Frontiers in Statistical Quality Control

More information about this series at http://www.springer.com/series/10821

Sven Knoth · Wolfgang Schmid
Editors

Frontiers in Statistical Quality Control 13

Springer

Editors
Sven Knoth
Department of Mathematics and Statistics
Helmut Schmidt University
Hamburg, Germany

Wolfgang Schmid
Department of Statistics
European University Viadrina
Frankfurt (Oder), Germany

Frontiers in Statistical Quality Control
ISBN 978-3-030-67855-5 ISBN 978-3-030-67856-2 (eBook)
https://doi.org/10.1007/978-3-030-67856-2

Mathematics Subject Classification (2010): 62-06, 62P30, 62L99, 62N05, 62K99

This Springer imprint is published by the registered company Springer Nature Switzerland AG
The registered company address is: Gewerbestrasse 11, 6330 Cham, Switzerland

Preface

The XIIIth International Workshop on *Intelligent Statistical Quality Control* took place in Hong Kong from August 12 to 14, 2019. The invitational workshop was jointly organized by Prof. K.-L. Tsui from the City University of Hong Kong, Prof. S. Knoth from the Helmut Schmidt University in Hamburg, Germany, and Prof. W. Schmid from the European University Viadrina in Frankfurt (Oder), Germany.

In line with the talks given at the workshop, the focus of the book is on major areas of Statistical Quality Control (SQC). It consists of 22 papers that were carefully selected and reviewed by the scientific program committee. The book is divided into two parts. The majority of the papers address Statistical Process Control (SPC), which is now often called Statistical Process Monitoring (SPM). Moreover, SPC is the subject of Part I, whereas Part II is devoted to selected topics of SQC (e.g., measurement uncertainty analysis and data quality).

Statistical Process Control

To evaluate the performance of a control chart, various measures have been proposed. In practice, the most popular criterion is the Average Run Length (ARL). However, it is often insufficient to summarize the run length behavior using the ARL, especially when the marginal distribution of the charting statistic is not the same for all time points. **Driscoll, Woodall, and Zou** propose the Conditional False Alarm Rate (CFAR) in such a situation. They emphasize that enforcing a constant CFAR allows dealing with varying sample sizes, population sizes, or other influential covariates appropriately. Moreover, they describe an unpretentious manner of implementing the corresponding procedures.

In applications, the in-control parameters are frequently unknown. **Goedhart** deals with the question of how the estimation of the unknown parameters using a reference sample influences control chart performance in Phase II. This introduces several challenges and tradeoffs regarding the design of a control chart. The author provides an overview and critical discussion of the present literature. He focuses on the Shewhart mean chart for independent samples and analyzes both the normal case and nonparametric approaches.

Knoth discusses the upper Exponentially Weighted Moving Average (EWMA) control chart for the mean of beta-distributed variables. He analyzes several ways to calculate the ARL of this chart. In particular, the Markov chain approximation and the Nyström procedure to approximate the solution of the ARL integral equation are investigated. The latter equation is dealt with, as well with collocation. In doing so, the ARL function is approximated using Chebyshev polynomials whose coefficients are determined by plugging them into the ARL integral equation. The Markov chain approach, in some cases, only provides a crude approximation whereas collocation yields reasonable results for all considered beta-distribution configurations.

Mahmood, Sanusi and Xie consider the zero-inflated Conway-Maxwell-Poisson (ZICOM-Poisson) distribution, which is applied to model over or under-dispersed zero-defect datasets. They introduce several Cumulative Sum (CUSUM) type control charts to detect an increasing shift in the rate parameter of the ZICOM-Poisson distribution. In a simulation study, all considered charts are compared with each other. As a measure of performance, the average number of observations to signal is used.

Yang and Lu deal with a skewed quality characteristic. They develop an average loss control chart for monitoring quality loss variation under skewed distributions. The statistical properties of the proposed chart are investigated, and the out-of-control behavior is analyzed using the ARL.

For a discrete quality characteristic, the ARL of a control chart for a parameter of interest is usually not a continuous function in the control limits. This is why the control limits cannot be determined such that the in-control ARL is equal to a prespecified value. Another problematic feature is caused by the skewness of most of the count distributions. Thus, a comparison of these charts is difficult. **Morais, Knoth, Cruz, and Weiß** introduce ARL-unbiased CUSUM charts for detecting both increases and decreases in the mean of binomial counts deploying randomization to achieve exact ARL values. They consider the case of independent samples and first-order autoregressive binomial counts. Explicit formulas (the Markov chains are exact models) for the ARLs of the introduced charts are derived.

Yashchin, Civil, Komatsu, and Zulpa investigate early warning systems (EWSs) for monitoring multi-stage data, in which downstream variables undergo changes associated with upstream process stages. In such applications, the EWS monitoring arm acts as a search engine that analyzes a number of data-streams for each monitored variable. The authors discuss principles of developing and managing targets, with examples from a supply chain operation.

Hryniewicz, Kaczmarek-Majer, and Opara address processes described by indirectly observed data, such as telehealth systems. The available data can be used to predict the characteristics of interest, which form a process to be monitored. If the process of interest takes only a finite number of observations, the prediction problem is related to a classification problem. The authors consider various classification methods, such as logistic regression and combined classifiers. In the present case, the result of the classification is a process taking the values of 0 and 1. The authors apply various control charts to these data considering autocorrelations.

Okhrin, Schmid, and Semeniuk present an overview of the literature on monitoring image processes. In most cases, the image is split into sub-images (regions of interests) and certain characteristics of these regions, such as the mean or the entropy, are monitored. The introduction of the regions of interest leads to a dimension reduction; nevertheless, the resulting process typically still has a high dimension (around 400). The authors discuss various control charts for the image characteristic assuming independent images and considering spatial correlations within an image. Thus, their approach is more general than in previous literature where spatial dependence has frequently not been directly considered. They compare several control procedures with each other using an extensive simulation study.

Otto monitors a spatio-temporal process, presenting several examples. In principle, a spatio-temporal process can be considered a multivariate time series with a specific autocorrelation matrix determined by the spatial structure. In most applications, it is assumed to be isotropic. The proposed control chart is based on a multivariate EWMA chart for time series, which has been studied in previous contributions. The author uses parallel multivariate control charts driven by the fact that spatial dependence decreases with an increasing distance between locations.

Huang, Jiang, and Shi consider the problem of monitoring warranty claims. To model warranty claims, various distributions are used, such as the Poisson distribution and the gamma-Poisson mixture model. The latter model typically results in a negative binomial distribution. However, if the gamma-distributed parameter is realized differently from the standard setup, the final count distribution deviates considerably from the latter one. For a specific setup driven by an application example, using the log-likelihood approach, a control chart for detecting an increase in the intensity is introduced. For this real data example on repair records of lifts installed for an high-speed rail, how the results can be applied is shown.

Megahed, Jones-Farmer, Cai, Rigdon, and Mohamed emphasize that computer acquisition of human and physical data is becoming more pervasive with continued technological advancements. Personal device data can be used as a proxy for human operations. The motivation behind their paper is to encourage the quality community to investigate relevant research problems that pertain to human operators. They describe three application areas: identification of physical human fatigue, capturing changes in a driver's safety performance, and human authentication for cyber-security applications.

Gan, Koh, and Ang describe an application of SPC in health monitoring. They propose a risk-adjusted control statistic, which is the ratio of the surgical outcome to the estimated probability of death. The main characteristic of this statistic is that the resulting penalty score is substantially higher if a patient with low risk dies, and the penalty score decreases sharply as the risk increases.

Wang and Zwetsloot employ functional data analysis to model and analyze health data recorded over time. In addition, SPM helps to detect an early event. They explore the usefulness of functional data analysis for prospective health surveillance and propose two strategies for monitoring using control charts. They apply their findings to monthly ovitrap index data.

Zhao, Yan, Holte, Kerani, and Mei deal with the detection of hot spots, which are defined as structured outliers that are sparse over the spatial domain but persistent over time. They propose a tensor decomposition method to uncover when and where the hot spots occur. The introduced method decomposes the tensor into three components: a smooth global trend, local hot spots, and residuals. A LASSO approach is used to estimate the model parameters and a CUSUM procedure is used for detecting hot spots.

Finally, **Sparks, Joshi, Paris, and Karimi** analyze EWMA control charts to detect increases in event frequencies to flag outbreaks promptly. These charts use the time between events, which is modeled by a Weibull distribution. The EWMA design is adaptive and deals with both homogeneous and non-homogeneous processes. An extensive discussion of an actual application completes this contribution.

Selected Topics from Statistical Quality Control

Suzuki, Takeshita, Ogawa, Lu, and Ojima discuss the evaluation of measurement methods. They focus on ordinal categorical variables. Using methods that can be applied to qualitative data, an analysis of a measurement precision experiment with measurements involving ordinal categorical variables is investigated.

Steiner, MacKay, and Fan consider the assessment of a binary measurement system with multiple operators when a gold measurement system is also available (for the assessment study). To model the data, it is assumed that some parts are more difficult to correctly classify than others. The assessment distinguishes between fixed and random operator effects. For each, a conditional and marginal model and their corresponding estimates of the parameters of interests are given.

Possolo deals with concepts, methods, and tools, evaluating measurement quality. The contribution provides an overview that is illustrated with examples of applying statistical methods that support measurement quality and guarantee the intercomparability of measurements made worldwide in all fields of commerce, industry, science, and technology, including medicine.

Bodnar and Elster propose a new statistical method for analyzing data from a key comparison when transfer standards are measured in two petals. Bayesian treatment of the model parameters and of the random effects is suggested. The latter can be viewed as potential laboratory effects that are assessed through the proposed analysis. While the prior for the labaratory effects naturally is assigned as a Gaussian distribution, the Berger and Bernardo reference prior is taken for the remaining model parameters.

Nishina analyzes how quality control activities can reduce the variability of outcomes in the value chain. He considers three variabilities: the variability before shipping to market, after shipping to market, and of the satisfaction with the market. He discusses various activities to reduce variability.

Benford's law is used worldwide to detect nonconformance or data fraud for numerical data. **Kössler, Lenz, and Wang** analyze five empirical numerical datasets of various sample sizes and evaluate the performance of Benford's law by applying various tests of goodness of fit.

The level of the workshop on *Intelligent Statistical Quality Control* is determined by the quality of its papers. We believe that this volume truly represents the frontiers of SQC. The editors would like to express their deep gratitude to the members of the scientific program committee, who carefully invited researchers from around the world and the reviewers of all submitted papers:

Sven Knoth, Germany
Fadel Megahed, USA
Wolfgang Schmid, Germany
Kwok L. Tsui, Hong Kong
Jiang Wei, China
William H. Woodall, USA

Additional Reviewers Include the Following:

Tomomichi Suzuki, Japan
Olgierd Hryniewicz, Poland
Emmanuel Yashchin, USA
Inez Maria Zwetsloot, Hong Kong
Stefan Steiner, Canada

Moreover, we thank Springer Heidelberg, for the continuing collaboration.

Hamburg, Germany
Frankfurt (Oder), Germany
September 2020

Sven Knoth
Wolfgang Schmid

Contents

Contributors

Janice J. Ang National University of Singapore, Singapore, Singapore

Olha Bodnar Unit of Statistics, School of Business, Örebro University, Örebro, Sweden

Miao Cai Saint Louis University, St. Louis, MO, USA

Aaron Civil IBM Corporation, Armonk, NY, USA

Camila Jeppesen Cruz Instituto Superior Técnico, Universidade de Lisboa, Lisbon, Portugal

Anne R. Driscoll Virginia Tech, Blacksburg, VA, USA

Steven E. Rigdon Saint Louis University, St. Louis, MO, USA

Clemens Elster Physikalisch-Technische Bundesanstalt, Abbestrasse 2-12, Berlin, Germany

Kevin Fan Department of Statistics and Actuarial Science, University of Waterloo, Waterloo, ON, Canada

Fah F. Gan National University of Singapore, Singapore, Singapore

Rob Goedhart IBIS UvA, Department of Operations Management, University of Amsterdam, Amsterdam, The Netherlands

Sarah E. Holte Division of Public Health Sciences, Fred Hutchinson Cancer Research Center, Seattle, WA, USA

Olgierd Hryniewicz Systems Research Institute, Polish Academy of Sciences, Warszawa, Poland

Wenpo Huang Hangzhou Dianzi University, Hangzhou, China

Wei Jiang Shanghai Jiao Tong University, Shanghai, China

R. Jock MacKay Department of Statistics and Actuarial Science, University of Waterloo, Waterloo, ON, Canada

L. Allison Jones-Farmer Miami University, Oxford, OH, USA

Aditya Joshi CSIRO Data61, Marsfield, NSW, Australia

Katarzyna Kaczmarek-Majer Systems Research Institute, Polish Academy of Sciences, Warszawa, Poland

Sarvnaz Karimi CSIRO Data61, Marsfield, NSW, Australia

Roxanne P. Kerani Department of Medicine, University of Washington, and Public Health-Seattle & King County, Seattle, WA, USA

Sven Knoth Department of Mathematics and Statistics, Faculty of Economics and Social Sciences, Helmut Schmidt University, Hamburg, Germany

Wei L. Koh National University of Singapore, Singapore, Singapore

Jeff Komatsu IBM Corporation, Armonk, NY, USA

Wolfgang Kössler Institut für Informatik, Humboldt Universität zu Berlin, Berlin, Germany

Hans-J. Lenz Institut für Statistik und Ökonometrie, Freie Universität Berlin, Berlin, Germany

Shan-Wen Lu Statistics Department, National Chengchi University, Taipei, Taiwan

Xiao-Nan Lu Department of Computer Science and Engineering, Faculty of Engineering, University of Yamanashi, Kofu, Yamanashi, Japan

Fadel M. Megahed Miami University, Oxford, OH, USA

Tahir Mahmood Department of Systems Engineering and Engineering Management, City University of Hong Kong, Kowloon, Hong Kong;
Department of Technology, School of Science and Technology, The Open University of Hong Kong, Kowloon, Hong Kong

Yajun Mei School of Industrial and Systems Engineering, Georgia Institute of Technology, Atlanta, GA, USA

Manar Mohamed University of Alabama at Birmingham, Birmingham, AL, USA

Manuel Cabral Morais CEMAT (Center for Computational and Stochastic Mathematics) and Department of Mathematics, Instituto Superior Técnico, Universidade de Lisboa, Lisbon, Portugal

Ken Nishina Aich Institute of Technology, Nagoya, Japan

Mayu Ogawa Department of Industrial Administration, Tokyo University of Science, Yamazaki, Noda, Chiba, Japan

Yoshikazu Ojima Department of Industrial Administration, Tokyo University of Science, Yamazaki, Noda, Chiba, Japan

Yarema Okhrin Department of Statistics, University of Augsburg, Augsburg, Germany

Karol R. Opara Systems Research Institute, Polish Academy of Sciences, Warszawa, Poland

Philipp Otto Institute of Cartography and Geoinformatics, Leibniz University Hannover, Hanover, Germany

Cecile Paris CSIRO Data61, Marsfield, NSW, Australia

Antonio Possolo NIST (National Institute of Standards and Technology), Gaithersburg, Maryland, USA

Ridwan A. Sanusi Department of Systems Engineering and Engineering Management, City University of Hong Kong, Kowloon, Hong Kong;
Department of Community Health Sciences, Rady Faculty of Health Sciences, University of Manitoba, Winnipeg, MB, Canada

Wolfgang Schmid Department of Statistics, European University Viadrina, Frankfurt(Oder), Germany

Ivan Semeniuk Department of Statistics, European University Viadrina, Frankfurt(Oder), Germany

Chengyou Shi Shanghai Jiao Tong University, Shanghai, China

Ross Sparks CSIRO Data61, Marsfield, NSW, Australia

Stefan H. Steiner Department of Statistics and Actuarial Science, University of Waterloo, Waterloo, ON, Canada

Tomomichi Suzuki Department of Industrial Administration, Tokyo University of Science, Yamazaki, Noda, Chiba, Japan

Jun-ichi Takeshita Research Institute of Science for Safety and Sustainability, National Institute of Advanced Industrial Science and Technology (AIST), Tsukuba, Ibaraki, Japan

Xing D. Wang Institut für Informatik, Humboldt Universität zu Berlin, Berlin, Germany

Zezhong Wang Department of Systems Engineering and Engineering Management, City University of Hong Kong, Kowloon, Hong Kong

Christian H. Weiß Department of Mathematics and Statistics, Faculty of Economics and Social Sciences, Helmut Schmidt University, Hamburg, Germany

William H. Woodall Virginia Tech, Blacksburg, VA, USA

Min Xie Department of Systems Engineering and Engineering Management, City University of Hong Kong, Kowloon, Hong Kong

Hao Yan School of Computing, Informatics, and Decision Systems Engineering, Arizona State University, Tempe, AZ, USA

Su-Fen Yang Statistics Department, National Chengchi University, Taipei, Taiwan

Emmanuel Yashchin IBM, Thomas J. Watson Research Ctr., Yorktown Heights, NY, USA

Yujie Zhao School of Industrial and Systems Engineering, Georgia Institute of Technology, Atlanta, GA, USA

Changliang Zou Nankai University, Tianjin, China

Paul Zulpa IBM Corporation, Armonk, NY, USA

Inez Maria Zwetsloot Department of Systems Engineering and Engineering Management, City University of Hong Kong, Kowloon, Hong Kong

Statistical Process Control

Use of Conditional False Alarm Metric in Statistical Process Monitoring

Anne R. Driscoll, William H. Woodall, and Changliang Zou

Abstract The conditional false alarm rate (CFAR) at a particular time is the probability of a false alarm for an assumed in-control process at that time conditional on no previous false alarm. Only the Shewhart control chart designed with known in-control parameters, or conditioned on the estimated parameters, has a constant conditional false alarm rate. Other types of charts, however, can have their control limits determined in order to have any desired pattern of CFARs. The important advantage of the use of this CFAR metric is when sample sizes, population sizes or other covariate information affecting chart performance vary over time. In these cases, the control limit at a particular time can be obtained through control of the CFAR value after the corresponding covariate value is known. This allows one to control the in-control performance of the chart without the need to model or forecast the covariate values. The approach is illustrated using the risk-adjusted Bernoulli cumulative sum (CUSUM) chart.

Keywords Changepoint approach · CUSUM chart · Dynamic control limits · Risk-adjusted CUSUM chart · Time-varying structure

1 Introduction

Recently, the conditional false alarm rate (CFAR) has proved useful in designing control charts, where this metric is defined as the probability of a false alarm for an assumed in-control process at a particular time given no previous false alarm. The

A. R. Driscoll · W. H. Woodall (✉)
Virginia Tech, Blacksburg, VA 24061, USA
e-mail: bwoodall@vt.edu

A. R. Driscoll
e-mail: adriscoll@vt.edu

C. Zou
Nankai University, Tianjin 300071, China
e-mail: nk.chlzou@gmail.com

S. Knoth and W. Schmid (eds.), *Frontiers in Statistical Quality Control 13*,
Frontiers in Statistical Quality Control,
https://doi.org/10.1007/978-3-030-67856-2_1

CFAR is analogous to the hazard function in reliability theory. This metric can be used to determine the control limits for any type of chart. It is particularly useful when the in-control (IC) parameter value varies over time due to a varying covariate such as the sample size. In Sect. 2 we discuss what we refer to as the static use of the CFAR. In these cases, the IC distribution of the observations is assumed to be known before monitoring begins. We discuss the dynamic use of the CFAR in Sect. 3. In the dynamic applications, the control limit at a particular time is not determined until the corresponding value of a time-varying covariate is known. We use the design of the risk-adjusted Bernoulli cumulative sum (CUSUM) chart as an example. We discuss implementation issues in Sect. 4. In Sect. 5, we give some other situations in which the CFAR could be used. Our conclusions are given in Sect. 6.

2 Static Use of CFAR

We consider a charting scheme consisting of two parts at time t: a detection statistic, $a(\{X_i\}_{i=1}^t)$ which is function of the observations X_is up to the time t, and a control limit, L. The time of an alarm, T, is the run length

$$T = \min\{t : a(\{X_t\}_{i=1}^t) \geq L\}.$$

The most commonly used criterion for evaluating the performance of a control chart is the average run length (ARL), say $E_{\text{IC}}(T)$. As recognized in the literature (e.g., Woodall and Montgomery 1999), it is often insufficient to summarize run length behavior by the ARL, especially when the marginal distribution of the charting statistic is not the same for all time points. In such situations, percentiles and the standard deviation of the run length (SDRL) would provide more information. To this end, the CFAR metric, which actually quantifies the uncertainty of run length, serves as a reasonable measure for the design and evaluation of control charts. Only the Shewhart control chart designed with known in-control parameters, or conditioned on the estimated parameters, has a constant CFAR. Other types of charts, however, can have their control limits determined in order to have any desired pattern of CFARs. The CFARs of popular exponentially weighted moving average (EWMA) and cumulative sum (CUSUM) methods are not constant over time, but the boundaries of these charts can be adjusted such that their CFARs can be controlled to be a specified constant. That is, we find a series of control limits L_t so that

$$\Pr\left(a(\{X_i\}_{i=1}^t) > L_t \mid a(\{X_i\}_{i=1}^s) < L_s, 1 \leq s < t\right) = \alpha, \quad \text{for } t > 1,$$
$$\Pr\left(a(X_1) > L_1\right) = \alpha,$$

for some pre-specified α.

This idea was first proposed by Margavio et al. (1995) and has been successfully utilized by D. Hawkins and his colleagues in the parametric change-point-based control charts with unknown IC parameters, such as Hawkins et al. (2003), Hawkins and Zamba (2005a, b), Zamba and Hawkins (2006, 2009). See also Zou et al. (2006, 2009) for its applications in profile monitoring.

In the use of CFAR above, it was typically assumed that one has a sequence of independent, normally distributed data with known IC parameter values. If the IC distribution of the variable or vector being monitored is known, as in Margavio et al. (1995), then the control limit values at all time periods can be determined at the beginning of monitoring in order to control the CFAR. One can use simulation or numerical methods to determine the control limits. We refer to this case as the static scenario. This is also the case for some nonparametric or distribution-free control charts of which the run length distributions are the same for every continuous distribution (or a broad family of distributions). Please refer to Zhou et al. (2009), Zou and Tsung (2010), Hawkins and Deng (2010) and Holland and Hawkins (2014) for formal use of CFAR in the design of univariate or multivariate distribution-free control charts. Morais and Pacheco (2012) evaluated the CFAR papers on several different types of control charts viewing the CFAR values over time as a hazard function. Similar approaches were taken by Nishina and Nishiyuki (2003) and Nishina et al. (2006).

It has been pointed out that the IC run length distribution is a geometric distribution with parameter α if the CFAR is controlled to be α, just like the run length distribution of the Shewhart chart. Some control charts perform quite well in terms of average run length (ARL), say achieving a desired IC ARL and having quite small out-of-control (OC) ARLs, but they may not be appealing if they have rather an unsatisfactory run length distributions. With some charts, the specified in-control (IC) ARL is attained with elevated probabilities of very short and very long runs, as compared with a geometric distribution. This is reflected in a much larger SDRL than that of a geometric distribution and an elevated probability of false alarms with short runs, which, in turn, hurt an operator's confidence in valid alarms. Too frequent and excessive early false alarms render these charts useless and thus, unacceptable in practice. The IC run length distribution is often considered to be satisfactory if it is close to the geometric distribution or more generally its variation is less than that of a geometric distribution. Controlling the CFAR at each time point is essentially equivalent to performing a formal hypothesis test at each time point, which automatically results in a geometric run length distribution. In this sense, an "approximately" steady CFAR could be a useful benchmark for evaluating the performance of a chart.

In most cases, it seems reasonable to determine control limits so that the CFAR is constant over time. Any desired pattern of CFAR values can be specified, however. A control chart with a "headstart", for example, will have higher CFARs at the start of monitoring. This would lead to quicker detection of process changes that occur early in the monitoring. It seems most reasonable to compare the OC performance of two competing methods only when the two corresponding sequences of CFARs are the same. If one considers, for instance, a given CUSUM chart with constant control limits compared to the CUSUM chart with the same IC ARL, but with DPCLs to have a constant CFAR, the latter method will more quickly detect an early shift in the process. The CFAR is also useful when two adaptive procedures are compared, such as the CUSUM or EWMA charts with adaptive tuning parameters to detect a non-abrupt deterioration of a monitored process. In such situations, maintaining a same CFAR will ensure that using OC ARLs and SDRLs as summary criteria is well-grounded.

3 Dynamic Use of CFAR

The static use of the CFAR is useful with the design of some methods, but is not always particularly impactful when applied to standard methods. In what we consider a major contribution, Shen et al. (2013) used simulation or Markov chain methodology to determine the control limits of an EWMA chart for monitoring with Poisson data when the area of opportunity varies over time. The control limits vary over time to maintain the CFAR at each time as a constant, and the control limit for a particular time is determined only after the corresponding area of opportunity is known. This procedure differs from the static use of CFAR in the sense that the control limits are determined online along with the process observations rather than determined before monitoring. That is, the control limits are data dependent. Benefiting from this unique feature, the performance of Shen et al. (2013)'s approach is better than that of competing methods because no model has to be specified for the area of opportunity. We refer to this case as the dynamic scenario of CFAR. Basically speaking, in this scenario, we usually have additional information, for instance, some covariates z_t, observed over time along with the process variable of interest X_t. At time point t, once we obtain the observations $(X_i, z_i)_{i=1}^t$, we find the control limit at time t via

$$\Pr\left(a(\{X_i\}_{i=1}^t) > L_t \mid a(\{X_i\}_{i=1}^s) < L_s, 1 \leq s < t; z_t\right) = \alpha.$$

The dynamic CFAR has been applied in a number of applications. Among others, Huang et al. (2016) applied it to cumulative sum (CUSUM) charts for monitoring the mean of a normal distribution when the sample size varies. Numerical methods were used to determine the control limits. Zhang and Woodall (2015) applied the CFAR approach to the design of the risk-adjusted Bernoulli CUSUM chart of Steiner et al. (2000) that is used to monitor surgical outcomes. In this application, the IC probability of an adverse surgical outcome varies widely from patient-to-patient. This is illustrated in Fig. 1 where the adverse outcome of interest is the 30-day mortality rate. The standard risk-adjusted Bernoulli CUSUM chart has constant control limits as shown in Fig. 2. The limits based on the CFAR, termed as dynamic probability control limits (DPCL), vary over time as shown in Fig. 2. The difference is that one needs to know the distribution of patient risk factors in order to obtain the constant limits whereas this information is not required to obtain the dynamic limits. The dynamic limit values are obtained one-by-one using simulation as the patient information becomes known.

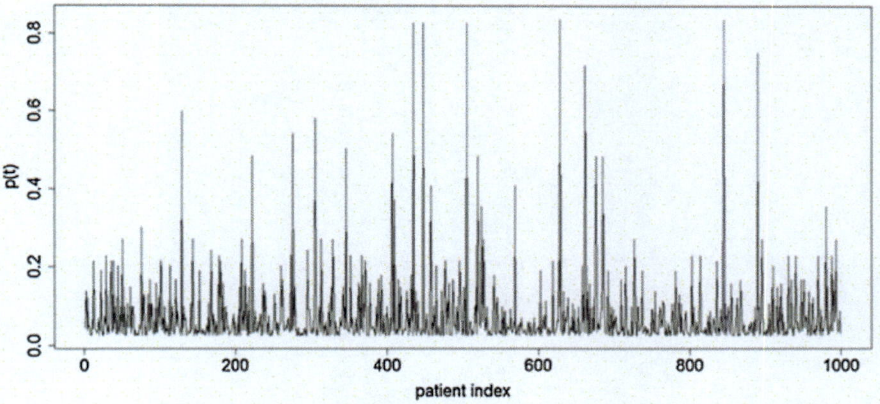

Fig. 1 Predicted mortality rates $p(t)$ of the first 1000 patients — Sequence 1 from Surgeon 1. *Reproduced from* Zhang and Woodall (2015). *Published with permission of © 2015 John Wiley & Sons, Ltd*

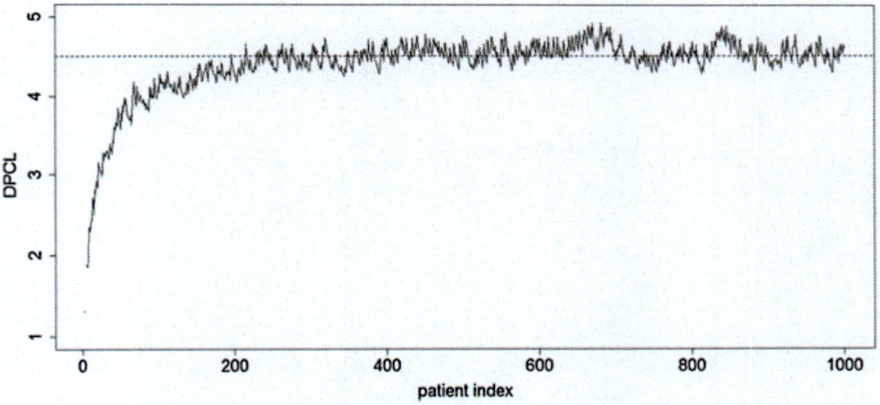

Fig. 2 Comparison of constant control limit (dashed line) and DPCLs (solid line) for comparable IC ARLs for population with all patients. *Reproduced from* Zhang and Woodall (2015). *Published with permission of © 2015 John Wiley & Sons, Ltd*

Figure 3 shows how the CFAR rate varies widely for the chart with a constant control limit. We have tight control of the CFAR values, however, when the dynamic control limits are used. If the risk population changes, the dynamic control limits will automatically adjust, as illustrated in Fig. 4. The control limits can be determined when one is interested in detecting improvements in quality in addition to detecting process deterioration. The resulting limits are shown in Fig. 5. It is more difficult to detect improvement because most patients are expected to survive more than thirty days after surgery.

Zhang and Woodall (2017b) showed that estimation error has less effect on the performance of the risk-adjusted Bernoulli CUSUM chart when the IC risk-adjustment

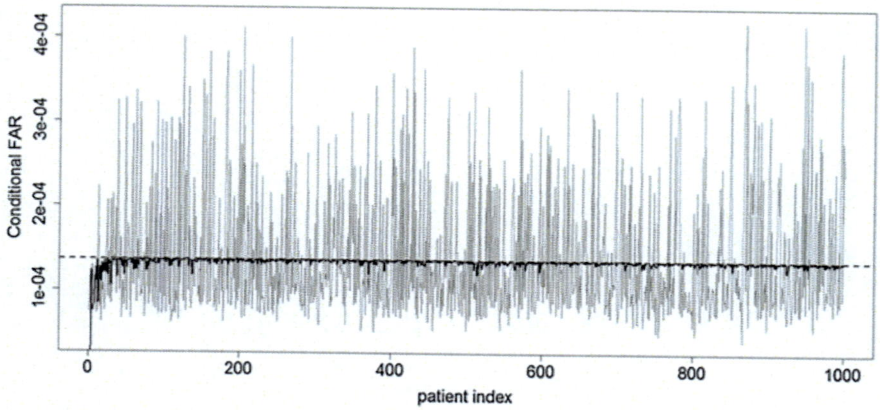

Fig. 3 Comparison of conditional false alarm rates of constant control limit (lighter line) and DPCLs (darker line) for comparable IC ARLs. *Reproduced from* Zhang and Woodall (2015). *Published with permission of © 2015 John Wiley & Sons, Ltd*

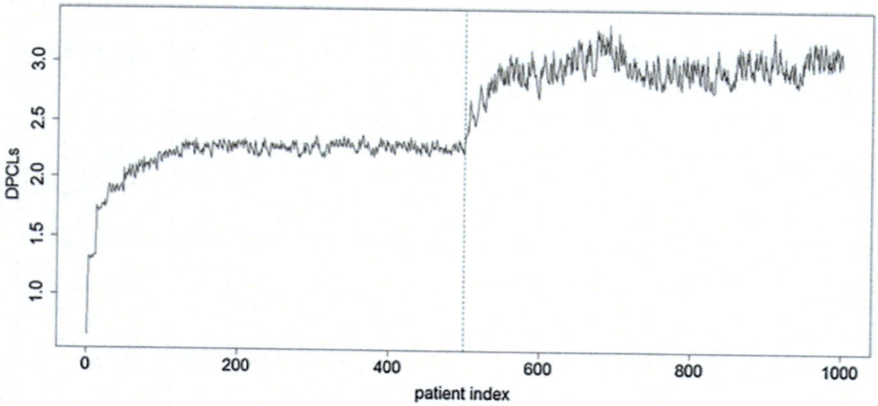

Fig. 4 Control limits when population shifts from lowest scores to highest scores after 500 patients. *Reproduced from* Zhang and Woodall (2015). *Published with permission of © 2015 John Wiley & Sons, Ltd*

model is estimated. In another application, Zhang et al. (2017) applied the dynamic control limit approach to the method proposed by Tang et al. (2015) which allows for surgical outcomes with more than two possible results. Recently, Aytaçoğlu and Woodall (2020) proposed to use the DPCLs for CUSUM charts when monitoring proportions with time varying sample sizes. Sogandi et al. (2019) generalizes Zhang and Woodall (2015)'s method to the case of multistage processess.

The CFAR metric has also been used in the self-starting chart context. Aminnayeri and Sogandi (2016) applied it to risk-adjusted monitoring, while Shen et al. (2016) extended it for monitoring Poisson random variables.

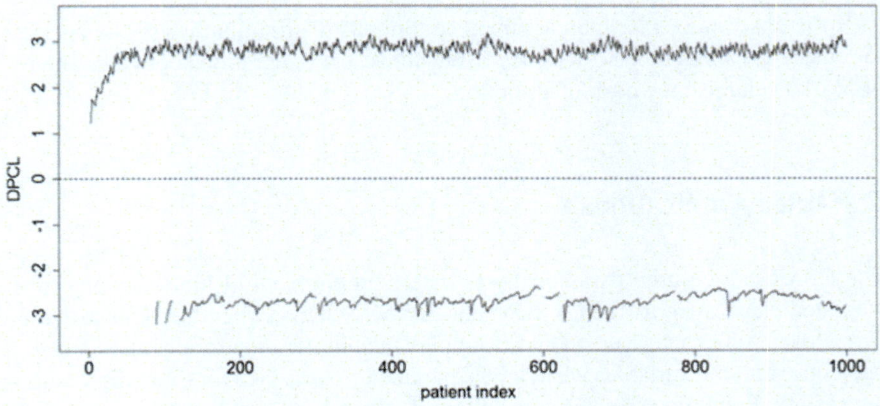

Fig. 5 Upper and lower DPCLs of the first 1,000 patients in Sequence 1 from Surgeon 1 with $\alpha = 0.001$. *Reproduced from* Zhang and Woodall (2017a). *Published with permission of © 2016 John Wiley & Sons, Ltd*

4 Implementation Methods

For a better understanding of the implementation of the CFAR approach to determine the DPCLs, we sketch below the general algorithm for the simulation-based method in order for a one-sided chart to have an approximately constant CFAR of α. The algorithm applies in both the static and dynamic cases.

1. Generate N random values from the IC model to obtain N values of the first control statistic.
2. Use the upper α-percentile of the empirical distribution of the control statistic as the first control limit. Stop if the first observation from the process being monitored leads to a control statistic value outside this limit.
3. Generate N values of the control statistic at the next time period by combining N values selected at random from the immediately previous simulated values of the control statistic that were below the control limit and N observations generated from the IC model.
4. Use the upper α-percentile of the empirical distribution of the control statistic as the control limit.
5. Repeats step (3) and (4) until the actual observed data lead to a value of the control statistic that falls outside the control limit.

In some cases, N will need to be large, e.g., 100,000 or 1,000,000. With discrete data, the percentile selected should be such that the CFAR is no larger than α. On some occasions, no such percentile will exist and thus there will be no control limit for that time period.

If the observations have an assumed continuous distribution, it may be possible to replace the simulation with a successive numerical integration approach. The use of Markov chains may be another option.

5 Other Applications

The CFAR can be used dynamically to determine the control limits for any time-weighted chart, such as the CUSUM and EWMA charts, when the IC distribution varies over time. A few unstudied examples include (i) monitoring a sequence of geometric random variables when the IC parameter p_0 varies over time; (ii) monitoring binomial random variables when either the sample size and/or the IC parameter p_0 varies over time; and (iii) monitoring exponential random variables when the IC mean β_0 varies over time. We strongly recommend that any CUSUM charts studied allow the possibility of a slack region, as discussed by Woodall and Faltin (2019). CUSUM charts based on an IC region, an indifference region, and an out-of-control region of parameter values can make control charting more practical and prevent an excessive number of alarms for process changes that are too small to be of practical importance. In such situations, the methods given in Woodall and Faltin (2019) combined with CFAR control could be a promising direction.

The dynamic use of CFAR has also successfully been used by Yang et al. (2017) to solve the problem of monitoring nonparametric profiles with time-varying sample sizes or random predictors. Traditional profile monitoring schemes, whose control limits are often determined before the monitoring begins, are constructed based on assumed knowledge of profile sample sizes and predictors; see Woodall (2007) for an overview. In practice, however, in some cases, the sample sizes or predictors are random and our foreknowledge about them is not available. Yang et al. (2017) proposed a kernel-based nonparametric profile monitoring scheme which integrates the multivariate exponentially weighted moving average procedure with the DPCLs.

Chen et al. (2016) employed the CFAR to construct an exactly distribution-free multivariate control chart for monitoring location parameters when only a small reference dataset is available. Although multivariate process monitoring has been extensively studied in the literature, designing distribution-free control schemes is still challenging because the multivariate generalization of signs or ranks usually does not have the distribution-free property over a wide class of distributions. The key idea in Chen et al. (2016) is to construct a series of conditionally distribution-free test statistics in the sense that their distributions are free of the underlying distribution given the empirical distribution functions up to the current time. The success of the proposed method, therefore, lies in the use of DPCLs to attain a specified CFAR.

We believe that the CFAR idea can be readily extended to other error rates, such as the false discovery rate (FDR) when we consider individual surveillance of high dimensional data streams. In many applications, it is often natural to assume that if an alarm with respect to a stream is made at the current time, then the process for this specific stream is usually halted. Only the streams with detection statistics within the threshold limits continue to be monitored at the next time. Du and Zou (2018) pointed out that since the ongoing streams monitored currently are dynamic in the sense that no alarm occurred previously, the conventional notion of FDR needs to be generalized to accommodate the dynamic nature of online monitoring.

6 Conclusions

We strongly encourage the use of the CFAR. Advantages include desired IC performance, approximately geometrically distributed IC run lengths, and no requirement of any information or assumptions about covariates such as sample sizes or patient risk factors. In addition, estimation error has less effect in cases when covariate information is used because no covariate distribution has to be estimated.

Acknowledgements The authors are grateful to an anonymous referee for insightful comments that have improved the article.

References

Aminnayeri, M., & Sogandi, F. (2016). A risk adjusted self-starting Bernoulli CUSUM control chart with dynamic probability control limits. *AUT Journal of Modeling and Simulation*, *48*(2), 103–110.

Aytaçoğlu B., Woodall, W. H. (2020). Dynamic probability control limits for CUSUM charts for monitoring proportions with time-varying sample sizes. *Quality and Reliability Engineering International, 36*, 592–603.

Chen, N., Zi, X., & Zou, C. (2016). A distribution-free multivariate control chart. *Technometrics*, *58*(4), 448–459.

Du, L., & Zou, C. (2018). On-line control of false discovery rates for multiple datastreams. *Journal of Statistical Planning and Inference*, *194*, 1–14.

Hawkins, D. M., & Deng, Q. (2010). A nonparametric change-point control chart. *Journal of Quality Technology*, *42*(2), 165–173.

Hawkins, D. M., & Zamba, K. (2005a). A change-point model for a shift in variance. *Journal of Quality Technology*, *37*(1), 21–31.

Hawkins, D. M., & Zamba, K. (2005b). Statistical process control for shifts in mean or variance using a changepoint formulation. *Technometrics*, *47*(2), 164–173.

Hawkins, D. M., Qiu, P., & Kang, C. W. (2003). The changepoint model for statistical process control. *Journal of Quality Technology*, *35*(4), 355–366.

Holland, M. D., & Hawkins, D. M. (2014). A control chart based on a nonparametric multivariate change-point model. *Journal of Quality Technology*, *46*(1), 63–77.

Huang, W., Shu, L., Woodall, W. H., & Tsui, K. L. (2016). CUSUM procedures with probability control limits for monitoring processes with variable sample sizes. *IIE Transactions*, *48*(8), 759–771.

Margavio, T. M., Conerly, M. D., Woodall, W. H., & Drake, L. G. (1995). Alarm rates for quality control charts. *Statistics and Probability Letters*, *24*(3), 219–224.

Morais, M. C., & Pacheco, A. (2012). A note on the aging properties of the run length of Markov-type control charts. *Sequential Analysis*, *31*(1), 88–98.

Nishina, K., & Nishiyuki, S. (2003). False alarm probability function of CUSUM charts. *Economic Quality Control*, *18*(1), 101–112.
Nishina, K., Kuzuya, K., & Ishii, N. (2006). Reconsidering control charts in Japan. *Frontiers in statistical quality control 8* (pp. 136–150). Springer.
Shen, X., Zou, C., Jiang, W., & Tsung, F. (2013). Monitoring Poisson count data with probability control limits when sample sizes are time varying. *Naval Research Logistics (NRL)*, *60*(8), 625–636.
Shen, X., Tsui, K. L., Zou, C., & Woodall, W. H. (2016). Self-starting monitoring scheme for Poisson count data with varying population sizes. *Technometrics*, *58*(4), 460–471.
Sogandi, F., Aminnayeri, M., Mohammadpour, A., & Amiri, A. (2019). Risk-adjusted Bernoulli chart in multi-stage healthcare processes based on state-space model with a latent risk variable and dynamic probability control limits. *Computers and Industrial Engineering*, *130*, 699–713.
Steiner, S. H., Cook, R. J., Farewell, V. T., & Treasure, T. (2000). Monitoring surgical performance using risk-adjusted cumulative sum charts. *Biostatistics*, *1*(4), 441–452.
Tang, X., Gan, F. F., & Zhang, L. (2015). Risk-adjusted cumulative sum charting procedure based on multiresponses. *Journal of the American Statistical Association*, *110*(509), 16–26.
Woodall, W. H. (2007). Current research on profile monitoring. *Production*, *17*(3), 420–425.
Woodall, W. H., & Faltin, F. (2019). Rethinking control chart design and evaluation. *Quality Engineering*, *31*(4), 596–605.
Woodall, W. H., & Montgomery, D. C. (1999). Research issues and ideas in statistical process control. *Journal of Quality Technology*, *31*(4), 376–386.
Yang, W., Zou, C., & Wang, Z. (2017). Nonparametric profile monitoring using dynamic probability control limits. *Quality and Reliability Engineering International*, *33*(5), 1131–1142.
Zamba, K., & Hawkins, D. M. (2006). A multivariate change-point model for statistical process control. *Technometrics*, *48*(4), 539–549.
Zamba, K., & Hawkins, D. M. (2009). A multivariate change-point model for change in mean vector and/or covariance structure. *Journal of Quality Technology*, *41*(3), 285–303.
Zhang, X., & Woodall, W. H. (2015). Dynamic probability control limits for risk-adjusted Bernoulli CUSUM charts. *Statistics in Medicine*, *34*(25), 3336–3348.
Zhang, X., & Woodall, W. H. (2017a). Dynamic probability control limits for lower and two-sided risk-adjusted Bernoulli CUSUM charts. *Quality and Reliability Engineering International*, *33*(3), 607–616.
Zhang, X., & Woodall, W. H. (2017b). Reduction of the effect of estimation error on in-control performance for risk-adjusted Bernoulli CUSUM chart with dynamic probability control limits. *Quality and Reliability Engineering International*, *33*(2), 381–386.
Zhang, X., Loda, J. B., & Woodall, W. H. (2017). Dynamic probability control limits for risk-adjusted CUSUM charts based on multiresponses. *Statistics in Medicine*, *36*(16), 2547–2558.
Zhou, C., Zou, C., Zhang, Y., & Wang, Z. (2009). Nonparametric control chart based on change-point model. *Statistical Papers*, *50*(1), 13–28.
Zou, C., & Tsung, F. (2010). Likelihood ratio-based distribution-free EWMA control charts. *Journal of Quality Technology*, *42*(2), 174–196.
Zou, C., Zhang, Y., & Wang, Z. (2006). A control chart based on a change-point model for monitoring linear profiles. *IIE Transactions*, *38*(12), 1093–1103.
Zou, C., Qiu, P., & Hawkins, D. (2009). Nonparametric control chart for monitoring profiles using change point formulation and adaptive smoothing. *Statistica Sinica*, *19*, 1337–1357.

Design Considerations and Trade-offs for Shewhart Control Charts

Rob Goedhart

Abstract When in-control parameters are unknown, they have to be estimated using a reference sample. The control chart performance in Phase II, which is generally measured in terms of the Average Run Length (ARL) or False Alarm Rate (FAR), will vary across practitioners due to the use of different reference samples in Phase I. This variation is especially large for small sample sizes. Although increasing the amount of Phase I data improves the control chart performance, others have shown that the amount required to achieve a desired in-control performance is often infeasibly high. This holds even when the actual distribution of the data is known. When the distribution of the data is unknown, it has to be estimated as well, along with its parameters. This yields even more uncertainty in control chart performance when parametric models are applied. With these issues in mind, choices have to be made in order to control the performance of control charts. We discuss several of these choices and their corresponding implications.

Keywords Control charts · Nonparametric · Parameter estimation

1 Introduction

In the field of statistical process monitoring (SPM) detecting changes in an underlying process is of major interest. To aid in the detect these changes, various statistical techniques have been developed, such as control charts. A common example of such a chart is the Shewhart control chart to monitor the mean of a variable, based on 3-sigma limits. When in-control parameters are known, this control chart yields a false alarm rate (FAR) of 0.27% for normally distributed data. This is equivalent to an in-control average run length (ARL) of 370.

R. Goedhart (✉)
IBIS UvA, Department of Operations Management, University of Amsterdam, Amsterdam, The Netherlands
e-mail: r.goedhart2@uva.nl

S. Knoth and W. Schmid (eds.), *Frontiers in Statistical Quality Control 13*, Frontiers in Statistical Quality Control,
https://doi.org/10.1007/978-3-030-67856-2_2

In practice, the in-control parameters μ_0 and σ_0 are generally unknown and need to be estimated from a Phase I reference sample. This requires an extensive Phase I analysis, as the obtained sample and its corresponding process estimates should be representative of the process. For example, the sample may contain contaminated data, and thus not be in-control itself. This would lead to a biased estimation of the process behavior, which in turn, affects the control chart performance in Phase II. A possible approach to address this problem is to use robust estimators, as described in Schoonhoven et al. (2011). A general overview of Phase I issues and considerations is given by Jones-Farmer et al. (2014).

While the use of robust estimation methods can provide a substantial improvement, it does not solve all the issues belonging to parameter estimation. Quesenberry (1993) recognized that, when parameters are estimated, consecutive false alarm events are dependent. As a consequence, the unconditional run length distribution is not geometric, as was previously assumed. He concluded that larger sample sizes are required in order to let the control chart behave as if parameters are known. Another important aspect of parameter estimation was not considered for this recommendation however, as shown by Saleh et al. (2015b) amongst others. They conclude that, as a result of *practitioner-to-practitioner* variation, the sample size requirements are much larger than suggested by Quesenberry (1993). For literature overviews on research on control chart performance when parameters are estimated, we refer to Jensen et al. (2006) and Psarakis et al. (2014).

In many situations, the sample size requirements for a sufficient control chart performance may be unfeasibly large. As an alternative, several researchers have proposed to adjust the control limits based on the sample size and the estimators used, such that a certain control chart performance criterion is satisfied. Albers and Kallenberg (2004a, b, 2005) provide two specific criteria as general directions for control limit adjustment, namely the *bias criterion* and the *exceedance probability* criterion. The first focuses on the unconditional run length distribution properties (i.e., averaged over all practitioners), while the latter aims to provide a certain minimum conditional performance for a large proportion of practitioners. Both of these approaches can also be combined with different performance measures, such as the FAR, ARL, or other similar run length characteristics.

After deciding on a design criterion and performance measure, one has to determine the estimation method to achieve it. This includes decisions on estimators to be used as mentioned earlier, but more importantly also on the accompanying parameter and distributional assumptions made, as well as the Phase I sample size. One of the most common assumptions in the literature of SPM is that process data are normally distributed, and the indicated control chart performance is then based on that assumption. However, data are often not normally distributed, which may substantially impact the actual performance of the control chart. To this end, more general models such as nonparametric methods have been developed. See for example Chakraborti et al. (2001, 2015) and Qiu (2018) for more information on nonparametric statistical process control. When sample sizes are small, parametric methods with appropriate distributional assumptions obviously perform better than nonparametric alternatives. However, in that case, the appropriateness of these distributional assumptions is also more difficult to validate, while deviations from these assumptions can have a substantial impact on performance.

This contribution is organized as follows. In the next section, we discuss the impact of parameter estimation, as well as several proposed initial countermeasures. In Sect. 3, we elaborate on different criteria to consider when constructing a control chart. After that, in Sect. 4, we elaborate on the distributional assumptions and corresponding implications on the control chart performance. In Sect. 5, we discuss the possible decisions regarding design parameters, such as sample size or strictness. Finally, in Sect. 6, we provide some concluding remarks.

2 Parameter Estimation

Consider the Shewhart $\bar{X}$ control chart to monitor the mean of a normally distributed variable through subgroups of size n, which has control limits

$$\begin{aligned} UCL &= \mu_0 + K\frac{\sigma_0}{\sqrt{n}}, \\ LCL &= \mu_0 - K\frac{\sigma_0}{\sqrt{n}}, \end{aligned} \tag{1}$$

where $K = \Phi^{-1}(1 - \alpha_0/2)$, with α_0 the nominal FAR. In practice, the values of μ_0 and σ_0 are generally unknown and need to be replaced by some estimates $\hat{\mu_0}$ and $\hat{\sigma_0}$ respectively. This results in the estimated control limits

$$\begin{aligned} \widehat{UCL} &= \hat{\mu_0} + K\frac{\hat{\sigma_0}}{\sqrt{n}}, \\ \widehat{LCL} &= \hat{\mu_0} - K\frac{\hat{\sigma_0}}{\sqrt{n}}. \end{aligned} \tag{2}$$

Many choices of estimators are possible, with the mean and median being common estimators for location, and the sample standard deviation and average moving range being common estimators for dispersion. A first property to consider for estimators is the efficiency of the estimators. This is done by Cryer and Ryan (1990), who show that the sample standard deviation is much more efficient than the average moving range when considering individual observations from normally distributed data. A similar conclusion was found by Mahmoud et al. (2010), who advise against the use of sample ranges to estimate dispersion. Another dimension to consider for estimators is the robustness, such that the estimation is less sensitive to contaminations in the Phase I data. Several robust estimation methods are considered in Schoonhoven et al. (2011), and Schoonhoven and Does (2012).

However, even when estimation is efficient and robust, there will still be estimation uncertainty. Quesenberry (1993) recognized that, due to parameter estimation, control charts with estimated parameters did not behave as they should according to the known parameter calculations. He suggested the use of a Phase I sample of at least $400/(n-1)$ subgroups of size n each, or around 300 observations when $n = 1$. However, for this recommendation he did not consider the variation between practitioners. Since different practitioners obtain different Phase I samples, their estimates and corresponding control limits will vary. As a consequence, the control chart performance in terms of the FAR or ARL will also vary. This effect is extensively discussed by Saleh et al. (2015b), and is often referred to as the *practitioner-to-practitioner* variability. This variation becomes less severe when the sample size increases, but the original sample size suggestions from Quesenberry (1993) are not sufficient to guarantee a performance equivalent to that of the known parameters situation. For S^2 and S charts Epprecht et al. (2015) conclude that the sample size requirements are often closer to several thousands.

Instead of waiting until a sufficient amount of Phase I data is obtained, another option is to make use of control charts that update the parameter estimates. Examples of such methods are the self-starting control charts of Hawkins (1987) and Quesenberry (1991). The advantage of such methods is that the monitoring Phase II can start early, and that the estimation error decreases over time as long as the process remains in-control. However, when data contaminations are present or when there is a shift or drift in the process mean, there is a possibility that these contaminations are incorporated in the updated parameters when they are not detected directly. This issue has been extensively investigated in Huberts et al. (2019). They conclude that the practitioner-to-practitioner variation reduces substantially when the process remains in-control or when the signals are correctly classified, and that the possible performance deterioration depends on the type of control chart and the level of data contamination.

3 Design Criteria

In many situations, the sample size requirements such as indicated in Epprecht et al. (2015) may not be available. To that end, several authors have proposed the use of adjusted control limits. Such an adjustment should be done to meet a certain performance criterion. In general, two directions are possible for control limit adjustment, which are the unconditional and conditional approach. Albers and Kallenberg (2004a, b) introduced their corresponding criteria as the *bias criterion* and the *exceedance probability criterion*, respectively.

3.1 Bias Criterion

Without parameter estimation, the run length (RL) distribution would be geometrical with parameter α_0. From this, several run length properties can be calculated, such as the ARL ($1/\alpha_0$) or other functions $g(\alpha_0)$ of the FAR. When parameters are estimated consecutive false alarm events are dependent, as pointed out by Quesenberry (1993). This means that the unconditional RL distribution (i.e., the RL distribution after averaging out the effects of estimation error) is no longer geometric. As a consequence, many properties of the unconditional RL distribution are different from their nominal value. This was shown for example by Chen (1997), who determined the average and standard deviation of the unconditional RL distribution when control limits are estimated. The bias criterion aims to adjust the control limits such that the control chart provides a specified in-control RL property (such as the FAR or ARL) in *expectation*.

In particular, the conditional false alarm rate (CFAR, conditional on the estimated control limits) can be written as

$$CFAR = 1 - P\left(\widehat{LCL} \leq \bar{X}_i \leq \widehat{UCL}\right), \tag{3}$$

where $\bar{X}_i$ is some in-control Phase II subgroup average at time period i. Conditional on the estimates $\widehat{LCL}$ and $\widehat{UCL}$, the (conditional) RL properties are geometric with parameter CFAR. The properties of the conditional RL distribution for these estimates can then be calculated as a function g of CFAR. For example, the conditional ARL (CARL) is equal to $CARL = g(CFAR) = 1/CFAR$. The unconditional equivalent of these properties is then equal to the expectation of these measures before the control limits are estimated. In general, for any RL property $g(CFAR)$, the bias criterion aims to provide a control chart performance equal to a nominal value $g(\alpha_0)$ in expectation, or more specifically

$$E\left(g(CFAR)\right) = g(\alpha_0). \tag{4}$$

Several researchers have proposed adjustments to the control limit coefficient K in (2) to achieve this for various performance measures. Examples of such adjustments for Shewhart type control charts can be found in Albers and Kallenberg (2004b, 2005), Tsai et al. (2005), Goedhart et al. (2016) and Diko et al. (2017).

3.2 Exceedance Probability Criterion

While it may seem logical to aim for a certain performance in expectation, there could still be a large probability of an unsatisfactory control chart performance for individual practitioners due to practitioner-to-practitioner variation. When the process is in-control, large FAR values (or low ARL values) are undesirable, as that would mean that the control chart produces many false signals. This could lead to a waste of time and effort invested in finding a non-existing special cause. As an alternative design criterion, Albers and Kallenberg (2004a, 2005) proposed the exceedance probability criterion, which aims to provide a specified minimum in-control performance with a specified large probability by focussing on the conditional performance. For example, when considering the CFAR as performance measure, this criterion can be denoted as

$$P(CFAR \leq \alpha_0) \geq 1 - p, \tag{5}$$

for some small probability p. Note that, since $CARL = 1/\alpha_0$ is a monotonically decreasing function of α_0, (5) is equivalent to $P(CARL \geq 1/\alpha_0) = 1 - p$, which would be the exceedance probability criterion for the CARL as performance measure. Note also that, when considering individual observations ($n = 1$), this objective for Shewhart control charts is equivalent to that of a tolerance interval (see e.g., Krishnamoorthy and Mathew 2009).

In order to satisfy the criterion, the control limit coefficient K in (2) should be adjusted. Various authors have proposed bootstrap, analytical and/or numerical (approximation) methods to determine the required control limit adjustments. Examples of these are given in Albers and Kallenberg (2004a, 2005), Jones and Steiner (2012), Gandy and Kvaløy (2013), Faraz et al. (2015), Saleh et al. (2015a), Goedhart et al. (2017b, a, 2018).

While this approach limits the possibility of an insufficient in-control performance, it does not remove the practitioner-to-practitioner variation. That means that the ARL values tend to be larger after the adjustment. Although this is beneficial in the in-control situation, one might expect that the out-of-control detection speed would suffer substantially. However, as concluded by Faraz et al. (2015), Saleh et al. (2015b) and Jardim et al. (2020) among others, the adjustments do not have too much of an adverse effect on the out-of-control performance. For that reason, the conditional approach of the exceedance probability criterion is suggested more often in recent literature.

4 Distributional Assumptions

After deciding on the desired design criterion, the next step is to actually achieve it. This requires appropriate modeling of the Phase I in-control distribution, where the functional form and corresponding underlying distributional assumptions of control limits form an important role. For example, the control limits as displayed in (2) are designed for normally distributed data. When the data distribution is skewed, these limits are no longer appropriate and the performance of the design criterion will not be met.

4.1 *Parametric Methods*

When the distribution of the data is known, one can determine the required control limits by using the corresponding probability limits. An S^2 chart using probability limits to monitor the standard deviation of a normally distributed variable is one such example of an approach since the monitoring characteristic follows a chi-squared distribution. The same approach can be used for other distributions of interest. For example, when considering the exceedance probability criterion, Krishnamoorthy and Mathew (2009) provide tolerance intervals for a wide range of distributions. However, in practice, the distribution of the data is generally unknown and has to be estimated as well. This was also indicated by Albers et al. (2004), who divided the total estimation error in two distinct factors: the *model error* and the *stochastic error*. The first is caused by inadequate model assumptions, while the latter is the result of estimation uncertainty such as considered in Sect. 3.

In order to reduce the model error, various parametric approaches are possible. The first option is to use data transformations or aggregations in order to make the normal theory more applicable. For example, a common practice in SPM is the use of subgroups. When considering the average of a subgroup, the normality assumptions should become more appropriate due to the central limit theorem (CLT). This approach is evaluated in Huberts et al. (2018) for various distributions. However, they conclude that for the applications in SPM, the CLT should be used with caution. Since the interest in SPM lies in the far end of the tails of the distribution, the convergence to normality as indicated by CLT is not (quick) enough when the data distribution is highly skewed or has wide tails. Other parametric alternatives are to use data transformations to make normal theory more applicable (e.g., Box and Cox 1964, Chou et al. 1998), or to use more general parametric models such as the Pearson system of distributions. However, Goedhart (2018) illustrates that for these methods, similar conclusions hold as for the use of the CLT. Because control charts are generally based on small tail probabilities, minor model deviations can cause major deviations in the obtained control chart performance. This leaves two alternative directions to control the control chart performance when the distribution is unknown: nonparametric methods and/or being more lenient in the in-control performance demands.

4.2 Nonparametric Methods

There is a wide range of literature available on nonparametric estimation methods in statistical process monitoring. For general information and detailed overview on this literature, we refer to Chakraborti et al. (2001, 2015), Chakraborti and Graham (2019) and Qiu (2018). The main advantage of such methods is that the model error is no longer an issue. Generally this means that the stochastic error increases, which results in larger sample size requirements. For example, when using the nonparametric tolerance intervals from Krishnamoorthy and Mathew (2009) to satisfy the exceedance probability criterion in (5) when $\alpha_0 = 0.0027$ and $p = 0.1$, the minimum sample size is already equal to $m = 1440$ individual observations. Adjustments based on extrapolation are available to lower this requirement somewhat (see Young and Mathew 2014, Goedhart et al. 2020).

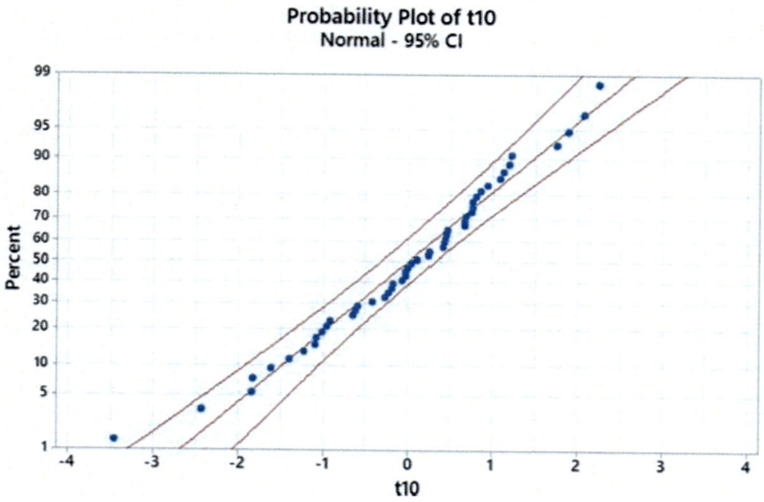

Fig. 1 Probability plot of $m = 50$ simulated observations from a t_{10} distribution

While this may seem like a major disadvantage of nonparametric methods, it is important to realize that when sample sizes are small, parametric assumptions are more difficult to validate in the first place. As a simple example, consider a Student's t-distribution with $df = 10$ degrees of freedom (t_{10}). This distribution has a mean equal to 0 and a standard deviation equal to 1.12. We simulate a random sample of $m = 50$ individual observations ($n = 1$) from this distribution, and find estimates of the mean and standard deviation of -0.008 and 1.150, respectively. For the Anderson–Darling test for normality, we find an AD value of 0.370 and a p-value of 0.412. A probability plot of the simulated sample is given in Fig. 1. In summary, there is no real reason to reject the normality assumption at this point. However, if we consider the actual quantiles of the t_{10} and the $N(0, 1.12)$ distributions as a comparison, the differences are quite large. In particular, the $1 - 0.00135$ $(1 - 0.0027/2)$ quantiles of these distributions equal 3.96 and 3.36, respectively. This already leads to control chart performance issues when the data distributions follow known theoretical functional forms such as the normal and t-distribution, but in practice, even this is not the case. As a consequence, it does not seem reasonable to expect an accurate estimate of the 0.27% (or even 0.135% for two-sided control charts) percentiles of an unknown probability distribution based on a small amount of data. This means that either more data should be collected, or lower performance demands should be placed on the control chart.

5 Sample Size and Strictness

Following the aspects discussed in the previous sections, decisions have to be made regarding the amount of Phase I data (m subgroups of size n) and the strictness of the design criteria (α_0, and possibly p).

When reliable data are available in abundance and/or easy to collect, this opens up the way for the use of nonparametric methods for a more robust control chart performance, even for small values of α_0 and p. At the same time, the effect of parameter estimation is then less severe in the first place, and distributional assumptions are easier to validate. When large amounts of data are not available, this is not as straightforward. As mentioned earlier, data requirements for nonparametric methods are generally considered large. On the other hand, when the available datasets are small, the appropriateness of a parametric model is difficult to validate, which may lead to undesirable control chart performances as well when model errors are substantial. An important consideration could be the reason behind the small sample. For example, it could be because of high costs associated with sampling, or because of the nature of the process. When sampling is difficult or costly in Phase I, this probably also holds for the monitoring part in Phase II. Although a larger value of α_0 means a lower in-control ARL when time units are measured in a number of observations, the impact could be limited when considering time instead. Therefore, lowering the desired (minimum) in-control ARL, which would lower the sample size requirements for a controlled control chart performance, may be required in such a situation.

6 Concluding Remarks

Generally, the in-control behavior of a process is unknown and requires estimation using a Phase I reference sample. This brings several challenges and trade-offs with it regarding the design of a control chart. Although adding additional assumptions improve the performance of statistical methods in theory when appropriate, they are often difficult to validate in practice. To this end, it may be advisable to be more conservative in the performance requirements instead. The issues considered also hold for control charts other than Shewhart type charts and are relevant to other statistical applications in a similar way.

References

Albers, W., & Kallenberg, W. C. M. (2004a). Are estimated control charts in control? *Statistics*, *38*(1), 67–79.

Albers, W., & Kallenberg, W. C. M. (2004b). Estimation in Shewhart control charts: Effects and corrections. *Metrika*, *59*(3), 207–234.

Albers, W., & Kallenberg, W. C. M. (2005). New corrections for old control charts. *Quality Engineering*, *17*(3), 467–473.

Albers, W., Kallenberg, W. C. M., & Nurdati, S. (2004). Parametric control charts. *Journal of Statistical Planning and Inference*, *124*(1), 159–184.

Box, G. E. P., & Cox, D. R. (1964). An analysis of transformations. *Journal of the Royal Statistical Society Series B (Methodological)*, *26*(2), 211–252.

Chakraborti, S., & Graham, M. A. (2019). *Nonparametric statistical process control*. New York: Wiley.

Chakraborti, S., Van der Laan, P., & Bakir, S. (2001). Nonparametric control charts: an overview and some results. *Journal of Quality Technology*, *33*(3), 304–315.

Chakraborti, S., Qiu, P., & Mukherjee, A. (2015). Editorial to the special issue: Nonparametric statistical process control charts. *Quality and Reliability Engineering International*, *31*(1), 1–2.

Chen, G. (1997). The mean and standard deviation of the run length distribution of $\bar{X}$ charts when control limits are estimated. *Statistica Sinica*, *7*(3), 789–798.

Chou, Y. M., Polansky, A. M., & Mason, R. L. (1998). Transforming nonnormal data to normality in statistical process control. *Journal of Quality Technology*, *30*(2), 133–141.

Cryer, J. D., & Ryan, T. P. (1990). The estimation of sigma for an X chart: MR/d_2 or S/c_4? *Journal of Quality Technology*, *22*(3), 187–192.

Diko, M. D., Goedhart, R., Chakraborti, S., Does, R. J. M. M., & Epprecht, E. K. (2017). Phase II control charts for monitoring dispersion when parameters are estimated. *Quality Engineering*, *29*(4), 605–622.

Epprecht, E. K., Loureiro, L. D., & Chakraborti, S. (2015). Effect of the amount of phase I data on the phase II performance of S^2 and S control charts. *Journal of Quality Technology*, *47*(2), 139–155.

Faraz, A., Woodall, W. H., & Heuchenne, C. (2015). Guaranteed conditional performance of the S^2 control chart with estimated parameters. *International Journal of Production Research*, *53*(14), 4405–4413.

Gandy, A., & Kvaløy, J. T. (2013). Guaranteed conditional performance of control charts via bootstrap methods. *Scandinavian Journal of Statistics*, *40*(4), 647–668.

Goedhart, R. (2018). Statistical Control of Shewhart Control Charts. Doctoral Dissertation, University of Amsterdam

Goedhart, R., Schoonhoven, M., & Does, R. J. M. M. (2016). Correction factors for Shewhart X and $\bar{X}$ control charts to achieve desired unconditional ARL. *International Journal of Production Research*, *54*(24), 7464–7479.

Goedhart, R., Schoonhoven, M., & Does, R. J. M. M. (2017a). Guaranteed in-control performance for the Shewhart X and $\bar{X}$ control charts. *Journal of Quality Technology*, *49*(2), 155–171.

Goedhart, R., da Silva, M. M., Schoonhoven, M., Epprecht, E. K., Chakraborti, S., Does, R. J. M. M., et al. (2017b). Shewhart control charts for dispersion adjusted for parameter estimation. *IISE Transactions*, *49*(8), 838–848.

Goedhart, R., Schoonhoven, M., & Does, R. J. M. M. (2018). On Guaranteed In-Control Performance for the Shewhart X and $\bar{X}$ Control Charts. *Journal of Quality Technology*, *50*(1), 130–132.

Goedhart, R., Schoonhoven, M., & Does, R. J. M. M. (2020). Nonparametric control of the conditional performance in statistical process monitoring. *Journal of Quality Technology, 52*(4), 355–369.

Hawkins, D. M. (1987). Self-starting CUSUM charts for location and scale. *The Statistician*, *36*(4), 299–316.

Huberts, L. C. E., Schoonhoven, M., Goedhart, R., Diko, M. D., & Does, R. J. M. M. (2018). The performance of control charts for large non-normally distributed datasets. *Quality and Reliability Engineering International*, *34*(6), 979–996.

Huberts, L. C. E., Schoonhoven, M., & Does, R. J. M. M. (2019). The effect of continuously updating control chart limits on control chart performance. *Quality and Reliability Engineering International*, *35*(4), 1117–1128.

Jardim, F. S., Chakraborti, S., & Epprecht, E. K. (2020). Two perspectives for designing a phase II control chart with estimated parameters: The case of the Shewhart chart. *Journal of Quality Technology, 52*(2), 198–217.

Jensen, W. A., Jones-Farmer, L. A., Champ, C. W., & Woodall, W. H. (2006). Effects of parameter estimation on control chart properties: a literature review. *Journal of Quality Technology, 38*(4), 349–364.

Jones, M. A., & Steiner, S. H. (2012). Assessing the effect of estimation error on risk-adjusted CUSUM chart performance. *International Journal for Quality in Health Care, 24*(2), 176–181.

Jones-Farmer, L. A., Woodall, W. H., Steiner, S. H., & Champ, C. W. (2014). An overview of phase I analysis for process improvement and monitoring. *Journal of Quality Technology, 46*(3), 265–280.

Krishnamoorthy, K., & Mathew, T. (2009). *Statistical tolerance regions: theory, applications, and computation*. New York, NY: Wiley.

Mahmoud, M. A., Henderson, G. R., Epprecht, E. K., & Woodall, W. H. (2010). Estimating the standard deviation in quality-control applications. *Journal of Quality Technology, 42*(4), 348–357.

Psarakis, S., Vyniou, A. K., & Castagliola, P. (2014). Some recent developments on the effects of parameter estimation on control charts. *Quality and Reliability Engineering International, 30*(8), 1113–1129.

Qiu, P. (2018). Some perspectives on nonparametric statistical process control. *Journal of Quality Technology, 50*(1), 49–65.

Quesenberry, C. P. (1991). SPC Q charts for start-up processes and short or long runs. *Journal of Quality Technology, 23*(3), 213–224.

Quesenberry, C. P. (1993). The effect of sample size on estimated limits for $\bar{X}$ and X control charts. *Journal of Quality Technology, 25*(4), 237–247.

Saleh, N. A., Mahmoud, M. A., Jones-Farmer, L. A., Zwetsloot, I. M., & Woodall, W. H. (2015a). Another look at the EWMA control chart with estimated parameters. *Journal of Quality Technology, 47*(4), 363–382.

Saleh, N. A., Mahmoud, M. A., Keefe, M. J., & Woodall, W. H. (2015b). The difficulty in designing Shewhart $\bar{X}$ and X control charts with estimated parameters. *Journal of Quality Technology, 47*(2), 127–138.

Schoonhoven, M., & Does, R. J. M. M. (2012). A robust standard deviation control chart. *Technometrics, 54*(1), 73–82.

Schoonhoven, M., Nazir, H. Z., Riaz, M., & Does, R. J. M. M. (2011). Robust location estimators for the $\bar{X}$ control chart. *Journal of Quality Technology, 43*(4), 363–379.

Tsai, T. R., Lin, J. J., Wu, S. J., & Lin, H. C. (2005). On estimating control limits of $\bar{X}$ chart when the number of subgroups is small. *International Journal of Advanced Manufacturing Technology, 26*(11–12), 1312–1316.

Young, D. S., & Mathew, T. (2014). Improved nonparametric tolerance intervals based on interpolated and extrapolated order statistics. *Journal of Nonparametric Statistics, 26*(3), 415–432.

On the Calculation of the ARL for Beta EWMA Control Charts

Sven Knoth

Abstract Accurate calculation of the Average Run Length (ARL) for exponentially weighted moving average (EWMA) charts might be a tedious task. The omnipresent Markov chain approach is a common and effective tool to perform these calculations — see Lucas and Saccucci (1990) and Saccucci and Lucas (1990) for its application in case of EWMA charts. However, Crowder (1987b) and Knoth (2005) provided more sophisticated methods from the rich literature of numerical analysis to solve the ARL integral equation. These algorithms lead to very fast implementations for determining the ARL with high accuracy such as Crowder (1987a), or the R package `spc` (Knoth 2019) with its functions `xewma.arl()` and `sewma.arl()`. Crowder (1987a) utilized the popular Nyström method (Nyström 1930) which fails for bounded random variables existing, for example, in the case of an EWMA chart monitoring the variance. For the latter, Knoth (2005) utilized the so-called collocation method. It turns out that the numerical problems are even more severe for beta distributed random variables, which are bounded from both sides, typically on (0, 1). We illustrate these subtleties and provide extensions from Knoth (2005) to achieve high accuracy in an efficient way.

Keywords Integral equation · Markov chain approximation · Collocation · Change point detection

1 Introduction

Compared to much more prominent continuous distributions such as the ubiquitous normal or the slightly less popular exponential, gamma or Weibull distributions, the beta distribution is sporadically used as vehicle for monitoring designs. There are a few cases such as Yousry et al. (1991) in a Bayes framework, Reynolds and Stoumbos

S. Knoth (✉)
Department of Mathematics and Statistics, Faculty of Economics and Social Sciences, Helmut Schmidt University Hamburg, Postfach 700822, 22008 Hamburg, Germany
e-mail: Sven.Knoth@hsu-hh.de

S. Knoth and W. Schmid (eds.), *Frontiers in Statistical Quality Control 13*,
Frontiers in Statistical Quality Control,
https://doi.org/10.1007/978-3-030-67856-2_3

(2005) providing a solid non-normal distribution example, Grigg and Spiegelhalter (2007) and Gan and Tan (2010), Loke and Gan (2012), Gan et al. (2012) dealing with risk-adjusted data in the field of health monitoring, where beta distribution is treated within the field of statistical process monitoring (SPM). In addition, the beta distribution is also used when handling order statistics in SPM, see Castagliola (2001) as particular example and Graham et al. (2012) for more general statements. However, only Loke and Gan (2012) explicitly monitored beta distributed variables to detect changes in the risk distribution within a large set of future patients. While Loke and Gan (2012) dealt with CUSUM (cumulative sum) charts, we analyze one-sided EWMA (exponentially weighted moving average) charts. In summary, we outline and evaluate numerical methods (Markov chain, Nyström, collocation) for calculating ARL (average run length) values for upper EWMA control charts.

In Sect. 2, the beta distribution is described in detail, and the EWMA chart for beta distributed data is introduced. Then, in Sect. 3, the aforementioned numerical procedures are presented, including some new recipe patterns. Utilizing these algorithms, we study their behavior for various shapes of beta distributions in Sect. 4. Conclusions are given in Sect. 5.

2 The Beta Distribution and the EWMA Control Chart

The beta distribution appeared in the realm of Karl Pearson's distributions introduced in Pearson (1895, 1916) as Pearson Type I and II distribution – see Lahcene (2013) for historical details. We follow standard literature about statistical distributions such as Johnson et al. (1995) or Forbes et al. (2011). For the sake of brevity, we use the standardized version. Hence, let X be a continuous random variable on (0, 1) ([0, 1] would be possible as well) having the following properties:

$$\text{PDF } f(x) = \frac{1}{B(\alpha,\beta)} x^{\alpha-1}(1-x)^{\beta-1} \quad \text{with} \quad B(\alpha,\beta) = \frac{\Gamma(\alpha)\Gamma(\beta)}{\Gamma(\alpha+\beta)} . \tag{1}$$

$$\text{CDF } F(x) = \frac{B(x;\alpha,\beta)}{B(\alpha,\beta)} = I_x(\alpha,\beta) \quad \text{with} \quad B(x;\alpha,\beta) = \int_0^x t^{\alpha-1}(1-t)^{b-1}\,dt . \tag{2}$$

PDF and CDF denote the probability density and the cumulative distribution function, respectively. Note that the two parameters α and β are positive. Moreover, $\Gamma()$ denotes the gamma function. The functions $B(,)$, $B(;,,)$, and $I_x(,)$ are the complete, incomplete, and regularized incomplete beta function, respectively. They are implemented in standard mathematical or statistical software packages for easy

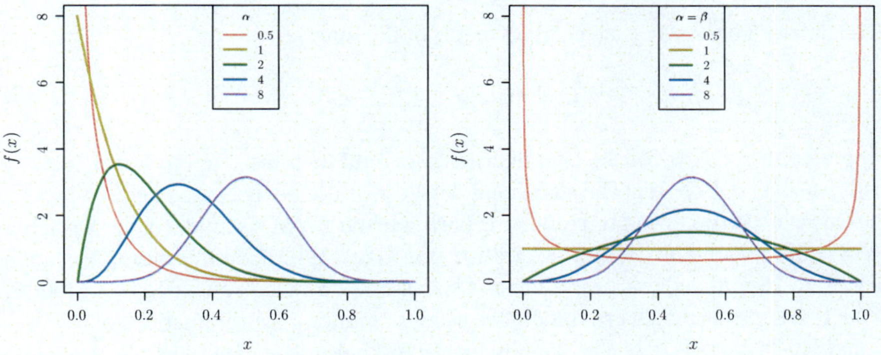

Fig. 1 Selected beta density functions, on the left β is fixed at 8

deployment. In the sequel, we write $X \sim \text{Beta}(\alpha, \beta)$ for declaring that X follows a beta distribution with parameters α and β. Based on the mentioned literature:

$$E(X) = \frac{\alpha}{\alpha + \beta} . \tag{3}$$

$$Var(X) = \frac{\alpha\beta}{(\alpha + \beta)^2(\alpha + \beta + 1)} . \tag{4}$$

$$Mode = \frac{\alpha - 1}{\alpha + \beta - 2} \quad \text{for } \alpha, \beta > 1 .$$

$$Skewness = \frac{2(\beta - \alpha)\sqrt{\alpha + \beta + 1}}{(\alpha + \beta + 2)\sqrt{\alpha\beta}} .$$

From (1), we conclude that $1 - X \sim \text{Beta}(\beta, \alpha)$. There are further interesting features and connections to other distributions, see again Johnson et al. (1995) or Forbes et al. (2011). In Fig. 1, we illustrate some typical beta distribution patterns choosing α and β from $\{0.5, 1, 2, 4, 8\}$. First, $\beta = 8$ is fixed and α varies, while in the second diagram, α and β coincide. The various shapes in Fig. 1 show the flexibility of the beta distribution. Besides the uniform ($\alpha = \beta = 1$) distribution, we recognize for $(\alpha, \beta) = (1, 8)$ a similar shape as the exponential distribution (which is not bounded to the right-hand side) and for $(\alpha, \beta) = (8, 8)$ a normal bell shape (again, a truncated version).

In order to detect changes in the mean of X, we deploy EWMA control charts, which were proposed by Roberts (1959). Hence, while observing sequentially $X_1, X_2, \ldots$, we apply the common EWMA smoothing.

$$\begin{aligned} Z_0 &= z_0 = \mu_0 := E_{\text{in-control}}(X) , \\ Z_t &= (1 - \lambda) Z_{t-1} + \lambda X_t , \quad t = 1, 2, \ldots \end{aligned}$$

An alarm is given, when Z_t is large, that is at the time

$$L = \inf\{t \geq 1 \colon Z_t > c_u\}\,.$$

Upper control charts for the beta distribution could be transferred to lower ones by using $1 - X \sim \mathrm{B}(\beta, \alpha)$. The two-sided design will not be discussed here.

One can also incorporate a reflecting barrier to the EWMA sequence, see Yashchin (1987), Crowder and Hamilton (1992), and Gan (1993). Furthermore, Yashchin (1989) re-formulated this procedure as a special case of weighted CUSUM of type 2. Note that such barriers were sometimes used for numeric approximations of conventional EWMA ARL (e.g., see Waldmann 1986), but not as genuine control scheme features. Contrary to the previous frameworks, our EWMA statistic Z_t exhibits the "natural" lower border 0. Therefore, it is less susceptible against inertia which is a pronounced problem for EWMA sequences without this reflecting border, as was demonstrated in Woodall and Mahmoud (2005). In the last paragraph of Sect. 4, we briefly discuss these patterns.

The upper control limit could also be written as the usual "mean + value × asymptotic standard deviation" of Z_t $(t \to \infty)$. For doing this, we recall that (in-control case)

$$\begin{aligned} E(Z_t) &= \mu_0\,, \\ Var(Z_t) &= \frac{\lambda}{2-\lambda}\big(1-(1-\lambda)^{2t}\big)Var(X) \to \frac{\lambda}{2-\lambda}Var(X)\,, \quad t \to \infty\,. \end{aligned}$$

Then we write

$$c_u = \mu_0 + U\sqrt{\frac{\lambda}{2-\lambda}}\sqrt{Var(X)}\,.$$

The design parameter c_u alias U is chosen to give a pre-defined in-control ARL (average run length). The latter is the most popular performance measure for control charts and was introduced in Page (1954). When all X_t follow the same beta distribution (either the in-control one or some disturbed version), it quantifies the expected number of observations (runs) until signal. Hence, we define $\ell := E(L)$. There are more sophisticated measures, but most of them are ARL related. In summary, the ARL measure is essential to evaluate and calibrate control charts. Hence, its calculation is an important prerequisite for utilizing control charts. Before presenting some methods to calculate the ARL, we emphasize that it is useful to deal with the ARL function $\ell(z)$ for an arbitrary EWMA starting value $Z_0 = z$. Besides the default value $\ell(z_0)$, it is also useful to consider $\ell(0)$ in the case of an increased mean (our detection objective), because it provides a worst-case bound for the out-of-control ARL.

3 Numerical Methods to Calculate EWMA ARL

First, we describe the most popular approach, namely the Markov chain approximation which was introduced into the control chart field by Brook and Evans (1972) and made popular for EWMA in Lucas and Saccucci (1990). It relies on the decomposition of the chart's continuation region $[0, c_u]$ into N subintervals of size $w = c_u/N$. Then we replace the original movement of the EWMA statistic Z_t by steps on the grid $\frac{w}{2}, \frac{3w}{2}, \ldots, \frac{(2N-1)w}{2}$. The probabilities to move from interval i to j are collected via

$$\begin{aligned} q_{ij} &= P\Big(Z_t \in \big((j-1)w, jw\big] \mid Z_{t-1} = (i-0.5)w\Big) \\ &= F\left(\frac{[j-(1-\lambda)(i-0.5)]w}{\lambda}\right) - F\left(\frac{[j-1-(1-\lambda)(i-0.5)]w}{\lambda}\right) \end{aligned}$$

in the matrix $\mathbb{Q} = (q_{ij}), i, j = 1, 2, \ldots, N$. Recall that $F(\cdot)$ denotes the CDF of the beta distribution. Another, slightly more complicated approach to fill this matrix was proposed by Hawkins (1992).

$$\begin{aligned} \tilde{q}_{ij} &= P\Big(Z_t \in \big((j-1)w, jw\big] \mid Z_{t-1} \in ((i-1)w, iw]\Big) \\ &\approx \Bigg[F\left(\frac{[j-(1-\lambda)(i-1)]w}{\lambda}\right) + 4F\left(\frac{[j-(1-\lambda)(i-0.5)]w}{\lambda}\right) + \\ &\quad F\left(\frac{[j-(1-\lambda)i]w}{\lambda}\right) - F\left(\frac{[j-1-(1-\lambda)(i-1)]w}{\lambda}\right) - \\ &\quad 4F\left(\frac{[j-1-(1-\lambda)(i-0.5)]w}{\lambda}\right) - F\left(\frac{[j-1-(1-\lambda)i]w}{\lambda}\right)\Bigg] /6 \,. \end{aligned}$$

Using standard results from Markov chain theory, we determine the solution of the following linear equation system:

$$(\mathbb{I} - \mathbb{Q})\boldsymbol{\ell} = \mathbf{1} \,,$$

where $\mathbb{I}$ denotes the identity matrix of size $N \times N$, $\mathbf{1}$ is a vector containing N ones, and $\boldsymbol{\ell}$ yields the discrete approximation of the ARL function $\ell(\cdot)$. After solving this equation, we need to pick the element that corresponds to z_0. The following code lines, written in R, demonstrate the simplicity and practicability of this approach.

```
ewmaU.arl <- function(lambda, cu, Alpha, Beta, z0, N=50) {
  i <- 1:N;  w <- cu/N
  qij <- function(i,j) {
    pbeta( w*(j  -(1-lambda)*(i-.5))/lambda, Alpha, Beta) -
    pbeta( w*(j-1-(1-lambda)*(i-.5))/lambda, Alpha, Beta) }
  QQ <- outer(i, i, qij)
  one <- rep(1, N);  II <- diag(1, N)
  ARL <- solve(II-QQ, one)
```

```
  arl <- ARL[round(z0/w)]
  arl
}
```

However, in the next section, we will see that quite large values of N are needed (for both q_{ij} and $\tilde{q}_{ij}$) to achieve decent accuracy.

Another algorithm to calculate the EWMA ARL is the Nyström method which was utilized the first time for EWMA by Crowder (1987b). It starts with the ARL integral equation,

$$\ell(x) = 1 + \int_0^{c_u} \ell(y) \frac{1}{\lambda} f\left(\frac{y - (1-\lambda)x}{\lambda}\right) dy\,,$$

where $f()$ is the PDF of the beta distribution. The core idea of the Nyström method is to replace the integral on the right-hand side by a quadrature. We obtain

$$\ell(z_i) = 1 + \sum_{j=1}^{N} w_j \ell(z_j) \frac{1}{\lambda} f\left(\frac{z_j - (1-\lambda)z_i}{\lambda}\right).$$

The quadrature weights w_j and nodes z_j, $j = 1, 2, \ldots, N$, are either very simple ones as for the Simpson rule or more sophisticated (and more effective) ones such as for the Gauß-Legendre quadrature. By utilizing the product midpoint rule, we recover the Markov chain equation system. The above system is solved in the usual way. Afterward, by applying the Nyström interpolation based on $\ell(z_j)$, $j = 1, 2, \ldots, N$,

$$\ell(z) = 1 + \sum_{j=1}^{N} w_j \ell(z_j) \frac{1}{\lambda} f\left(\frac{z_j - (1-\lambda)z}{\lambda}\right),$$

we approximate the ARL function for every $0 \le z \le c_u$. As we see later, the Nyström method works well only for a few of our considered parameter situations.

A further numerical procedure to solve (approximately, but with high accuracy) the aforementioned integral equation is collocation. An early treatment of collocation for the ARL integral equation was given by Gianino et al. (1990). However, its first "indispensable" application was Knoth (2005), where the ARL of S^2 EWMA was calculated. The basic principle is to approximate the function $\ell(x)$ by, for example, polynomials. Then the original problem is reduced to one of determining a finite number of polynomial coefficients. We use, as in Knoth (2005), Chebyshev polynomials defined for the interval $[0, c_u]$. Using Chebyshev polynomials of order up to $N-1$, namely $T_0^*()$, ..., $T_{N-1}^*()$, we write

$$\sum_{j=1}^{N} c_j T_{j-1}^*(z_i) = 1 + \sum_{j=1}^{N} c_j \int_{(1-\lambda)z_i}^{\min\{(1-\lambda)z_i+\lambda, c_u\}} T_{j-1}^*(y) \frac{1}{\lambda} f\left(\frac{y-(1-\lambda)z_i}{\lambda}\right) dy\,.$$

By evaluating the above equation at reasonably chosen $z = z_i$ (here the roots of the Chebyshev polynomial of order N) and determining the definite integrals (deploying quadrature again), we obtain a linear equation system. Note that we are able to treat the special support of the beta variable X by adjusting the integral limits. Later we see that this procedure, which was used in the same way for the upper EWMA S^2 chart, works well for most of the considered parameter configurations. But for some other, it deteriorates. Below we introduce a setup that accounts more effectively for the bounded support of the beta distribution. In fact, a piece-wise design helps a lot for these subtle configurations. We construct these subintervals starting from

$$\left(\frac{c_u - \lambda}{1 - \lambda}, c_u\right] ,$$

which is the only subinterval allowing an alarm signal within the next observation. The other subintervals are built in the same way, allowing a signal in at least two, three, etc. observations. Some arithmetics give the number of subintervals K and their shape $[a_k, b_k]$, $k = 1, 2, \ldots, K$:

$$\begin{aligned} K &= \left\lceil \frac{\log(1 - c_u)}{\log(1 - \lambda)} \right\rceil . \\ b_k &= 1 - \frac{1 - c_u}{(1 - \lambda)^{K-k}} , \quad (b_K = c_U) . \\ a_k &= \max\left\{0, 1 - \frac{1 - c_u}{(1 - \lambda)^{K-k+1}}\right\} , \quad (a_1 = 0) . \\ b_k - a_k &= \frac{\lambda(1 - c_u)}{(1 - \lambda)^{K-k+1}} , \quad \text{(subinterval widths)} . \end{aligned}$$

Eventually, we obtain a more complicated linear equation system, for $z_{ki} \in [a_k, b_k]$, $k = 1, 2, \ldots, K$ and $i = 1, 2, \ldots, N$:

$$\sum_{j=1}^{N} c_{kj} T^*_{k,j-1}(z_{ki}) = 1 + \sum_{l} \sum_{j=1}^{N} c_{lj} \int_{\ldots}^{\ldots} T^*_{l,j-1}(y) \frac{1}{\lambda} f\left(\frac{y - (1 - \lambda)z_{ki}}{\lambda}\right) dy \quad (5)$$

The integral limits are now more involved. By accounting for the restricted supports of $T^*_{l,\cdot}()$, the lower limit is given by $\max\{(1 - \lambda)z_{ki}, a_l\}$ and the upper one by $\max\{(1 - \lambda)z_{ki} + \lambda, b_l\}$. Moreover, the summation over $l = 1, 2, \ldots, K$ could be reduced. It ends already at $\min\{k + 1, K\}$ and starts possibly much later than at $l = 1$. For the considered configurations we provide the number of intervals in Table 1, where the EWMA constant is set to $\lambda = 0.1$, and the target in-control ARL to $\ell^! = 500$.

Both collocation setups rely on reliable quadrature procedures. Knoth (2005) showed that substitutions improve the standard Gauß-Legendre quadrature a lot. One idea is to substitute $x^2 = y - (1 - \lambda)z_{ki}$ to make the quadrature robust for $\alpha < 1$ (density $f()$ not bounded at 0). In case of the piece-wise design, we can apply this

Table 1 Some example configurations, alarm thresholds c_u (aka U), and number K of subintervals for the piece-wise collocation design; $\lambda = 0.1$, $\ell^! = 500$

(α, β)	(0.5,8)	(1,8)	(2,8)	(4,8)	(8,4)	(8,2)	(8,1)	(8,0.5)
c_u	0.114208	0.178124	0.276491	0.412137	0.738949	0.862811	0.936718	0.974214
U	3.162384	2.939243	2.764543	2.627252	2.409846	2.270129	2.097802	1.886420
K	2	2	4	6	13	19	27	35
(α, β)	(8,8)	(4,4)	(2,2)	(1,1)	(0.5,0.5)			
c_u	0.570147	0.596072	0.628256	0.664641	0.700625			
U	2.521399	2.512603	2.500164	2.486026	2.473469			
K	9	9	10	11	12			

idea also for $\beta < 1$ by setting $x^2 = (1-\lambda)z_{ki} + \lambda - y$ to deal with $f()$ that is now not bounded at 1. The latter is applied for $l = k+1$ in (5). Another transformation used here is to perform partial integration when computing the terms of (5).

$$
\begin{aligned}
I_{kil} &= \int_{\max\{(1-\lambda)z_{ki},a_l\}}^{\max\{(1-\lambda)z_{ki}+\lambda,b_l\}} T^*_{l,j-1}(y)\frac{1}{\lambda} f\left(\frac{y-(1-\lambda)z_{ki}}{\lambda}\right) dy \\
&= T^*_{l,j-1}(y) F\left(\frac{y-(1-\lambda)z_{ki}}{\lambda}\right)\Bigg|_{\max\{(1-\lambda)z_{ki},a_l\}}^{\max\{(1-\lambda)z_{ki}+\lambda,b_l\}} \\
&\quad - \int_{\max\{(1-\lambda)z_{ki},a_l\}}^{\max\{(1-\lambda)z_{ki}+\lambda,b_l\}} t^*_{l,j-1}(y) F\left(\frac{y-(1-\lambda)z_{ki}}{\lambda}\right) dy\,.
\end{aligned}
$$

Again, $T^*_{l,j-1}()$ denote the modified Chebyshev polynomials and $t^*_{l,j-1}()$ stand for their first derivatives. The latter can be determined easily by applying the rule $t_j(x) = j/(1-x^2)\big(T_{j-1}(x) - xT_j(x)\big)$ for $j > 1$ (plus $t_0(x) \equiv 0$ and $t_1(x) \equiv 1$). The versions without "$*$" correspond to the ordinary Chebyshev polynomials on $[-1, 1]$. It holds that $t^*_{l,j-1}(y) = 2/(b_l - a_l)\, t_{l,j-1}\big((2y - a_l - b_l)/(b_l - a_l) - 1\big)$ for $y \in [a_l, b_l]$ and, of course, $T^*_{l,j-1}(y) = T_{j-1}\big((2y - a_l - b_l)/(b_l - a_l) - 1\big)$. More sophisticated modifications are possible,but the latter works well enough for practical purposes.

In the following section, we compare the two Markov chain designs, the Nyström procedure with both Gauß-Legendre and Simpson rule nodes, and collocation without and with piece-wise setup of the base functions.

4 Comparison Study

We start with the most popular ARL calculation technique, the Markov chain approximation following Lucas and Saccucci (1990). We consider two (α, β) configurations, one $(1, 8)$ works well and another one $(8, 1)$ that leads to problematic behavior. The

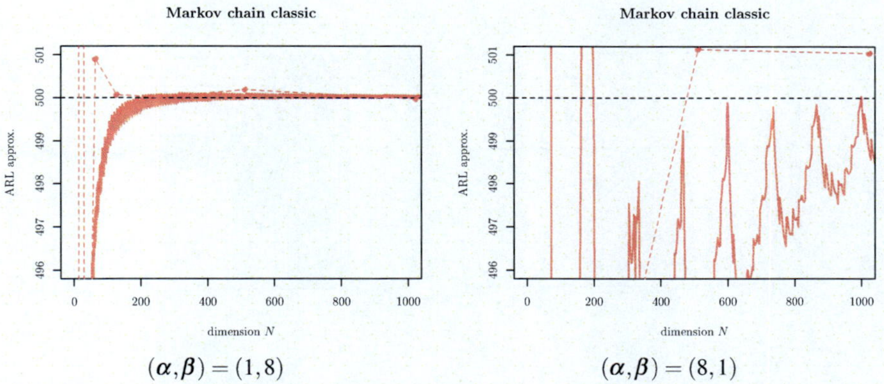

Fig. 2 Classic Markov chain approximation of an upper EWMA chart for two beta distributions. True ℓ is 500, $\lambda = 0.1$, N runs from 10 to 1000 by 1

EWMA smoothing constant is set to $\lambda = 0.1$ throughout this section. Figure 2 illustrates the resulting ARL approximation patterns, augmented with some potentially improving numerical extensions (dashed lines). The conclusions are twofold. On the one hand, we observe combinations where the Markov chain approximation is a reasonable tool to calculate EWMA ARL in case of beta distributions. On the other hand, we discover others where it does not stabilize for N up to 1000. In particular, most of the ARL approximations for (8, 1) in Figs. 2 and 3 are far away from the correct value 500 so that parts of the profiles are not visible.

Applying convergence acceleration techniques (see Brezinski 1985, for some more theoretical thoughts) might help. Popular attempts in the statistical process monitoring literature are Brook and Evans (1972), Lucas (1982), Lucas and Saccucci (1990), and Hawkins (1992). The latter deploys Richardson extrapolation and yields

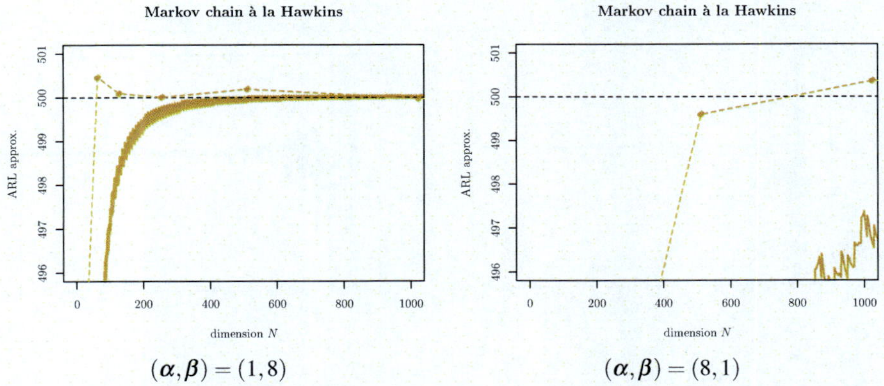

Fig. 3 Markov chain approximation following Hawkins (1992) of an upper EWMA chart. True ℓ is 500, $\lambda = 0.1$, N runs from 10 to 1000 by 1

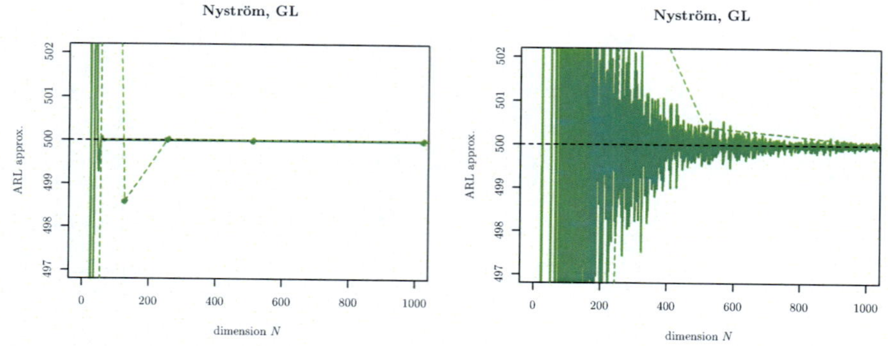

$(\alpha, \beta) = (8, 8)$ $(\alpha, \beta) = (2, 8)$

Fig. 4 Nyström method with Gauß-Legendre nodes for an upper EWMA chart. True ℓ is 500, $\lambda = 0.1$, N runs from 10 to 1000 by 1

some impressive results for a couple of designs (see Yashchin 2019, for a recent example). We added to Figs. 2, 3, 4, 5, 6, and 7 corresponding numbers (for $N = 8, 16, 32, \ldots, 1024$, drawn as dashed lines) to provide some illustrations for the beta distribution. As mentioned in Knoth (2006) for CUSUM control charts monitoring normal variance, these techniques are effective only for well behaving convergence patterns — compare here the two cases (1, 8) and (8, 1), for example.

Applying the more advanced Markov chain design relying on ideas of Hawkins (1992) does not lead to improvement, as illustrated in Fig. 3. Below we will see examples where the design of Hawkins performs slightly better than the original Markov chain setup. For both Markov chain frameworks, the Richardson extrapolation yields some improvements for the raw Markov chain results. However, the collocation results, presented below, are substantially closer to the true ARL values.

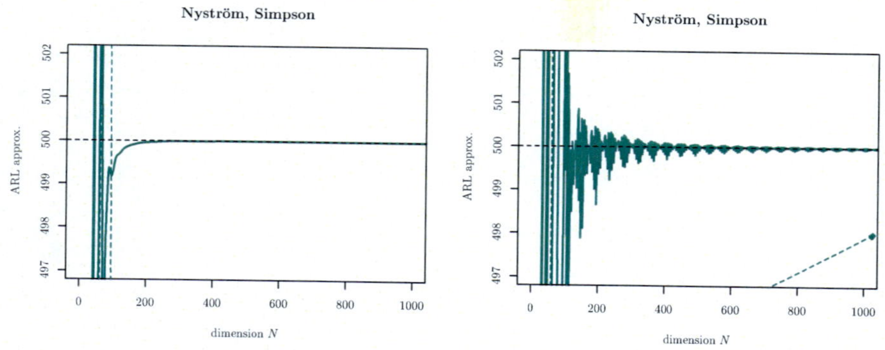

$(\alpha, \beta) = (8, 8)$ $(\alpha, \beta) = (2, 8)$

Fig. 5 Nyström method with Simpson rule nodes for an upper EWMA. True ℓ is 500, $\lambda = 0.1$, N runs from 10 to 1000 by 1

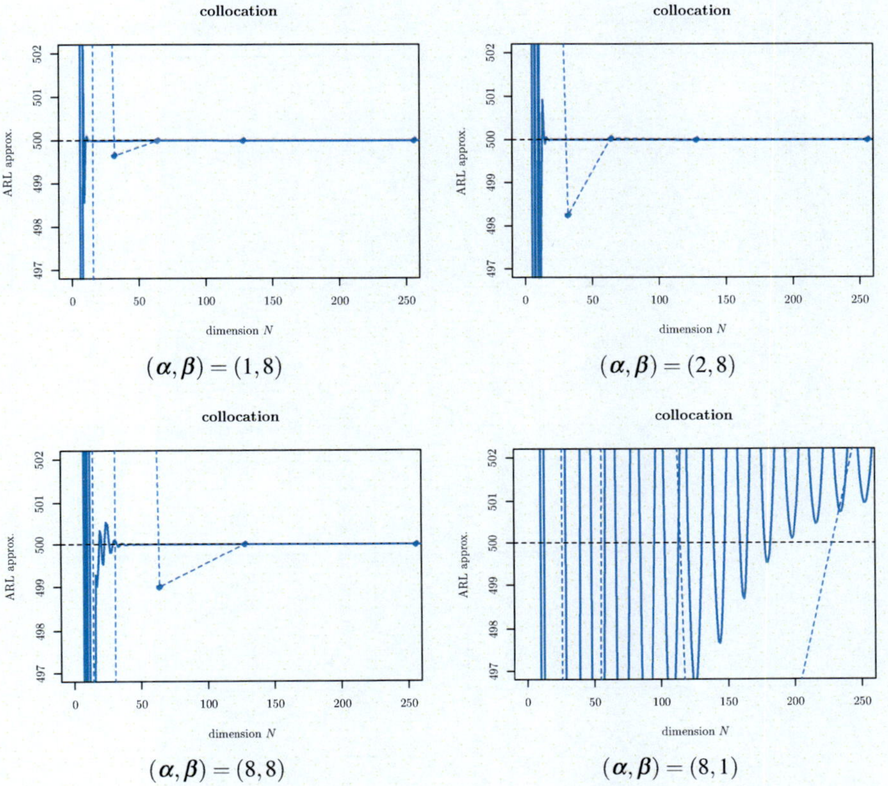

Fig. 6 Collocation for an upper EWMA chart. True ℓ is 500, $\lambda = 0.1$, N runs from 5 to 250 by 1

After performing a similar study for solving the ARL integral equation with the Nyström method as in Crowder (1987b), we show excellent results for (8, 8) and for (2, 8), where the convergence patterns are already unstable (see Fig. 4). At least, for (8, 8), the Nyström method works excellently as for EWMA monitoring the normal mean. However, (2, 8) exhibits a worse behavior. The parameter pairs $(\alpha, \beta) =$ (4, 8), (8, 4), and (4, 4) yield more stable patterns, but not as good as (8, 8). For all other pairs from Table 1, the Nyström method with Gauß-Legendre nodes collapses completely, meaning highly oscillating patterns without suitable regularities. Hence, this method is applicable only for approximately bell-shaped beta distributions. Next, we look at Nyström with simpler node designs based on the Simpson rule. Figure 5 indicates that it converges slower than Gauß-Legendre in case of (8, 8). But it behaves considerably better for (2, 8) than the more complicated Gauß-Legendre Nyström, which is quite surprising. Nonetheless, it does not converge for half of the pairs in Table 1 so that it is not the method to go.

Initially, we were assuming that collocation deals with all the above problems in a reasonable way. Applying collocation to the four presented parameter combinations,

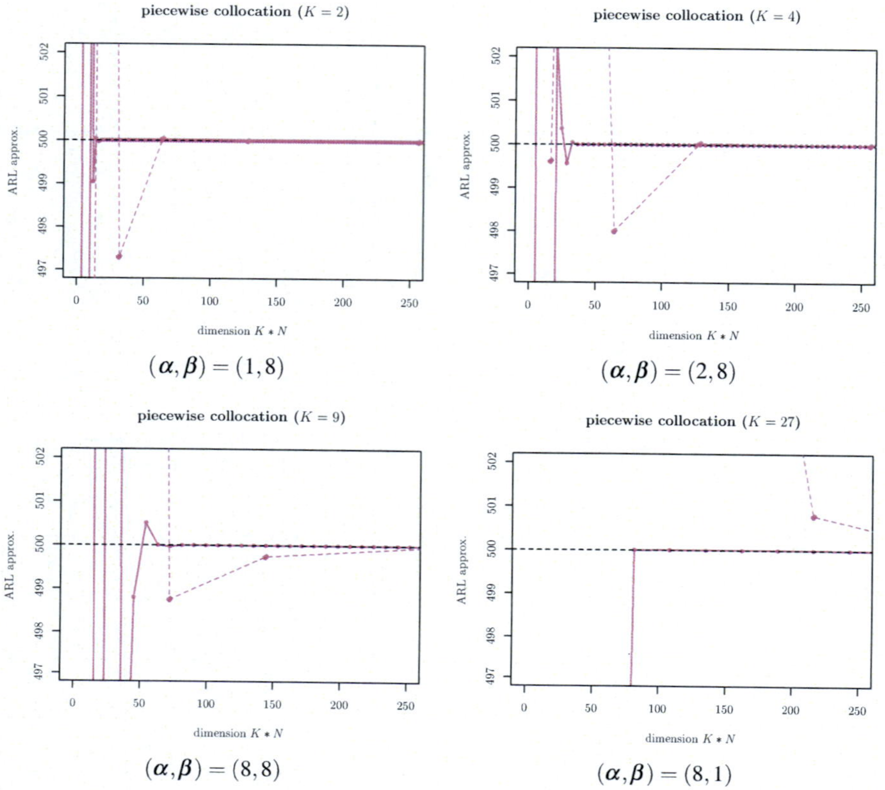

Fig. 7 Collocation, piece-wise basis, for an upper EWMA chart. True ℓ is 500, $\lambda = 0.1$, N runs from 2 to 300 by 1, but $K * N \leq 360$

we obtain the following patterns. Because the collocation procedure is considerably slower than the competitors for fixed N, we deal only with $5 \leq N \leq 250$, which leads to reasonable accurate ARL evaluations. Except for (8, 1), collocation seems to be a reliable method to calculate the ARL of an EWMA control chart monitoring beta variables. Already for small values of N, the ARL approximation is stabilized. However, the intractable parameter pair (8, 1) not only requires high values of N to stabilize, but the stabilized value differs also from the true value. Because of the reduced smoothness of the ARL function at some points (see Fig. 8), the collocation utilizing polynomials defined on the complete range $[0, c_u]$ deteriorates. Therefore, the piece-wise defined polynomials – see previous section – might help.

Our last scheme, collocation with piece-wise defined base functions, is illustrated in Fig. 7. This time, the matrix dimension is the product of the polynomial order N and the number of subintervals K. The intricate pair (8, 1) is now under control. In general, the piece-wise collocation algorithm works sufficiently well. Below we check the accuracy of both collocation designs and the Markov chain approach with

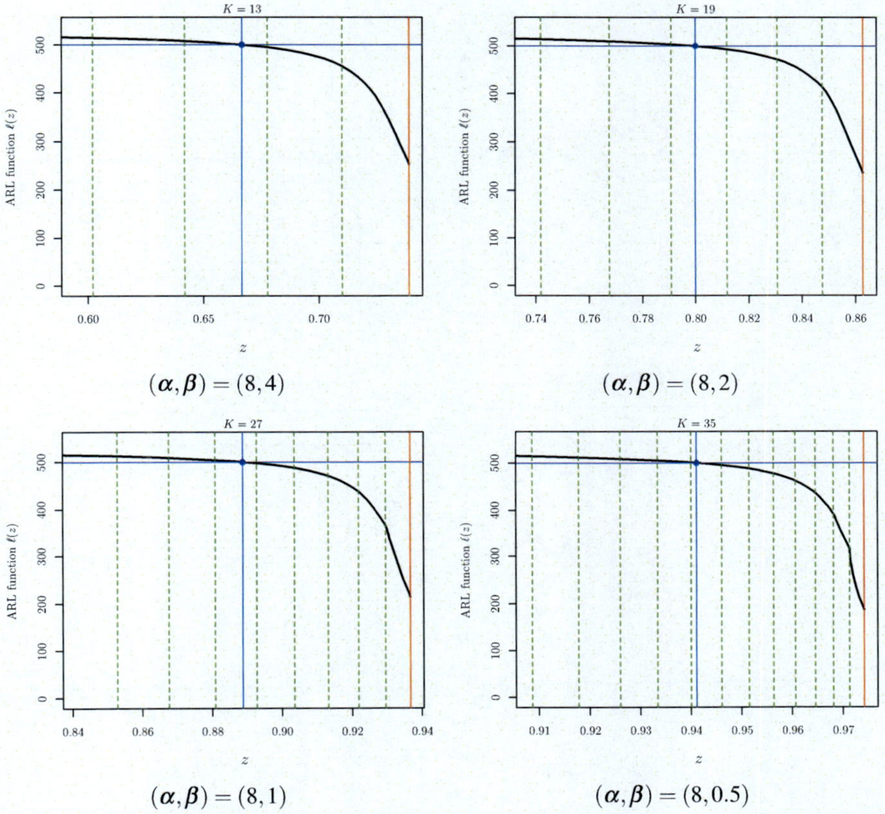

Fig. 8 Beta EWMA ARL function $\ell(z)$ for $2z_0 - c_u \le z \le c_u$

higher resolutions while confirming the results with the Monte Carlo studies. In addition, we explore the most demanding (α, β) configurations, where at least one of the components is equal to 0.5.

Before validating the accuracy, we provide some $\ell(z)$ profiles which might explain the difficulties we have to deal with. In particular, we see cusps in the $\ell()$ profiles for $\beta = 1$ and $\beta = 0.5$. It turns out that these cusps are at the interval borders we used for our piece-wise design, that is, at $b_{K-1}, b_{K-2}, \ldots$ $(b_K = c_u)$. Hence, it seems reasonable to approximate $\ell()$ on the corresponding intervals separately. It would probably be sufficient to utilize fewer than K intervals, because $\ell()$ gets progressively smoother along these interval borders. In addition, for smaller β, the descent of $\ell(z)$ from $z = z_0$ to $z = c_u$ becomes steeper. For upper control charts of beta distributions with small β, the upper control limit c_u is quite close to the upper limit of the support of the monitored variable.

As a final confirmation, we explored for the following (α, β) configurations the "high accuracy" behavior of the Markov chain approximation (N goes up to 5000)

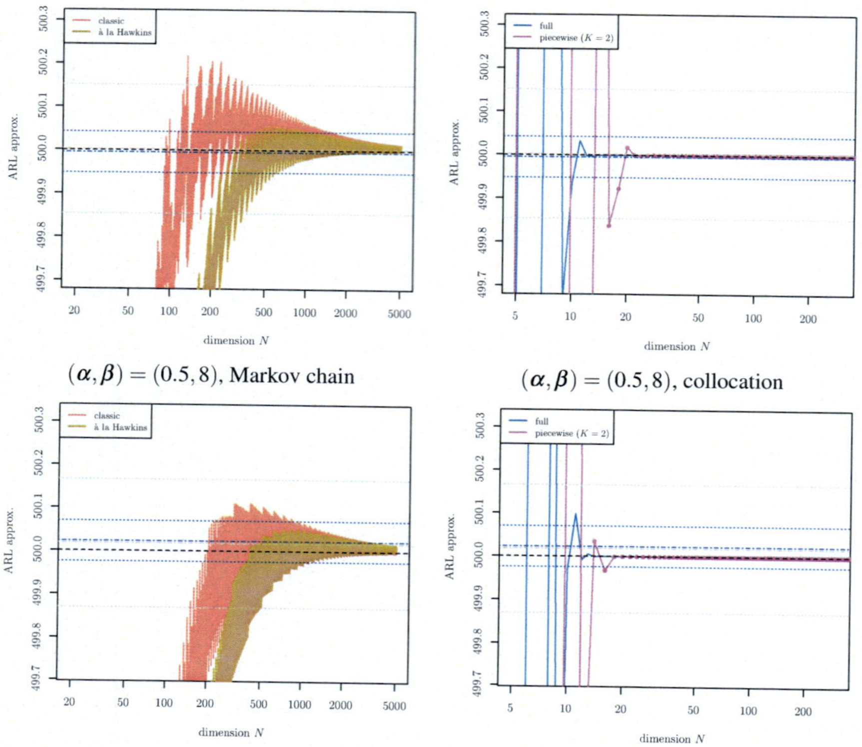

$(\alpha, \beta) = (0.5, 8)$, Markov chain

$(\alpha, \beta) = (0.5, 8)$, collocation

$(\alpha, \beta) = (1, 8)$, Markov chain

$(\alpha, \beta) = (1, 8)$, collocation

Fig. 9 High accuracy study for peculiar setups – Markov chain approximation and collocation methods. The Monte Carlo results added (mean + three times standard error for 10^8 (gray) and 10^9 (blue) replicates)

and both collocation setups (N similar to the above figures). The computations were carried out for (0.5, 8), (1, 8), (8, 1), (8, 0.5) and the symmetric pairs (2, 2), (1, 1), (0.5, 0.5). In order to judge the spread of the more or less fluctuating approximations, we add the Monte Carlo results utilizing 10^8 and 10^9 replications by plotting their averages and lines at average plus/minus three times the corresponding standard errors. For the first two pairs, (0.5, 8) and (1, 8), in Fig. 9, we conclude that the Markov chain approximation works reasonably well, though it requires large values of N to achieve high accuracy. But both collocation designs demonstrate that high accuracy could be achieved already for $N = 21$ and $K \times N = 38$, respectively. This is in contrast to the Markov chain approximation, where even for the five largest values of $N \in \{4996, \ldots, 5000\}$, the range of the ARL approximations is larger than 0.007, while for collocation we achieved the final five digits after the decimal point quite early. Hence, for configurations with α considerably smaller than β, both collocation procedures deliver high accuracy with little computation time. The Markov chain might be used, but be sure to have the matrix dimension of at least

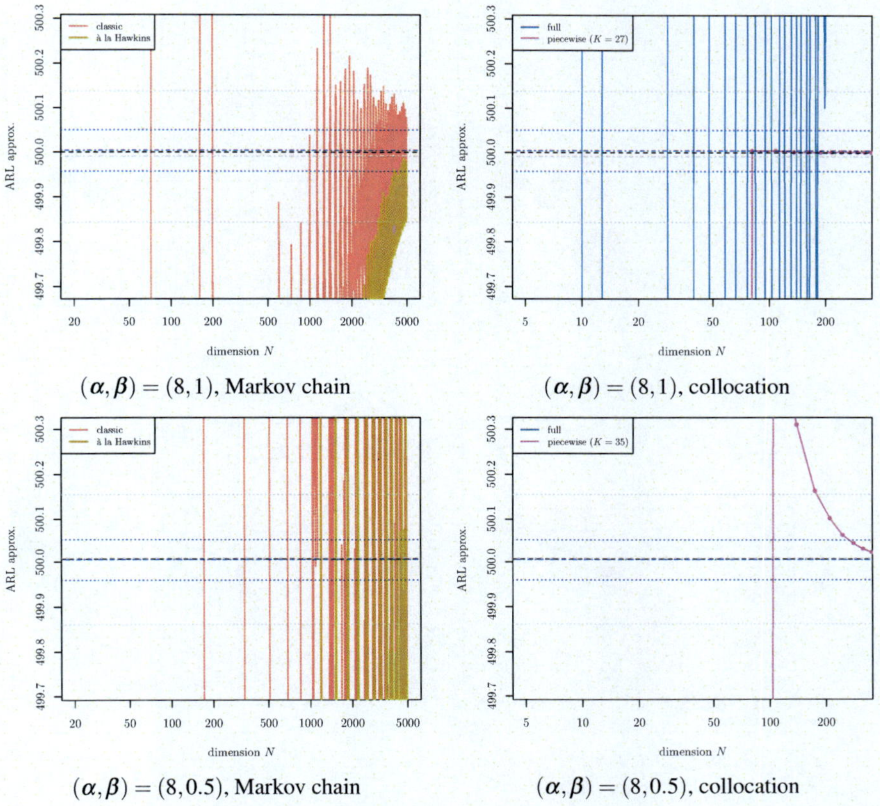

Fig. 10 High accuracy study for further peculiar setups – Markov chain approximation and collocation methods. The Monte Carlo results added (mean + three times standard error for 10^8 (gray) and 10^9 (blue) replicates)

$N = 500$ to approximate the ARL within 0.02% accuracy (first digit after the decimal point).

Turning to the mirrored designs, namely (8, 0.5) and (8, 1), Fig. 10 indicates that the quality of three of our algorithms degrades considerably. Only the piece-wise collocation approach yields satisfactory results. Of course, it needs larger dimensions than for "nicer" beta distributions because of the large number of considered subintervals, K. For the less demanding case, (8, 1), the Markov chain results will work in a pinch. They seem to behave better than collocation with the full base functions. For the most difficult case, $(8, 0.5)$, we observe that only the piece-wise collocation design delivers reliable results. The other (and simpler) collocation procedure departs from the true value so that it is not visible on that scale. And the Markov chain results vary heavily so that a final approximation result is difficult to pick, even for dimensions around 5000 (Fig. 11).

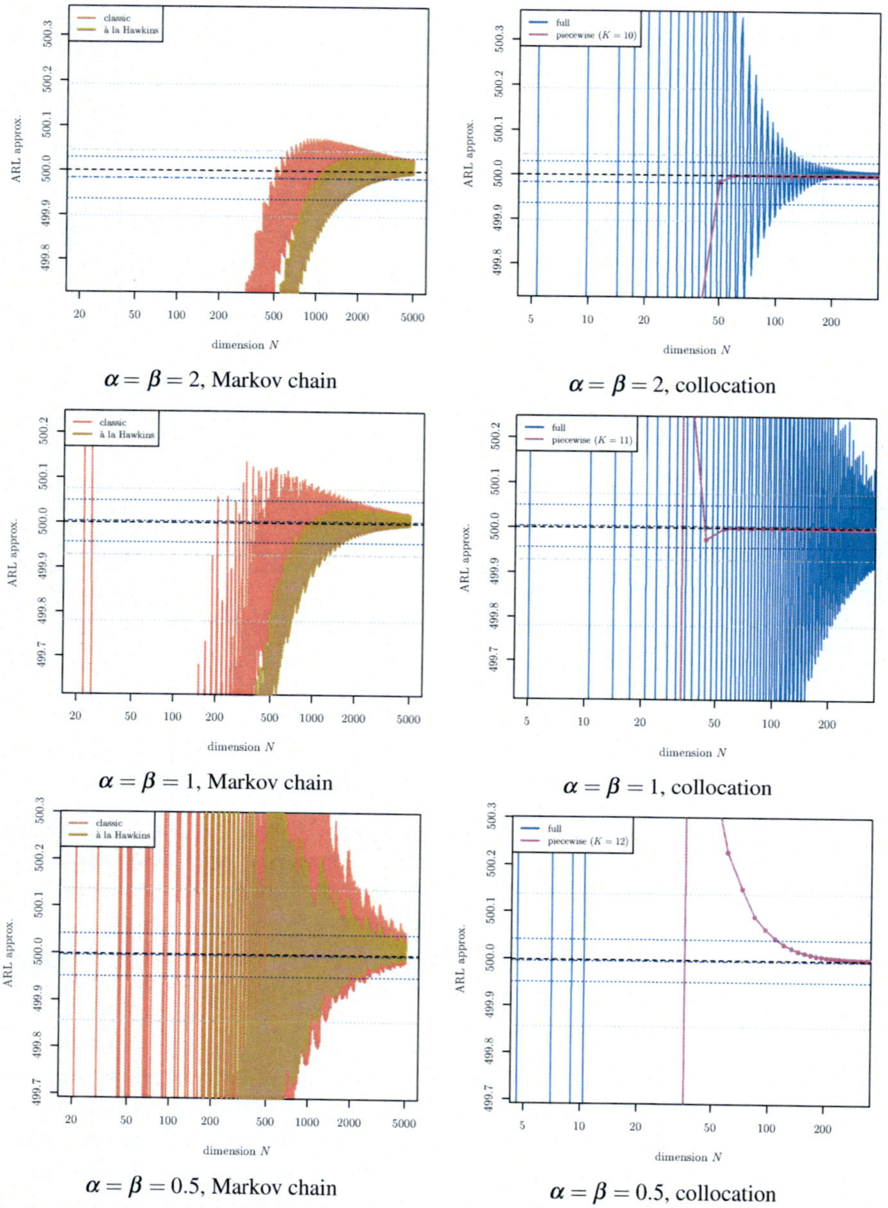

Fig. 11 High accuracy study for symmetric setups – Markov chain approximation and collocation methods. The Monte Carlo results added (mean + three times standard error for 10^8 (gray) and 10^9 (blue) replicates)

Finally, we look at three symmetric setups, (2, 2), (1, 1), and (0.5, 0.5). The most inoffensive patterns could be seen, not surprisingly, for (2, 2). Only the full collocation behaves strangely (spread and systematic deviation). It is clearly dominated by the piece-wise design. Fortunately, the number of subintervals K is quite small (≤ 12) so that the polynomial order N could be larger. Looking at the beta configuration (1, 1) that corresponds to the uniform distribution, we see a further worsening of the full collocation approach. The Markov chain behaves sufficiently well. Yet again, piece-wise collocation provides the best performance. Its computation could be accelerated further because the quadratures could be replaced by explicit formulas deploying the simple shape of the density of X in this case. Similar to the problematic non-symmetric example (8, 0.5), the symmetric (0.5, 0.5) is quite difficult in terms of ARL calculation. Surprisingly, the Markov chain and piece-wise collocation work out better for the symmetric version. Recall that the corresponding density of X is unbounded at $x = 0$ and $x = 1$.

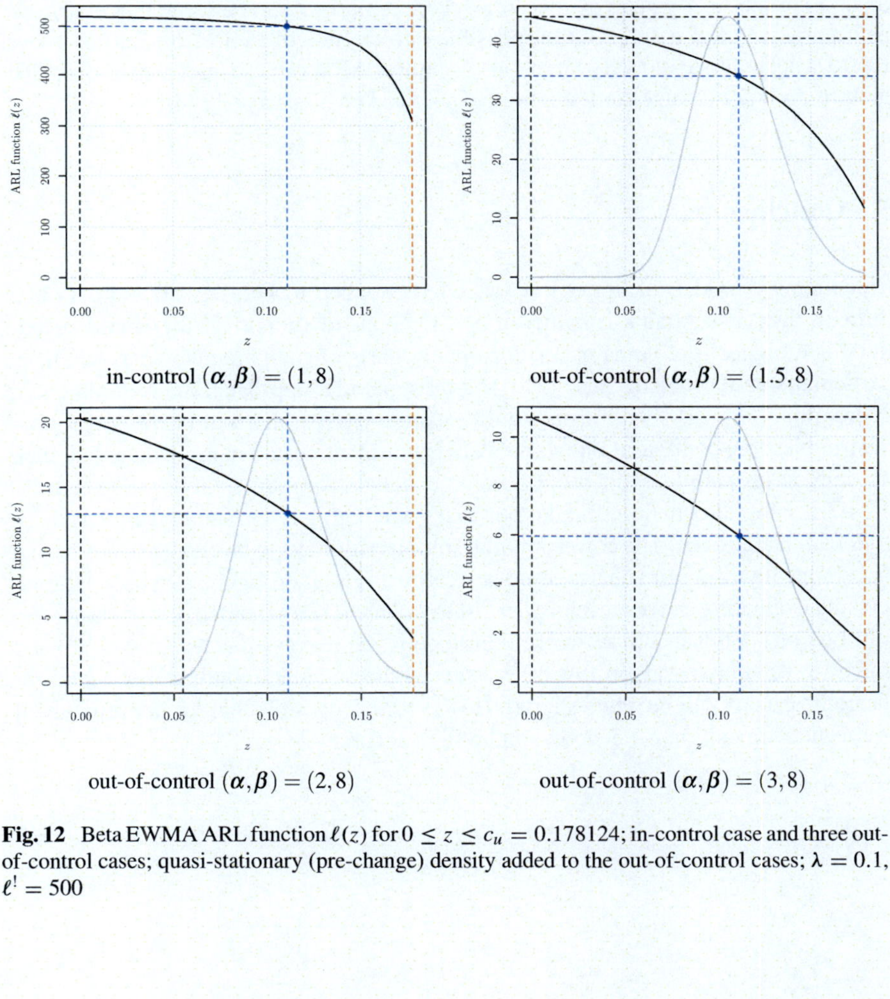

Fig. 12 Beta EWMA ARL function $\ell(z)$ for $0 \leq z \leq c_u = 0.178124$; in-control case and three out-of-control cases; quasi-stationary (pre-change) density added to the out-of-control cases; $\lambda = 0.1$, $\ell^! = 500$

In summary, piece-wise collocation yields for all considered beta distributions reasonable ARL approximations. For most of our examples, the simple collocation is satisfactory as well. The Markov chain approximation works always, but it needs a large matrix dimension to provide high accuracy.

Finally, we look at the full ARL function $\ell()$ to illustrate potential limitations of our EWMA control chart design. In Fig. 12, we consider the numerically manageable in-control case (1, 8) and the out-of-control cases $\alpha \in \{1.5, 2, 3\}$ with unchanged $\beta = 8$. Hence, the mean increases from $1/9 \approx 0.11$ to $1.5/9.5 \approx 0.16$, $2/10 = 0.2$, and $3/11 \approx 0.27$, respectively. First we notice that in the in-control case, $\ell(z)$ does not change much for $0 \le z \le 1.2 z_0$. For the out-of-control case $\alpha = 1.5$, however, $\ell(0) = 44.7$ is considerably larger than the typically reported $\ell(z_0) = 34.2$. Concluding from the conditional steady-state distribution (the bell-shaped curve in the out-of-control diagrams of Fig. 12; it is a proxy for the distribution of the EWMA statistic just before the change happens), the extreme $\ell(0)$ appears to be less important and might be replaced by $\ell(z_{0.001}) = 40.8$, where $z_{0.001} \approx 0.055$ is the 0.001 quantile of the steady-state distribution. There remains a gap between these two values, namely 40.8 and 34.2. But the possible introduction of a reflecting barrier to our EWMA design to avoid very large out-of-control ARL values is not imperative. The picture does not change for the more pronounced changes $\alpha \in \{2, 3\}$.

5 Conclusions

Calculating the ARL of upper EWMA charts applied to beta distributed data is a difficult task. Essentially, it is driven by the bounded support of the beta distribution. The biggest problems occur, for upper charts, if the parameter β is small, in particular smaller than or close to 1. The most popular approach, the Markov chain approximation, works well if only crude accuracy (typically within 1%, except for the troublesome configurations) is needed. If one is interested in more than that, then in many cases simple collocation does an excellent job while solving numerically the ARL integral equation. In all other cases, we highly recommend the piece-wise collocation approach. The corresponding subintervals have to be chosen accordingly. Recall that lower control charts could be dealt with by reverting the two beta parameters and evaluating the resulting upper control chart. The two-sided design, however, will be considered in future work. Fortunately, the piece-wise design would need much fewer subintervals in this case, because the so-called continuation region of the control chart will be much shorter. However, it is possible that further refinement of this decomposition is needed to attain high accuracy.

References

Brezinski, C. (1985). Convergence acceleration methods: The past decade. *Journal of Computational and Applied Mathematics*, *12–13*, 19–36. https://doi.org/10.1016/0377-0427(85)90005-6.

Brook, D., & Evans, D. A. (1972). An approach to the probability distribution of CUSUM run length. *Biometrika*, *59*(3), 539–549. https://doi.org/10.2307/2334805.

Castagliola, P. (2001). An $(\tilde{X}/R)$-EWMA control chart for monitoring the process sample median. *International Journal of Reliability, Quality, and Safety Engineering*, *8*(2), 123–135. https://doi.org/10.1142/S0218539301000414.

Crowder, S. V. (1987a). Average run lengths of exponentially weighted moving average control charts. *Journal of Quality Technology*, *19*(3), 161–164. https://doi.org/10.1080/00224065.1987.11979055.

Crowder, S. V. (1987b). A simple method for studying run-length distributions of exponentially weighted moving average charts. *Technometrics*, *29*(4), 401–407. https://doi.org/10.1080/00401706.1987.10488267.

Crowder, S. V., & Hamilton, M. D. (1992). An EWMA for monitoring a process standard deviation. *Journal of Quality Technology*, *24*(1), 12–21. https://doi.org/10.1080/00224065.1992.11979369.

Forbes, C., Evans, M., Hasting, N., & Peacock, B. (2011). *Statistical Distributions* (4th ed.). New York: Wiley.

Gan, F. F. (1993). Exponentially weighted moving average control charts with reflecting boundaries. *Journal of Statistical Computation and Simulation*, *46*(1–2), 45–67. https://doi.org/10.1080/00949659308811492.

Gan, F. F., & Tan, T. (2010). Risk-adjusted number-between failures charting procedures for monitoring a patient care process for acute myocardial infarctions. *Health Care Management Science*, *13*(3), 222–233. https://doi.org/10.1007/s10729-010-9125-8.

Gan, F. F., Lin, L., & Loke, C. K. (2012). Risk-adjusted cumulative sum charting procedures. In H. J. Lenz, W. Schmid, & P. T. Wilrich (Eds.), *Frontiers in Statistical Quality Control 10* (pp. 207–225). Heidelberg: Physica (Springer). https://doi.org/10.1007/978-3-7908-2846-7.

Gianino, A. B., Champ, C. W., & Rigdon, S. E. (1990). Solving integral equations by the collocation method. In *ASA Proceedings of the Statistical Computing Section, American Statistical Association* (pp. 101–102)

Graham, M. A., Mukherjee, A., & Chakraborti, S. (2012). Distribution-free exponentially weighted moving average control charts for monitoring unknown location. *Computational Statistics and Data Analysis*, *56*(8), 2539–2561. https://doi.org/10.1016/j.csda.2012.02.010.

Grigg, O., & Spiegelhalter, D. (2007). A simple risk-adjusted exponentially weighted moving average. *Journal of the American Statistical Association*, *102*(477), 140–152. https://doi.org/10.1198/016214506000001121.

Hawkins, D. M. (1992). Evaluation of average run lengths of cumulative sum charts for an arbitrary data distribution. *Communications in Statistics - Simulation and Computation*, *21*(4), 1001–1020. https://doi.org/10.1080/03610919208813063.

Johnson, N. L., Kotz, S., & Balakrishnan, N. (1995). *Continuous Univariate Distributions* (2nd ed., Vol. 2)., Wiley Series in Probability and Statistics New York: Wiley.

Knoth, S. (2005). Accurate ARL computation for EWMA-S^2 control charts. *Statistics and Computing*, *15*(4), 341–352. https://doi.org/10.1007/s11222-005-3393-z.

Knoth, S. (2006). Computation of the ARL for CUSUM-S^2 schemes. *Computational Statistics and Data Analysis*, *51*(2), 499–512. https://doi.org/10.1016/j.csda.2005.09.015.

Knoth, S. (2019). spc: Statistical Process Control - Collection of Some Useful Functions. R package version 0.6.1.

Lahcene, B. (2013). On Pearson families of distributions and its applications. *African Journal of Mathematics and Computer Science Research*, *6*(5), 108–117. https://doi.org/10.5897/AJMCSR2013.0465.

Loke, C. K., & Gan, F. F. (2012). Joint monitoring scheme for clinical failures and predisposed risks. *Quality Technology and Quantitative Management*, *9*(1), 3–21. https://doi.org/10.1080/16843703.2012.11673274.

Lucas, J. M. (1982). Combined Shewhart-CUSUM quality control schemes. *Journal of Quality Technology*, *14*(2), 51–59. https://doi.org/10.1080/00224065.1982.11978790.

Lucas, J. M., & Saccucci, M. S. (1990). Exponentially weighted moving average control schemes: Properties and enhancements. *Technometrics*, *32*(1), 1–12. https://doi.org/10.1080/00401706.1990.10484583.

Nyström, E. J. (1930). Über die praktische Auflösung von Integralgleichungen mit Anwendungen auf Randwertaufgaben. *Acta Mathematica*, *54*(1), 185–204. https://doi.org/10.1007/BF02547521.

Page, E. S. (1954). Control charts for the mean of a normal population. *Journal of the Royal Statistical Society: Series B (Statistical Methodology)*, *16*(1), 131–135. https://doi.org/10.1111/j.2517-6161.1954.tb00154.x.

Pearson, K. (1895). X. Contributions to the mathematical theory of evolution.–II. Skew variation in homogeneous material. *Philosophical Transactions of the Royal Society A: Mathematical, Physical and Engineering Sciences*, *186*, 343–414. https://doi.org/10.1098/rsta.1895.0010.

Pearson, K. (1916). IX. Mathematical contributions to the theory of evolution.–XIX. Second supplement to a memoir on skew variation. *Philosophical Transactions of the Royal Society A: Mathematical, Physical and Engineering Sciences*, *216*, 429–457. https://doi.org/10.1098/rsta.1916.0009.

Reynolds, M. R., & Stoumbos, Z. G. (2005). Should exponentially weighted moving average and cumulative sum charts be used with Shewhart limits? *Technometrics*, *47*(4), 409–424. https://doi.org/10.1198/004017005000000382.

Roberts, S. W. (1959). Control chart tests based on geometric moving averages. *Technometrics*, *1*(3), 239–250. https://doi.org/10.1080/00401706.1959.10489860.

Saccucci, M. S., & Lucas, J. M. (1990). Average run lengths for exponentially weighted moving average control schemes using the Markov chain approach. *Journal of Quality Technology*, *22*(2), 154–162. https://doi.org/10.1080/00224065.1990.11979227.

Waldmann, K. H. (1986). Bounds for the distribution of the run length of geometric moving average charts. *Journal of the Royal Statistical Society: Series C (Applied Statistics)*, *35*(2), 151–158. https://doi.org/10.2307/2347265.

Woodall, W. H., & Mahmoud, M. A. (2005). The inertial properties of quality control charts. *Technometrics*, *47*(4), 425–436. https://doi.org/10.1198/004017005000000256.

Yashchin, E. (1987). Some aspects of the theory of statistical control schemes. *IBM Journal of Research and Development*, *31*(2), 199–205. https://doi.org/10.1147/rd.312.0199.

Yashchin, E. (1989). Weighted cumulative sum technique. *Technometrics*, *31*(1), 321–338. https://doi.org/10.1080/00401706.1989.10488555.

Yashchin, E. (2019). Gradient analysis of Markov-type control schemes and its applications. *Communications in Statistics – Simulation and Computation*, online: 1–23. https://doi.org/10.1080/03610918.2019.1687718

Yousry, M. A., Sturm, G. W., Feltz, C. J., & Noorossana, R. (1991). Process monitoring in real time: Empirical Bayes approach - discrete case. *Quality and Reliability Engineering International*, *7*(3), 123–132. https://doi.org/10.1002/qre.4680070303.

Flexible Monitoring Methods for High-yield Processes

Tahir Mahmood, Ridwan A. Sanusi, and Min Xie

Abstract In recent years, advancement in technology brought a revolutionary change in the manufacturing processes. Therefore, manufacturing systems produce a large number of conforming items with a small amount of non-conforming items. The resulting dataset usually contains a large number of zeros with a small number of count observations. It is claimed that the excess number of zeros may cause over-dispersion in the data (i.e., when variance exceeds mean), which is not entirely correct. Actually, an excess amount of zeros reduce the mean of a dataset which causes inflation in the dispersion. Hence, modeling and monitoring of the products from high-yield processes have become a challenging task for quality inspectors. From these highly efficient processes, produced items are mostly zero-defect and modeled based on zero-inflated distributions like zero-inflated Poisson (ZIP) and zero-inflated Negative Binomial (ZINB) distributions. A control chart based on the ZIP distribution is used to monitor the zero-defect process. However, when additional over-dispersion exists in the zero-defect dataset, a control chart based on the ZINB distribution is a better alternative. Usually, it is difficult to ensure that data is over-dispersed or under-dispersed. Hence, a flexible distribution named zero-inflated Conway–Maxwell–Poisson (ZICOM-Poisson) distribution is used to model over or under-dispersed zero-defect dataset. In this study, CUSUM charts are designed based

T. Mahmood (✉) · R. A. Sanusi · M. Xie
Department of Systems Engineering and Engineering Management, City University of Hong Kong, Tat Chee Avenue, Kowloon, Hong Kong
e-mail: tmahmood@ouhk.edu.hk

R. A. Sanusi
e-mail: rasanusi2-c@my.cityu.edu.hk

M. Xie
e-mail: minxie@cityu.edu.hk

T. Mahmood
Department of Technology, School of Science and Technology, The Open University of Hong Kong, Ho Man Tin, Kowloon, Hong Kong

R. A. Sanusi
Department of Community Health Sciences, Rady Faculty of Health Sciences, University of Manitoba, Winnipeg, MB, Canada

S. Knoth and W. Schmid (eds.), *Frontiers in Statistical Quality Control 13*,
Frontiers in Statistical Quality Control,
https://doi.org/10.1007/978-3-030-67856-2_4

on the ZICOM-Poisson distribution. These provide a flexible monitoring method for quality practitioners. A simulation study is designed to access the performance of the proposed monitoring methods and their comparison. Moreover, a real application is presented to highlight the importance of the stated proposal.

Keywords Conway-Maxwell–Poisson · High-yield process · Over-dispersion · Random shocks · Zero-inflated

1 Introduction

In most of the manufacturing processes, the quality of the process is determined based on the fraction of nonconforming items produced in the production batches. Such count data is discrete in nature and commonly modeled by the count models such as Poisson, Geometric, Bernoulli, Binomial, and Negative Binomial (NB) distributions (Mahmood 2020). The most common Poisson distribution is an equi-dispersed distribution (i.e., where mean and variance are equal) and the charts based on it were known by c and u control charts (Montgomery 2009). The Bernoulli and Binomial distributions are considered as under-dispersed models (i.e., where the variation is smaller than the mean) and, the p and np charts are designed on these distributions (Gan 1990). Further, the negative Binomial and Geometric distributions are considered as over-dispersed models (i.e., where the variation is more substantial than the mean), and the conventional Geometric (g) control chart was designed on the basis of geometric distribution (for more details see Xie et al. 2000 and Riaz et al. 2017). In real scenarios, it is a challenging task to find the adequate model among the discussed above. Therefore, Shmueli et al. (2005) provide a revived flexible model known as Conway–Maxwell–Poisson (COM-Poisson) distribution, which can fit over-dispersed data, as well as, efficiently fit under-dispersed and equi-dispersed data.

For the monitoring purpose, Sellers (2012) proposed a Shewhart structure based on the COM-Poisson distribution and showed that it is a flexible and generalized structure of the p, c, and u charts. In the study, the structure is based on the k-sigma limits, which provides false conclusions in the case of an asymmetric COM-Poisson distribution (Saghir et al. 2013). Alternatively, Saghir et al. (2013) suggest another Shewhart structure based on the probability limits approach. Also, Saghir and Lin (2014c) introduced an exponentially weighted moving average (EWMA) and a generalized EWMA (GEWMA) structures based on COM-Poisson distribution. Under the COM-Poisson distribution, Alevizakos and Koukouvinos (2019), Saghir and Lin (2014b), and Saghir and Lin (2014a), respectively, proposed a double EWMA (DEWMA), a cumulative sum (CUSUM), and a multivariate Shewhart-type schemes. For more details on these charts, interested readers may read Saghir and Lin (2015) and Ali et al. (2016).

In most of the industrial processes, count dataset consists of a large number of zeros and is termed as zero-defect or high-yield dataset. Moreover, in healthcare studies, this type of dataset is referred to as rare health-related events. The above-stated count models provide biased and inadequate estimates while fitting the high-yield dataset. Therefore, to overwhelm the effect of zero excess, zero-inflated version of the ordinary models are derived. Lambert (1992) introduced a zero-inflated Poisson (ZIP) distribution, McCullagh and Nelder (1983) derived a zero-inflated Negative Binomial (ZINB) distribution, Chang and Gan (1999) proposed a zero-inflated geometric (ZIG) model, and Barriga and Louzada (2014) presented zero-inflated COM-Poisson (ZICOM-Poisson) distribution. Further, Sellers and Raim (2016) and Sim et al. (2018) discussed more properties of the ZICOM-Poisson distribution.

The standard u chart produces a high false alarm rate, even when the exact probability limits of the ordinary Poisson distribution are used to monitor the high-yield processes. Therefore, Xie and Goh (1993) initiated a Shewhart structure based on the ZIP distribution, which was further extended by Xie et al. (1995) and Chang and Gan (1999). In monitoring literature, several studies are developed on the methods for the ZIP, ZIG, and ZINB models, which are comprehensively reviewed in Mahmood and Xie (2019). Specifically, He et al. (2012, 2014) proposed CUSUM structures based on likelihood ratio statistics for the ZIP process. He et al. (2012) proposed $p - \lambda$ CUSUM chart and He et al. (2014) developed CRL-ZTP CUSUM chart using conforming run length (CRL) idea, for detecting increasing shifts in the ZIP process parameters. Motivated by He et al. (2012, 2014), we design similar structures based on the ZICOM-Poisson distribution to monitor an increasing shift in the mean and zero-inflation parameter of the process. To the best of our knowledge, there may not exist a single monitoring study based on the ZICOM-Poisson distribution. Hence, this study is designed to incorporate the following prime objectives:

1. To provide the generalized structure of the CUSUM charts proposed by He et al. (2012, 2014), which would be flexible for the over/under/equi-dispersed datasets.
2. To design a simulation-based comparative study between the suggested CUSUM structures.
3. Implementation of the proposed CUSUM charts on the real-data scenario.

The rest of this article is arranged as follows: In Sect. 2, we describe the zero-inflated COM-Poisson model. In Sect. 3, the structure of the proposed control charts is given, and the performance evaluations are discussed in Sect. 4. In Sect. 5, we provide the comparative analysis of the proposed charts. In Sect. 6, we present a case study using a LED dataset. Finally, Sect. 7 includes summary, conclusions, and recommendations drawn from the proposed study.

2 The Zero-Inflated Conway-Maxwell-Poisson (ZICOM-Poisson) Distribution

The Conway–Maxwell–Poisson (COM-Poisson) distribution was originated by Conway and Maxwell (1962). It was revived by Shmueli et al. (2005). The probability mass function of the COM-Poisson distribution for a random variable Y is defined as follows:

$$Pr(Y = y) = \frac{\lambda^y}{(y!)^\nu Z(\lambda, \nu)}; \; y = 0, \; 1, \; 2, \; \ldots, \tag{1}$$

where $\lambda > 0$ is the usual rate parameter of the Poisson distribution, ν is the dispersion parameter and $Z(\lambda, \nu) = \sum_{j=0}^{\infty} \lambda^j/(j!)^\nu$ normalizes the distribution. For more details about the normalizing factor, see Gillispie and Green (2015).

In the ZICOM-Poisson process, outcomes emanate from the two processes such as the first process models zeros with a proportion p (zero-inflation) and $p/z(\lambda, \nu)$ (zeros coming from COM-Poisson distribution); and another process models the nonzero counts from the zero-truncated COM-Poisson distribution (for details on zero-truncated COM-Poisson distribution, see Chou et al. (2015). Thus, the ZICOM-Poisson model for a random variable Y can be formulated as follows:

$$Pr\,(Y = y) = \begin{cases} (1 - p) + \frac{p}{Z(\lambda,\nu)} & , \quad if \; y = 0 \\ p\frac{\lambda^y}{(y!)^\nu Z(\lambda,\nu)} & , \quad if \; y > 0 \end{cases}. \tag{2}$$

The mean and variance of the ZICOM-Poisson distribution are, respectively, given by

$$E\,(Y) = (1 - p)\left(\lambda^{1/\nu} - \frac{\nu - 1}{2\nu}\right), \tag{3}$$

and

$$Var\,(Y) = (1 - p)\left(\frac{1}{\nu}\lambda^{1/\nu} + p\left(\lambda^{1/\nu} - \frac{\nu - 1}{2\nu}\right)^2\right). \tag{4}$$

The ZICOM-Poisson distribution is the generalized form of many under-dispersed ($\nu > 1$) and over-dispersed ($\nu < 1$) models. Specifically, the ZICOM-Poisson distribution reduces to the ZIP model when ($\nu = 1$) and to the ZIG distribution when ($\nu = 0$).

3 Monitoring Methods Based on ZICOM-Poisson Distribution

A standard structure of a control chart consists of two decision lines named as upper control limit (UCL) and lower control limit (LCL). A process is declared stable or in an in-control (IC) state when sample points lie within decision lines. Else, it is declared unstable or in an out-of-control (OOC) state when sample points exceed decision lines (Mahmood et al. 2019). One of the traditional control charts is the

Shewhart chart, proposed by Shewhart (1926). It is efficient in detecting large shifts in process parameters. However, the EWMA (Roberts 1959) and the CUSUM (Page 1954) charting structures are efficient in detecting small to moderate shifts in the process parameters. More discussion on the CUSUM charts, one may read Faisal et al. (2018) and Abbas et al. (2020). This study designs a CUSUM structure for the timely detection of small to moderate increasing shifts in the rate (λ) and the zero-inflation (p) parameters of the ZICOM-Poisson distribution. The proposed CUSUM structures are given in the following subsections.

3.1 The $p - \lambda$ CUSUM Chart

The $p - \lambda$ CUSUM chart is the combination of two CUSUM charts (i.e., p CUSUM chart and λ CUSUM chart) to detect an increasing shift in the parameters (i.e., λ and/or p) of the ZICOM-Poisson distribution.

3.1.1 The p CUSUM Chart

The p CUSUM chart is designed to detect an increasing shift in the zero-inflation parameter of the ZICOM-Poisson distribution. As a Phase II method, we assume that the IC parameters of the ZICOM-Poisson distribution are p_0, λ_0, and υ and are known. The p CUSUM chart is designed to detect a shift from p_0 to p_1 (where $p_1 > p_0$). Based on the likelihood ratio method, the p CUSUM statistic is obtained by;

$$B_i = \max\left(0,\ b_i + B_{i-1}\right); i = 1, 2, \ldots \tag{5}$$

where the initial value of the statistic is set at zero (i.e., $B_0 = 0$) and the reference value b_i is obtained by

$$b_i = \begin{cases} ln\left(\frac{1-p_1+(p_1/Z(\lambda_0,\upsilon))}{1-p_0+(p_0/Z(\lambda_0,\upsilon))}\right) & y_i = 0 \\ ln\left(\frac{p_1}{p_0}\right) & y_i > 0 \end{cases}. \tag{6}$$

If $B_i > h_b$, an upward shift in p is signaled, where h_b is the control limit that is selected to achieve the desired IC performance.

3.1.2 The λ CUSUM Chart

The λ CUSUM chart is designed to monitor an increasing shift in the rate parameter of the ZICOM-Poisson distribution. The λ CUSUM chart is designed to detect a shift from λ_0 to λ_1 (where $\lambda_1 > \lambda_0$). The λ CUSUM statistic is obtained by

$$L_i = \max\left(0,\ M_i + L_{i-1}\right); i = 1, 2, \ldots, \quad (7)$$

where $\lambda_0 = 0$. The reference value M_i is based on the log-likelihood ratio and defined as

$$M_i = \begin{cases} ln\left(\frac{1-p_0+(p_0/Z(\lambda_1,\nu))}{1-p_0+(p_0/Z(\lambda_0,\nu))}\right) & y_i = 0 \\ y_i \ln\left(\frac{\lambda_1}{\lambda_0}\right) + \ln\left(\frac{Z(\lambda_0,\nu)}{Z(\lambda_1,\nu)}\right) & y_i > 0 \end{cases}. \quad (8)$$

If $L_i > h_l$, an upward shift in λ is signaled, where h_l is the control limit that is selected to achieve the desired IC performance. It is noted that the $p - \lambda$ CUSUM chart based on the ZICOM-Poisson distribution converts to the $p - \lambda$ CUSUM chart for the ZIP model (He et al. 2012) when the dispersion parameter equals to one (i.e., $\nu = 1$).

3.2 *The CRL-ZTCOMP CUSUM Chart*

The CRL-ZTCOMP CUSUM chart is also a combination of CRL CUSUM chart and ZTCOMP CUSUM chart. This chart is also designed to detect an increasing shift in the parameters (i.e., λ and/or p) of the ZICOM-Poisson distribution.

3.2.1 The CRL CUSUM Chart

The conforming run length (CRL) is the number of conforming products between successive nonconforming products. Therefore, a CRL value is observed when a nonconforming product is seen. The random variable CRL follows a geometric distribution under the assumption of an independently and identically distributed (i.i.d) dataset. The CRL CUSUM statistic is obtained by;

$$C_i = \max\left(0,\ k - CRL_i - C_{i-1}\right); i = 1, 2, \ldots \quad (9)$$

where the initial value is selected as zero (i.e., $C_0 = 0$) and the reference value k is calculated by using the same formula as Bourke (1991)

$$k = \frac{ln\,(p_1/p_0)}{ln\,[(1-p_1)\,/\,(1-p_0)]} + 1. \quad (10)$$

Where p_0 is the IC zero-inflation parameter and p_1 is the shifted parameter to be detected quickly. If $C_i > h_c$, an upward shift in p is signaled, where h_c is the control limit that is selected to achieve the desired IC performance.

3.2.2 The ZTCOMP CUSUM Chart

The ZTCOMP CUSUM chart is developed to monitor an increasing shift in the rate parameter of the ZTCOM-Poisson distribution. It detects a shift from λ_0 to λ_1 (i.e., $\lambda_1 > \lambda_0$). The ZTCOMP CUSUM statistic is obtained by

$$T_i = \max\left(0,\ N_i + T_{i-1}\right); i = 1, 2, \ldots \tag{11}$$

where initial value is assumed to be zero (i.e., $T_0 = 0$) and the reference value N_i is based on the log-likelihood ratio statistic of zero-truncated COM-Poisson distribution (for details, see Chou et al. 2015), which is defined as

$$N_i = y_i \ln\left(\frac{\lambda_1}{\lambda_0}\right) + \ln\left(\frac{Z\left(\lambda_0, \nu\right)}{Z\left(\lambda_1, \nu\right)}\right) + \ln\left(\frac{Z\left(\lambda_1, \nu\right)\left[Z\left(\lambda_0, \nu\right) - 1\right]}{Z\left(\lambda_0, \nu\right)\left[Z\left(\lambda_1, \nu\right) - 1\right]}\right). \tag{12}$$

If $T_i > h_t$, an upward shift in λ is signaled, where h_t is the control limit that is selected to achieve the desired IC performance. It is worthy of note that, when the dispersion parameter equals to one (i.e., $\nu = 1$), the CRL-ZTCOMP CUSUM chart reduces to the CRL-ZTP CUSUM chart proposed by He et al. (2014). Next, we discuss the IC design along with the OOC performance of both proposed CUSUM structures.

4 Performance Evaluations

This section includes the definition of the performance measure used to compute the competency of the proposed charts. IC and OOC parameters settings and the development of control limits are also discussed in this section.

4.1 Performance Measure

Saghir and Lin (2014b) suggested the average number of observations to signal (ANOS) as a performance measure, to evaluate the performance of CUSUM charts based on the COM-Poisson distribution. Similarly, we have adopted ANOS measures to assess and compare the performance ability of the proposed charts. The ANOS is defined as the expected number of items inspected from the beginning of the process until a signal is highlighted by a control chart. The ANOS is further categorized as $ANOS_0$: ANOS when parameters of the ZICOM-Poisson distribution are IC or under the null hypothesis, and $ANOS_1$: ANOS when parameters of the ZICOM-Poisson distribution are OOC or under the alternative hypothesis. In the CRL-ZTCOMP CUSUM chart, the ANOS for CRL CUSUM chart is calculated by adopting the formula of Bourke (1991);

$$ANOS = \frac{1}{r} E(Number\ of\ TNS\ to\ signal), \quad (13)$$

where TNS is termed as terminal nonconformity sequence and define as a run of the conforming items followed by a nonconforming item. r is the probability of nonconformities which is defined as $r = p\,(Z\,(\lambda, \nu) - 1/Z\,(\lambda, \nu))$ in a ZICOM-Poisson process. For the ZTCOMP CUSUM chart, the same formula is used, but TNS is replaced with the number of observations in a ZICOM-Poisson process. Further, the decision is made such that on the fixed $ANOS_0$, a chart having the minimum $ANOS_1$ is declared as the best chart as compared to all others under consideration.

4.2 IC and OOC Parameters for the Simulation Study

In this study, we have carried out an extensive simulation study with 10^6 iterations to evaluate the performance ability of the $p - \lambda$ CUSUM chart and the CRL-ZTCOMP CUSUM chart. An illustration is provided in Fig. 1 to express the relationship between the probability of observing zero counts and different combinations of ZICOM-Poisson parameters.

The probability of zero counts against ZICOM-Poisson parameters λ ranging from 1 to 10, $p_0 = 0.1$ and 0.2, with $\nu = 0.5$ are portrayed in Fig. 1a; with $\nu = 1$ are presented in Fig. 1b, and with $\nu = 2$ are plotted in Fig. 1c. It is revealed that the probability of a zero count approaches $1 - p$ as λ increases. It is clearly seen that the λ have some effect on the probability of observing a zero count given the dispersion parameter ν. When $\nu = 0.5$ than $\lambda \leq 3$ has effect on the probability of observing a zero count. Similarly, when $\nu = 1$ and 2 than $\lambda \leq 6$ and $\lambda \leq 8$ have effect on the probability of observing a zero count, respectively. Hence, for the brevity in the simulation study, two IC values are selected for the zero-inflation parameter of the ZICOM-Poisson distribution such as $p_0 = 0.1$ and $p_0 = 0.2$. Similarly, two IC values of the rate parameter of ZICOM-Poisson distribution are considered such as $\lambda_0 = 4$ and $\lambda_0 = 6$. The dispersion parameter (ν) plays a vital role in the ZICOM-Poisson distribution. The ZICOM-Poisson model covers under-dispersed models (when ν is set above one), over-dispersed models (when ν is set below one), and equi-dispersed model (when ν is equals to one). Hence, to provide proper coverage to all models, we consider three IC values of the dispersion parameter such as $\nu = 0.5$, $\nu = 1$, and $\nu = 2$.

It is also noted from Fig. 1 that, as p and λ increases, the frequency of non-zero counts being observed also increases. Therefore, this study is designed to monitor an increasing shift in the zero-inflation and rate parameters of the ZICOM-Poisson distribution with fixed dispersion parameter. Hence, two shift sizes of each parameter p and λ are pre-determined for the CUSUM charts to have fast detections such as $p_1 = 1.25p_0$, $p_1 = 1.5p_0$, $\lambda_1 = \lambda_0 + 1$, and $\lambda_1 = \lambda_0 + 2$. Moreover, several shifts are introduced in the parameters to evaluate the OOC performance of the $p - \lambda$ and CRL-ZTCOMP CUSUM charts, which are provided below:

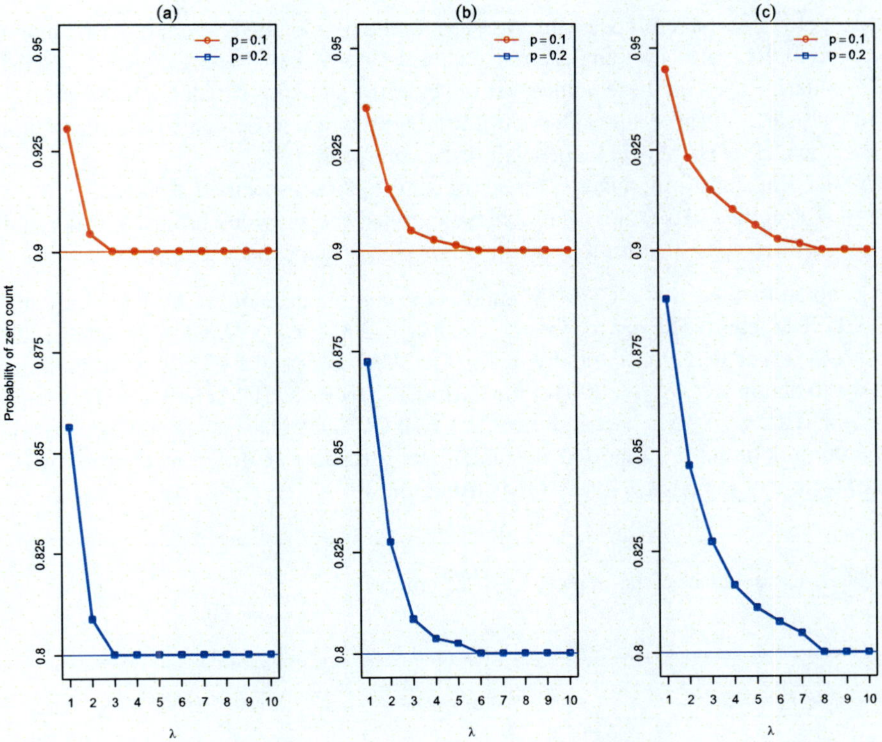

Fig. 1 An illustration of the relationship between the probability of observing zero counts and different combinations of ZICOM-Poisson parameters; **a** $\nu = 0.5$, **b** $\nu = 1$, and **c** $\nu = 2$

- For the zero-inflation parameter: when $p_0 = 0.1$, shifts are introduced as $p = 0.2, 0.3, 0.4$, and 0.5, while for $p_0 = 0.2$, shifts are considered as $p = 0.25, 0.30, 0.40,$, and 0.50.
- For rate parameter, shifts are considered as $\lambda_0 + 1$, $\lambda_0 + 2$, $\lambda_0 + 3$, and $\lambda_0 + 4$.

4.3 Derivation of Control Limits

As mentioned above that the $p - \lambda$ CUSUM chart has the control limits h_b and h_l, while the CRL-ZTCOMP CUSUM chart has decision lines h_c and h_t. The procedure to find the control limits for the stated charts is illustrated in the following steps:

(i) Generate an IC dataset of fixed sample size $n = 10000$ from ZICOM-Poisson distribution. This can be easily done using the R package of Sellers et al. (2017).

(ii) Estimates the reference value b and M for the $p - \lambda$ CUSUM chart. Also, obtain the CRL value (N) and reference values (k) for the CRL-ZTCOMP CUSUM chart. Based on these values, get the charting statistics for all CUSUM charts.
(iii) Use an arbitrary value as a control limit for the respective control chart and plot the CUSUM statistic value against the control limit.
(iv) Repeat steps i–iii, a large number of runs to obtain specified $ANOS_0$.
(v) If specified $ANOS_0$ is not achieved, adjust the previous arbitrary value and repeat steps i–iv until the specified $ANOS_0$ is achieved.

It is noted that the $p - \lambda$ CUSUM chart is the combination of p CUSUM chart and λ CUSUM chart. Hence, to obtain $ANOS_0 = 200$ for $p - \lambda$ CUSUM chart, the $ANOS_0$ is set at 352 for the individual p CUSUM chart and λ CUSUM chart. Similarly, to obtain $ANOS_0 = 200$ for the CRL-ZTCOMP CUSUM chart, the $ANOS_0$'s for the CRL CUSUM chart and the ZTCOMP CUSUM chart is set at 370. Further, the control limits are reported in Table 1 with respect to different choices of IC parameters of the ZICOM-Poisson distribution.

Table 1 Control limits of the proposed CUSUM structures

λ_1	ν_0	p_0	λ_0	$p_1 = 1.25p_0$				$p_1 = 1.5p_0$			
				h_b	h_l	h_c	h_t	h_b	h_l	h_c	h_t
$\lambda_0 + 1$	0.5	0.1	4	1.100	2.060	31.982	2.074	1.760	2.060	24.234	2.065
	0.5	0.1	6	1.100	2.050	31.900	2.040	1.750	2.050	24.400	2.060
	0.5	0.2	4	1.530	2.730	20.600	2.700	2.380	2.740	14.600	2.700
	0.5	0.2	6	1.560	2.700	20.600	2.710	2.330	2.730	14.500	2.710
	1.0	0.1	4	1.090	1.330	30.450	1.380	1.760	1.340	23.200	1.340
	1.0	0.1	6	1.090	1.220	30.600	1.230	1.760	1.220	23.600	1.234
	1.0	0.2	4	1.530	1.850	19.493	1.851	2.330	1.870	14.220	1.850
	1.0	0.2	6	1.530	1.700	20.400	1.690	2.330	1.700	14.500	1.698
	2.0	0.1	4	1.050	0.730	25.000	0.689	1.700	0.750	20.000	0.691
	2.0	0.1	6	1.050	0.640	29.000	0.620	1.700	0.650	23.300	0.620
	2.0	0.2	4	1.410	1.100	16.200	1.100	2.220	1.100	12.220	1.100
	2.0	0.2	6	1.500	0.900	18.260	0.930	2.290	0.930	13.300	0.930
$\lambda_0 + 2$	0.5	0.1	4	1.080	1.160	32.000	1.150	1.750	1.200	24.160	1.410
	0.5	0.1	6	1.120	1.390	32.100	1.370	1.750	1.170	24.160	1.396
	0.5	0.2	4	1.540	2.100	20.500	2.100	2.350	2.000	14.550	2.000
	0.5	0.2	6	1.550	2.254	20.620	2.254	2.330	2.254	14.500	2.254
	1.0	0.1	4	1.080	1.700	30.000	1.710	1.740	1.700	23.200	1.700
	1.0	0.1	6	1.080	1.760	31.200	1.750	1.750	1.760	23.700	1.760
	1.0	0.2	4	1.530	2.480	19.420	2.460	2.340	2.475	14.230	2.470
	1.0	0.2	6	1.530	2.340	20.420	2.340	2.360	2.340	14.500	2.340
	2.0	0.1	4	1.030	1.230	26.000	1.150	1.700	1.220	20.300	1.150
	2.0	0.1	6	1.050	1.060	29.000	1.040	1.700	1.080	22.000	1.050
	2.0	0.2	4	1.450	1.650	16.100	1.580	2.240	1.650	12.200	1.600
	2.0	0.2	6	1.500	1.480	18.280	1.480	2.290	1.500	13.300	1.480

5 Results and Discussions

For brevity, the OOC study is only reported for $p_0 = 0.2$. The findings of $p - \lambda$ CUSUM and CRL-ZTCOMP CUSUM charts, at fixed dispersion parameter $\nu = 0.5$, are presented in Table 2. The few results for $p_1 = 1.25p_0$ are discussed below:

- When λ_0 is set at IC value 4 and $\lambda_1 = \lambda_0 + 1$, a shift in the zero-inflation parameter (i.e., $p = 0.30$) results in 73.32% and 62.00% decrease in the $ANOS_0$ values of the $p - \lambda$ CUSUM and CRL-ZTCOMP CUSUM charts, respectively. Further, a change in the zero-inflation parameter (i.e., $p = 0.50$) results in 90.69% and 78.96% decrease in the $ANOS_0$ values of the $p - \lambda$ CUSUM and CRL-ZTCOMP CUSUM charts, respectively, for the parameter setting $\lambda_0 = 6$ and $\lambda_1 = \lambda_0 + 2$.
- For fixed $p_0 = 0.2$, a shift in the rate parameter (i.e., $\lambda = 6$) under parameter setting $\lambda_0 = 4$, and $\lambda_1 = \lambda_0 + 2$ results in 96.75% and 96.74% reduction in the $ANOS_0$ values of the $p - \lambda$ CUSUM and CRL-ZTCOMP CUSUM charts, respectively. However, a shift $\lambda = 9$ with parameter setting $\lambda_0 = 6$ and $\lambda_1 = \lambda_0 + 1$ results in 97.40% and 97.41% decrease in the $ANOS_0$ values of the $p - \lambda$ CUSUM and CRL-ZTCOMP CUSUM charts, respectively.
- For fixed $\lambda_0 = 4$ and $\lambda_1 = \lambda_0 + 1$, a shift in zero-inflation parameter (i.e., $p = 0.30$) and rate parameter (i.e., $\lambda = 7$) results in 3.44 and 5.19 $ANOS_1$ values of the $p - \lambda$ and CRL-ZTCOMP CUSUM charts, respectively. Moreover, for $\lambda_0 = 6$ and $\lambda_1 = \lambda_0 + 2$, a shift in both parameters (i.e., $p = 0.40$ and $\lambda = 8$) results in 3.29 and 6.46 $ANOS_1$ values of the $p - \lambda$ CUSUM and CRL-ZTCOMP CUSUM charts, respectively.

Table 3 consists of the results for $p - \lambda$ CUSUM and CRL-ZTCOMP CUSUM charts at fixed dispersion parameter $\nu = 1$. Some findings for $p_1 = 1.5p_0$ are discussed below:

- At fixed $\lambda_0 = 4$ and $\lambda_1 = \lambda_0 + 1$, a shift $p = 0.30$ results in 86.97% and 75.79% reduction in the $ANOS_0$ values for $p - \lambda$ CUSUM and CRL-ZTCOMP CUSUM charts, respectively. Further, on the parameter setting $\lambda_0 = 6$ and $\lambda_1 = \lambda_0 + 2$, a change $p = 0.50$ results in 91.52% and 80.85% decrease in the $ANOS_0$ values of the $p - \lambda$ CUSUM and CRL-ZTCOMP CUSUM charts, respectively.
- At fixed $p_0 = 0.2$, $\lambda_0 = 4$, and $\lambda_1 = \lambda_0 + 2$, a shift $\lambda = 5$ results in 65.09 and 66.85 $ANOS_1$ values for $p - \lambda$ CUSUM and CRL-ZTCOMP CUSUM charts, respectively. However, on the parameter setting $\lambda_0 = 6$ and $\lambda_1 = \lambda_0 + 1$, a change $\lambda = 10$ results in 10.35 and 15.97 $ANOS_1$ values of the $p - \lambda$ CUSUM and CRL-ZTCOMP CUSUM charts, respectively.
- At fixed $\lambda_0 = 4$ and $\lambda_1 = \lambda_0 + 1$, shifts $p = 0.30$ and $\lambda = 8$ results in 94.85% and 91.98% reduction in the $ANOS_0$ values of the $p - \lambda$ CUSUM and CRL-ZTCOMP CUSUM charts, respectively. Further, in the case where $\lambda_0 = 6$, and $\lambda_1 = \lambda_0 + 2$, a shift in both parameters (i.e., $p = 0.50$ and $\lambda = 8$) results in 93.95% and 85.57% decrease in the $ANOS_0$ values of the $p - \lambda$ CUSUM and CRL-ZTCOMP CUSUM charts, respectively.

Table 2 The ANOS profile of the proposed CUSUM structures with fixed dispersion parameter $\nu = 0.5$

p_0	λ_0	λ	$\lambda_1 = \lambda_0 + 1$				$\lambda_1 = \lambda_0 + 2$			
			$p_1 = 1.25p_0$		$p_1 = 1.5p_0$		$p_1 = 1.25p_0$		$p_1 = 1.5p_0$	
			$p-\lambda$	CRL-ZTCOMP	$p-\lambda$	CRL-ZTCOMP	$p-\lambda$	CRL-ZTCOMP	$p-\lambda$	CRL-ZTCOMP
0.20	4	4	199.90	201.07	201.94	200.36	201.72	200.36	200.12	199.69
0.25	4	4	90.86	112.26	94.50	113.29	90.95	111.16	92.48	111.46
0.30	4	4	53.32	76.39	52.37	74.57	52.93	75.63	52.38	73.77
0.40	4	4	27.97	51.97	25.83	47.23	27.41	51.19	25.44	46.78
0.50	4	4	18.84	42.73	16.85	37.66	18.76	42.01	16.62	37.23
0.20	6	6	200.85	200.87	201.11	199.83	199.86	199.98	199.99	199.98
0.25	6	6	92.47	111.52	92.40	111.95	89.21	106.20	88.97	109.15
0.30	6	6	53.70	75.77	52.16	73.65	52.70	74.90	50.98	72.30
0.40	6	6	28.29	51.73	25.63	47.26	27.39	50.74	25.18	46.51
0.50	6	6	19.11	42.46	16.87	37.66	18.60	42.08	16.51	37.08
0.20	4	5	16.26	16.26	16.47	16.51	18.66	18.40	18.49	18.43
0.20	4	6	6.82	6.88	6.80	6.82	6.49	6.52	6.51	6.48
0.20	4	7	5.24	5.19	5.17	5.18	5.14	5.10	5.11	5.09
0.20	4	8	4.94	5.01	4.99	5.00	5.07	5.00	5.09	5.00
0.20	6	7	16.65	16.69	16.93	16.44	19.01	18.79	18.92	18.42
0.20	6	8	7.06	7.13	7.17	7.09	6.57	6.54	6.42	6.49
0.20	6	9	5.18	5.22	5.14	5.23	5.14	5.10	5.15	5.10
0.20	6	10	5.04	5.01	5.04	5.01	5.00	5.00	4.98	5.00
0.30	4	4	6.85	76.30	53.32	74.33	52.96	75.82	51.07	73.68
0.30	4	5	5.40	16.33	10.86	16.47	12.15	18.33	12.31	18.42
0.30	4	6	4.58	6.86	4.58	6.88	4.39	6.50	4.30	6.47
0.30	4	7	3.44	5.19	3.48	5.18	3.38	5.10	3.43	5.11
0.30	4	8	2.73	5.01	3.31	5.00	3.31	5.00	3.34	5.00
0.30	6	6	53.82	76.53	51.77	73.99	52.69	74.95	51.20	72.41
0.30	6	7	11.12	16.42	11.28	16.52	12.69	18.42	12.56	18.16
0.30	6	8	4.73	7.14	4.74	7.08	4.34	6.52	4.34	6.45
0.30	6	9	3.47	5.22	3.47	5.24	3.42	5.11	3.37	5.10
0.30	6	10	3.37	5.01	3.37	5.01	3.34	5.00	3.36	5.00
0.20	4	6	6.99	6.87	6.93	6.84	6.47	6.50	6.59	6.47
0.25	4	6	5.54	6.90	5.45	6.86	5.26	6.52	5.21	6.55
0.30	4	6	4.89	6.89	4.65	6.90	4.33	6.45	4.34	6.47
0.40	4	6	3.53	6.87	3.44	6.86	3.28	6.50	3.28	6.53
0.50	4	6	2.94	6.87	2.74	6.85	2.64	6.48	2.61	6.52
0.20	6	8	7.03	7.06	7.05	7.04	6.53	6.46	6.54	6.50
0.25	6	8	5.64	7.11	5.67	7.10	5.25	6.47	5.18	6.49
0.30	6	8	4.73	7.10	4.74	7.08	4.39	6.49	4.33	6.47
0.40	6	8	3.53	7.17	3.54	7.12	3.29	6.46	3.25	6.46
0.50	6	8	2.82	7.09	2.83	7.10	2.63	6.49	2.63	6.51

Table 3 The ANOS profile of the proposed CUSUM structures with fixed dispersion parameter $\nu = 1$

p_0	λ_0	λ	$\lambda_1 = \lambda_0 + 1$				$\lambda_1 = \lambda_0 + 2$			
			$p_1 = 1.25p_0$		$p_1 = 1.5p_0$		$p_1 = 1.25p_0$		$p_1 = 1.5p_0$	
			$p - \lambda$	CRL-ZTCOMP	$p - \lambda$	CRL-ZTCOMP	$p - \lambda$	CRL-ZTCOMP	$p - \lambda$	CRL-ZTCOMP
0.20	4	4	201.96	199.45	201.19	199.30	201.17	200.75	199.54	199.15
0.25	4	4	94.76	113.82	94.19	115.61	92.62	114.57	92.84	118.02
0.30	4	4	54.92	78.35	53.65	76.71	54.66	76.98	52.50	75.93
0.40	4	4	28.90	52.00	26.22	48.25	28.45	51.43	25.98	47.45
0.50	4	4	19.31	42.70	17.33	38.23	18.93	42.36	17.25	38.12
0.20	6	6	200.06	199.64	201.98	200.23	199.98	201.63	201.12	199.94
0.25	6	6	93.63	114.89	93.52	116.22	92.93	115.01	94.01	114.12
0.30	6	6	54.48	78.26	52.71	77.44	53.97	78.27	52.54	76.07
0.40	6	6	28.69	52.66	26.34	48.86	28.29	52.48	26.61	48.22
0.50	6	6	19.31	43.86	17.33	38.72	19.11	43.27	17.05	38.29
0.20	4	5	63.48	64.32	63.64	63.67	67.03	65.73	65.09	66.85
0.20	4	6	31.55	32.21	31.57	31.94	30.40	30.36	30.29	31.37
0.20	4	7	20.62	20.96	20.76	21.09	19.01	19.29	18.96	19.73
0.20	4	8	15.62	15.72	15.50	15.96	14.03	14.37	13.95	14.74
0.20	6	7	76.74	74.35	75.16	75.96	78.08	77.14	76.52	77.51
0.20	6	8	39.90	38.81	39.52	39.68	37.42	37.96	37.28	37.73
0.20	6	9	26.12	25.59	26.30	26.55	23.61	24.21	23.77	24.26
0.20	6	10	20.04	19.21	19.65	19.92	17.55	17.76	17.75	17.93
0.30	4	4	55.06	77.90	53.30	76.55	54.44	76.77	52.03	75.89
0.30	4	5	35.07	51.79	33.93	49.83	34.27	50.40	33.28	50.37
0.30	4	6	20.48	31.30	20.13	30.46	19.56	29.39	19.12	29.69
0.30	4	7	13.77	20.88	13.85	20.82	12.49	19.28	12.16	19.80
0.30	4	8	10.42	15.91	10.35	15.97	9.28	14.49	9.18	14.70
0.30	6	6	54.16	78.53	53.33	77.20	53.74	78.09	53.59	76.05
0.30	6	7	39.43	56.44	37.99	55.59	38.49	56.64	37.31	54.41
0.30	6	8	25.20	36.81	24.62	37.01	23.97	35.30	23.27	34.94
0.30	6	9	17.33	25.57	17.25	26.47	15.67	23.84	15.72	23.61
0.30	6	10	13.25	19.24	13.11	19.99	11.72	17.73	11.65	17.87
0.20	4	6	31.86	31.94	31.65	31.78	30.48	30.48	30.09	31.27
0.25	4	6	24.85	31.60	24.79	31.45	23.68	29.96	23.37	30.64
0.30	4	6	20.29	31.06	20.28	30.38	19.15	29.54	18.90	29.68
0.40	4	6	14.88	29.49	14.48	28.51	13.75	27.82	13.47	27.45
0.50	4	6	11.54	28.03	11.09	27.01	10.73	26.46	10.24	25.98
0.20	6	8	39.94	38.55	39.35	39.51	37.60	38.01	37.87	37.57
0.25	6	8	31.00	38.29	30.58	38.88	29.36	37.30	28.87	36.59
0.30	6	8	25.17	36.94	24.49	37.33	23.68	35.51	23.31	34.78
0.40	6	8	17.57	34.32	17.13	33.93	16.56	32.63	16.01	31.39
0.50	6	8	13.61	32.14	12.89	30.87	12.65	30.84	12.17	28.86

At fixed dispersion parameter $\nu = 2$, findings of the $p - \lambda$ CUSUM and CRL-ZTCOMP CUSUM charts are reported in Table 4. Few results for $p_1 = 1.25p_0$ are discussed below:

- On the parameter setting $\lambda_0 = 4$ and $\lambda_1 = \lambda_0 + 1$, a change $p = 0.25$ may cause 53.15% and 39.87% decrease in the $ANOS_0$ values of the $p - \lambda$ CUSUM and CRL-ZTCOMP CUSUM charts, respectively. Further, a change in the zero-inflation parameter (i.e., $p = 0.30$) may result in 72.51% and 60.33% reduction in the $ANOS_0$ values of the $p - \lambda$ CUSUM and CRL-ZTCOMP CUSUM charts, respectively, for the parameter setting $\lambda_0 = 4$, and $\lambda_1 = \lambda_0 + 2$.
- For the parameter setting $p_0 = 0.2$, a change $\lambda = 7$ may result in 51.56 and 54.58 $ANOS_1$ values of the $p - \lambda$ CUSUM and CRL-ZTCOMP CUSUM charts, respectively, for the parameter setting $\lambda_0 = 4$ and $\lambda_1 = \lambda_0 + 2$. However, at $\lambda_0 = 6$ and $\lambda_1 = \lambda_0 + 1$, a shift $\lambda = 6$ results in 36.49 and 55.06 $ANOS_1$ values of the $p - \lambda$ CUSUM and CRL-ZTCOMP CUSUM charts, respectively.
- When $\lambda_0 = 4$ and $\lambda_1 = \lambda_0 + 1$, shifts in both parameters (i.e., $p = 0.30$ and $\lambda = 6$) may cause 81.80% and 72.39% reduction in the $ANOS_0$ values of the $p - \lambda$ CUSUM and CRL-ZTCOMP CUSUM charts, respectively. Further, for the fixed $\lambda_0 = 6$ and $\lambda_1 = \lambda_0 + 2$, shifts $p = 0.30$ and $\lambda = 8$ result in 79.45% and 69.54% decrease in the $ANOS_0$ values of the $p - \lambda$ CUSUM and CRL-ZTCOMP CUSUM charts, respectively.

In conclusion, increasing shift in rate parameter has a severe effect on the performance ability of both charts, but the chart show relatively same performance against these kinds of changes. When shifts are introduced in the zero-inflation parameter, the $p - \lambda$ CUSUM outperforms the CRL-ZTCOMP CUSUM chart. Moreover, when shifts are presented in both parameters, again, the $p - \lambda$ CUSUM has better detection ability as compared to the CRL-ZTCOMP CUSUM chart.

6 An Example

In a real scenario, we apply the CUSUM charts based on the ZICOM-Poisson distribution to the light-emitting diode (LED) packaging industry dataset adopted from He et al. (2012, 2014). Similarly, we have used the first 96 IC observations (after excluding four values which are greater than 10) and obtained estimates as $\hat{p}_0 = 0.1268$, $\hat{\lambda}_0 = 3.003$, and $\hat{\nu} = 0.6643$. It is clearly seen that $\hat{\nu} < 1$, which is the evidence that the dataset is also suffering from over-dispersion and hence, the ZIP distribution is not an appropriate model for the LED dataset. Further, we designed CUSUM structures by fixing $ANOS_0 = 200$, $p_1 = 1.5\hat{p}_0$, and $\lambda_1 = \hat{\lambda}_0 + 2$. The $p - \lambda$ CUSUM is plotted in Fig. 2 with its control limits $h_b = 1.95$ and $h_l = 2$. However, the CRL-ZTCOMP CUSUM chart with limits $h_c = 20$ and $h_t = 2.09$ is plotted in Fig. 3.

The p CUSUM chart based on the ZICOM-Poisson showed 26 OOC signals with indexes 162–164 and 174–196, while the λ CUSUM chart based on the ZICOM-Poisson model revealed 39 OOC points with indexes 100–110 and 129–156. The

Table 4 The ANOS profile of the proposed CUSUM structures with fixed dispersion parameter $\nu = 2$

p_0	λ_0	λ	$\lambda_1 = \lambda_0 + 1$				$\lambda_1 = \lambda_0 + 2$			
			$p_1 = 1.25p_0$		$p_1 = 1.5p_0$		$p_1 = 1.25p_0$		$p_1 = 1.5p_0$	
			$p - \lambda$	CRL-ZTCOMP	$p - \lambda$	CRL-ZTCOMP	$p - \lambda$	CRL-ZTCOMP	$p - \lambda$	CRL-ZTCOMP
0.20	4	4	200.43	199.44	200.85	199.83	201.66	200.15	199.52	201.10
0.25	4	4	93.89	119.92	95.69	123.07	95.95	119.31	94.79	123.13
0.30	4	4	56.51	80.04	55.74	82.83	56.69	79.83	54.61	81.20
0.40	4	4	29.22	51.86	27.67	49.85	29.76	51.90	27.60	49.93
0.50	4	4	19.80	41.37	18.19	39.15	20.11	41.14	18.09	38.48
0.20	6	6	199.82	201.97	201.42	200.37	201.04	201.03	199.99	200.71
0.25	6	6	96.06	118.73	96.69	119.38	96.17	115.80	91.15	117.76
0.30	6	6	56.25	79.44	54.87	78.58	55.26	79.74	53.34	78.40
0.40	6	6	29.44	52.75	26.77	49.24	29.19	52.26	26.35	48.70
0.50	6	6	19.86	42.81	17.43	38.77	19.72	42.64	17.26	38.74
0.20	4	5	113.50	112.96	112.38	113.82	114.49	114.73	111.59	115.38
0.20	4	6	72.78	74.73	72.00	74.38	73.51	75.64	71.12	75.20
0.20	4	7	53.30	55.30	53.27	54.77	51.56	54.58	50.66	54.85
0.20	4	8	42.12	43.71	41.59	43.65	40.25	43.45	39.56	43.82
0.20	6	7	126.25	130.87	126.89	129.29	128.48	130.11	125.86	130.02
0.20	6	8	88.16	90.41	87.87	90.74	88.21	89.34	86.17	88.40
0.20	6	9	65.37	68.47	66.78	67.67	65.61	66.44	63.83	65.61
0.20	6	10	52.32	55.43	53.52	54.76	51.58	53.59	50.99	52.33
0.30	4	4	55.61	80.78	55.38	82.37	56.47	80.90	54.56	82.87
0.30	4	5	45.21	66.86	43.77	66.73	44.44	67.15	43.48	66.43
0.30	4	6	36.49	55.06	35.69	54.42	35.94	54.56	34.58	54.16
0.30	4	7	30.37	46.83	29.67	45.76	29.04	46.67	28.24	45.73
0.30	4	8	25.16	39.58	24.97	39.15	24.01	39.34	23.28	38.92
0.30	6	6	55.95	80.42	54.47	78.64	56.55	79.87	51.98	77.54
0.30	6	7	49.03	71.63	47.72	70.05	48.83	70.14	45.91	68.60
0.30	6	8	42.17	63.12	41.12	60.77	41.70	61.61	39.23	59.48
0.30	6	9	36.03	55.13	35.35	53.64	36.04	53.13	33.54	51.73
0.30	6	10	31.12	48.62	30.69	46.64	30.30	46.49	29.22	44.45
0.20	4	6	72.25	73.10	72.06	74.50	72.75	75.09	70.83	75.08
0.25	4	6	49.91	65.05	49.33	64.29	49.42	65.04	48.27	64.27
0.30	4	6	36.57	55.48	35.81	54.49	36.00	55.28	34.99	54.61
0.40	4	6	22.97	43.54	21.83	41.61	22.07	42.98	20.75	41.40
0.50	4	6	16.54	37.03	15.38	35.21	15.81	36.92	14.51	34.80
0.20	6	8	86.81	91.51	88.14	89.58	88.94	89.75	85.47	88.50
0.25	6	8	58.83	76.12	58.45	74.37	58.84	75.33	57.00	73.19
0.30	6	8	42.27	62.90	40.81	61.06	41.31	61.23	39.42	59.85
0.40	6	8	25.33	47.64	23.72	44.74	25.07	46.53	22.91	43.71
0.50	6	8	17.85	40.63	16.20	37.06	17.63	39.89	15.70	36.37

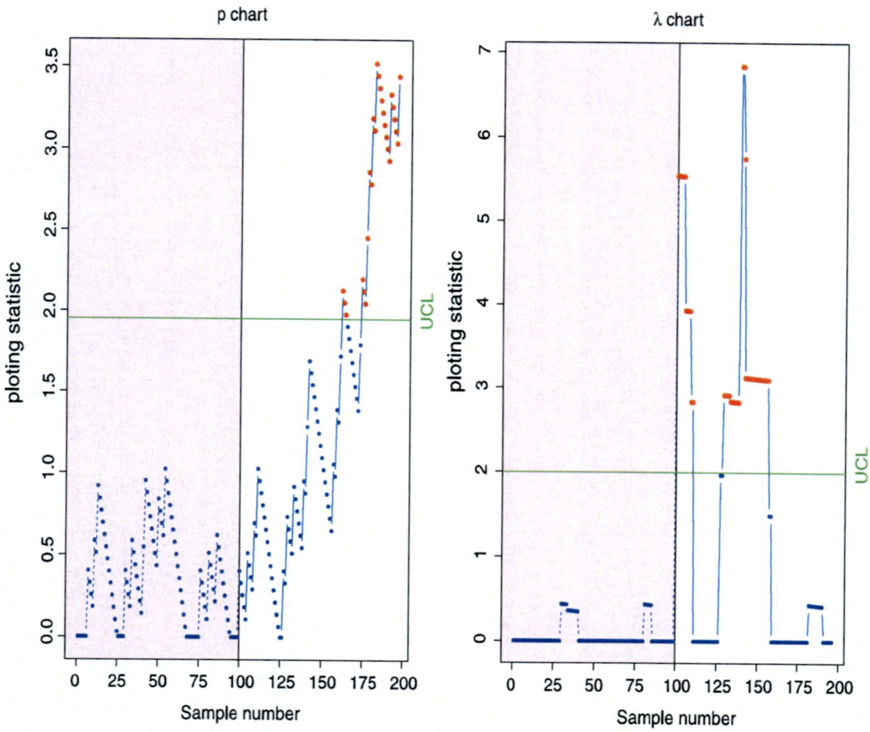

Fig. 2 The $p - \lambda$ charts based on the ZICOM-Poisson distributions for the LED dataset

CRL CUSUM chart based on the ZICOM-Poisson has captured 8 OOC signals with indexes 26 and 28–34. However, the ZTCOMP CUSUM chart reveals 8 OOC signals with indexes 13–15 and 18–22. Hence, it is revealed that the CUSUM structures based on the ZICOM-Poisson distribution are the general setups, which can also be able to cover underlying dispersion (i.e., over or under) of the dataset. Specifically, the $p - \lambda$ CUSUM chart provides a detailed diagnosis as compared to the CRL-ZTCOMP CUSUM chart.

7 Summary, Conclusions, and Recommendations

Most of the high-yield processes produce datasets which contains a large number of zeros with a small amount of count observations. Such datasets are often known as zero-defect and rare health-related datasets. Some researchers claimed that the superfluous zeros might cause over-dispersion in the data (i.e., when variance exceeds mean), which may not be wholly accurate. Sometimes, a surplus zero counts may reduce the mean of a dataset which causes inflation in the dispersion. Hence, zero-

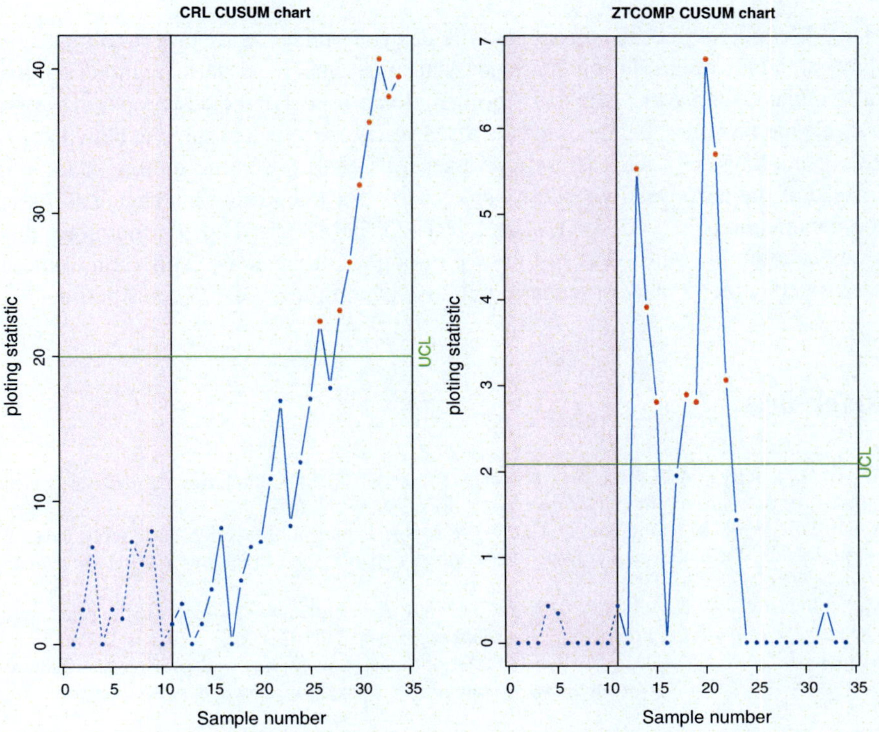

Fig. 3 The CRL-ZTCOMP CUSUM charts for the LED dataset

inflated models (i.e., ZIP, ZIG, and ZINB) were designed to fit high-yield datasets. However, in practice, it is difficult to ensure that the monitoring dataset follows the over-dispersed (i.e., ZINB), under-dispersed (i.e., ZIG), or equi-dispersed (i.e., ZIP) distribution. Therefore, the ZICOM-Poisson distribution is an alternative model, which is not only flexible to data dispersion but also a generalized form of all the above-stated distributions. From the last two decades, several monitoring methods were designed based on the zero-inflated models (e.g., ZIP, ZIG, and ZINB distributions). However, there may not exist any monitoring study based on the ZICOM-Poisson distribution, to the best of our knowledge. Hence, following He et al. (2012, 2014), we design CUSUM structures based on the ZICOM-Poisson distribution.

We design a simulation study, and the results reveal that the rate parameter (λ) has a significant effect on the performance of the CUSUM structures as compared to the zero-inflation parameter (p). It is also noted that both CUSUM structures (i.e., $p - \lambda$ CUSUM and CRL-ZTCOMP CUSUM) show almost equal performance when shifts are introduced in (λ). However, under the variations in p, the $p - \lambda$ CUSUM outperforms the CRL-ZTCOMP CUSUM chart. Further, behavior of the stated charts is also observed under the shifts in both parameters and results revealed that the $p - \lambda$ CUSUM has better detection ability as compared to the CRL-ZTCOMP CUSUM chart. It is noted that the current study is designed under zero-state condition, one may extend it by considering the steady-state approach, head starts method (i.e., the false initial response mechanism). Another good prospect of this research is to design a single monitoring scheme for the CUSUM structures to detect shifts in both parameters simultaneously. That is, to have a separate chart (with one plotting statistic), instead of the proposed two side-by-side charts, for monitoring increasing shifts in the parameters of $p - \lambda$ CUSUM and CRL-ZTCOMP CUSUM. Furthermore, this study is compiled on the basis of known in-control parameters; estimation of these parameters under unknown scenario will also be a future research contribution.

References

Abbas, N., Abujiya, M. R., Riaz, M., Mahmood, T., et al. (2020). Cumulative sum chart modeled under the presence of outliers. *Mathematics*, *8*(2), 269.

Alevizakos, V., & Koukouvinos, C. (2019). A double exponentially weighted moving average control chart for monitoring com-poisson attributes. *Quality and Reliability Engineering International*, *35*(7), 2130–2151.

Ali, S., Pievatolo, A., & Göb, R. (2016). An overview of control charts for high-quality processes. *Quality and Reliability Engineering International*, *32*(7), 2171–2189.

Barriga, G. D., & Louzada, F. (2014). The zero-inflated conway-maxwell-poisson distribution: Bayesian inference, regression modeling and influence diagnostic. *Statistical Methodology*, *21*, 23–34.

Bourke, P. D. (1991). Detecting a shift in fraction nonconforming using run-length control charts with 100% inspection. *Journal of Quality Technology*, *23*(3), 225–238.

Chang, T., & Gan, F. (1999). Charting techniques for monitoring a random shock process. *Quality and Reliability Engineering International*, *15*(4), 295–301.

Chou, Y. C., Chuang, H. H. C., & Shao, B. B. (2015). Information initiatives of mobile retailers: a regression analysis of zero-truncated count data with underdispersion. *Applied Stochastic Models in Business and Industry*, *31*(4), 457–463.

Conway, R. W., & Maxwell, W. L. (1962). A queuing model with state dependent service rates. *Journal of Industrial Engineering*, *12*(2), 132–136.

Faisal, M., Zafar, R. F., Abbas, N., Riaz, M., & Mahmood, T. (2018). A modified cusum control chart for monitoring industrial processes. *Quality and Reliability Engineering International*, *34*(6), 1045–1058.

Gan, F. (1990). Monitoring observations generated from a binomial distribution using modified exponentially weighted moving average control chart. *Journal of Statistical Computation and Simulation*, *37*(1–2), 45–60.

Gillispie, S. B., & Green, C. G. (2015). Approximating the conway-maxwell-poisson distribution normalization constant. *Statistics*, *49*(5), 1062–1073.

He, S., Huang, W., & Woodall, W. H. (2012). Cusum charts for monitoring a zero-inflated poisson process. *Quality and Reliability Engineering International*, *28*(2), 181–192.

He, S., Li, S., & He, Z. (2014). A combination of cusum charts for monitoring a zero-inflated poisson process. *Communications in Statistics-Simulation and Computation*, *43*(10), 2482–2497.

Lambert, D. (1992). Zero-inflated poisson regression, with an application to defects in manufacturing. *Technometrics*, *34*(1), 1–14.

Mahmood, T. (2020). Generalized linear model based monitoring methods for high-yield processes. *Quality and Reliability Engineering International*, *36*(5), 1570–1591.

Mahmood, T., & Xie, M. (2019). Models and monitoring of zero-inflated processes: The past and current trends. *Quality and Reliability Engineering International*, *35*(8), 2540–2557.

Mahmood, T., Wittenberg, P., Zwetsloot, I. M., Wang, H., & Tsui, K. L. (2019). Monitoring data quality for telehealth systems in the presence of missing data. *International Journal of Medical Informatics*, *126*, 156–163.

McCullagh, P., & Nelder, J. (1983). *Generalized linear models*. London: Chapman and Hall.

Montgomery, D. C. (2009). *Statistical quality control* (Vol. 7). New York: Wiley.

Page, E. S. (1954). Continuous inspection schemes. *Biometrika*, *41*(1/2), 100–115.

Riaz, M., Abbas, N., & Mahmood, T. (2017). A communicative property with its industrial applications. *Quality and Reliability Engineering International*, *33*(8), 2761–2763.

Roberts, S. (1959). Control chart tests based on geometric moving averages. *Technometrics*, *1*(3), 239–250.

Saghir, A., & Lin, Z. (2014a). Control chart for monitoring multivariate com-poisson attributes. *Journal of Applied Statistics*, *41*(1), 200–214.

Saghir, A., & Lin, Z. (2014b). Cumulative sum charts for monitoring the com-poisson processes. *Computers and Industrial Engineering*, *68*, 65–77.

Saghir, A., & Lin, Z. (2014c). A flexible and generalized exponentially weighted moving average control chart for count data. *Quality and Reliability Engineering International*, *30*(8), 1427–1443.

Saghir, A., & Lin, Z. (2015). Control charts for dispersed count data: an overview. *Quality and Reliability Engineering International*, *31*(5), 725–739.

Saghir, A., Lin, Z., Abbasi, S. A., & Ahmad, S. (2013). The use of probability limits of com-poisson charts and their applications. *Quality and Reliability Engineering International*, *29*(5), 759–770.

Sellers, K., Lotze, T., & Raim, A. (2017). Compoissonreg: Conway-maxwell poisson (com-poisson) regression. https://CRAN.R-project.org/package=COMPoissonRegRpackage version 04 1:380

Sellers, K. F. (2012). A generalized statistical control chart for over-or under-dispersed data. *Quality and Reliability Engineering International*, *28*(1), 59–65.

Sellers, K. F., & Raim, A. (2016). A flexible zero-inflated model to address data dispersion. *Computational Statistics and Data Analysis*, *99*, 68–80.

Shewhart, W. A. (1926). Quality control charts. *The Bell System Technical Journal*, *5*(4), 593–603.

Shmueli, G., Minka, T. P., Kadane, J. B., Borle, S., & Boatwright, P. (2005). A useful distribution for fitting discrete data: revival of the conway-maxwell-poisson distribution. *Journal of the Royal Statistical Society: Series C (Applied Statistics)*, *54*(1), 127–142.

Sim, S. Z., Gupta, R. C., & Ong, S. H. (2018). Zero-inflated conway-maxwell poisson distribution to analyze discrete data. *The International Journal of Biostatistics*, *14*(1), 20160,070.

Xie, M., & Goh, T. (1993). Spc of a near zero-defect process subject to random shocks. *Quality and Reliability Engineering International*, *9*(2), 89–93.

Xie, M., Goh, T., & Kuralmani, V. (2000). On optimal setting of control limits for geometric chart. *International Journal of Reliability, Quality and Safety Engineering*, *7*(1), 17–25.

Xie, W., Xie, M., & Goh, T. (1995). Control charts for processes subject to random shocks. *Quality and Reliability Engineering International*, *11*(5), 355–360.

An Average Loss Control Chart Under a Skewed Process Distribution

Su-Fen Yang and Shan-Wen Lu

Abstract In the global market the quality of products is a crucial factor separating competitive companies within numerous industries. These firms may employ a loss function to measure the loss caused by a deviation of the quality variable from the target value. From the view of Taguchi's philosophy, monitoring this deviation from the process target value is important, but in practice many quality data have distributions that are not normal but skewed. This paper thus develops an average loss control chart for monitoring quality loss variation under skewed distributions. We investigate the statistical properties of the proposed control chart and measure the out-of-control process detection performance of the proposed loss control charts by using the average run length. The average loss control chart illustrates the best performance in detecting of out-of-control loss location for a left-skewed process distribution and performs better than the existing median loss control chart.

Keywords Loss function · Non-normal distribution · Control chart · Run length

1 Introduction

Control charts are commonly used tools in process change detection for improving the quality of manufacturing and service processes. In the past few years, more and more statistical process control techniques have been applied to the service industry, with control charts also becoming an effective tool to enhance service quality. There have been a few studies in this area of the literature, including Tsung et al. (2008), Ning et al. (2009), Yang et al. (2012), Yang and Yang (2013), Yang and Wu (2017a, b), Yang and Jiang (2019). In practice, many service quality data follow non-normal distributions. For example, the service time of a local bank branch is a

S.-F. Yang (✉) · S.-W. Lu
Statistics Department, National Chengchi University, Taipei, Taiwan
e-mail: yang@mail2.nccu.tw

S.-W. Lu
e-mail: ollymeow@hotmail.com

S. Knoth and W. Schmid (eds.), *Frontiers in Statistical Quality Control 13*,
Frontiers in Statistical Quality Control,
https://doi.org/10.1007/978-3-030-67856-2_5

critical service quality characteristic, and efficiently monitoring the location and/or dispersion of service data is an important issue to bank managers. Bank service data that have been analyzed tend to have a right-skewed distribution as shown in Yang and Wu (2017a, b). Some other examples of service quality data are fatigue symptoms of breast cancer patients Ho et al. (2014), passenger counts of Taipei's mass rapid transit (MRT) system on a weekday basis Yang and Yang (2013), and health care costs Zhou et al. (2008). The commonly used Shewhart variables control charts, whose statistical properties depend on a normality assumption, clearly may not be suitable for monitoring service data when the variables exhibit non-normal or unknown distributions. Furthermore, McCracken and Chakraborti (2013) note that normality is often an elusive assumption, and discuss some available nonparametric schemes for jointly monitoring location and scale in overviewing control charts for joint monitoring mean and variance.

Product, service quality, and productivity loss are all crucial competitive factors of companies in numerous industries, and the loss function is a popular method for measuring the loss caused by variations in product or service quality. Taguchi (1986) proposed that target values are vital during process specification, while Sullivan (1984) emphasized the importance of monitoring deviations from the target value. Because increases in the difference between the mean and the target and/or variability are the sources of out-of-control loss, it is crucial to monitor the loss variation of a manufacturing or service process.

Scant research has been done to deal with monitoring process loss location. Existing loss-function-based control charts are based on the assumption that the in-control mean of a process quality variable equals the target value; see, for example, Zhang and Wu (2006) and Wu et al. (2009). However, in practice, the in-control process mean may not actually be the process target, and diagnosing the source of an out-of-control signal is crucial for correcting an out-of-control process loss location. Yang (2013a, b), Yang and Lin (2014) and Yang et al. (2017) proposed loss-based control charts in order to monitor the loss location that arises when quality variables deviate from target values.

A major drawback of loss-based control charts is that almost all of them are based on the assumption that the quality variable has a normal distribution. This paper focuses on discussing a loss-based control chart under non-normal distributions. We note that the sample median is more robust than the sample average for estimating the population location as the former is less affected by extreme values Graham et al. (2011). Motivated by this, Yang et al. (2017) considered using the median loss to express the quality loss function under a non-normal distribution. For this reason, the resulting loss-based control chart is called the median loss (ML) control chart throughout their paper. Their ML chart and the optimal variable sampling intervals median loss (VSI ML) chart both illustrate the best out-of-control detection performance for the left-skewed distributed process among the considered left-skewed, symmetric, and right-skewed distributions. Even under a normal distribution, they illustrated that the resulting out-of-control detection performance of the VSI ML chart is better than the VSI average loss (AL) chart in Yang (2013b) and the weighted loss (WL) control charts in Yang and Lin (2014), except for very small shifts in process mean. However, the properties of average loss (ALSN) control chart were not discussed for a non-normal distributed process. Here in this present study, we consider that both the sample size and sampling interval are fixed and will examine whether the ALSN control chart has better out-of-control detection performance than that of the ML chart under a skew-normal distributed process. Hence, we proceed to derive the ALSN control chart and discuss the out-of-control detection performances of the ALSN control chart either when the process distribution is left-skewed, symmetric, or right-skewed, respectively.

The paper is organized as follows. Section 2 introduces the sampling distribution of the median loss for a quality variable, X, with a skew-normal distribution. Section 3 illustrates the control limits of the ML chart for various sample sizes and out-of-control detection performances for small to moderate shifts in mean and variance. Section 4 derives the distribution of sample average loss, constructs the ALSN control chart, and measures its out-of-control detection performance for small to moderate shifts in mean and variance. Section 5 compares the out-of-control detection performance between the proposed ALSN chart and the ML chart in Yang et al. (2017) by considering the process with left-skewed, symmetric, and right-skewed normal distribution, respectively. Section 6 summarizes the findings and provides a recommendation.

2 The ML Control Chart

2.1 The Skew-Normal Distribution

We let the random variable X have a skew-normal distribution with location parameter $\xi_0 \in (-\infty, \infty)$, scale parameter $a_0 \in (0, \infty)$, and shape parameter $b \in (-\infty, \infty)$, i.e., $X \sim SN(\xi_0, a_0, b)$. From Azzalini (1985), the probability density function (pdf) of X is:

$$f_X(x) = \frac{2}{a_0}\varphi\left(\frac{x-\xi_0}{a_0}\right)\Phi\left(b\frac{x-\xi_0}{a_0}\right), \quad x \in (-\infty, \infty), \tag{1}$$

where $\varphi(\cdot)$ and $\Phi(\cdot)$ are respectively the pdf and cumulative distribution function (cdf) of the standard normal distribution.

In (1) we know that if $b = 0$, then the skew-normal distribution will reduce to the normal distribution with mean ξ_0 and standard deviation a_0. The distribution is right-skewed if $b > 0$ and is left-skewed if $b < 0$. The plot of the pdfs for $b = -2, 0, 3$ is shown in Fig. 1.

The cumulative distribution function (cdf) of the skew-normal random variable X is:

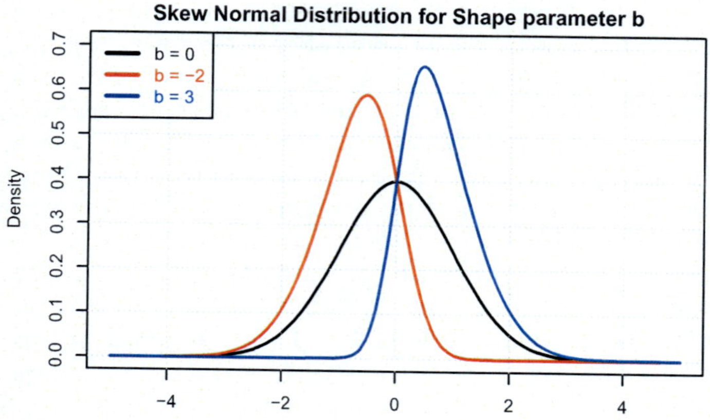

Fig. 1 The pdfs for different b

$$F_X(x) = \Phi\left(\frac{x-\xi_0}{a_0}\right) - \frac{1}{\pi}\int_0^b \frac{\exp\left[-\frac{1}{2}\left(\frac{x-\xi_0}{a_0}\right)^2(1+y^2)\right]}{1+y^2}\,dy\,, \quad x \in (-\infty, \infty)\,. \tag{2}$$

2.2 The Loss Function

The loss function is defined as $L = k(X-T)^2$. Let $X_i, i = 1, 2, \ldots, n$, be a random sample from the in-control distribution of $SN(\xi, a, b)$. The sample median loss depends on the sample size being odd or even. We only consider the case where the sample size is an odd value for easier derivation of the distribution of the sample median loss.

Denote the sample statistic of median loss as $ML = (X-T)^2_{\left(\left(\frac{n+1}{2}\right)\right)}$, where ML is the loss value separating the higher half from the lower half of a loss data sample. For a loss dataset, this may be thought of as the "middle" loss value.

Referring to Yang et al. (2017), the derived cdf of ML is as follows.

$$\begin{aligned} F_{ML}(t) &= \int_0^t f_M(u)\,du \\ &= \frac{n!}{\left[\left(\frac{n-1}{2}\right)!\right]^2}\int_0^t F_{(X-T)^2}(u)^{\frac{n-1}{2}}\left[1-F_{(X-T)^2}(u)\right]^{\frac{n-1}{2}} f_{(X-T)^2}(u)\,du \\ &= \frac{n!}{\left[\left(\frac{n-1}{2}\right)!\right]^2} B\left(\frac{1}{a_0\sqrt{t}}\left[\varphi\left(\frac{\sqrt{t}+T-\xi_0}{a_0}\right)\Phi\left(b\frac{\sqrt{t}+T-\xi_0}{a_0}\right)\ldots\right.\right. \end{aligned}$$

$$+\varphi\left(\frac{-\sqrt{t}+T-\xi_0}{a_0}\right)\Phi\left(b\frac{-\sqrt{t}+T-\xi_0}{a_0}\right)\Bigg], \frac{n+1}{2}, \frac{n+1}{2}\Bigg), \; t>0, \tag{3}$$

where $B(x, a, b) = \int_0^x t^{a-1}(1-t)^{b-1}\,dt$ is an incomplete beta function.

We determine the ML control chart based on the cdf of ML in order to monitor the changes in the loss location or, equivalently, to monitor the shifts in the in-control population mean (or the deviation of $\mu_0 - T$) and/or standard deviation.

2.3 The Design of a Median Loss Control Chart

We first establish the ML control chart with a specified false alarm rate α by using Eq. (3). The upper control limit (UCL) and the lower control limit (LCL) of the ML chart are obtained by taking the inverse cdf of ML – that is:

$$UCL = F_{ML}^{-1}(1-\alpha/2)\,, \quad LCL = F_{ML}^{-1}(\alpha/2)\,. \tag{4}$$

The process is deemed to be out-of-control if the monitoring statistic ML is smaller than LCL or larger than UCL; otherwise, the process is considered to be in-control.

The expectation (μ_0) and variance (σ_0^2) of in-control X are $\mu_0 = \xi_0 + a_0\frac{b}{\sqrt{1+b^2}}\sqrt{\frac{2}{\pi}}$ and $\sigma_0^2 = a_0^2\left[1-\frac{2b^2}{\pi(1+b^2)}\right]$, respectively. Let δ_3 denote the dispersion parameter that satisfies $\mu_0 - T = \delta_3\sigma_0$. For a skewed distribution, we set $\delta_3 > 0$.

Table 1 gives the control limits of the ML chart for various combinations of $n = 5, 11$, $\delta_3 = 0, 1, 2$, and $b = -500, -2, 0, 2$ and 500 under ARL$_0$ = 370.4, $\mu_0 = 0$,

Table 1 Control limits of the ML chart with ARL$_0$ =370.4

n	b	δ_3		
		0	1	2
		(LCL, UCL)	(LCL, UCL)	(LCL, UCL)
5	–500	(0.006, 3.573)	(0.021, 4.958)	(0.157, 10.331)
	–2	(0.004, 3.707)	(0.016, 6.190)	(0.176, 12.086)
	0	(0.004, 3.754)	(0.012, 6.868)	(0.198, 13.099)
	2	(0.004, 3.707)	(0.009, 7.546)	(0.283, 14.040)
11	500	(0.006,3.573)	(0.003, 8.354)	(0.618, 15.135)
	–500	(0.036, 1.661)	(0.132, 4.264)	(0.800, 9.290)
	–2	(0.027, 2.192)	(0.102, 4.374)	(0.814, 9.463)
	0	(0.028, 2.268)	(0.075, 4.498)	(0.796, 9.713)
	2	(0.027, 2.192)	(0.054, 4.542)	(0.855, 9.802)
	500	(0.036, 1.661)	(0.020, 4.729)	(0.907, 10.078)

and $\sigma_0 = 1$. From Table 1 we can see that the widths of the control limits become narrower when n increases and b and δ_3 are fixed, and the widths of the control limits become wider when δ_3 increases and n and b are fixed. When $\delta_3 = 0$, the widths of the control limits are the widest for a symmetric ($b = 0$) distributed quality variable. When δ_3 >0, the widths of the control limits become wider under an increasing b or when the distribution of the quality variable changes from left-skewed, to normal, to right-skewed.

3 Performance Measurement of the ML Chart

We fix ARL_0 at a desired level, for example 370.4, while for an out-of-control process average run length (ARL_1) being smaller is better. Here, ARL_0 for the ML chart is:

$$\text{ARL}_0 = 1/\big(1 - P(LCL < ML < UCL \mid \text{in-control } ML)\big) . \tag{5}$$

Suppose that X^* is the quality characteristic for the out-of-control process, and $X^* \sim SN(\xi^*, a^*, b)$ has mean $\mu_1 = \mu_0 + \delta_1\sigma_0, \delta_1 \neq 0$, and standard deviation $\sigma_1 = \delta_2\sigma_0, \delta_2 \geq 1$. The power $(1 - \beta)$ is the probability that the out-of-control median loss statistic (ML^*) is larger than UCL or smaller than LCL – that is:

$$1 - \beta = 1 - P(LCL < ML^* < UCL) = F_{ML^*}(LCL) + 1 - F_{ML^*}(UCL) .$$

Hence, we obtain:

$$\text{ARL}_1 = \frac{1}{1 - \beta} = \frac{1}{F_{ML^*}(LCL) + 1 - F_{ML^*}(UCL)} , \tag{6}$$

where $F_{ML^*}(\cdot)$ is the cdf of the out-of-control median loss statistic.

Table 2 illustrates the out-of-control detection performance of the ML chart for the shifts in mean and standard deviation, $\delta_1 = 1.0, 2.0, \delta_2 = 1.0, 2.0$, the dispersion parameter, $\delta_3 = 0, 1, 2$, $\text{ARL}_0 = 370.4$, $n = 5$, $\mu_0 = 0$, $\sigma_0 = 1$, and the quality variable with the left half normal ($b = -500$), left-skewed ($b = -2$), symmetric ($b = 0$), right-skewed ($b = 2$), and right half normal ($b = 500$) distributions. In Table 2 we can see, whether $b = -500, -2, 0, 2$, or 500, that ARL_1 decreases when δ_1 and/or δ_2 are far away from $\delta_1 = 0$ and/or $\delta_2 = 1$ under a specified δ_3 (≥ 0); the ARL_1 of the ML chart decreases when δ_3 rises for a specified combination of ($\delta_1 > 0, \delta_2 > 0, b$); and the ARL_1s of the ML chart with the left-skewed distributed ($b < 0$) quality variable are all smaller than those of the quality variable with symmetric ($b = 0$) and right-skewed ($b > 0$) distributions. These findings suggest that the ML chart has the best performance for the left-skewed distributed quality variable.

Table 2 ARL_1 of the ML chart ($n = 5$)

δ_1	δ_2	b	δ_3		
			0	1	2
1	1	–500	13.869	2.059	2.027
		–2	22.527	4.764	4.615
		0	24.152	8.146	8.113
		2	25.040	14.131	14.131
		500	22.726	22.207	22.207
2	2	–500	1.415	1.067	1.065
		–2	1.618	1.164	1.158
		0	1.829	1.313	1.312
		2	2.027	1.588	1.588
		500	2.308	2.308	2.308
1	2	–500	2.010	1.577	1.715
		–2	2.757	2.393	2.579
		0	3.317	3.355	3.528
		2	4.176	4.882	4.889
		500	6.860	6.867	6.452
2	2	–500	1.326	1.168	1.190
		–2	1.523	1.335	1.351
		0	1.757	1.554	1.563
		2	2.052	1.874	1.875
		500	2.519	2.522	2.522

4 The Average Loss Control Chart

4.1 The Distribution of Average Loss

The Taguchi loss function is defined as $L = k(X - T)^2$. Without loss of generality, we set $k = 1$. In order to design an average loss control chart, suppose that a sequence of random samples $X_1, X_2, \ldots, X_n$ of size n are taken from $SN(\xi_0, a_0, b)$. We further define the sample average loss (AL) as:

$$AL = \frac{1}{n}\sum_{i=1}^{n}(X_i - T)^2 = \frac{n-1}{n}S_X^2 + (\bar{X} - T)^2 \,. \tag{7}$$

The first step to construct the ALSN chart is to find the distribution of AL when X follows a skew-normal distribution. Since the exact distribution of AL is not available, we use Edgeworth expansion (for example, see Hall 1992) to approximate the distribution of AL.

Edgeworth (1905) derived Edgeworth expansion that relates the cdf of a random variable having expectation zero and variance 1 to the cumulative density function (cdf) of the standard normal distribution using Chebyshev–Hermite polynomials.

Since the in-control X_i follows $SN(\xi_0, a_0, b)$, we can obtain the first and the second moments of $L_i = (X_i - T)^2$ by using the Chebyshev–Hermite polynomials. Hence, the expectation and the standard deviation of L (μ_L and σ_L) can be obtained by the moments of L.

If we define $Z_n = \sqrt{n}(AL - \mu_L)/\sigma_L$, then we can approximate the pdf of Z_n by Edgeworth expansion:

$$f_{Z_n}(z) \approx \varphi(z) - \frac{1}{\sqrt{n}}\left(\frac{1}{6}\lambda_3 \Phi^{(4)}(z)\right) + \frac{1}{n}\left(\frac{1}{24}\lambda_4 \Phi^{(5)}(z) + \frac{1}{72}\lambda_3^2 \Phi^{(7)}(z)\right), \quad (8)$$

where $\Phi^{(r)}(z) = (-1)^{r-1} He_{r-1}(z)\varphi(z)$, $He_{r-1}(z)$ is the Chebyshev–Hermite polynomial, and λ_r is the rth cumulant of $(L - \mu_L)/\sigma_L$ (see Hall 1992).

We can therefore obtain the cdf of AL by the following.

$$\begin{aligned} F_{AL}(t) &= P(AL \le t) = P\left(Z_n \le \frac{\sqrt{n}(t-\mu_L)}{\sigma_L}\right) = F_{Z_n}\left(\frac{\sqrt{n}(t-\mu_L)}{\sigma_L}\right) \\ &\approx \Phi\left(\frac{\sqrt{n}(t-\mu_L)}{\sigma_L}\right) - \frac{1}{\sqrt{n}}\left(\frac{1}{6}\lambda_3 \Phi^{(3)}\left(\frac{\sqrt{n}(t-\mu_L)}{\sigma_L}\right)\right) \\ &+ \frac{1}{n}\left(\frac{1}{24}\lambda_4 \Phi^{(4)}\left(\frac{\sqrt{n}(t-\mu_L)}{\sigma_L}\right) + \frac{1}{72}\lambda_3^2 \Phi^{(6)}\left(\frac{\sqrt{n}(t-\mu_L)}{\sigma_L}\right)\right). \end{aligned} \quad (9)$$

The accuracy of this approximation is examined by the Pearson χ^2 goodness-of-fit test. We find when the number of random samples m is 2000 or 1000 and the sample size $n = 11$ that the test reveals that the approximated cdf has no significant difference from the cdf using Monte Carlo simulation.

4.2 The Design of an Average Loss Control Chart

Using Eq. (9), the upper control limit (UCL) and lower control limit (LCL) of an average loss control chart with false alarm rate α are expressed as follows.

$$UCL = F_{AL}^{-1}(1 - \alpha/2), \quad LCL = F_{AL}^{-1}(\alpha/2). \quad (10)$$

We let the ALSN control chart represent the average loss control chart throughout the paper and estimate the control limits using Monte Carlo simulation.

Table 3 lists the control limits of the ALSN chart with $\text{ARL}_0 = 370.4$ for various combinations of $n = 5, 11, \delta_3 = 0, 1, 2, b = -500, -2, 0, 2, 500, \mu_0 = 0$, and $\sigma_0 = 1$. From the table we can see that the widths of the control limits become narrower when n increases and b and δ_3 are fixed, and the widths of the control limits become

Table 3 Control limits of the ALSN chart

n	b	δ_3		
		0	1	2
		(LCL, UCL)	(LCL, UCL)	(LCL, UCL)
5	–500	(0.095, 4.468)	(0.000, 4.718)	(0.950, 9.275)
	–2	(0.083, 3.984)	(0.167, 5.647)	(0.991, 11.009)
	0	(0.024, 3.665)	(0.114, 6.431)	(1.062, 12.117)
	2	(0.083, 3.984)	(0.040, 7.377)	(1.190, 13.315)
	500	(0.095, 4.468)	(0.000, 8.324)	(1.112, 14.469)
11	–500	(0.282, 3.220)	(0.595, 3.739)	(2.147, 7.954)
	–2	(0.262, 2.902)	(0.568, 4.312)	(2.023, 8.869)
	0	(0.224, 2.692)	(0.509, 4.806)	(1.958, 9.523)
	2	(0.262, 2.902)	(0.443, 5.413)	(1.984, 10.265)
	500	(0.282, 3.220)	(0.288, 6.020)	(1.913, 11.000)

wider when δ_3 increases and n and b are fixed. When $\delta_3 = 0$, the width of the control limits is the widest for a symmetric ($b = 0$) distributed quality variable. When $\delta_3 > 0$, the widths of the control limits become wider under an increasing b or for the distribution of quality variable changing from left-skewed, to normal, to right-skewed.

5 Performance Measurement of the ALSN Control Chart

To measure the detection performance of the proposed ALSN control chart, we let the out-of-control mean and standard deviation be $\mu_1 = \mu_0 + \delta_1\sigma_0$, $\delta_1 \neq 0$, and $\sigma_1 = \delta_2\sigma_0$, $\delta_2 \geq 1$, where $\delta_1 = 1.0, 2.0$, $\delta_2 = 1.0, 2.0$, $\delta_3 = 0, 1, 2$, and $b = -500, -2.0, 2, 500$. We estimate the ARL_1s using Monte Carlo simulation.

Table 4 illustrates the out-of-control detection performance of the ALSN chart for the changes in mean and standard deviation, $\delta_1 = 1.0, 2.0$, $\delta_2 = 1.0, 2.0$, the dispersion parameter, $\delta_3 = 0, 1, 2$, $\mathrm{ARL}_0 = 370.4$, $n = 5$, $\mu_0 = 0$, $\sigma_0 = 1$, and the quality variable with the left half normal ($b = -500$), left-skewed ($b = -2$), symmetric ($b = 0$), right-skewed ($b = 2$), and right half normal ($b = 500$) distributions. In this table we can see whether $b = -500, -2, 0, 2$, or 500 that ARL_1 decreases when δ_1 and/or δ_2 are far away from $\delta_1 = 0$ and/or $\delta_2 = 1$ under a specified $\delta_3 \geq 0$); the ARL_1 of the ALSN chart decreases when δ_3 rises for only mean changes ($\delta_1 = 1, 2$), but is almost same for $\delta_3 = 1$ and $\delta_3 = 2$; and the ARL_1s of the ALSN chart with the left-skewed distributed ($b < 0$) quality variable are all smaller than those of the quality variable with symmetric ($b = 0$) and right-skewed ($b > 0$) distributions, except for $\delta_1 = 1$ and $\delta_2 = 1$. These findings suggest that the ALSN chart has better performance for the left-skewed distributed quality variable.

Table 4 ARL_1 of the ALSN chart ($n = 5$)

δ_1	δ_2	b	δ_3		
			0	1	2
1	1	–500	190.23	1.754	1.604
		–2	29.969	3.002	2.976
		0	13.037	4.718	4.707
		2	12.993	7.874	7.611
		500	10.049	11.795	11.147
2	1	–500	1.578	1.012	1.009
		–2	1.415	1.039	1.038
		0	1.346	1.095	1.088
		2	1.544	1.234	1.200
		500	1.903	1.504	1.413
2	1	–500	1.953	1.150	1.253
		–2	1.701	1.416	1.621
		0	1.653	1.712	1.964
		2	1.908	2.155	2.416
		500	2.031	2.496	2.836
2	2	–500	1.125	1.024	1.034
		–2	1.160	1.068	1.093
		0	1.183	1.139	1.169
		2	1.300	1.263	1.287
		500	1.423	1.442	1.453

We further compare the ARL_1s between the proposed ALSN chart and the existing ML chart for a process with a skew-normal distribution. From the resulting Tables 2 and 4 we can see under $b = -500, -2, 0, 2, 500$, respectively, that the ARL_1s of the ALSN chart performs better than those of the ML chart whether the process has small or moderate changes in location and/or dispersion.

6 Conclusions

In this paper we propose a new ALSN control chart to monitor the changes in process loss location or in the deviation of process mean and target and/or variance when the distribution of a process is not symmetric but left-skewed or right-skewed. We also develop the numerical approaches for calculating control limits and ARL of the ALSN control chart are developed. Through numerical analyses, the proposed ALSN chart shows reasonable and reliable detection ability compared to the ML chart. Furthermore, the proposed ALSN chart illustrates best out-of-control detection performance for the left-skewed distributed quality variable. We thus recommend the application of the proposed ALSN chart for process loss location monitoring. In the future, we suggest to study the exponentially weighted moving average ALSN control chart, adaptive control schemes and the effect of contamination by outliers.

Acknowledgements The research was partially supported by a MOST 106-2118-M-004-001-MY2 research grant, Taiwan.

References

Azzalini, A. (1985). A class of distributions which includes the normal ones. *Scandinavian Journal of Statistics*, *12*(2), 171–178.

Edgeworth, F. Y. (1905). The law of error. *Transactions of the Cambridge Philosophical Society*, *20*, 36–65.

Graham, M. A., Chakraborti, S., & Human, S. W. (2011). A nonparametric exponentially weighted moving average signed-rank chart for monitoring location. *Computational Statistics and Data Analysis*, *55*(8), 2490–2503.

Hall, P. (1992). *The bootstrap and edgeworth expansion*. New York: Springer.

Ho, R. T. H., Fong, T. C. T., & Cheung, I. K. M. (2014). Cancer-related fatigue in breast cancer patients: Factor mixture models with continuous non-normal distributions. *Quality of Life Research*, *23*(10), 2909–2916.

McCracken, A. K., & Chakraborti, S. (2013). Control charts for joint monitoring of mean and variance: An overview. *Quality Technology and Quantitative Management*, *10*(1), 17–36.

Ning, X., Shang, Y., & Tsung, F. (2009). Statistical process control techniques for service processes: A review. In *6th International Conference on Service Systems and Service Management 2009, Institute of Electrical and Electronics Engineers* (pp 927–931).

Sullivan, L. P. (1984). Reducing variability: A new approach to quality. *Quality Progress*, *17*(7), 15–21.

Taguchi, G. (1986). *Introduction to quality engineering: Designing quality into products and processes*. Toyko: Asian Productivity Organization.

Tsung, F., Li, Y., & Jin, M. (2008). Statistical process control for multistage manufacturing and service operations: a review and some extensions. *International Journal of Services Operations and Informatics*, *3*(2), 191–204.

Wu, Z., Wang, P., & Wang, Q. (2009). A loss function-based adaptive control chart for monitoring the process mean and variance. *The International Journal of Advanced Manufacturing Technology*, *40*(9), 948–959.

Yang, C. C., & Yang, S. F. (2013). Optimal variable sample size and sampling interval 'mean squared error' chart. *The Service Industries Journal*, *33*(6), 652–665.

Yang, S. F. (2013a). Using a new VSI EWMA average loss control chart to monitor changes in the difference between the process mean and target and/or the process variability. *Applied Mathematical Modelling*, *37*(16–17), 7973–7982.

Yang, S. F. (2013b). Using a single average loss control chart to monitor process mean and variability. *Communications in Statistics - Simulation and Computation*, *42*(7), 1549–1562.

Yang, S. F., & Jiang, T. A. (2019). Service quality variation monitoring using the interquartile range control chart. *Quality Technology and Quantitative Management*, *16*(5), 613–627.

Yang, S. F., & Lin, L. Y. (2014). Monitoring and diagnosing process loss using a weighted-loss control chart. *Quality and Reliability Engineering International*, *30*(7), 951–959.

Yang, S. F., & Wu, S. H. (2017a). A double sampling scheme for process mean monitoring. *IEEE Access*, *5*, 6668–6677.

Yang, S. F., & Wu, S. H. (2017b). A double sampling scheme for process variability monitoring. *Quality and Reliability Engineering International*, *33*(8), 2193–2204.

Yang, S. F., Cheng, T. C., Hung, Y. C., & Cheng, S. W. (2012). A new chart for monitoring service process mean. *Quality and Reliability Engineering International*, *28*(4), 377–386.

Yang, S. F., Zhou, R., & Lu, S. W. (2017). A median loss control chart for monitoring quality loss under skewed distributions. *Journal of Statistical Computation and Simulation*, *87*(17), 3241–3260.

Zhang, S., & Wu, Z. (2006). Monitoring the process mean and variance using a weighted loss function cusum scheme with variable sampling intervals. *IIE Transactions*, *38*(4), 377–387.

Zhou, X. H., Lin, H., & Johnson, E. (2008). Non-parametric heteroscedastic transformation regression models for skewed data with an application to health care costs. *Journal of the Royal Statistical Society: Series B (Statistical Methodology)*, *70*(5), 1029–1047.

ARL-Unbiased CUSUM Schemes to Monitor Binomial Counts

Manuel Cabral Morais, Sven Knoth, Camila Jeppesen Cruz, and Christian H. Weiß

Abstract Counted output, such as the number of defective items per sample, is often assumed to have a marginal binomial distribution. The integer and asymmetrical nature of this distribution and the value of its target mean hinders the quality control practitioner from dealing with a chart for the process mean with a pre-stipulated in-control average run length (ARL) and the ability to swiftly detect not only increases but also decreases in the process mean. In this paper we propose *ARL-unbiased* cumulative sum (CUSUM) schemes to rapidly detect both increases and decreases in the mean of independent and identically distributed as well as first-order autoregressive (AR(1)) binomial counts. Any shift is detected more quickly than a false alarm is generated by these schemes and their in-control ARL coincide with the pre-specified in-control ARL. We use the R statistical software to provide compelling illustrations of all these CUSUM schemes.

Keywords ARL-unbiased design · CUSUM · Binomial AR(1) model · Statistical process control

M. C. Morais (✉)
CEMAT (Center for Computational and Stochastic Mathematics) and Department of Mathematics, Instituto Superior Técnico, Universidade de Lisboa, Av. Rovisco Pais, Lisbon, Portugal
e-mail: maj@math.ist.utl.pt

S. Knoth · C. H. Weiß
Department of Mathematics and Statistics, Faculty of Economics and Social Sciences, Helmut Schmidt University, Hamburg, Postfach 700822, 22008, Hamburg, Germany
e-mail: knoth@hsu-hh.de

C. H. Weiß
e-mail: weissc@hsu-hh.de

C. J. Cruz
Instituto Superior Técnico, Universidade de Lisboa, Av. Rovisco Pais, Lisbon, Portugal
e-mail: camilacruz@tecnico.ulisboa.pt

S. Knoth and W. Schmid (eds.), *Frontiers in Statistical Quality Control 13*, Frontiers in Statistical Quality Control,
https://doi.org/10.1007/978-3-030-67856-2_6

1 Introduction

Shewhart quality control charts have a widely known limitation: they are not particularly swift when it comes to the detection of small-to-moderate shifts. The reason being that these charts only use the last observed value of their control statistics to decide whether or not a signal should be triggered.

The rationale of the cumulative sum (CUSUM) control statistic, introduced by Page (1954), is building up by addition of differences between a score X_t and a reference value k. If this control statistic exceeds the decision interval value h, a signal is triggered by the CUSUM control chart and the parameter is deemed out-of-control.

CUSUM charts, like most quality control charts, are usually assessed by determining their average run length (ARL) profile. The ARL is undisputedly the most popular performance metric and represents the average number of samples taken before a signal is triggered by the chart.

The reference value k of the CUSUM chart should be chosen to be between the target process mean and the process mean level that the CUSUM chart is to detect promptly, observed Lucas (1985). Another recommendation can be discerned in Lucas (1985): after k is selected, the decision interval h is chosen in order to give a sufficiently large in-control ARL.

CUSUM control charts/schemes:

- give an indication of small-to-moderate shifts earlier than their Shewhart counterparts (see e.g., Ewan and Kemp 1960; Lucas 1985; Gan 1993; Hawkins and Olwell 1998, pp. 7–8; Montgomery 2009, p. 402);
- can be related to Wald sequential probability ratio tests (SPRT) (Johnson and Leone 1962);
- and their (asymptotic) optimality properties for detecting a change in distribution have been thoroughly discussed (Lorden 1971; Pollak 1985).

As expertly put by Lucas (1985), two-sided CUSUM schemes to detect either an increase or a decrease in the process mean are obtained by running simultaneously two one-sided CUSUM charts for this parameter. The standard upper and lower one-sided CUSUM charts have control statistics given by

$$S_t^+ = \max\{0,\ S_{t-1}^+ + (X_t - k^+)\} \tag{1}$$

$$S_t^- = \max\{0,\ S_{t-1}^- + (k^- - X_t)\} \tag{2}$$

(respectively), where

- $\{X_t : t \in \mathbb{N}\}$ is the output process;
- the starting values are $S_0^+ = S_0^- = 0$, that is, the fast initial response (FIR) feature Lucas and Crosier (1982) has not been adopted; k^+ and k^- are the reference values of each one-sided chart.

If the control statistic S_t^+ (resp. S_t^-) exceeds the decision interval value h^+ (resp. h^-), then a signal is prompted at sample t by the upper (resp. lower) one-sided CUSUM charts at sample t.

1.1 A Handful of CUSUM Charts and Schemes for Independent Counts

Counts of nonconforming items (resp. defects) arise frequently in statistical process control (SPC) and are assumed to have a marginal binomial (resp. Poisson) distribution.

As far as we have investigated, Ewan and Kemp (1960, Sects. 13–14) were the first authors to propose the use of CUSUM charts to detect persistent shifts in the mean of independent and identically distributed (i.i.d.) Poisson output from a target value μ_0 to an out-of-control level μ_1. Ewan and Kemp (1960, Sect. 15) go on to add that some of the results they obtained for the Poisson variate can also be used to formulate schemes to control the fraction nonconforming, provided this proportion is very small so that the Poisson distribution may be used to describe the sampling distribution of the number of nonconforming items in a fixed size sample.

Lucas (1985) suggests that the reference value should be selected to be close to $k = (\mu_1 - \mu_0)/[\ln(\mu_1) - \ln(\mu_0)]$ and when $k \geq 1$ the reference value will usually be rounded to the nearest integer. If a two-sided CUSUM scheme is put to use and its constituent upper and lower one-sided charts are meant to swiftly detect sustained shifts from μ_0 to the off-target levels μ_1^+ ($\mu_1^+ > \mu_0$) and μ_1^- ($\mu_1^- < \mu_0$), then the reference values should be equal to

$$k^* = \frac{\mu_1^* - \mu_0}{\ln(\mu_1^*) - \ln(\mu_0)}, \quad * = +, -. \tag{3}$$

Lucas (1985) advocates the use of two-sided CUSUM schemes for the mean of i.i.d. Poisson output, explains how to obtain its RL performance metrics, states under which conditions the ARL can be obtained in terms of the ARL of the one-sided CUSUM constituent charts, and yet does not provide tables for the ARL of those two-sided schemes.

We are convinced that Gan (1993) is a seminal paper when it comes to the use of CUSUM charts in the detection of sustained shifts in the fraction of nonconforming items.

Gan (1993) and Hawkins and Olwell (1998, p. 123) suggest that the reference value of the upper and lower one-sided CUSUM chart for i.i.d. binomial output should be selected as close as possible of

$$k^* = \frac{n \times \ln[(1 - p_0)/(1 - p_1^*)]}{\ln[(1 - p_0)p_1^*/(1 - p_1^*)p_0]}, \quad * = +, -. \tag{4}$$

Recall that np_0 is the nominal expected number of nonconforming items per random sample of size n, and np_1^+ (resp. np_1^-) denotes the corresponding out-of-control value that we want to quickly detect by making use of the upper (resp. lower) one-sided CUSUM for this sort of output.

Let us remind the reader that the reference values in (3) and (4) were obtained by capitalizing on the relation between the SPRT and the CUSUM procedure, as described in detail by Hawkins and Olwell (1998, pp. 139–140, 145–147), for the general exponential family with a single parameter and for the Poisson and binomial families.

Hawkins and Olwell (1998, p. 108) refer the routines *ANYARL* and *ANYGETH* to deal with one-sided CUSUM charts for i.i.d. binomial, Poisson, and negative binomial output. *ANYARL* returns the ARL of the chart for a given h and *ANYGETH* attempts to find a suitable h for the target in-control ARL. Hawkins and Olwell (1998, pp. 108–110, 124–126) go on to illustrate the use of these two routines for i.i.d. Poisson and binomial counts.

1.2 A Few CUSUM Charts and Schemes for Autocorrelated Counts

Autocorrelation among successive counts is a more realistic assumption while dealing for instance with very high sampling rates (Rakitzis et al. 2017). Moreover, it has been amply shown in the literature that we need indeed specific charts/schemes to monitor autocorrelated counts such as the ones mentioned in the comprehensive review in Weiß (2018, Chap. 8) or found in Weiß (2009a, Chap. 20), Weiß and Testik (2009) propose an upper one-sided CUSUM chart to monitor the mean of first-order integer-valued autoregressive (INAR(1)) Poisson counts. Yontay et al. (2013) discuss upper and lower one-sided CUSUM charts and two-sided CUSUM schemes for Poisson INAR(1) output.

Weiß and Testik (2012) discuss CUSUM charts to control counts modeled by the integer-valued counterpart to the usual first-order autoregressive conditional heteroskedasticity models, INARCH(1). Additionally, Rakitzis et al. (2017) introduce separate upper and lower one-sided charts to monitor the mean of binomial AR(1) and beta-binomial AR(1) counts, and provide practical guidelines for the statistical design of these charts. Ottenstreuer et al. (2018) proposed a combined Shewhart-CUSUM scheme with a switching limit to monitor not only i.i.d. and INAR(1) Poisson counts, but also Gaussian output.

1.3 On ARL-Unbiased Charts for Discrete Output

While dealing with a control chart/scheme, the in-control ARL should never be smaller than any out-of-control ARL. This behavior of the ARL profile means, according to Knoth and Morais (2013, 2015), that the chart satisfies what Ramalhoto and Morais (1995, 1999) called the primordial criterion and Pignatiello et al. (1995) and Acosta-Mejía and Pignatiello (2000) expertly termed an *ARL-unbiased* chart.

On being questioned about ARL-unbiased charts for parameters of discrete distributions, we are bound to reply that the SPC literature is scarce and we ought to mention:

- the attempts to derive ARL-unbiased charts that resulted in *nearly* or *approximately* ARL-unbiased designs, such as the np-chart proposed by Acosta-Mejía and Pignatiello (1999), the geometric chart found in Zhang et al. (2004), and the cumulative count conforming (CCC) chart under group inspection (CCC_G) promoted by Zhang et al. (2012);
- the control charts that are indeed ARL-unbiased, specifically the EWMA$-p$ chart (to monitor the variance for process data with non-normal or unknown distributions) proposed by Yang and Arnold (2015), the c- and np- charts derived by Paulino et al. (2016) and Morais (2016), the geometric and CCC_G charts found in Morais (2017), the c-chart for Poisson INAR(1) counts proposed by Paulino et al. (2019), the thinning-based exponentially weighted moving average (TEWMA) found in Morais et al. (2018) for the mean of i.i.d. Poisson counts, and the EWMA chart proposed by Morais and Knoth (2020) for this same mean.

Having all this in mind, we review the derivation of the ARL of the one-sided CUSUM charts and the two-sided CUSUM scheme (Sect. 2), propose ARL-unbiased versions of the two-sided CUSUM schemes for the mean of i.i.d. binomial counts (Sect. 3) and of binomial AR(1) counts (Sect. 4), and wrap up the paper briefly discussing some related topics worthy of future research (Sect. 5).

We use the R statistical software to provide instructive illustrations of all these schemes and to assess their in-control and out-of-control performance.

2 The ARL of CUSUM Charts and Schemes for i.i.d. Counts

Throughout this section let us assume that:

- $\{X_t : t \in \mathbb{N}\}$ is a sequence of i.i.d. counts;
- the reference value k^* and the upper control limit h^* are, for $* = +, -$, positive rational numbers written as $k^* = a^*/b^*$ and $h^* = c^*/b^*$, where $a^*, b^*, c^* \in \mathbb{N}$.

2.1 Upper and Lower One-Sided CUSUM Charts

$\{S_t^* : t \in \mathbb{N}_0\}$ constitutes a Markov chain with state space $\mathscr{S}^* = \{0, 1/b^*, 2/b^*, \ldots, c^*/b^*, (c^*+1)/b^*, \ldots\}$.

The transition probabilities referring to the upper one-sided CUSUM chart, $p_{i^+,j^+} = P(S_t^+ = j^+ \mid S_{t-1}^+ = i^+)$, $i^+, j^+ \in \mathscr{S}^+$, are equal to:

- for $i^+ \in \mathscr{S}^+$ and $j^+ = 0$,

$$\begin{aligned} p_{i^+,j^+} &= P(S_t^+ = 0 \mid S_{t-1}^+ = i^+) \\ &= P[i^+ + (X_t - k^+) \leq 0] \\ &= P(X_t \leq k^+ - i^+); \end{aligned} \tag{5}$$

- for $i^+ \in \mathscr{S}^+$ and $j^+ \in \mathscr{S}^+\backslash\{0\}$,

$$\begin{aligned} p_{i^+,j^+} &= P(S_t^+ = j^+ \mid S_{t-1}^+ = i^+) \\ &= P[i^+ + (X_t - k^+) = j^+] \\ &= P(X_t = k^+ + j^+ - i^+). \end{aligned} \tag{6}$$

As for the lower one-sided CUSUM chart, the transition probabilities $p_{i^-,j^-} = P(S_t^- = j^- \mid S_{t-1}^+ = i^-)$, $i^-, j^- \in \mathscr{S}^-$ are given by:

- for $i^- \in \mathscr{S}^-$ and $j^- = 0$,

$$\begin{aligned} p_{i^-,j^-} &= P(S_t^- = 0 \mid S_{t-1}^- = i^-) \\ &= P[i^- + (k^- - X_t) \leq 0] \\ &= P(X_t \geq k^- + i^-); \end{aligned} \tag{7}$$

- for $i^- \in \mathscr{S}^-$ and $j^- \in \mathscr{S}^-\backslash\{0\}$,

$$\begin{aligned} p_{i^-,j^-} &= P(S_t^- = j^- \mid S_{t-1}^- = i^-) \\ &= P[i^- + (k^- - X_t) = j^-] \\ &= P(X_t = k^- - j^- + i^-). \end{aligned} \tag{8}$$

Let $u^* \in \{0, 1/b^*, 2/b^*, \ldots, c^*/b^*\}$ be the initial value of the control statistic S_t^*, $* = +, -$. If $S_t^* > h^* = c^*/b^*$ then a signal is triggered at sample t by the associated one-sided CUSUM chart and its run length is given by

$$RL^{*,u^*}(p) = \min\{t \in \mathbb{N} : S_t^* > h^* = c^*/b^* \mid S_0^* = u^*\},\ * = +, -. \tag{9}$$

Capitalizing on the work by Brook and Evans (1972), we can derive the ARL of the one-sided CUSUM chart with control statistic S_t^* as the expected value of a time to absorption of a Markov chain with transient (or in-control) states

$\mathcal{T}^* = \{0, 1/b^*, 2/b^*, \ldots, c^*/b^*\}$ and absorbing (out-of-control) state corresponding to $S_t^* > h^*$. By doing so, we get

$$ARL^{*,u^*} = \underline{\mathbf{e}}_{u^*}^\top \times (\mathbf{I}^* - \mathbf{Q}^*)^{-1} \times \underline{\mathbf{1}}^*, \; * = +, -, \tag{10}$$

where

- $\underline{\mathbf{e}}_{u^*}^\top$ is the $(u^* + 1)$-th vector of the orthogonal basis for $\mathbb{R}^{(c^*+1)}$;
- $\mathbf{I}^*$ represents the identity matrix with rank $(c^* + 1)$;
- $\mathbf{Q}^* = [p_{i^*,j^*}]_{i^*,j^* \in \mathcal{T}^*}$ is the sub-stochastic matrix governing the transitions between the transient (i.e., in-control) states of the absorbing Markov chain;
- $\underline{\mathbf{1}}^*$ is a column-vector with $(c^* + 1)$ ones.

2.2 The Two-Sided CUSUM Scheme

$\{(S_t^+, S_t^-) : t \in \mathbb{N}_0\}$ is a bivariate Markov chain with state space $\mathcal{S} = \mathcal{S}^+ \times \mathcal{S}^-$ and transition probabilities

$$p_{(i^+,i^-)(j^+,j^-)} = P(S_t^+ = j^+, S_t^- = j^- \mid S_{t-1}^+ = i^+, S_{t-1}^- = i^-).$$

A double subscript was obviously used to index any state; $(0, 0)$ represents a two-sided CUSUM scheme with both constituent charts CUSUM at zero, whereas (i^+, i^-) refers to a two-sided CUSUM scheme with upper (resp. lower) one-sided CUSUM chart in state i^+ (resp. i^-).

Keeping in mind that the two control statistics of the two-sided CUSUM control scheme are $S_t^+ = \max\{0, \; S_{t-1}^+ + (X_t - k^+)\}$ and $S_t^- = \max\{0, \; S_{t-1}^- + (k^- - X_t)\}$ the probabilities of transitioning from state (i^+, i^-) to state (j^+, j^-) can be easily derived and written in terms of indicator functions:

- for $j^+ = 0$ and $j^- = 0$,

$$\begin{aligned} &p_{(i^+,i^-)(j^+,j^-)} \\ &\quad = P(S_t^+ = 0, \; S_t^- = 0 \mid S_{t-1}^+ = i^+, \; S_{t-1}^- = i^-) \\ &\quad = P[i^+ + (X_t - k^+) \leq 0, \; i^- + (k^- - X_t) \leq 0] \\ &\quad = P(k^- + i^- \leq X_t \leq k^+ - i^+) \times \mathbb{1}_{[0,k^+-k^-]}(i^+ + i^-); \end{aligned} \tag{11}$$

- for $j^+ = 0$ and $j^- \in \mathcal{S}^- \backslash \{0\}$,

$$\begin{aligned}
&p_{(i^+,i^-)(j^+,j^-)} \\
&\quad = P(S_t^+ = 0,\ S_t^- = j^- \mid S_{t-1}^+ = i^+,\ S_{t-1}^- = i^-) \\
&\quad = P[i^+ + (X_t - k^+) \le 0,\ i^- + (k^- - X_t) = j^-] \\
&\quad = P(X_t \le k^+ - i^+,\ X_t = k^- - j^- + i^-) \\
&\quad = P(X_t = k^- - j^- + i^-) \times \mathbb{1}_{(-\infty, k^+ - k^-]}(i^+ + i^- - j^-); \qquad (12)
\end{aligned}$$

- for $j^+ \in \mathscr{S}^+ \backslash \{0\}$ and $j^- = 0$,

$$\begin{aligned}
&p_{(i^+,i^-)(j^+,j^-)} \\
&\quad = P(S_t^+ = j^+,\ S_t^- = 0 \mid S_{t-1}^+ = i^+,\ S_{t-1}^- = i^-) \\
&\quad = P[i^+ + (X_t - k^+) = j^+,\ i^- + (k^- - X_t) \le 0] \\
&\quad = P(X_t = k^+ + j^+ - i^+,\ X_t \ge k^- + i^-] \\
&\quad = P(X_t = k^+ + j^+ - i^+) \times \mathbb{1}_{(-\infty, k^+ - k^-]}(i^+ + i^- - j^+); \qquad (13)
\end{aligned}$$

- for $j^+ \in \mathscr{S}^+ \backslash \{0\}$ and $j^- \in \mathscr{S}^- \backslash \{0\}$,

$$\begin{aligned}
&p_{(i^+,i^-)(j^+,j^-)} \\
&\quad = P(S_t^+ = j^+,\ S_t^- = j^- \mid S_{t-1}^+ = i^+,\ S_{t-1}^- = i^-) \\
&\quad = P[i^+ + (X_t - k^+) = j^+,\ i^- + (k^- - X_t) = j^-] \\
&\quad = P(X_t = k^+ + j^+ - i^+,\ X_t = k^- - j^- + i^-) \\
&\quad = P(X_t = k^+ + j^+ - i^+) \times \mathbb{1}_{\{k^+ - k^-\}}(i^+ + i^- - j^+ - j^-). \qquad (14)
\end{aligned}$$

Once we derived these transition probabilities, we can extend the use of the Markov chain approach and obtain the ARL function using the procedure described by Lucas and Crosier (1982). Thus, we shall consider an absorbing Markov chain with a set of transient (or in-control) states $\mathscr{T} = \mathscr{T}^+ \times \mathscr{T}^-$ and a single absorbing state comprising all the out-of-control states of the original Markov chain characterized by (i^+, i^-), with $i^+ > h^+$ or $i^- > h^-$.[1] We can find the ARL of the two-sided CUSUM scheme with no head start using only one matrix inversion,

$$ARL = \underline{\mathbf{e}}_0^\top \, (\mathbf{I} - \mathbf{Q})^{-1} \, \underline{\mathbf{1}}, \qquad (15)$$

where

- $\underline{\mathbf{e}}_0^\top$ is the first vector of the orthogonal basis for $\mathbb{R}^{(c^+ + 1) \times (c^- + 1)}$;
- $\mathbf{I}$ represents the identity matrix with rank $(c^+ + 1) \times (c^- + 1)$;
- $\mathbf{Q} = [p_{(i^+,i^-)(j^+,j^-)}]_{(i^+,i^-),(j^+,j^-) \in \mathscr{T}}$;
- $\underline{\mathbf{1}}$ is a column-vector with $(c^+ + 1) \times (c^- + 1)$ ones.

[1] Notice that the transient states can be ordered as follows: $(0, 0)$, $(0, 1/b^-)$, $\ldots$, $(0, c^-/b^-)$, $(1, 0)$, $(1, 1/b^-)$, $\ldots$, $(1, c^-/b^-)$, $\ldots$, $(c^+/b^+, 0)$, $(c^+/b^+, 1)$, $\ldots$, $(c^+/b^+, c^-/b^-)$.

2.3 *Relating the ARL of One-Sided CUSUM Charts and the Two-Sided CUSUM Scheme*

According to Lucas (1985), the ARL of the two-sided CUSUM scheme with initial state $(S_0^+, S_0^-) = (u^+, u^-)$, ARL^{u^+,u^-}, can be obtained directly from the ARL of two one-sided charts, under certain conditions. Indeed, if

$$h^+ + k^+ - k^- \geq u^+ + u^- \tag{16}$$
$$h^- + k^+ - k^- \geq u^+ + u^- \tag{17}$$
$$k^+ - k^- \geq h^+ - h^- \tag{18}$$
$$k^+ - k^- \geq h^- - h^+, \tag{19}$$

then

$$ARL^{u^+,u^-} = \frac{ARL^{+,u^+} ARL^{-,0} + ARL^{+,0} ARL^{-,u^-} - ARL^{+,0} ARL^{-,0}}{ARL^{+,0} + ARL^{-,0}}, \tag{20}$$

where ARL^{+,u^+} (resp. ARL^{-,u^-}) is the ARL of the upper (resp. lower) one-sided CUSUM chart with initial state $u^+ \in \mathscr{T}^+$ (resp. $u^- \in \mathscr{T}^-$). When $(S_0^+, S_0^-) = (u^+, u^-) = (0, 0)$, the conditions (16)–(19) can be written as

$$k^+ - k^- \geq \max\{-h^+, -h^-, h^+ - h^-, h^- - h^+\} = |h^+ - h^-| \tag{21}$$

and the ARL of the two-sided CUSUM scheme with no head start reads as the familiar formula

$$\frac{1}{ARL} = \frac{1}{ARL^+} + \frac{1}{ARL^-}, \tag{22}$$

where $ARL \equiv ARL^{0,0}$, $ARL^+ \equiv ARL^{+,0}$, and $ARL^- \equiv ARL^{-,0}$.

3 The ARL-Unbiased Two-Sided CUSUM Scheme for i.i.d. Binomial Output

The ARL-unbiased two-sided CUSUM scheme to monitor i.i.d. binomial counts triggers a signal at sample t with:

- probability one if $S_t^+ > h^+$ or $S_t^- > h^-$;
- probability γ^+ (resp. γ^-) if $S_t^+ = h^+$ (resp. $S_t^- = h^-$).

Since we are dealing with a discrete control statistic, randomization probabilities are essential to bring the in-control ARL to a pre-specified in-control ARL value, say $ARL^\star$. They also play a major role in achieving a maximum of the ARL function at $p = p_0$.

3.1 The Control Limits and Randomization Probabilities

Randomizing the emission of a signal means considering an altered sub-stochastic matrix $\mathbf{Q} \equiv \mathbf{Q}(p, \gamma^+, \gamma^-)$ in (15). As a matter of fact, when there is a transition from state $(i^+, i^-) \in \mathscr{T}$ to state:

- (j^+, h^-), for $j^+ \in \mathscr{T}^+ \backslash \{h^+\}$, $(1-\gamma^-) \times p_{(i^+,i^-)(j^+,h^-)}$ replaces the entry $p_{(i^+,i^-)(j^+,h^-)}$;
- (h^+, j^-), for $j^- \in \mathscr{T}^- \backslash \{h^-\}$, $(1-\gamma^+) \times p_{(i^+,i^-)(h^+,j^-)}$ takes the place of the entry $p_{(i^+,i^-)(h^+,j^-)}$;
- (h^+, h^-), $(1-\gamma^+) \times (1-\gamma^-) \times p_{(i^+,i^-)(h^+,h^-)}$ is a replacement for $p_{(i^+,i^-)(h^+,h^-)}$.

The remaining entries of the sub-stochastic matrix $\mathbf{Q}(p, \gamma^+, \gamma^-)$ continue to be equal to the transition probabilities $p_{(i^+,i^-)(j^+,j^-)}$, for $(i^+, i^-), (j^+, j^-) \in (\mathscr{T}^+ \backslash \{h^+\}) \times (\mathscr{T}^- \backslash \{h^-\})$.

Expectedly, the ARL function of the ARL-unbiased two-sided CUSUM scheme for i.i.d. binomial output is obtained by using (15) with this altered sub-stochastic matrix $\mathbf{Q}$.

The randomization of the emission of a signal when S_t^+ (resp. S_t^-) is equal to h^+ (resp. h^-) decreases the ARL of a two-sided CUSUM scheme with the very same control limits but no randomization probabilities. Consequently, we want to obtain an in-control ARL larger than the pre-specified value, so that complementing control limits with randomization probabilities brings the in-control ARL down to $ARL^\star$.

This simple fact has to be taken into account by any search procedure we may use to obtain the control limits and the associated randomization probabilities of an ARL-unbiased chart/scheme with dependent control statistics. That is the case of the search procedure proposed by Paulino et al. (2019), which comprises two main steps: identifying the grid of control limits; obtaining admissible randomization probabilities.

Apart from its first step, the search procedure we used to derive the control limits and randomization probabilities of the ARL-unbiased two-sided CUSUM scheme does not differ much from the description found in Paulino et al. (2019). The differences are essentially due to the fact that, unlike Paulino et al. (2019), we are dealing with a two-sided scheme and a rather time-consuming search procedure.

In order to reduce the search procedure run time, we considered integer reference values and control limits. However, when it was not possible to derive an ARL-unbiased two-sided CUSUM scheme with in-control ARL equal to $ARL^\star$ by considering integer reference values, they were rounded to one decimal place, in particular to multiples of 0.5, 0.2, or 0.1. Consequently, the set of possible values for the control limits is $\{d \times l : d \in \mathbb{N}\}$ with l either equal to 1, 0.5, 0.2, or 0.1.

Given (22), if we consider initial control limits h_0^+ and h_0^- such that

$$ARL^*(p_0) > 2 \times ARL^\star, \; * = +, -, \tag{23}$$

we may ensure that

$$ARL(p_0) > ARL^\star. \tag{24}$$

If that is not the case, we have to fix the value of h_0^- and increase h_0^+ until (24) is fulfilled.

Once this condition is satisfied, we have to check whether $ARL(p)$ achieves a maximum, $ARL(p^\star)$, at a point $p^\star$ that lies to the left or to the right of p_0. Bearing in mind that, for a fixed h^+, an increase (resp. decrease) of h^- forces $p^\star$ to move left (resp. right), we have to proceed with the search of the control limits and consider two separate cases: $p^\star > p_0$ and $p^\star < p_0$. The goal in both of them is to obtain, for a fixed h^+, the largest h^- such that the maximum point of the ARL function lies to the right of p_0. As a result we must adopt the following procedure, when $p^\star > p_0$ (resp. $p^\star < p_0$).

- For $h^- = h_0^-, h_0^- + l, \dots$ (resp. $h^- = h_0^-, h_0^- - l, \dots$), consider a value of h^+ large enough ($h^+ \geq h_0^+$) such that (24) is met and search for the smallest (resp. largest) h^-, say h, so that $p^\star$ moves to the left (resp. right). Under these circumstances, h and $h - l$ (resp. h and $h + l$) define the two candidate values for the h^-. Moreover, for each of these candidates, identify the smallest possible value of h^+, say H_h and H_{h-l} (resp. H_h and H_{h+l}), which guarantee that (24) is valid; these are the candidate values for h^+.

Note that the largest of these two candidates to be h^+ has to be increased by one unit to handle some *exotic* cases.

As for the obtention of the randomization probabilities, the search procedure relies on the pairs of (h^+, h^-) previously identified, intervals of admissible randomization probabilities, approximate values of the derivative of the ARL function at the target value of the parameter, and on a secant rule to attain a zero derivative, at least approximately.

3.2 Preliminary Results

Figure 1 allows us to compare the performances of two ARL-unbiased two-sided CUSUM schemes for binomial i.i.d. counts:

- one considering integer reference values and control limits;
- the other one with rational values for k^* and h^*, $* = +, -$.

Their in-control ARL were brought exactly to its desired value $ARL^\star = 370.4$.

This figure leads to the conclusion that the scheme with rational reference values and control limits performs slightly better than the scheme associated with integer values. This improvement comes with a price: the search procedure run time is substantially higher.

Figure 2 refers to the comparison of:

- the ARL-unbiased version of the np-chart (Morais 2016), with control limits L and U and randomization probabilities γ_L and γ_U;

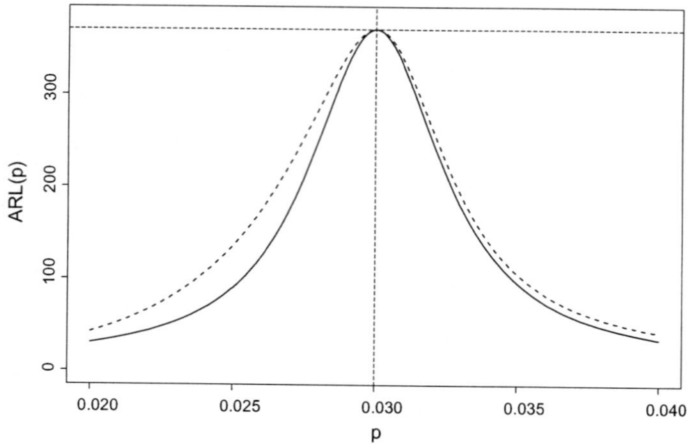

Fig. 1 ARL-unbiased curves to two-sided CUSUM schemes with integer and rational reference values—$n = 60$, $p_0 = 0.03$, $p_1^{\pm} = p_0 \pm 0.01$, $(k^-, k^+, h^-, h^+, \gamma^-, \gamma^+) = (1, 2, 3, 18, 0.028753, 0.323484)$ (dashed line), $(1.5, 2.1, 10, 12.4, 0.182710, 0.864451)$ (solid line)

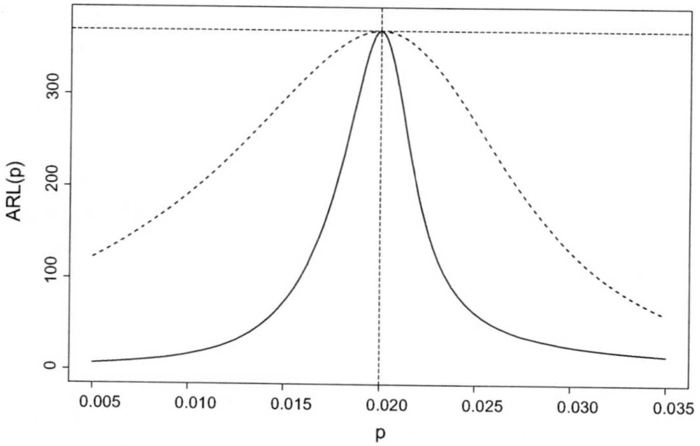

Fig. 2 ARL-unbiased curves of a *np*-chart and a two-sided CUSUM scheme—$n = 90$, $p_0 = 0.02$, $p_1^{\pm} = p_0 \pm 0.01$, $(L, U, \gamma_L, \gamma_U) = (0, 7, 0.012851, 0.084593)$ (dashed line), $(k^-, k^+, h^-, h^+, \gamma^-, \gamma^+) = (1, 2, 3, 18, 0.020530, 0.204149)$ (solid line)

- the ARL-unbiased two-sided CUSUM scheme we have just proposed.

It is apparent that the former is considerably outperformed by the latter even for values of p very close to p_0. Thus, replacing an ARL-unbiased np-chart by its two-sided CUSUM counterpart certainly pays off when we are monitoring the fraction nonconforming of binomial i.i.d. counts.

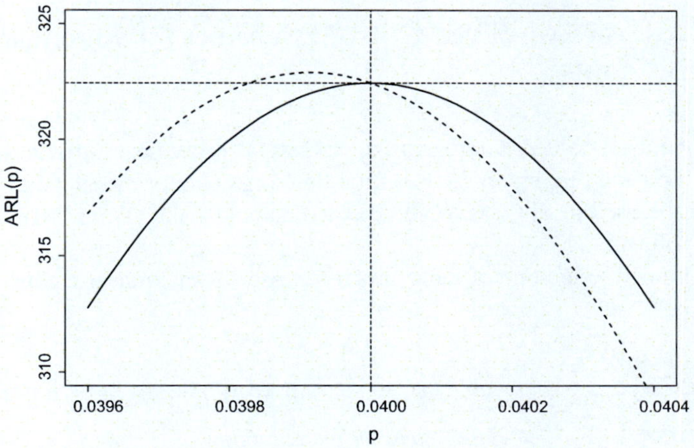

Fig. 3 ARL profiles of a standard two-sided CUSUM scheme and an ARL-unbiased two-sided CUSUM scheme—$n = 200$, $p_0 = 0.04$, $p_1^{\pm} = p_0 \pm 0.01$, $ARL^\star = 322.4388$, $(k^-, k^+, h^-, h^+) = (7, 9, 15, 17)$ (dashed line), $(k^-, k^+, h^-, h^+, \gamma^-, \gamma^+) = (7, 9, 15, 18, 0.165153, 0.647683)$ (solid line)

In Cruz (2016, pp. 27–33) the reader can find: further examples of the comparisons described above; ARL estimates obtained via Monte Carlo simulation that virtually coincide with the values of the ARL function; and a brief study of the impact of the choice of p_1^+ and p_1^- on the ARL performance of the ARL-unbiased two-sided CUSUM scheme.

As for the bias reduction of the ARL profile of a standard two-sided CUSUM scheme for i.i.d. binomial output, it is not significant because these schemes tend to have ARL functions with a maximum very close to the target value p_0, as shown by Fig. 3.

But keep in mind that due to the *graininess* of the ARL of the CUSUM charts for discrete output, a phenomenon thoroughly reported by Hawkins and Olwell (1998, pp. 107–110) achieving a desired in-control ARL might be difficult without randomization even if we consider rational reference values and control limits.

In fact what we did to produce Fig. 3 was to obtain a standard two-sided CUSUM scheme with its in-control ARL as close as possible to 370.4 (using integer reference values and control limits); the in-control ARL of the resulting scheme was 322.4388. Then we set $ARL^\star = 322.4388$ and obtained a matched in-control ARL-unbiased two-sided CUSUM scheme, to make a fair comparison between the two ARL profiles.

Figure 3 leads us to add that the standard two-sided CUSUM scheme tends to be in average quicker than its ARL-unbiased counterpart when it comes to the detection of upward shifts in the fraction nonconforming. This comes as no surprise, after all the derivative of the ARL function of the ARL-biased scheme is negative at $p = p_0$, whereas the one of the ARL-unbiased scheme is virtually null.

4 The ARL of Two-Sided CUSUM Schemes for Binomial AR(1) Counts

First-order integer-valued autoregressive (INAR(1)) processes, introduced by McKenzie (1985) and relying on the binomial thinning operator (Steutel and VanHarn 1979), can be used in SPC, as Weiß (2009a, Chap. 20) and (Weiß 2018, Chap. 8) thoroughly illustrated.

Let us now describe one of those processes and for that purpose consider from now on:

- $n \in \mathbb{N}, \quad p \in (0, 1)$;
- $\rho \in [\max\{-p/(1-p), -(1-p)/p\}, 1], \quad \beta = p(1-\rho), \quad \alpha = \beta + \rho$.

Then the process $\{X_t : t \in \mathbb{N}_0\}$ defined by the recursion

$$X_t = \alpha \circ X_{t-1} + \beta \circ (n - X_{t-1}),\ t \in \mathbb{N}, \tag{25}$$

is said to be a binomial AR(1) process if:

- $X_0 \sim \text{binomial}(n, p)$;
- $\circ$ represents the binomial thinning operator (recall that $\alpha \circ X \mid X \sim$ binomial (X, α));
- all thinning operations are performed independently of each other, and the thinnings at time t are independent of $\{\ldots, X_{t-2}, X_{t-1}\}$.

Weiß (2009b) not only added that $E(X_t) = np$ and $V(X_t) = np(1-p)$, but also that a binomial AR(1) process is a stationary Markov chain with state space $\{0, 1, \ldots, n\}$, binomial(n, p) marginal distribution and transition probabilities $p_{ij} = P(X_t = j \mid X_{t-1} = i)$ given by

$$p_{ij} = \sum_{m=\max(0,j+i-n)}^{min(j,i)} \binom{i}{m} \alpha^m (1-\alpha)^{i-m} \times \binom{n-i}{j-m} \beta^{j-m} (1-\beta)^{n-i+m-j}. \tag{26}$$

The binomial AR(1) process has been used to describe the number of counts in random samples of fixed size n, for example, by Weiß (2009b, c) and Rakitzis et al. (2017).

Throughout this section, we assume that:

- in the absence of assignable causes, $p = p_0$ and $\rho = \rho_0$;
- the purpose of using a control chart/scheme is to monitor shifts from p_0 to $p = p_0 + \delta_p$ or from ρ_0 to $\rho = \rho_0 + \delta_\rho$, where $\delta_p \in (-p_0, 1 - p_0)$ and $\delta_\rho \in (\max\{-p_0/(1-p_0), -(1-p_0)/p_0\} - \rho_0, 1 - \rho_0)$.

4.1 Overall ARL Functions

The ARL functions of upper one-sided CUSUM charts and two-sided CUSUM schemes for Poisson INAR(1) counts were derived by Weiß and Testik (2009) and Yontay et al. (2013), respectively.

By following closely Yontay et al. (2013), we are able to derive the ARL function of two-sided CUSUM schemes to monitor binomial AR(1) counts.

Firstly, note that $\{(X_t, S_t^+, S_t^-) : t \in \mathbb{N}_0\}$ is a trivariate Markov chain with state space $\{0, 1, \ldots, n\} \times \mathscr{S}^+ \times \mathscr{S}^-$ and transition probabilities

$$
\begin{aligned}
&p_{(b,i^+,i^-)(a,j^+,j^-)} \\
&\quad = P\left(X_t = a, S_t^+ = j^+, S_t^- = j^- \mid X_{t-1} = b, S_{t-1}^+ = i^+, S_{t-1}^- = i^-\right),
\end{aligned} \tag{27}
$$

with $a, b \in \{0, 1, ..., n\}$, $i^+, j^+ \in \mathscr{S}^+$, and $i^-, j^- \in \mathscr{S}^-$. These transition probabilities were derived by Cruz (2016, p. 49) and are written below in terms of indicator functions:

- For $j^+ = 0$ and $j^- = 0$,

$$
\begin{aligned}
&p_{(b,i^+,i^-)\,(a,j^+,j^-)} \\
&\quad = P\left(X_t = a,\ S_t^+ = 0,\ S_t^- = 0 \mid X_{t-1} = b,\ S_{t-1}^+ = i^+,\ S_{t-1}^- = i^-\right) \\
&\quad = P\left(X_t = a,\ i^+ + X_t - k^+ \le 0,\ i^- + k^- - X_t \le 0 \mid X_{t-1} = b\right) \\
&\quad = P\left(X_t = a,\ X_t \le k^+ - i^+,\ X_t \ge i^- + k^- \mid X_{t-1} = b\right) \\
&\quad = P\left(X_t = a \mid X_{t-1} = b\right) \times \mathbb{1}_{[i^- + k^-,\, k^+ - i^+]}(a);
\end{aligned} \tag{28}
$$

- for $j^+ = 0$ and $j^- \in \mathscr{S}^- \backslash \{0\}$,

$$
\begin{aligned}
&p_{(b,i^+,i^-)\,(a,j^+,j^-)} \\
&\quad = P\left(X_t = a,\ S_t^+ = 0,\ S_t^- = j^- \mid X_{t-1} = b,\ S_{t-1}^+ = i^+,\ S_{t-1}^- = i^-\right) \\
&\quad = P\left(X_t = a,\ i^+ + X_t - k^+ \le 0,\ i^- + k^- - X_t = j^- \mid X_{t-1} = b\right) \\
&\quad = P\left(X_t = a,\ X_t \le k^+ - i^+,\ X_t = i^- + k^- - j^- \mid X_{t-1} = b\right) \\
&\quad = P\left(X_t = a \mid X_{t-1} = b\right) \times \mathbb{1}_{[0,\, k^+ - i^+] \cap \{i^- + k^- - j^-\}}(a);
\end{aligned} \tag{29}
$$

- for $j^+ \in \mathscr{S}^+ \backslash \{0\}$ and $j^- = 0$,

$$
\begin{aligned}
&p_{(b,i^+,i^-)\,(a,j^+,j^-)} \\
&\quad = P\left(X_t = a,\ S_t^+ = j^+,\ S_t^- = 0 \mid X_{t-1} = b,\ S_{t-1}^+ = i^+,\ S_{t-1}^- = i^-\right) \\
&\quad = P\left(X_t = a,\ i^+ + X_t - k^+ = j^+,\ i^- + k^- - X_t \le 0 \mid X_{t-1} = b\right) \\
&\quad = P\left(X_t = a,\ X_t = j^+ - i^+ + k^+,\ X_t \ge i^- + k^- \mid X_{t-1} = b\right) \\
&\quad = P\left(X_t = a \mid X_{t-1} = b\right) \times \mathbb{1}_{[i^- + k^-,\, n] \cap \{j^+ - i^+ + k^+\}}(a);
\end{aligned} \tag{30}
$$

- for $j^+ \in \mathscr{S}^+ \backslash \{0\}$ and $j^- \in \mathscr{S}^- \backslash \{0\}$,

$$
\begin{aligned}
&p_{(b,i^+,i^-)\,(a,j^+,j^-)} \\
&= P\left(X_t = a,\ S_t^+ = j^+,\ S_t^- = j^- \mid X_{t-1} = b,\ S_{t-1}^+ = i^+,\ S_{t-1}^- = i^-\right) \\
&= P\left(X_t = a,\ i^+ + X_t - k^+ = j^+,\ i^- + k^- - X_t = j^- \mid X_{t-1} = b\right) \\
&= P\left(X_t = a,\ X_t = j^+ - i^+ + k^+,\ X_t = i^- + k^- - j^- \mid X_{t-1} = b\right) \\
&= P\left(X_t = a \mid X_{t-1} = b\right) \times \mathbb{1}_{\{j^+ - i^+ + k^+\}\cap\{i^- + k^- - j^-\}}(a). \qquad (31)
\end{aligned}
$$

Secondly, let us consider once more an absorbing Markov chain, in this case with set of transient states equal to $\{0, 1, ..., n\} \times \mathscr{T}^+ \times \mathscr{T}^-$ and a single absorbing state combining all the original states (a, i^+, i^-), with $i^+ > h^+$ or $i^- > h^-$.[2] If $X_0 = x \in \{0, 1, ..., n\}$, $S_0^+ = u^+$, and $S_0^- = u^-$ then the ARL function of the two-sided CUSUM scheme for binomial AR(1) counts equals

$$
ARL^{x,u^+,u^-}(p, \rho) = \underline{\mathbf{e}}^\top_{x,u^+,u^-} \times (\mathbf{I} - \mathbf{Q})^{-1} \times \underline{\mathbf{1}}, \qquad (32)
$$

where

- $\underline{\mathbf{e}}^\top_{x,u^+,u^-}$ is the $(x \times (c^+ + 1) \times (c^- + 1) + u^+ \times (c^- + 1) + u^- + 1)$-th vector of the orthogonal basis for $\mathbb{R}^{(n+1)\times(c^+ +1)\times(c^- +1)}$;
- $\mathbf{I}$ represents the identity matrix with rank $(n + 1) \times (c^+ + 1) \times (c^- + 1)$;
- $\mathbf{Q} \equiv \mathbf{Q}(p, \rho) = [p_{(b,i^+,i^-)(a,j^+,j^-)}]_{(a,i^+,i^-),(b,j^+,j^-)\in\{0,1,...,n\}\times\mathscr{T}^+\times\mathscr{T}^-}$;
- $\underline{\mathbf{1}}$ is a column-vector with $(n + 1) \times (c^+ + 1) \times (c^- + 1)$ ones.

Thirdly, since the value of X_0 is usually unknown, it is plausible to rely on $X_1 \equiv X_1(p, \rho) \sim \text{binomial}(n, p)$ and define the *overall ARL* (Weiß and Testik 2009) as:

$$
\begin{aligned}
ARL(p, \rho) = 1 + &\sum_{(x,u^+,u^-)\in\{0,1,...,n\}\times\mathscr{T}^+\times\mathscr{T}^-} ARL^{x,u^+,u^-}(p, \rho) \\
&\times P(X_1 = x,\ S_1^+ = u^+,\ S_1^- = u^- \mid S_0^+ = 0,\ S_0^- = 0). \qquad (33)
\end{aligned}
$$

The probabilities $P(X_1 = x,\ S_1^+ = u^+,\ S_1^- = u^- \mid S_0^+ = 0,\ S_0^- = 0)$ are denoted by $p_{(\bullet,0,0)\,(x,u^+,u^-)}$ and are taken from Cruz (2016, pp. 50–51):

- for $u^+ = 0$ and $u^- = 0$,

$$
\begin{aligned}
&p_{(\bullet,0,0)(x,u^+,u^-)} \\
&= P\left(X_1 = x,\ S_1^+ = 0,\ S_1^- = 0 \mid S_0^+ = 0,\ S_0^- = 0\right) \\
&= P\left(X_1 = x,\ X_1 - k^+ \le 0,\ k^- - X_1 \le 0\right) \\
&= P\left(X_1 = x,\ X_1 \le k^+,\ X_1 \ge k^-\right) \\
&= P(X_1 = x) \times \mathbb{1}_{[k^-,\,k^+]}(x); \qquad (34)
\end{aligned}
$$

- for $u^+ = 0$ and $u^- \in \mathscr{T}^-\backslash\{0\}$,

[2]The transient states can be ordered as follows: $(0, 0, 0), (0, 0, 1/b^-), \ldots, (0, 0, c^-/b^-)$, $(0, 1, 0), (0, 1, 1/b^-), \ldots, (0, 1, c^-/b^-), \ldots, (n, c^+/b^+, 0), (n, c^+/b^+, 1), \ldots, (n, c^+/b^+, c^-/b^-)$.

$$
\begin{aligned}
p_{(\bullet,0,0)(x,u^+,u^-)} \\
&= P\left(X_1 = x,\ S_1^+ = 0,\ S_1^- = u^- \mid S_0^+ = 0,\ S_0^- = 0\right) \\
&= P\left(X_1 = x,\ X_1 - k^+ \leq 0,\ k^- - X_1 = u^-\right) \\
&= P\left(X_1 = x,\ X_1 \leq k^+,\ X_1 = k^- - u^-\right) \\
&= P(X_1 = x) \times \mathbb{1}_{[0,\, k^+] \cap \{k^- - u^-\}}(x); \qquad (35)
\end{aligned}
$$

- for $u^+ \in \mathscr{T}^+ \backslash \{0\}$ and $u^- = 0$,

$$
\begin{aligned}
p_{(\bullet,0,0)(x,u^+,u^-)} \\
&= P\left(X_1 = x,\ S_1^+ = u^+,\ S_1^- = 0 \mid S_0^+ = 0,\ S_0^- = 0\right) \\
&= P\left(X_1 = x,\ X_1 - k^+ = u^+,\ k^- - X_1 \leq 0\right) \\
&= P\left(X_1 = x,\ X_1 = u^+ + k^+,\ X_1 \geq k^-\right) \\
&= P(X_1 = x) \times \mathbb{1}_{[k^-,\, n] \cap \{u^+ + k^+\}}(x); \qquad (36)
\end{aligned}
$$

- for $u^+ \in \mathscr{T}^+ \backslash \{0\}$ and $u^-\ in \mathscr{T}^- \backslash \{0\}$,

$$
\begin{aligned}
p_{(\bullet,0,0)(x,u^+,u^-)} \\
&= P\left(X_1 = x,\ S_1^+ = u^+,\ S_1^- = u^- \mid S_0^+ = 0,\ S_0^- = 0\right) \\
&= P\left(X_1 = x,\ X_1 - k^+ = u^+,\ k^- - X_1 = u^-\right) \\
&= P\left(X_1 = x,\ X_1 = u^+ + k^+,\ X_1 = k^- - u^-\right) \\
&= P(X_1 = x) \times \mathbb{1}_{\{u^+ + k^+\} \cap \{k^- - u^-\}}(x). \qquad (37)
\end{aligned}
$$

To derive an ARL-unbiased two-sided CUSUM scheme we proceed as in Sect. 3.

We are still dealing with the two discrete control statistics S_t^*, $* = +, -$, thus the randomization probabilities play a vital role in achieving an in-control ARL equal to $ARL^\star$ and eliminating the bias of the ARL function.

Unsurprisingly, randomizing the emission of a signal means modifying some entries of the sub-stochastic matrix $\mathbf{Q}$ in (32), in particular the entries associated with states (a, j^+, h^-), (a, h^+, j^-), and (a, h^+, h^-), where $a \in \{0, 1, ..., n\}$, $j^+ \in \mathscr{T}^+$, and $j^- \in \mathscr{T}^-$. When there is a transition from state $(b, i^+, i^-) \in \{0, 1, ..., n\} \times \mathscr{T}^+ \times \mathscr{T}^-$, to state:

- (a, j^+, h^-), for $j^+ \in \mathscr{T} \backslash \{h^+\}$, $(1 - \gamma^-) \times p_{(b,i^+,i^-)(a,j^+,h^-)}$ takes the place of $p_{(b,i^+,i^-)(a,j^+,h^-)}$;
- (a, h^+, j^-), for $j^- \in \mathscr{T} \backslash \{h^-\}$, $(1 - \gamma^+) \times p_{(b,i^+,i^-)(a,h^+,j^-)}$ stands in for $p_{(b,i^+,i^-)(a,h^+,j^-)}$;
- (a, h^+, h^-), $(1 - \gamma^+) \times (1 - \gamma^-) \times p_{(b,i^+,i^-)(a,h^+,h^-)}$ replaces $p_{(b,i^+,i^-)(a,h^+,h^-)}$.

The remaining entries of the altered matrix $\mathbf{Q} \equiv \mathbf{Q}(p, \rho, \gamma^+, \gamma^-)$ are equal to the transition probabilities $p_{(b,i^+,i^-)(a,j^+,j^-)}$, where $(a, j^+, j^-) \in \{0, 1, ..., n\} \times (\mathscr{T}^+ \backslash \{h^+\}) \times (\mathscr{T}^- \backslash \{h^-\})$.

The overall ARL function of the ARL-unbiased two-sided CUSUM scheme for the mean of binomial AR(1) counts, $ARL(p, \rho, \gamma^-, \gamma^+)$ is given by

$$\begin{aligned}
&1 + \sum_{(x,u^+,u^-):u^+\neq h^+,\, u^-\neq h^-} ARL^{x,u^+,u^-} \times p_{(\bullet,0,0)\,(x,u^+,u^-)} \\
&+(1-\gamma^+) \times \sum_{(x,h^+,u^-):u^-\neq h^-} ARL^{x,h^+,u^-} \times p_{(\bullet,0,0)\,(x,h^+,u^-)} \\
&+(1-\gamma^-) \times \sum_{(x,u^+,h^-):u^+\neq h^+} ARL^{x,u^+,h^-} \times p_{(\bullet,0,0)\,(x,u^+,h^-)} \\
&+(1-\gamma^-)(1-\gamma^+) \times \sum_{(x,h^+,h^-)} ARL^{x,h^+,h^-} \times p_{(\bullet,0,0)\,(x,h^+,h^-)}, \qquad (38)
\end{aligned}$$

where $(x, u^+, u^-) \in \{0, 1, ..., n\} \times \mathscr{T}^+ \times \mathscr{T}^-$, $(x, u^-) \in \{0, 1, ..., n\} \times \mathscr{T}^-$, $(x, u^+) \in \{0, 1, ..., n\} \times \mathscr{T}^+$ in those summations; and $ARL^{x,u^+,u^-} \equiv ARL^{x,u^+,u^-}(p, \rho, \gamma_{h^+}, \gamma_{h^-})$ is obtained from (32) with $\mathbf{Q}(p, \rho)$ replaced by $\mathbf{Q}(p, \rho, \gamma^+, \gamma^-)$.

The control limits and randomization probabilities of the ARL-unbiased two-sided CUSUM scheme, for the mean of binomial AR(1) counts, are derived using the search procedure described in Sect. 3.

4.2 Further Preliminary Results

Since the mean of the binomial AR(1) process is equal to np regardless of the value of the autocorrelation parameter ρ, we could be tempted to resort to an ARL-unbiased two-sided CUSUM scheme designed for i.i.d. binomial counts to monitor the mean of such autocorrelated process.

However, the scheme that takes into account the autocorrelation structure of the counts and the one that ignores it has contrasting performances, as shown by Fig. 4. The two-sided CUSUM scheme that disregards the autocorrelation structure is ARL-biased and has in-control ARL smaller than the other two-sided CUSUM scheme.

Figure 5 refers to the comparison of the ARL profiles of:

- the ARL-unbiased modified np-chart (Cruz 2016, pp. 42–48), with control limits L and U and randomization probabilities γ_L and γ_U;
- the ARL-unbiased two-sided CUSUM scheme for binomial AR(1) counts.

Unsurprisingly, the latter scheme outperforms its Shewhart counterpart.

It is important to note that Monte Carlo simulation was used by Cruz (2016, p. 57) to provide ARL estimates and verify the results obtained for $ARL(p, \rho_0, \gamma^-, \gamma^+)$. Another relevant finding, it was not possible to derive a two-sided CUSUM scheme whose ARL profile achieves a maximum at $\rho = \rho_0$. As a matter of fact these schemes have a very poor performance when it comes to the detection of shifts in ρ, as previously reported by Weiß (2009b) while discussing the use of several other charts to control binomial AR(1) counts.

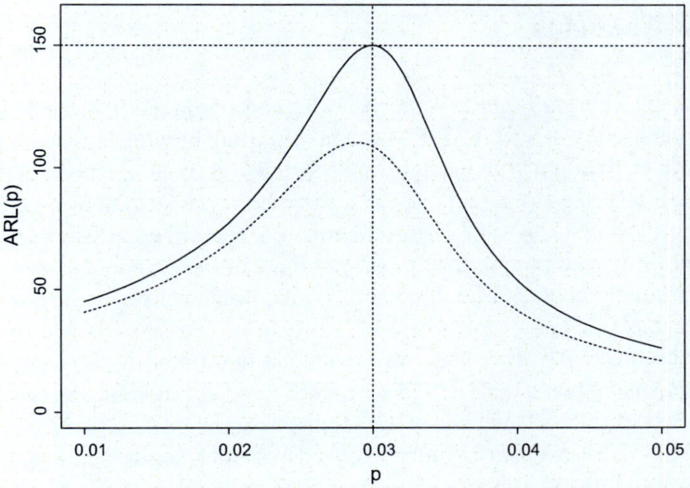

Fig. 4 $ARL(p, \rho_0, \gamma^-, \gamma^+)$ curve of the ARL-unbiased two-sided CUSUM schemes for AR(1) and i.i.d. binomial counts in the presence of autocorrelation—$n = 30$, $p_0 = 0.03$, $\rho_0 = 0.2$, $p_1^{\pm} = p_0 \pm 0.01$, $(k^-, k^+, h^-, h^+, \gamma^-, \gamma^+) = (1, 1, 31, 12, 0.406215, 0.079627)$ (solid line), $(1, 1, 28, 10, 0.493716, 0.540321)$ (dashed line)

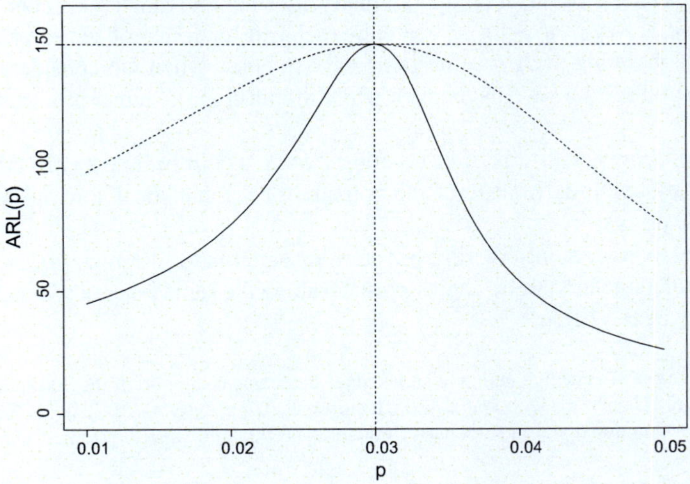

Fig. 5 ARL profiles of the ARL-unbiased modified *np*-chart and two-sided CUSUM scheme (resp. dashed and solid lines)—$n = 30$, $p_0 = 0.03$, $p_1^{\pm} = p_0 \pm 0.01$, $\rho_0 = 0.2$, $(L, U, \gamma_L, \gamma_U) = (0, 5, 0.013757, 0.575849)$ (dashed line), $(k^-, k^+, h^-, h^+, \gamma^-, \gamma^+) = (1, 1, 31, 12, 0.406215, 0.079627)$ (solid line)

5 Final Thoughts

We have come a long way since Walter Shewhart proposed the p-chart, by proposing an ARL-unbiased two-sided CUSUM scheme for i.i.d. binomial counts.

Since we are dealing with discrete control statistics, randomization probabilities are essential to bring the in-control ARL down to a pre-specified in-control ARL and to eliminate the bias of the ARL function. The search procedure, used to obtain the control limits and randomization probabilities of the two constituent one-sided CUSUM charts to monitor i.i.d. binomial counts, was inspired by the one proposed by Paulino et al. (2019).

As for binomial AR(1) counts, we concluded that ignoring the autocorrelation structure and using two-sided CUSUM schemes for i.i.d. binomial output would lead to biased ARL profiles. Thus, it is of the utmost importance to derive ARL-unbiased two-sided CUSUM schemes for autocorrelated binomial counts. Once again this was done by adapting the search procedure described by Paulino et al. (2019).

We ought to note that ARL-unbiased two-sided CUSUM schemes to monitor the mean of i.i.d. and INAR(1) Poisson counts have been proposed and thoroughly discussed by Clara (2016, Chap. 3).

The search procedure used to obtain ARL-unbiased two-sided CUSUM schemes for binomial AR(1) counts is computationally intensive and rather inefficient, namely, because we are dealing with sparse sub-stochastic matrices of considerable size. Moreover, obtaining such schemes was only possible when we considered $ARL^\star$ smaller than the values we have taken while monitoring the mean of i.i.d. binomial counts.

As a consequence, a direction for future work would be improving this search procedure by taking advantage of the potential of the statistical software R to deal with sparse arrays.

Another topic certainly worthy of future research is to minimize the number of states included in the Markov chain approach along the same lines as Woodall (1984) and Yontay et al. (2013).

Acknowledgements The first author acknowledges the financial support of the Portuguese FCT – Fundação para a Ciência e a Tecnologia, through the project UID/Multi/04621/2019 of CEMAT/IST-ID, Center for Computational and Stochastic Mathematics, Instituto Superior Técnico, Universidade de Lisboa.

References

Acosta-Mejía, C. A. (1999). Improved p-charts to monitor process quality. *IIE Transactions*, *31*, 509–516.

Acosta-Mejía, C. A., & Pignatiello, J. J, Jr. (2000). Monitoring process dispersion without subgrouping. *Journal of Quality Technology*, *32*, 89–102.

Brook, D., & Evans, D. A. (1972). An approach to the probability distribution of CUSUM run length. *Biometrika*, *59*, 539–549.

Clara, C. J. (2019). Esquemas CUSUM com ARL sem viés para processos i.i.d. e INAR(1) com marginais de Poisson. (On ARL-unbiased two-sided CUSUM schemes for i.i.d. and INAR(1) Poisson counts.) M.Sc. thesis, Instituto Superior Técnico, Universidade de Lisboa.

Cruz, C. J. (2019). Cartas com ARL sem viés para processos i.i.d. e AR(1) com marginais binomiais. (On ARL-unbiased charts for i.i.d. and AR(1) binomial counts.) M.Sc. thesis, Instituto Superior Técnico, Universidade de Lisboa.

Ewan, W. D., & Kemp, K. W. (1960). Sampling inspection of continuous processes with no autocorrelation between successive results. *Biometrika*, *47*, 363–380.

Gan, F. F. (1993). An optimal design of CUSUM control charts for binomial counts. *Journal of Applied Statistics*, *20*, 445–460.

Hawkins, D. M., & Olwell, D. H. (1998). *Cumulative sum control charts and charting for quality improvement*. New York: Springer.

Johnson, N. L., & Leone, F. C. (1962). Cumulative sum control charts - mathematical principles applied to their construction and use - Part III. *Industrial Quality Control*, *19*, 22–28.

Knoth, S., & Morais, M. C. (2013). On ARL-unbiased control charts. In S. Knoth, W. Schmid, & R. Sparks (Eds.), *Proceedings of the XIth International Workshop on Intelligent Statistical Quality Control* (pp. 31–50). Sydney, Australia, 20–23 August 2013.

Knoth, S., & Morais, M. C. (2015). On ARL-unbiased control charts. In S. Knoth & W. Schmid (Eds.), *Frontiers in statistical quality control* (Vol. 11, pp. 95–117). Switzerland: Springer International Publishing.

Lorden, G. (1971). Procedures for reacting to a change in distribution. *Annals of Mathematical Statistics*, *42*, 1897–1908.

Lucas, J. M. (1985). Counted data CUSUM's. *Technometrics*, *27*, 129–144.

Lucas, J. M., & Crosier, R. B. (1982). Fast initial response (FIR) for cumulative sum quality control schemes. *Technometrics*, *24*, 199–205.

McKenzie, E. (1985). Some simple models for discrete variate time series. *Water Resources Bulletin*, *21*, 645–645.

Montgomery, D. C. (2009). *Introduction to statistical quality control* (6th ed.). New York: Wiley.

Morais, M. C. (2016). An ARL-unbiased np-chart. *Economic Quality Control*, *31*, 11–21.

Morais, M. C. (2017). ARL-unbiased geometric and CCC_G control charts. *Sequential Analysis*, *36*, 513–527.

Morais, M. C. & Knoth, S. (2020). Improving the ARL profile and the accuracy of its calculation for Poisson EWMA charts. *Quality and Reliability Engineering International*, *36*, 876–889.

Morais, M. C., Knoth, S., & Weiß, C. H. (2018). An ARL-unbiased thinning-based EWMA chart to monitor counts. *Sequential Analysis*, *37*, 487–510.

Ottenstreuer, S., Weiß, C. H., & Knoth, S. (2019). Combined Shewhart-CUSUM chart with switching limit. *Quality Engineering*, *31*, 255–268.

Page, E. S. (1954). Continuous inspection scheme. *Biometrika*, *41*, 100–115.

Paulino, S., Morais, M. C., & Knoth, S. (2016). An ARL-unbiased c-chart. *Quality and Reliability Engineering International*, *32*, 2847–2858.

Paulino, S., Morais, M. C., & Knoth, S. (2019). On ARL-unbiased c-charts for INAR(1) Poisson counts. *Statistical papers*, *60*, 1021–1038.

Pignatiello, J. J., Jr., Acosta-Mejía, C. A., & Rao, B. V. (1995). The performance of control charts for monitoring process dispersion. In *4th Industrial Engineering Research Conference*, May 24–25, 1995, Nashville, TN (pp. 320–328).

Pollak, M. (1985). Optimal detection of a change in distribution. *Annals of Statistics*, *13*, 206–227.

R Core Team (2013). R: A Language and Environment for Statistical Computing. R Foundation for Statistical Computing. Vienna, Austria. http://www.R-project.org

Rakitzis, A. C., Weiß, C. H., & Castagliola, P. (2017). Control charts for monitoring correlated counts with a finite range. *Applied Stochastic Models in Business and Industry*, *33*, 733–749.

Ramalhoto, M. F. & Morais, M. (1995). Cartas de controlo para o parâmetro de escala da população Weibull tri-paramétrica. (Control charts for the scale parameter of the Weibull population.) *Actas do II Congresso Anual da Sociedade Portuguesa de Estatística* (Proceedings of the Annual Congress of the Portuguese Statistical Society) (pp. 345–371).

Ramalhoto, M. F., & Morais, M. (1999). Shewhart control charts for the scale parameter of a Weibull control variable with fixed and variable sampling intervals. *Journal of Applied Statistics*, *26*, 129–160.

Steutel, F. W., & Van Harn, K. (1979). Discrete analogues of self-decomposability and stability. *Annals of Probability*, *7*, 893–899.

Weiß, C. H. (2009a). *Categorical Time Series Analysis and Applications in Statistical Quality Control*. Ph.D. Thesis, Fakultät für Mathematik und Informatik der Universität Würzburg. dissertation.de - Verlag im Internet GmbH.

Weiß, C. H. (2009b). Controlling correlated processes with binomial marginals. *Journal of Applied Statistics*, *36*, 399–414.

Weiß, C. H. (2009c). Jumps in binomial AR(1) processes. *Statistics and Probability Letters*, *79*, 2012–2019.

Weiß, C. H. (2018). *An introduction to discrete-valued time series*. Chichester: Wiley Inc.

Weiß, C. H., & Testik, M. C. (2009). CUSUM monitoring of first-order integer-valued autoregressive processes of Poisson counts. *Journal of Quality Technology*, *41*, 389–400.

Weiß, C. H., & Testik, M. C. (2012). Detection of abrupt changes in count data time series: Cumulative sum rerivations for INARCH(1) models. *Journal of Quality Technology*, *44*, 249–264.

Woodall, W. H. (1984). On the Markov chain approach to the two-sided CUSUM procedure. *Technometrics*, *26*, 41–46.

Yang, S.-F., & Arnold, B. C. (2015). Monitoring process variance using an ARL-unbiased EWMA-p control chart. *Quality and Reliability Engineering International*, *32*, 1227–1235.

Yontay, P., Weiß, C. H., Testik, M. C., & Bayindir, Z. P. (2013). A two-sided cumulative sum chart for first-order integer-valued autoregressive process of Poisson counts. *Quality and Reliability Engineering International*, *29*, 33–42.

Zhang, L., Govindaraju, K., Bebbington, M., & Lai, C. D. (2004). On the statistical design of geometric control charts. *Quality Technology & Quantitative Management*, *2*, 233–243.

Zhang, C. W., Xie, M., & Jin, T. (2012). An improved self-starting cumulative count of conforming chart for monitoring high-quality processes under group inspection. *International Journal of Production Research*, *50*, 7026–7043.

Statistical Aspects of Target Setting for Attribute Data Monitoring

Emmanuel Yashchin, Aaron Civil, Jeff Komatsu, and Paul Zulpa

Abstract We consider early warning systems (EWS) for monitoring multi-stage data, in which downstream variables undergo changes associated with upstream process stages. In such applications, the EWS monitoring arm acts as a search engine that analyses a number of data streams for each monitored variable, as the problems of change detection and identification of the change-causing stage are handled jointly. Given massive amounts of data involved in analysis, it is important to achieve an acceptable balance between false alarms and sensitivity requirements, by focusing on changes of practical significance. The role of the target-setting arm of EWS is to ensure and maintain this balance via suitable selection of control scheme parameters. In this paper, we discuss principles of developing and managing targets, with examples from a supply chain operation.

Keywords Statistical process control · Robust methods · Trimmed mean · Search engines · Hypothesis testing · Practical significance

1 Introduction

An early warning system (EWS) can be viewed as a kind of search engine that analyzes available data on a periodic basis, selects cases meriting engineering attention and produces supplemental information that is instrumental for problem diagnosis and alarm prioritization, see Yashchin (2018). The main objective of an EWS is to make sure that (a) all unfavorable conditions are detected reasonably early and (b) the rate of false alarms is acceptably low. To achieve this objective, we rely on the basic knowledge of processes driving the data and operating conditions that are communicated to the EWS: for example, we expect the users to specify, for every monitored

E. Yashchin (✉)
IBM, Thomas J. Watson Research Ctr., Box 218, Yorktown Heights, NY 10598, USA
e-mail: yashchi@us.ibm.com

A. Civil · J. Komatsu · P. Zulpa
IBM Corporation, Armonk, NY, USA

S. Knoth and W. Schmid (eds.), *Frontiers in Statistical Quality Control 13*,
Frontiers in Statistical Quality Control,
https://doi.org/10.1007/978-3-030-67856-2_7

process, its target, acceptable / unacceptable deviations of the process mean from the target, acceptable false alarm rate, and so forth. This requires non-trivial effort, especially in the presence of massive or highly intensive data streams. The target setting procedures are meant to facilitate this process by helping users in populating the files governing the monitoring process based on the available data and other information about the process. In this article, we describe target setting procedures for the attribute data monitoring arm of an EWS.

We assume that the data monitoring stream consists of pairs $(n_i, x_i), i = 1, 2, \ldots$, where i is the index of a vintage, n_i is the number of items (we will call them *parts*) tested for it, and x_i is the number of items failed. For every vintage, some concomitant information is also available, but the details are not important for this article. The EWS monitors the Binomial mean p in some setting, e.g., see Gan (1993), Hawkins and Olwell (1998), Civil et al. (2013a, b), Kenett and Zacks (2014). We use the three-zone approach that calls for specifying the *acceptable* and *unacceptable* levels p_0 and p_1, respectively ($p_1 > p_0$), e.g., see Woodall (1986), Yashchin (1985, 2012). Such a setup has several advantages: of key importance is the fact that it enables one to focus the search engine on detecting changes of *practical* (as opposed to *statistical*) significance. This property is highly valuable in large-scale monitoring systems, as it guarantees a-priori that detected unfavorable conditions are of interest to at least some of the users. Conventional statistical process control (SPC) systems based on the Western Electric rules or similar anomaly detection systems focus instead on detecting changes from some target value, and thus they do not scale well, as they tend to produce too many alarms that are of no practical interest. We refer to the process of determining (p_0, p_1) as target-setting.

Once the acceptable/unacceptable values are available, one can convert the information in the observed vintages to the corresponding values of the control scheme $S_i, i = 1, 2, \ldots,$ for the purpose of monitoring; for example,

$$S_0 = s_0\,, \quad S_i = \max[0, \gamma S_{i-1} + w_i(\hat{P}_i - k)]\,, \quad i = 1, 2, \ldots\,, \tag{1}$$

where s_0 is the scheme headstart, $\hat{P}_i$ is the vintage-based estimate of p, w_i is the corresponding weight, the *reference value* k is given by the formula

$$k = \frac{p_1 - p_0}{\ln p_1 - \ln p_0} \approx (p_0 + p_1)/2\,, \tag{2}$$

(assuming that the Poisson approximation is adequate) and γ is the evidence-damping parameter (typically chosen in the range [0.7, 1]). This version of weighted Geometric Cusum $GC(\gamma)$ can be applied in the conventional setting (when the newly arriving information affects only the last vintage data) or in the setting involving dynamically changing observations (DCO, see, Yashchin 2010). In practical applications, one will typically use (1) in conjunction with supplemental rules depending on the mode of deployment; these rules are essential for ensuring acceptable operating characteristics of the monitoring scheme. In the context of this article, we will focus on the case where $\hat{P}_i$ is the sample proportion of failures in the i-th vintage and w_i is the corresponding sample size, n_i. However, the methods described in this paper can be applied in more general situations, as we will discuss in the last section.

In many situations, the acceptable/unacceptable levels are dictated by business requirements. In this case, the selection (p_0, p_1) could be related, for example, to the practical (and possibly, financial) consequences of the process operating at these failure rates, and it does not have to reflect the capability of the process to produce parts having the acceptable failure rate. Disregarding capability, however, could come at a cost of tolerating constantly underperforming parts that will be prominently featured in red light on dashboards and displays, making it difficult to track the quality improvement process. Given that there can be many parts that exhibit such behavior, dashboard management could become exceedingly complicated, as messages delivered by its red color component become less actionable.

Furthermore, considering part capability might also facilitate detection of unfavorable conditions of "drift" type that originate at some point and then get progressively worse. Therefore, in this paper we emphasize the approach based on estimating the ability of various processes to operate at the levels compatible with the historical capability demonstrated by the process itself, its peers and similar processes. Of course, the proven capability could turn out to be inferior to that dictated by the business needs, and an ongoing effort will be required to bring the processes to desired performance levels—however, dashboards based on process capability (or influenced by it) could help in guiding actions geared towards process improvement.

In practice, the targets used by the monitoring system are influenced by many factors, including business—driven necessity (e.g., financial implications of keeping targets at specified levels), continuous improvement objectives (e.g. directives to reduce failure rate by 10% every year) or contractual obligations (e.g. related to vendors or customers)—and process capability is only one of them. It is quite an important component, though—and typically a target-setting system (which in our context is essentially a capability analysis system) is run in parallel with the monitoring system and plays a key (advisory, not direct) role in driving the quality improvement activities.

In the next section, we introduce basic robust estimation procedures for the proportion of defectives, which reflects the process capability. In Section 3 we describe the target-derivation process. In Section 4 we discuss an example related to implementation of the target-setting system in the IBM Supply Chain organization. In Section 5 we discuss broader aspects of the target-setting process.

2 Robust Point Estimation of *p*

One of the key statistical issues is to evaluate, for every part, the inherent capability of the part manufacturer. In the problem of monitoring by variables, capability is often measured in terms of indices such as C_p or C_{pk}. In the case of the inspection / attribute data, we typically do not have rigidly defined specs, so another approach is needed. For simplicity, we focus on the prevailing situation where the monitored proportions are low, so that one can rely on adequacy of the Poisson approximation to the distribution of the observed counts.

In practice, the pass/fail data is typically contaminated with outliers. It is commonly observed that the data for a given part (we refer to it as Part A) comes from several vintages, and some of them tend to be either unusually good or unusually bad. It is therefore necessary to neutralize the effect of the outlier vintages. When the sample sizes $\{n_i\}$ are all the same, we could use the *trimmed mean* of the sample proportions, e.g., see Huber and Ronchetti (2009). In the case where the sample sizes vary, this approach (called *weighted trimmed mean*, see, NIST Handbook 2019) is unadvisable, both from the statistical and business perspectives. Leaving aside the statistical aspects of standard trimming, note that it can introduce issues in relationship with vendors, whose parts are supposed to deliver a certain level of performance. Suppose, for simplicity, that we apply trimming by removing the two vintages with largest estimated p_i (we call them top vintages) and two vintages with smallest estimated p_i (bottom vintages), and then compute a weighted average of the proportions in the remaining vintages. In cases where the removed bottom vintages have a much lower overall sample size than the top vintages, a vendor can legitimately protest that his capability estimate is biased against him since the resulting trimmed weighted average will contain fewer failures than what he would normally be capable of delivering. Furthermore, since we have no control of the sample sizes corresponding to the trimmed observations, it is difficult to control the variance of the resulting estimates. Instead, we use the procedure which we call *weight-trimmed mean* introduced in the next section.

2.1 Weight-Trimmed Estimate for p

In this section, we describe an alternative procedure for obtaining $\hat{p}_0$, the robust estimate of the underlying proportion of defective parts of type A. Suppose that for the Part A we have data for N vintages (e.g. corresponding to weeks). Accordingly, given are

(i) Sequence of sample sizes $n_1, n_2, \ldots, n_N$ corresponding to these vintages,
(ii) Corresponding numbers of defective parts, $X_1, X_2, \ldots, X_N$ and
(iii) Proportions of defective parts, $P_i = X_i/n_i, i = 1, 2, \ldots, N$.

Note that here we use the upper-case notation X_i to emphasize that we treat the numbers of defective parts as random variables (as usual, the lower-case letters will refer to *realizations* of the random variables). A natural way to estimate the overall proportion of defective parts is to take the weighted average of P_i - we denote this estimator $\hat{p}_w$:

$$\hat{p}_w = \frac{\sum_{i=1}^{N} n_i P_i}{\sum_{i=1}^{N} n_i} = \frac{\sum_{i=1}^{N} X_i}{\sum_{i=1}^{N} n_i}, \tag{3}$$

To obtain a robust estimate, we apply the following process:

(a) Arrange the estimates P_i in increasing order. Denote the resulting sequence of estimates, the corresponding numbers of defective parts and sample sizes by

$$\{P_{(1)}, P_{(2)}, \ldots, P_{(N)}\}, \{X_{(1)}, X_{(2)}, \ldots, X_{(N)}\}, \{n_{(1)}, n_{(2)}, \ldots, n_{(N)}\} \quad (4)$$

(b) Set the *lower* trimming level α_1 (for example $\alpha_1 = 0.1$). Discard the proportion α_1 of the total sample size (and related numbers of defective parts) from *below*. Typically, this would require discarding few initial data points and a portion of data for the vintage at a boundary. In this case, the estimated proportion of defective parts for this boundary vintage remains the same as before, but its sample size is adjusted downward to satisfy the rate of trimming, α_1.
(c) Similarly, set the *upper* trimming level α_2 (for example, $\alpha_2 = 0.05$). Discard the proportion α_2 of the total sample size and related numbers of failed parts from *above*. Typically, this would require discarding few points corresponding to very high fallout rates and a portion of data for the boundary vintage. As before, the estimated proportion of defective items for this boundary vintage remains the same as before, but its sample size is adjusted downward to satisfy the upper rate of trimming α_2.
(d) Deliver a robust estimate $\hat{p}_0$ of the proportion of defective parts of type A computed as a weighted average based on the remaining (non-discarded) sample size.

More details on this process can be found in Civil et al. (2013a, b). In practice, it is often convenient to use *symmetric trimming*, i.e., $\alpha_1 = \alpha_2 = \alpha$ (e.g., one can use $\alpha = 0.1$, which leads to 10% trimming from each side, i.e., 20% of the overall sample size is trimmed). Consequently, for the two-sided trimming only values $\alpha < 0.5$ are permissible.

Geometric interpretation of the procedure (a)–(d) is shown in Fig. 1. One can see that both "usual" weighted and "robust weighted" estimates of p can be represented as slopes on the weighted cumulative sum plot.

2.2 Confidence Bounds for *p*

Once the weight-trimmed robust point estimate $\hat{p}_0$ is available, we can proceed to obtain the confidence bounds for the underlying failure rate p. A technique that we use is somewhat similar to the one used in derivation of the binomial confidence bounds based on the properties of the Beta distribution, see Johnson et al. (2005) Two significant differences are that (a) the confidence procedure is based on the robust estimate $\hat{p}_0$ and not on $\hat{p}_w$, and (b) the Beta-distribution is computed based on fractional number of defective parts and it uses a special procedure to compute the effective samples size, as shown below.

The two-sided $\beta * (100\%)$ confidence bounds (L, U) for p are computed as follows:

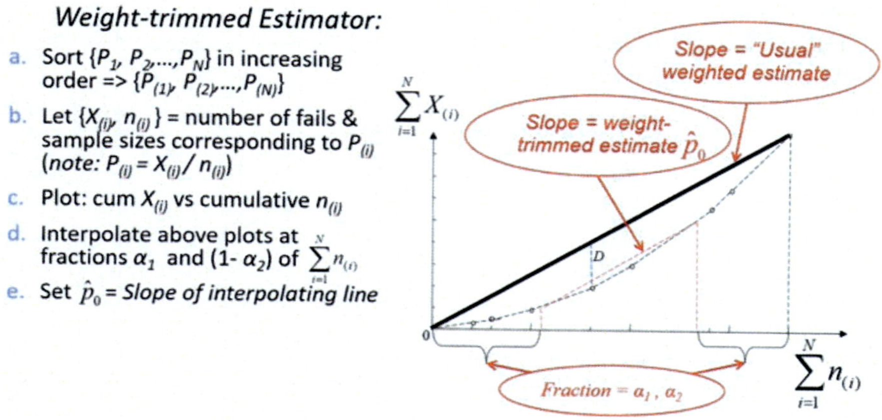

Fig. 1 Obtaining the robust (weight-trimmed) estimate $\hat{p}_0$ of the proportion of defective parts

(a) Compute the effective sample size

$$n_{eff} = (1 - u\alpha) \sum_{i=1}^{N} n_i, \tag{5}$$

where α is the trimming fraction and u is an appropriately chosen coefficient. The value n_{eff} reflects the degree of loss of statistical power that is associated with the use of robust estimate instead of the more efficient (but non-robust) weighted estimate, $\hat{p}_w$.

(b) Define the equivalent number of fails by

$$X_{eff} = n_{eff} * \hat{p}_0 \tag{6}$$

(note that both n_{eff} and X_{eff} are, in general, non-integers).

(c) Compute the lower and upper confidence bounds (L, U) by solving the equations

$$(1 - \beta)/2 = F(x, X_{eff}, n_{eff} + 1 - X_{eff}), \tag{7a}$$

$$1 - (1 - \beta)/2 = F(x, X_{eff} + 1, n_{eff} - X_{eff}), \tag{7b}$$

where $F(x, a, b)$ is the cumulative distribution function (cdf) of the Beta distribution with parameters (a, b). Note that the above formulas are given in terms of the coverage probabilities as opposed to the more commonly used escape probabilities. In the case $X_{eff} = 0$, the lower limit is set to 0; in this case, U is set to β * (100%) upper confidence bound computed by solving (7b) with the LHS = β. Similarly, if $X_{eff} = n_{eff}$, the upper limit is set to 1 and L is set to β * (100% lower confidence bound computed by solving (7a) with the LHS = $(1 - \beta)$. Equations (7) correspond to the classical Clopper–Pearson method (see

Clopper and Pearson 1934; Brown et al. 2001, 2002) with the exception that the counts are allowed to be non-integer.

Applying a non-zero value of u in (5) enables one to adjust the confidence bounds to achieve the nominal coverage probability β*(100%) or to regulate the lower and upper escape rates corresponding to the confidence bounds. For a given data set, a suitable value of u can be determined by setting $p = \hat{p}_0$ and performing simulation analysis conditional on $\{n_i\}$; this value is then applied to adjust the escape probabilities of the bounds (7). Typically, $u > 0$, since the process of weight-trimming is associated with some loss of information, as reflected by n_{eff}—however, this is not universally true, especially for very low p and small sample sizes. Even for the non-robust unbiased estimate $\hat{p}_w$ the coverage probabilities can turn out to be above the nominal level, calling for $u < 0$, and this remains the case for the trimmed versions as well. Note, however, that in cases of this type there will typically be asymmetry between the left and right escape rates, and one might prefer maintaining the nominal escape rate of $[(1 - \beta)/2]$ *100% only for the upper bound.

2.3 *Bias and Robustness Issues*

The methods of Sects. 2.1, 2.2 are useful in many practical situations—however, one needs to be aware of several properties that can complicate the estimation process, especially in situations involving very low proportions of defectives and presence of numerous vintages with low sample sizes. Generally, (symmetrically) trimmed mean procedures estimate the underlying mean of a symmetrically distributed population, provided that the mean exists. However, when the distribution of the population is skewed, the trimmed mean (of any kind) is just a measure of its central tendency. In our case of Binomial population with p estimated exclusively based on the observed number of trials and failures, the weight-trimmed mean is biased as an estimate of the mean, p; typically, p is small and one can see that is biased downwards, due to the positive skewness of the Binomial distribution. Asymptotically, when N is fixed and $\sum_{i=1}^{N} n_i$ tends to infinity, $\hat{p}_0 \to p$. However, for other regimes of increasing the total sample size (e.g., when the number of vintages N also increases rapidly enough), the weight-trimmed mean is not a consistent estimator of p. There are ways to turn it into a consistent estimator—however, this involves additional computational effort, and in many practical situations the bias is not sufficiently large to justify this investment, especially when targets are computed on a massive scale. As will be shown in Sect. 3, we do not use $\hat{p}_0$ directly to declare it as a measure of capability. For example, if the history of part A includes $N = 10$ vintages with the sample sizes $n_1 = n_2 = \ldots = n_N = 100$, and we observe no failures, we will not declare $p = 0$ as the proven part capability. Confidence bounds are a much more important indicator, and in our experience the confidence bounds, as described in Sect. 2.2, provide coverage that is reasonably close to the nominal β *(100%) even when the estimator $\hat{p}_0$ is biased.

Of special importance is the ability of $\hat{p}_0$ to serve as a useful estimator of p in the presence of outliers. For example, let us assume that

$$\text{Probability of defective unit} = \begin{cases} p\,, & \text{with probability } (1-\epsilon)\,, \\ p+a \gg p\,, & \text{with probability } \epsilon\,, \end{cases} \tag{8}$$

where a is a relatively large disturbance and ϵ is some small number. Our objective is to estimate p, which serves as the basis for declaring the part capability. The disturbance model (8) can emerge in two major settings: (a) it can affect the individual items independently of the vintage to which they belong or (b) it can affect the whole vintage. Such disturbances are quite common in industrial data, and they can adversely affect the conventional estimators; however, the proposed robust estimator $\hat{p}_0$, with all its bias-related issues, tends to provide useful estimates of part capability under both modes of disturbance. Selection of the trimming proportions determines the degree of robustness; in our applications, the default values are $\alpha_1 = \alpha_2 = 0.1$.

Next, we briefly discuss the issue of bias. In some cases, one may want to reduce the bias so as to obtain better Root Mean Square Error (RMSE) properties or to improve the coverage probability (by focusing, for example, on the escape rate of the lower confidence bound L). This activity should always be performed by *conditioning* on the observed sample sizes, $\{n_1, n_2, \ldots, n_N\}$, as they form a set of ancillary statistics in our estimation process. Denote

$$\psi\,(p|n_1, n_2, \ldots, n_N) = E\left(\hat{p}_0 \middle| p; n_1, n_2, \ldots, n_N\right)\,, \tag{9}$$

where the function $\psi(p)$ also implicitly depends on the trimming proportions (α_1, α_2). Clearly, our weight-trimmed mean is an unbiased estimator of $\psi(p)$ and not of p. However, once this function is estimated (for example, for every p we can do it via simulation analysis, where we explore the averages of $\hat{p}_0$ based on B replications), we would be able to construct an estimator $\hat{p}_0^{(1)}$ by solving the equation, in p:

$$\psi\,(p|n_1, n_2, \ldots, n_N) = \hat{p}_0\,. \tag{10}$$

This estimator can then be used as a basis for the confidence bounds in lieu of $\hat{p}_0$, using the methods of Sect. 2.2. We could represent $\hat{p}_0^{(1)} = \hat{p}_0 + b^{(1)}$, where $b^{(1)}$ is a positive bias-correction term. The above method of bias correction would lead to consistent estimation under most asymptotic conditions involving growth of sample sizes and number of vintages involved - however, it is computationally expensive, as it relies on combination of simulation and root finding, i.e., a *stochastic approximation* problem, see Bouleau and Lepingle (1994). Another way of bias reduction involves manipulating (α_1, α_2): for example, instead of trimming 10% on each side, bias reduction can be achieved by trimming 8% from the upper side and 12% from the lower side. Conditional on the sample sizes, one can establish the best combination of (α_1, α_2) satisfying the conditions like $\alpha_1 + \alpha_2 = 0.1, \alpha_1/\alpha_2 < 1.5$; note that we need to keep the trimming proportions balanced, as one of our primary goals is to ensure robustness. This method also works well, but requires nontrivial computational investment, and is difficult to implement on a massive scale. This type of estimation is most useful in situations where historical part performance records (involving part A and similar ones) exist, suggesting that a certain a-priori combination (say, $\alpha_1 = 0.12, \alpha_2 = 0.08$) yields more realistic evaluations of proven capability.

The easiest way to achieve bias correction is via parametric bootstrap which involves simulating B replications under the assumption $p = \hat{p}_0$ (as usual, conditional on the observed sample sizes), providing an estimate of the negative bias, $-b^{(2)}$, under these conditions, and then correcting for this bias by setting $\hat{p}_0^{(2)} = \hat{p}_0 + b^{(2)}$. Note, however, that using this correction can lead to increase in variance of the estimator that is not compensated enough by reduction in bias, and result in the increase of the overall RMSE—so the usefulness of this correction depends on the configuration of the sample sizes and the relevant values of p. Based on bootstrap replications, one could decide to apply a more moderate correction term, like $0.5{*}b^{(2)}$ to obtain a better RMSE and/or confidence bound coverage properties.

As an example, consider the case $N = 10$, with the sample sizes $n_1 = n_2 = n_3 = n_4 = 50$, $n_5 = n_6 = \ldots = n_{10} = 1000$, and let us assume that the observed counts of defectives were $x_1 = x_2 = x_3 = x_4 = 0$, $x_5 = 1$, $x_6 = 3$, $x_7 = x_8 = x_9 = 1$, $x_{10} = 0$. Based on these data, using the weigh-trimming with $\alpha_1 = \alpha_2 = 0.1$, yields $\hat{p}_0 = 0.0010$ with 95% confidence bounds (0.0004, 0.0022). To decide whether bias-correction or escape rate corrections to the confidence bounds should be applied, let us set $p = \hat{p}_0 = 0.0010$ and generate 10000 replications conditional on the observed sample sizes. This experiment yields the bias-correction term $b^{(2)} = 0.00008$, suggesting $\hat{p}_0^{(2)} = \hat{p}_0 + b^{(2)} = 0.0011$. With this correction, the RMSE = 0.00042 remains the same, but it is now solely due to variance, not bias - so there appears to be improvement in this respect. The estimated escape probabilities are 0.0132 (left) and 0.0235 (right), so the confidence interval tends to overcover overall, as the total estimated escape probability is 0.0367 instead of the nominal 0.05. The average half-width of the confidence interval is 0.00084. The bias-corrected estimator's escape probabilities are more symmetric, 0.0204 (left) and 0.0222 (right), with the average half-width of the interval 0.00087. Since in our target-setting decisions the upper confidence bound plays a somewhat more important role, we might want to emphasize the closer to nominal right-side escape rate of the original estimator; with this rate, the mean and standard deviation (in parentheses) of the upper bound are 0.0020 (0.00057). For the bias-corrected estimate, these are 0.0021 (0.00058). We also consider midways of the confidence bounds, as these are used as *measures of proven capability*, see Sect. 3: for the original estimator the mean and standard deviation are 0.0012 (0.0004) and for the bias-corrected one 0.0013 (0.0004), i.e., they are quite close. Therefore, there is no strong benefit of applying the bias correction in this case.

Computations show that the bias correction term of the approach in (10) is similar in magnitude: $b^{(1)}$=0.00009, i.e., $\hat{p}_0^{(1)} = \hat{p}_0 + b^{(1)} = 0.0011$. In our context, its properties are similar to those of $\hat{p}_0^{(2)}$. Note that both bias-corrected estimates are close to = 0.0011. Finally, compare the above results to those of the weight-trimmed estimator with asymmetric trimming proportions, $\alpha_1 = 0.12, \alpha_2 = 0.08$. The point estimate corresponding to our data is 0.0011 and the estimated bias term is fairly low, $b^{(3)}$=0.00003. We could have reduced the bias even more by skewing the trim proportions further (say, to $\alpha_1 = 0.13, \alpha_2 = 0.07$)—however, such shift comes at the expense of robustness property, and so it is undesirable. A better way to reduce bias would be to apply other techniques discussed above, on top of the asymmetry in the trim proportions. One can see, however, that using asymmetric trim proportions does not markedly improve the properties of the confidence bounds. The escape probabilities are 0.0169 (left) and 0.0222 (right), with the average half-width of the CIs 0.00086. The mean and standard deviation of the upper bound are 0.0021 (0.00058), and for the mid-CI values they are 0.0012 (0.0004). In summary, for our data set this bias-correction method does not appear to be preferable to the ones discussed earlier.

In a way similar to bias-correction, one can also adjust the confidence bounds so as to achieve the nominal coverage probability β*(100%) by using a non-zero value of u in (3), as discussed in Sect. 2.2—this will lead to an increase of the escape probabilities for both upper and lower confidence bounds. One will probably not want to apply a large value of u when computing bounds based on $\hat{p}_0$, as the upper bound's escape probability is already close to the preferred 0.025. Since our confidence bounds are over-covering, let us explore the value $u = -0.5$. For $\hat{p}_0$, the escape probabilities increase to 0.0156 (left) and 0.0254 (right), with the average half-width of the CIs 0.00081. The upper bound and mid-point statistics do not change much, however: 0.002 (0.00057) and 0.0012 (0.0004). For the bias-corrected estimator $\hat{p}_0^{(2)}$, the bounds become more symmetric and almost achieve the nominal escape rates: 0.0236 (left) and 0.0232 (right), with the average half-width of the CIs 0.00081. However, the half-width of the interval and the midpoint statistics do not change much: 0.002 (0.00057) and 0.0013 (0.0004).

In summary, applying bias-correction methods to our data set improves somewhat the properties of $\hat{p}_0$ as a point estimator—however this does not translate to substantial improvement in the properties of confidence bounds and related quantities that are of primary importance for our target-setting routine.

3 Target Derivation Process

In this section, we discuss the procedures and statistical considerations employed in the process of assigning capability-based targets to parts. Of key importance is the concept of a *yardstick* (denoted by Y in what follows) which reflects, for any given part, the performance of its peers, according to the rules described in Sect. 3.1.

The four major steps in the scheme are (a) obtaining a robust estimate of the underlying failure rate p for the part A based on the data related exclusively to part A, (b) obtaining the lower and upper β*(100%) confidence bounds (L, U) for p based on $\hat{p}_0$, (c) establishing a yardstick Y based on performance of peers of part A

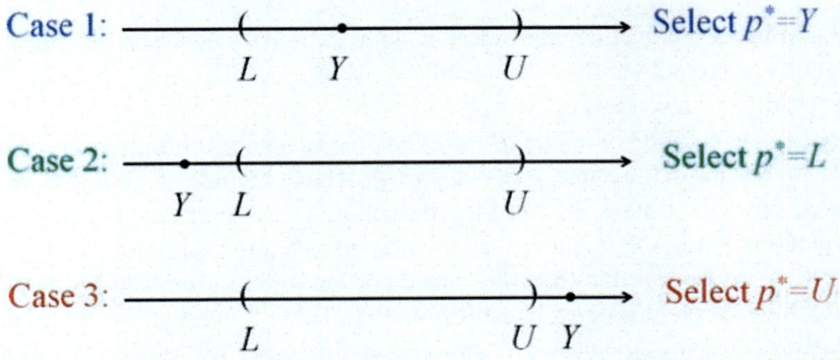

Fig. 2 Setting the target p^* for Part A based on the yardstick Y and confidence bounds (L, U) for p

and (d) delivering the target p^* as function of Y, L and U. The scheme for deriving target for a given part A is shown in Fig. 2. Three cases are shown: when Y falls within the confidence bounds (Case 1), $Y < L$ (Case 2) and $Y > U$ (Case 3).

As illustrated in Fig. 2, the target is selected so that it is always consistent with the process capability as characterized by the bounds (L, U). Our goal here is to (a) minimize intervention as much as possible (so that parts corresponding to the same yardstick generally get the same target as long as their performance is, to some degree, consistent with the yardstick, see Case 1) and (b) encourage positive behavior of process owners by moving them towards better performance while remaining consistent with their capability (Case 2) and not "penalizing" them too much for demonstrating performance that is significantly better than the yardstick (Case 3). Note that the estimate $\hat{p}_0$ is not playing a direct role in setting p^*.

The process described above generally results in meaningful and manageable initial set of targets. One appealing property of this process is that it creates naturally groups of parts that share the same target, facilitating the process of target management and vendor communication.

3.1 The Processing Parameters

There are three major parameters that govern the capability-based target-setting. We will refer to the target recommended for a given part A as p^*_A or simply p^* when it is clear what part is being discussed. The parameters are as follows:

1. *Trimming proportion* α (typically, $\alpha = 0.1$; non-symmetric proportions α_1, α_2 can also be applied). This parameter controls the estimation of the process capability. For every given part (say, part A) our objective is to establish the failure rate of the part that the process is capable of delivering. The data for part A consists of a sequence of failure rates and sample sizes, aggregated by date. We need to eliminate extreme cases from the data, trimming away vintages (dates) for which

the failure rates are either extremely good or extremely bad. The remaining data is then used to deliver the robust estimate $\hat{p}_0$ of the underlying failure rate p for the part A, as described in Sect. 2.

2. *Confidence level* β (typically, $\beta = 0.9$) of the two-sided confidence bounds for the underlying failure rate p of the part A. These bounds, denoted (L, U), are based on the estimate $\hat{p}_0$ and they determine the interval of the failure rates containing values of p that can be considered as being "supported by the data". We show in Sect. 4 that they play an important role in establishing p^*, as they reflect the inherent process capability.
3. *Minimal sample size needed to establish a yardstick*. We refer to this value as *minyards* (typically, *minyards* = 100). In the process of establishing p^*, we take into account not only the capability of a given part A, but also the performance of other parts in the hierarchy. Based on the performance of peers, we establish a *yardstick* value Y. Once we have Y, the target p^* is computed as a function of Y, L and U, see Fig. 2. However, the yardstick computation requires us to borrow data from the parts in higher levels of hierarchy if the sample sizes in lower levels of hierarchy are too small. So, *minyards* = 100 tells us that we should not use data from higher levels of hierarchy if we managed to accumulate a sample of size 100 using lower levels. A more detailed explanation will be provided in Sect. 3.2.

3.2 The Yardstick

As noted earlier, the part A's target is based on (a) the confidence bounds (L, U) for the failure rate obtained exclusively from the data for the part A, and (b) the yardstick Y that incorporates information from the part hierarchy. Without loss of generality, here and in what follows we assume that every part has up to six layers of hierarchy that determine its "siblings", "cousins" and more distant "relatives". Generally, we compute yardsticks for every layer of hierarchy, and we will refer to them as $\{Y_1, Y_2, \ldots, Y_6\}$. The overall yardstick Y is computed sequentially, based on these yardsticks:

$$\begin{array}{ll} \text{Step 1.} & Y = a_1 Y_1 \\ \text{Step 2.} & Y = a_1 Y_1 + a_2 Y_2 \\ \ldots & \\ \text{Step K.} & Y = a_1 Y_1 + a_2 Y_2 + \ldots + a_K Y_K \,, \end{array} \tag{11}$$

stopping at some step K corresponding to the overall sample size accumulated in computing the yardstick. The stopping step K is governed by selection of the parameter *minyards*, see Sect. 3.1. Below is an example of how this computation is performed.

Suppose that in the input file the first categorization field is GRP_ID, the second is *Subcommodity* (like Channel/Optics) and the third is *Commodity* (like Adapters). Suppose that the minimal sample size needed to obtain the yardstick is set to *min-yards* =100. We compute the robust estimates of failure proportions for every part in GRP_ID and the corresponding confidence bounds (L, U), and then compute the Y_1 as the average of midpoints of confidence bounds $(L + U)/2$ of these parts. Next, we check whether the sample size of GRP_ID is 100 or more. If yes, then the yardstick for the Part A can be established based on GRP_ID (i.e., on Y_1) alone - in this case, the last two fields of the corresponding row of the output file (see Fig. 5) will be (1 1), indicating that the 1st tier was sufficient to produce a yardstick, and this tier got 100% of the weight. In other words, we have no need to incorporate the *subcommodity* (tier 2) or *commodity* (tier 3) data to establish targets.

Now consider the case where tier 1 contains too little data for a yardstick; for example, overall sample size for the tier GRP_ID corresponding to our part A is only 60, less than the minimum of 100 required. We will still compute the yardstick Y_1 based on GRP_ID - but this yardstick will only carry 60% of the weight in establishing the overall yardstick for the Part A. We will then need to incorporate subcommodity data (tier 2) - and if the sample size for the subcommodity is high enough (greater than 160, indicating 100 "new" members) then the yardstick based on the subcommodity will receive the remaining 40% of the weight. So, the final yardstick for the Part A will be based on 60% of the GRP_ID data (as represented by Y_1) and 40% of the subcommodity data (as represented by Y_2). The last two fields of the output file row for the part A (Fig. 5) will then be (2 0.4), indicating that the 2-nd tier was touched, and it got 40% of the weight.

If the *subcommodity* tier does not have enough sample size (let us say its sample size is 130, corresponding to just 70 (= 130-60) additional items), then *subcommodity* will get only 70% of weight (of the 40%), resulting in the overall weight for *subcommodity* of 0.4*0.7 = 0.28. So, we are in a situation where GRP_ID gets 60% of weight, *subcommodity* (represented by its own yardstick Y_2) gets 28% of weight, and we need to incorporate *commodity* data to get the remaining 100-60-28 = 12% of weight, So, in this case the last two fields of the output line for the part A will be (3 0.12), indicating that the 3rd tier data (*commodity*, represented by its own yardstick Y_3) was used and its weight in establishing the yardstick Y is 12%. Note, however, that if the commodity data is missing, then the subcommodity data will get the remaining 40% instead of 28%, so that the sum of weights is always 1.

A scheme for constructing a yardstick Y is shown in Fig. 3. In this case, the part A (a power cable for a type of desktop computer) has an immediate group of peers represented by several other models of desktop computers (Level 1); it also belongs to the class "Power Cables" (Level 2) that is nested in the class "Connectors" (Level 3).

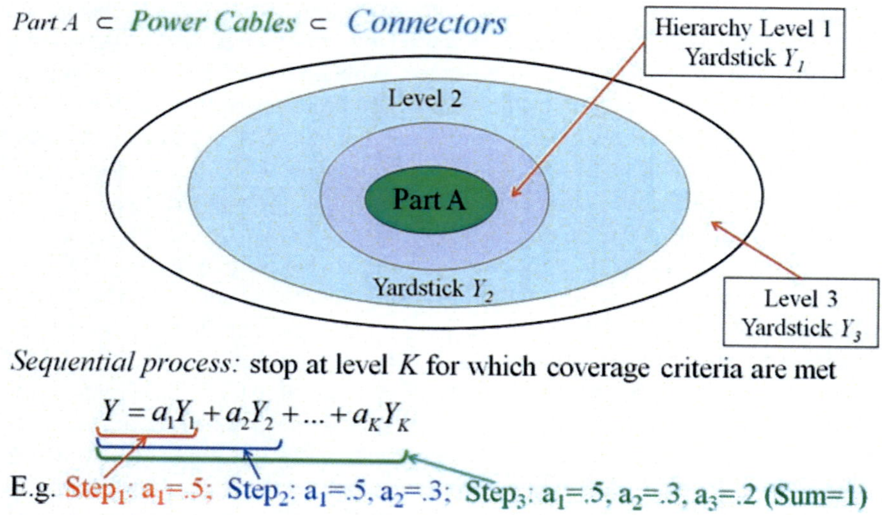

Fig. 3 The scheme of constructing the yardstick Y for the part A based on part hierarchy

3.3 Decisions

Targets delivered as a result of the process described above can be considered for use in the setup files that drive activities of a monitoring system, such as an EWS. These targets could either be accepted outright or further modified in accordance with business needs or quality management policy in place. For example, some quality improvement programs could mandate a systematic adjustment of the current targets downward (say, by 10% per year), even if the historic data does not support such process capability. Another possible policy is to ensure that the targets of underperforming parts are moved by some minimum amount towards their respective yardsticks. For example, suppose that for a given part A, the estimated proportion of defectives is $\hat{p}_0$, the yardstick is $Y < \hat{p}_0$ and the target delivered by the process is $p^* > Y$. Then one might want to set the final target p^*_{fin} as follows:

$$p^*_{fin} = \min\ [p^*, w\hat{p}_0 + (1-w)Y], \tag{12}$$

where w is some suitable weight, say, 0.8. This policy guarantees that

$$(p^*_{fin} - Y)/(\hat{p}_0 - Y) \le w\,. \tag{13}$$

In whatever way the target is chosen, let us assume that this is the value designated as "acceptable level", i.e. if the process operates at this level, the probability of a (false) alarm should be low. In what follows, we will use the notation (p_{acc}, p_{unacc}) to denote the acceptable and unacceptable levels based on a wide range of considerations, as opposed to the earlier notation (p_0, p_1), since in that notation we were mostly focused on the capability-based targets.. For simplicity, let us assume that the acceptable level $p_{acc} = p^*$ is delivered as the output of the described process, i.e., it depends mostly of the proven part capability and peer performance. Next, we need to decide on the unacceptable level, p_{unacc}. Generally, any number above p^* could serve as the unacceptable level. However, it is not desirable to set it too close to the acceptable level. The reason is that in order to resolve between the levels (p_{acc}, p_{unacc}) that are close, we will need the detection thresholds to be quite high. Our ability to detect p_{unacc} will indeed be about the best achievable for the data stream intensity at hand - however, the detection speed will still be low in practical terms; on the other hand, the high detection threshold will hamper our ability to detect changes of larger magnitude quickly enough - so, we will need to rely more on the supplemental rules. The best policy is to set p_{unacc} to the level that it is important to detect quickly. In many practical situations, the policy

$$p_{unacc} = c_u * p_{acc} \tag{14}$$

where c_u is in the interval [1.5, 2.5], is reasonable. For example, $c_u = 1.75$ is often used for the supply chain data. This choice is partially motivated by the fact that the factor 1.75 typically reflects a non-trivial change in the monitored defect rate that engineers are willing to recognize as being of practical importance. Furthermore, in many cases it also ensures separation by at least one standard deviation of the acceptable defect rate - and, with the choice of k given by (2), such situations tend to produce detection schemes that are more comfortable to work with - i.e., the resulting alarm thresholds tend to be neither too high nor too low. For example, suppose that $p_{acc} = 0.01$ and the sample size corresponding to the individual point (vintage) is $n = 175$. Then the standard deviation of the observed proportion of defectives is approximately $\sigma = sqrt[(0.01 * 0.99)/175] = 0.0075$ - and so, by setting $p_{unacc} = 1.75 * p_{acc}$ we are achieving a 1*σ separation between p_{acc} and p_{unacc}.

The policy (14) enables one to establish targets on a massive scale without having to deal with the reality of prevalent vintage sample sizes. Under some conditions, however, targets that take them into account are more suitable. For example, consider the case of ultra-low failure rates: $p_{acc} = 1e - 6 =$ 1ppm (part per million), and assume that the prevalent vintage sample size is $n = 10000$. In this case, $\sigma \approx$ 10ppm, i.e., the gap of 0.75ppm between the unacceptable and acceptable levels suggested by (12) is 0.075*σ, leading to the resolvability issue described above. In such situations, a formula

$$p_{unacc} = p_{acc} + f_u * \sqrt{\frac{p_{acc} * (1 - p_{acc})}{n}} \tag{15}$$

with f_u in the interval [0.5, 1] is likely to lead to a monitoring procedure with more appealing operating characteristics. In the above case, the policy (15) would call for setting p_{unacc} to a value between 6ppm and 11ppm.

The above arguments illustrate a target setting approach for which the starting point is establishing p_{acc} based on the concept of proven part capability. In some cases, however, one could choose to start with establishing p_{unacc}, and then invert the relations (14)–(15) to obtain p_{acc}. This approach could be used when the primary purpose of the monitoring scheme is to flag parts that directly endanger business objectives or contractual obligations. For example, an organization could maintain a list of thresholds for failure rates and focus on flagging parts for which failure rates exceed these thresholds in some sense. When using this approach, it is important to understand the meaning of such thresholds, as it can vary from one organization to another, even within the same company. In some cases, thresholds are meant to represent the behavior of the process mean, and in this case they can be interpreted as p_{unacc}. In other cases they represent a form of a "spec", indicating that p_{unacc} should be chosen way below the threshold, using a statistical argument in conjunction with the threshold definition.

4 Example of Implementation

In this section we describe implementation of a target-setting system in the IBM Integrated Supply Chain, which monitors quality of parts as they progress from their manufacturing plants (typically owned by the vendors) through the IBM's own manufacturing and assembly plants, and to the field, as part of computer systems. Parts undergo many tests, typically with the pass/fail outcome. A monitoring system called Quality Early Warning System (QEWS) is used to analyze performance of parts in relation to various tests. The key parts of QEWS include the data preparation module, the search engine, user interface (dashboard) and the target-setting system. The target-setting involves many factors, as mentioned earlier, and the segment of the system we focus on is related to the part capability analysis.

Inputs. The basic data file contains the following fields:

1. Analysis ID (e.g., ADP1CHO1PU0001, for an adapter part in a channel-optic family)
2. Part number (e.g., 0000000E0807)
3. Test identifier (e.g., 000)
4. Date (e.g., date of manufacturing)
5. Number of parts tested
6. Number of parts failed
7. Fields 7–12 give up to 6 layers of hierarchy. The part's number can be considered as the 0-th level of hierarchy.

A typical record of the input data file is shown in Fig. 4. One can see that in this case levels 1, 3 and 4 of the hierarchy are missing (only levels 2, 5 and 6 are there). For records of this type, we often refer to "Channel/Optics" as the name of *subcommodity* and to "Adapters" as *commodity*. The 2nd level of the hierarchy groups part numbers by some criterion (in this case, GRP_CHO1 represents a group of "very similar" parts).

Processing the file using the parameters described in Sect. 3.1, yields the output file. Its lines contain targets and related information for every part number, see Fig. 5.

ADP1CHO1PU0001|0000000E0807|000|2012-12-21|622|2||GRP_CHO1|||Channel/Optics|Adapters

Fig. 4 Typical data record of the target-setting procedure. The trailing six fields (three of them empty) establish the part hierarchy

	A	B	C	D	E	F	G	H	I	J	K	L	M	N	O	P	Q	R	S	T
1	ADP1CHO1PU0001	0000000E0807	0	4/19/2013	0.00177	30	9129	20	16	123	232	301.5	1528	0.00112	0.00267	146934	0.007607	0.00267	2	1
2	ADP1CHO1PU0005	0000010N9824	0	2/22/2013	0.00747	13	162	3	1	3.25	5.5	13.5	69	0.00059	0.03099	146934	0.007607	0.007607	2	1
3	ADP1CHO1PU001R	0000046M6067	0	4/5/2013	0	22	233	0	1	1.5	3	5.5	113	0	0.00983	146934	0.007607	0.007607	2	1
4	ADP1CHO1PU001V	0000000E1731	0	4/5/2013	0	1	9	0	9	9	9	9	9	0	0.22574	146934	0.007607	0.007607	2	1
5	ADP1CHO1PU001Y	0000000Y3263	0	4/5/2013	0.009953	19	522	9	2	6.25	23.5	32.75	110	0.00402	0.02053	146934	0.007607	0.007607	2	1
6	ADP1CHO1PU002L	0000000Y3328	0	4/5/2013	0	14	48	0	1	2	2	4.5	8	0	0.04684	146934	0.007607	0.007607	2	1
7	ADP1CHO1PU0025	0000010N8620	0	1/25/2013	0	1	4	0	4	4	4	4	4	0	0.43766	141364	0.015372	0.015372	5	0.96
8	ADP1CHO1PU003P	0000000E0806	0	4/19/2013	0.005271	30	63792	352	264	798	1450	2923	6774	0.00481	0.00577	146934	0.007607	0.00577	2	1
9	ADP1CHO1PU003Q	0000000E0808	0	4/5/2013	0.005211	27	443	6	1	4	12.5	23.25	60	0.0011	0.01517	146934	0.007607	0.007607	2	1
10	ADP1CHO1PU0035	0000000Y3272	0	4/12/2013	0.015477	16	617	12	1	9	18	54	130	0.00828	0.02643	146934	0.007607	0.00828	2	1
11	ADP1CHO1PU0058	0000039Y9192	0	4/5/2013	0.003207	20	2129	11	1	2	4	14	819	0.00149	0.00606	146934	0.007607	0.00606	2	1
12	ADP1CHO1PU005E	0000041Y8581	0	12/21/2012	0	3	7	0	1	1	1.5	2.5	4	0	0.28031	136947	0.016223	0.016223	5	0.93
13	ADP1CHO1PU005H	0000044X1925	0	11/2/2012	0	1	2	0	2	2	2	2	2	0	0.68377	146934	0.007607	0.007607	2	1
14	ADP1CHO1PU006K	0000046M6142	0	4/19/2013	0.003087	26	666	3	1	3.5	10	26.5	118	0.00057	0.00955	146934	0.007607	0.007607	2	1

Fig. 5 The output file of the target-setting procedure

The fields of the output file are as follows (the letter in parentheses refers to the spreadsheet field of Fig. 5):

(A) Analysis ID
(B) Part number (PN)
(C) Test identifier
(D) Date of the last data file row for this PN
(E) Estimated proportion of defectives, $\hat{p}_0$
(F) Number of vintages (dates) for which data are available
(G) Total number of tested parts for this PN
(H) Total number of failed parts for this PN
(I) n_{min}, minimal sample size observed for various dates
(J) $n_{0.25}$, lower quartile of sample sizes observed for various dates
(K) $n_{0.5}$, median of sample sizes observed for various dates
(L) $n_{0.75}$, upper quartile of sample sizes observed for various dates
(M) n_{max}, maximal sample size observed for various dates
(N) Lower confidence bound L for p based on $\hat{p}_0$ (and the two-sided β *(100%) confidence interval)
(O) Upper confidence bound U for p based on $\hat{p}_0$
(P) Overall number of tested parts in the process of establishing a yardstick (note that in Fig. 5, many of these numbers are the same, as the corresponding parts belong to the same group having a large overall sample size, and so these parts share a common yardstick)
(Q) Estimated proportion of defectives based on peer performance (i.e., yardstick).
(R) Target proportion p^* of defectives for this PN (it is based on the part capability and it can be considered for the role of p_0)
(S) Top tier of part hierarchy used in establishing the yardstick
(T) Weight given to the top tier of part hierarchy when establishing the yardstick

For example, consider the leading record of the output file in Fig. 5. It is related to the performance of PN = E0807 with respect to the test '000'. The robust estimate of the failure rate, based on 30 vintages (date rows) in the data file was 0.00177.

The data file contained information on 9329 parts, of which 20 failed the test '000' (note that the regular estimate of the failure rate is $\hat{p}_w = 20/9329 = 0.00214$, which is somewhat higher than the robust estimate based on 10% trimming). The sample sizes corresponding to rows of the data file varied considerably: $n_{min} = 16$, $n_{0.25} = 123$, $n_{0.5} = 232$, $n_{0.75} = 301.5$, $n_{max} = 1528$. The 90% confidence bounds for the failure rate are $L = 0.00112$, $U = 0.00267$. The yardstick (reflecting the peer performance) was 0.007607, which means that the part E0807 tended to perform considerably better than the yardstick. The yardstick was computed based on 146934 peer parts. The top tier of the hierarchy used to compute the yardstick was 2. In other words, parts belonging to the group GRP_CHO1 provided enough information to compute the yardstick. Finally, the last element indicates that the information based on the 2nd level of hierarchy (i.e., GRP_CHO1) received the weight 1 (highest possible), when establishing the yardstick.

In contrast, the part 10N8620 corresponding to row 7 of the output file belongs to a smaller group of parts and so, the yardstick could not be computed based on the 2nd level of the hierarchy. Information had to be borrowed from the level 5 of the hierarchy, i.e., we had to incorporate performance data of all parts for which the subcommodity is "Channel/Optics". Furthermore, the information of level 5 played a dominant part in establishing the yardstick, receiving the weight of 0.96.

5 Discussion

In practical implementations, EWSs typically handle many data streams for which targets have been assigned by some process based on statistical and business requirements. The target-setting system handles a much wider set of issues than just determining the process capability, but the capability-centered target generation sub-system is running in parallel with the monitoring system, and both often use shared data sources. If for a given part A the current acceptable level is, say, 0.0001, i.e., 100 defective parts per million (ppm) and the capability analysis based on the presented methodology suggests that the vendor is capable of delivering 10 ppm, this will not typically lead to automatic tightening of requirements for the part manufacturer, as such policy could lead to undesirable consequences involving costs, behaviors and business relations, and in some cases it would not be in line with contractual obligations. Knowledge of the underlying capability, however, is valuable, and it can be used in many ways, including product planning, vendor selection and quality monitoring. Of special importance is the process of assigning targets to new parts—and this is where capability-based targets can be quite useful. Once the initial targets are assigned, the target-management system will govern their future development, using its own logic and procedures. Below, we discuss several additional aspects of the capability-based target setting.

1. It is important to select the window of vintages (for every part) that are included in the target-setting computation. One will often decide that, say, no more than

32 weeks of the most recent vintages will be included. Such a requirement is motivated by the desire to maintain the quality level inside the window roughly constant, so that the estimated capability can be viewed as representative for the whole window. The robust methods help us to eliminate the effect of outliers - but they can produce misleading results in the presence of trends and other temporal effects, and limiting the window size helps to reduce their impact.

2. Of special importance is the definition of vintage. For example, consider the process in which we focus on the chip failure rate, when the chip is installed into a printed circuit board and subjected to various tests as part of a sub-system. One can define the vintage as (a) date (or timestamp) of the chip installation into the board, (b) date of the initial system test involving the chip as part of the board, (c) date of the board production (usually available on a board's barcode), and so forth. The vintages are selected to represent operations that can have a negative impact on the chip performance, see Yashchin (2018). The EWS is typically applied to various types of vintages (in parallel), and so should be the capability-based target analysis. This can result in different failure rate capability assessments based on the definition of vintages, which complicates the process of the target assignment (after all, only one set of targets should be assigned to the chip, independently of the definition of vintage) - however, it also presents the opportunity of a deeper study of the underlying chip capability. If the estimated capability-based target differs strongly for various definitions of vintages, this suggests that the chip's failures might be affected by external factors rather than its own inherent flaws. For example, consider the case where the vintage corresponds to the chip mounting date and we concluded that the chip is capable of 1000 ppm. However, when the chip failures were organized by the board manufacturer dates, the estimated target dropped to 100 ppm. A closer examination could reveal that a large fraction of failed chips was mounted on boards of a single vintage, and the impact of this vintage was automatically reduced through the trimming process, resulting in delivering 100 ppm as a robust estimate. One can then legitimately suspect that the cause of the failures might not be related to defects inside the chip, but rather caused by defective boards. Under such conditions, one might accept 100 ppm (rather than 1000 ppm) as a more realistic estimate of the inherent chip capability.
3. When numerous vintages with small sample sizes are present, it might be beneficial to aggregate vintages into larger sub-groups, i.e., to use *temporal aggregation*, e.g., see Zwetsloot and Woodall (2019). This could enhance statistical performance of the robust techniques discussed in this paper and present a more realistic picture of the underlying process capability, especially for very low p.
4. In large-scale implementations, it is important to explore the connection between target-setting and ability of the monitoring system to detect unfavorable conditions early. If the targets are set primarily based on business requirements, it can happen that they are quite lax relative to the process capability. This can lead to delayed detection of unfavorable trends: some of them would undoubtedly be considered as being of no practical importance, at least initially. Back-testing of targets against the previously known unfavorable conditions can help in making the case for stricter targets. Furthermore, one can also decide to use two sets

of targets: one for driving business decisions, and another for early detection of unfavorable trends. In this way, one can detect, for example, origination of drifts in the process level way before these drifts reach the magnitude where they turn into a business issue.

5. Confidence bounds for p play a crucial role in our target-setting scheme; the Clopper–Pearson procedure (generalized for non-integer values of tested/failed counts) described in Sect. 2.2 has been used in the IBM Supply Chain quality management system for several years and it was well-received by the users. Of course, this procedure is known to be somewhat conservative, e.g., see Brown et al. (2001, 2002)—however, this did not appear to affect the quality of the target-setting process: one needs to keep in mind that the confidence bounds are used primarily as an intermediate tool for establishing proven part capability, so consistent conservatism in coverage probability can generally be tolerated. The procedures can be easily adapted to use other methods for confidence bound derivation, such as the mid-P or Jeffrey's methods. We do not expect significant practical consequences from using such intervals, once the default values of β are adjusted appropriately to match the coverage probabilities of the Clopper - Pearson procedure.
6. In this article, we focused on the problem of target-setting for p using attribute data: only the counts of tested and failed items for various vintages played a role in the robust estimation process. Note, however, that monitoring schemes for p can also use other types of data, e.g., variables data, see Knoth and Steinmetz (2013). Specifically, if a failure of an item can be associated with measured values of its characteristics, then one can typically obtain an estimate $\hat{P}_i$ for the i-th vintage failure rate using a sample of these values; such estimates often have properties that are superior to those based on the pass/fail data alone. The monitoring scheme (1) can be directly applied to variables-based sequence of estimates, with weights $\{w_i\}$ inversely proportional to the variances of $\{\hat{P}_i\}$. One would typically need to take special precautions to ensure low bias of the estimates $\hat{P}_i$ and obtain control schemes with good statistical properties. Other schemes, e.g., based on the generalized likelihood ratio statistics, can also be used for a similar purpose.

The process of target-setting is crucial for such situations as well, and the procedures described in this article can be adapted to yield high-quality robust estimates of p based on a set of vintages. Specifically, the process illustrated in Fig. 1 can be deployed, with $n_i = w_i$, in conjunction with low-bias $\{\hat{P}_i\}$. As in the present article, the resulting robust estimates could require separate bias-correction, which can be done using methods described in Sect. 2.3. The concepts of *yardstick* and *proven capability* are applicable in this general context, and they can be used in a similar way. One technical issue is obtaining confidence bounds for p based on the robust estimate, see Fig. 2. Such bounds can be obtained in several standard ways (including bootstrap and delta-method), depending on the concrete situation and relationship between p and the basic parameters driving the data.

Acknowledgements We would like to thank John Maresca, John Wargo and Susan Zichitella (IBM Supply Chain organization), Naoki Abe, Yasuo Amemiya, Chid Apte, Jayant Kalagnanam and Bonnie Ray (IBM Research) for their support of developing the Quality Early Warning System (QEWS) that incorporates, among other things, the approach to target-setting described in this article. We deeply appreciate the contribution of David L. Jensen and Brian F. White (IBM Research) to software development. We would also like to acknowledge and thank the Editor and the Referee for their work on this article.

References

Bouleau, N., & Lepingle, D. (1994). *Numerical methods for stochastic processes*. New York: Wiley.

Brown, L. D., Cai, T. T., & DasGupta, A. (2001). Interval estimation for a binomial proportion. *Statistical Science*, *16*(2), 101–133.

Brown, L. D., Cai, T. T., & DasGupta, A. (2002). Confidence Intervals for a binomial proportion and asymptotic expansions. *Annals of Statistics*, *30*(1), 160–201.

Civil, A., Komatsu, J. G., Wargo, J. M., Yashchin, E., & Zulpa, P. (2013). Advanced statistical detection of emerging trends. *US Patent Application Publication US 2013/0041625 A1*

Civil, A., Komatsu, J. G., Wargo, J. M., Yashchin, E., & Zulpa, P. (2013). Trend-based target setting for process control. *US Patent Application Publication US 2013/0030863 A1*

Clopper, C. J., & Pearson, E. S. (1934). The use of confidence or fiducial limits illustrated in the case of the binomial. *Biometrika*, *26*(4), 404–413.

Gan, F. F. (1993). An optimal design of CUSUM control charts for binomial counts. *Journal of Applied Statistics*, *20*(4), 445–460.

Hawkins, D. M., & Olwell, D. H. (1998). *Cumulative sum charts and charting for quality improvement*. New York: Springer.

Huber, P. J., & Ronchetti, E. M. (2009). *Robust statistics* (2nd ed.). New York: Wiley.

Johnson, N. L., Kemp, A. W., & Kotz, S. (2005). *Univariate discrete distributions* (3rd ed.). New York: Wiley.

Kenett, R., & Zacks, S. (2014). *Modern industrial statistics: With applications in R, MINITAB and JMP* (2nd ed.). New York: Wiley.

Knoth, S., & Steinmetz, S. (2013). EWMA *p* charts under sampling by variables. *International Journal of Production Research*, *51*(13), 3795–3807.

National Institute of Standards and Technology (NIST) Handbook (2019). Dataplot, weighted trimmed mean. https://www.itl.nist.gov/div898/software/dataplot/refman2/auxillar/weigtmea.htm.

Woodall, W. H. (1986). The design of Cusum quality control charts. *Journal of Quality Technology*, *18*(2), 99–102.

Yashchin, E. (1985). On the analysis and design of Cusum-Shewhart control schemes. *IBM Journal of Research and Development*, *29*(4), 377–391.

Yashchin, E. (2010). Computational and monte-carlo aspects of systems for monitoring reliability data. In Y. Lechevallier & G. Saporta (Eds.), *Proceedings of the COMPSTAT 2010* (pp. 253–262). Paris: Springer.

Yashchin, E. (2012). On detection of changes in categorical data. *Quality Technology and Quantitative Management*, *9*(1), 79–96.

Yashchin, E. (2018). Statistical monitoring of multi-stage processes. In S. Knoth & W. Schnid (Eds.), *Frontiers in statistical quality control* (Vol. 12, pp. 185–209). Cham, Switzerland: Springer.

Zwetsloot, I. M. & Woodall, W. H. (2019). A review of some sampling and aggregation strategies for basic statistical process monitoring. *Journal of Quality Technology* (with discussion, published online). https://doi.org/10.1080/00224065.2019.1611354.

MAV Control Charts for Monitoring Two-State Processes Using Indirectly Observed Binary Data

Olgierd Hryniewicz, Katarzyna Kaczmarek-Majer, and Karol R. Opara

Abstract Processes described by indirectly observed data naturally arise in applications, such as telehealth systems. The available data can be used to predict the characteristics of interest, which form a process to be monitored. Its randomness is largely related to the classification (diagnosis) errors. To minimize them, one can use ensembles of predictors or try to benefit from the availability of heterogeneous sources of data. However, these techniques require certain modifications to the control charts, which we discuss in this paper. We consider three methods of classification: classical—based on the full set of attributes, and two combined—based on the number of positive evaluations yielded by an ensemble of inter-correlated classifiers. For monitoring the results of classification, we use a moving average control chart for serially dependent binary data. The application of the proposed procedure is illustrated with a real example of the monitoring of patients suffering from bipolar disorder. This monitoring procedure aims to detect a possible change in a patient's state of health.

Keywords Moving average control chart · Indirectly observed binary data · Correlated classification data · Bipolar disorder disease

1 Introduction

In Statistical Process Control (SPC) a monitored process is assumed to be in two states: in-control (stable) and out-of-control (non-stable, deteriorated). The in-control state is defined by the probability distribution of process quality characteristics. When

O. Hryniewicz (✉) · K. Kaczmarek-Majer · K. R. Opara
Systems Research Institute, Polish Academy of Sciences, Newelska 6, 01-447 Warszawa, Poland
e-mail: hryniewi@ibspan.waw.pl

K. Kaczmarek-Majer
e-mail: K.Kaczmarek@ibspan.waw.pl

K. R. Opara
e-mail: Karol.Opara@ibspan.waw.pl

S. Knoth and W. Schmid (eds.), *Frontiers in Statistical Quality Control 13*,
Frontiers in Statistical Quality Control,
https://doi.org/10.1007/978-3-030-67856-2_8

the type of this distribution is known (e.g., the normal distribution) the in-control state is defined by the values of parameters of this distribution. The probability distribution of quality characteristics in the out-of-control state may not be defined. However, in practice this distribution is assumed to be the same as in the in-control state, but with different parameters. For example, having a shifted expected value (in this case we talk about a shifted process level). When the type of a probability distribution is not known, one can use non-parametric SPC procedures, but this approach is rather seldom applied in practice.

Dozens of control charts (the main tool of SPC) have been proposed for different distributions of process quality characteristics, different types of observed data, and different types of decision rules. For nearly all these charts, it is assumed that important quality characteristics are directly observable. However, sometimes measurements of these quality characteristics are either practically impossible (e.g., lifetimes of produced electronic elements) or too costly (e.g., in destructive testing). Furthermore, in many medical contexts, for example, in the mental state monitoring of bipolar disorder patients, measurements of the patient's psychiatric condition (the psychiatric assessment) shall not be performed too often due to the patient's well-being. In such cases, one has to find observable characteristics that are related to the quality characteristics of interest, and either monitor the stability of these characteristics or build a prediction model and monitor predicted values of quality characteristics of interest. An example of this second approach is described in Hryniewicz (2015).

In all the cases mentioned above, it is assumed that the probability distributions that describe states of the monitored process can be, directly or indirectly, identified. This is the case for all industrial, financial, etc., processes, where results of measurements allow us to identify all these distributions. There exist, however, processes where the state of a process cannot be objectively determined. A good example of such a process is human health which varies in time. It is very difficult, or sometimes even impossible, to define the state of health or the state of illness. A possible "objective" way to do so is to describe a person by a set of measurable characteristics (e.g., temperature, blood pressure, pulse rate, etc.), and define some limits on the values of these characteristics. A person is considered as "healthy" if all observed values of these characteristics are inside these limits. The state of "illness" is a complement of "healthy". Note, however, that the transition between states defined in such a way is usually not abrupt, and this creates serious problems when some decision rules have to be established. Despite these difficulties, the application of classical SPC procedures for monitoring of such health-related processes is possible, if we define considered states in term of stability or instability of measured processes. Unfortunately, in some important cases, the state of a process is evaluated to a great extent subjectively. This is frequently encountered in the case of psychiatric diseases, such as bipolar disorder, considered an example in this paper. Moreover, for such diseases, objectively measured health-related characteristics may not exist. Therefore, to evaluate a patient's state, psychiatrists have to rely on observations of symptoms (sometimes loosely-related), called context variables.

In this paper, we consider monitoring a two-state process whose states cannot be evaluated objectively. We assume that the state of a monitored process can be, in some way, predicted, but considerable errors are possible. In the case of two-state processes, the prediction process boils down to the problem of binary classification. We consider two cases when classifiers used in the monitoring procedure are built using complete and incomplete data. In the case of complete data, all vectors of context variables from a training set have their assigned labels. Thus, for building a classifier we can use the methodology known in machine learning as supervised learning. In the case of incomplete data, we know all vectors of context variables, but only for some of them, we know their respective labels. In this case, for building a classifier we can use the methodology known as semi-supervised learning, where unknown labels in the training set are estimated using an iterative estimation process. The results of a classification in the training set can be used for the construction of a control chart for monitoring of an indirectly observed process.

This paper is an extended version of the paper (Hryniewicz et al. 2019) published in the proceedings of the international conference ISQC 2019 held in Hong Kong. In particular, it contains new results to be used when process data in the training phase are incomplete, and the respective classifiers have to be built using a semi-supervised learning algorithm. This situation takes place in the case of the real-life example considered in this paper. The paper is organized as follows. In the second section, we present methods used for the indirect evaluation of the state of the monitored process using supervised and semi-supervised learning algorithms for binary classification. In the third section, we propose a control chart based on the data obtained using considered classification procedures. The fourth section is dedicated to the real problem of monitoring patients suffering from bipolar disorder (BD). The state of a monitored patient is evaluated indirectly by monitoring his/her smartphone activity. The practical part of the research described in this paper is still in its initial stage, so in the last section of the paper, we will present problems that require further investigation.

2 Indirect Evaluation of States Using State-Related Observational Data

Let's consider a two-state process whose states are denoted by 0 and 1. By convention, we label the state 0 as "negative", and the state 1 as "positive". Note, that connotations attributed to these labels in certain contexts may be misleading, as, e.g., the label "negative" may not have the meaning "undesirable". We assume that the actual state of the process is not directly observable, and we can know only its predicted value, described by a binary random variable Y (0 or 1). We estimate it using a set of observable attributes (predictors) $(X_1, \ldots, X_m)$, and a function $\hat{Y} = f(X_1, \ldots, X_m)$, called a "*classifier*". Now, let us assume that we observe n sets of predictors $(X_{1,j}, \ldots, X_{m,j})$, $j = 1, \ldots, n$ together with their known labels

Table 1 Confusion matrix

	Predicted negative	Predicted positive
Actual negative	True negative (TN)	False positive (FP)
Actual positive	False negative (FN)	True positive (TP)

$Y_j, j = 1, \ldots, n$. These data are often called *training data set*. The information contained in the training data set is now used for building the classifier, i.e., for the estimation of the unknown function $\hat{Y} = f(X_1, \ldots, X_n)$. This estimated function can be further used for the prediction of labels assigned to observed sets of predictors.

The procedure described in the previous paragraph solves a problem of "*classification*", which is known in different communities under such names, for example, "discrimination" or "supervised learning". It is a special case of a more general problem of "statistical learning", whose theoretical foundations can be found, e.g., in monographs by Duda et al. (2000) and Hastie et al. (2008). In these books, one can also find the description of many algorithms whose aim is to classify objects described by sets of observable predictors. A comprehensive view on the application aspects of such classification algorithms can be found in the book by Witten et al. (2011).

Many different classifiers have been already proposed in the literature. They differ in their statistical properties and computational complexity. If we do not consider such additional information as, for example, costs of misclassification, the whole information about the quality of classifiers is contained in the so-called confusion matrix, presented in Table 1 (Japkowicz and Shah 2015). The most popular measures of the quality of classifiers are built using the information contained in this matrix. The most frequently used quality measure of a classifier is its *Accuracy*, defined as $= (TN + TP)/(N + P)$. It estimates the probability of correct classification. However, in certain circumstances (e.g., when classes are unbalanced) this measure does not let to discriminate the quality of different classifiers. Other popular and important measures, such as *Precision*, defined as $= TP/(TP + FP)$, *Sensitivity* (*Recall*), defined as $= TP/(TP + FN)$, and *Specificity*, defined as $= TN/(FP + TN)$, describe these features of binary classifiers which are related to classification errors of different types. For example, high values of *Precision* in statistical terms are equivalent to low values of type I classification error when "Positives" are considered as the relevant class. Similarly, high values of *Sensitivity* in statistical terms are equivalent to low values of type II classification error. When the quality of the classification of "Negatives" is also worthy of consideration, one has to take into account the value of *Specificity*. There exist also certain aggregate quality measures of classifiers, For example, the *F1 score* (or *F1 measure*) is defined as the harmonic average of *Precision* and *Sensitivity*. Low values of this measure indicate that a classifier has a large value of at least one of type I or type II errors.

Let us consider the problem of the classifier's quality in the context of process monitoring. When the monitored process is in the desirable state 0, then we are interested in a low rate of false alarms, triggered by the occurrence of "false positives", indicating that the actual state of the process is 1. Thus, we are interested in a high value of *Specificity*. On the other hand, we are also interested in the generation of an immediate alarm when the monitored process switches to the undesirable state 1. This happens when "true positives" occur. Thus, we are interested in a high value of *Sensitivity* (*Recall*).

In classical applications of SPC, it is assumed that after the process deterioration has been revealed, the process is reverted to its "in-control" state. However, in practice, this might be not straightforward. To correct a deteriorated process, some measures are undertaken, and we hope that after a certain time the process will revert to its "in-control" state. In our context, it means that we want to avoid a signal of false improvement, so we need a high value of *Sensitivity* (*Recall*). On the other hand, we expect an immediate signal that the process has reverted to its in-control state, so our classifier should have a high value of *Specificity*. So in both considered cases, high values of *Specificity* and *Sensitivity* (*Recall*) are required for the used classifier. These two requirements can be summarized by the requirement of the high value of the aggregate characteristics, called *Informedness*, defined as *Informedness* = *Specificity* + *Sensitivity* −1, see Powers (2011). When costs of misclassification can be considered, one can assign different weights to these two measures of quality. In this research, however, we do not make such a distinction.

We have already noticed that the quality measures mentioned above have a probabilistic interpretation. The question arises then how to estimate these probabilities. The best way to do so is to use a separate test set of labeled observations. When the total number of observations is limited, another approach, named cross-validation, is advisable. In this approach, the set of all observations is randomly split into two sets: a training one (for building a classifier) and a test one (for evaluation of the classifier's quality). The procedure is repeated several times in such a way that every observation is included in one of the test sets. The results of the quality evaluation are averaged. A comprehensive description of evaluation procedures of this type can be found in the book by Japkowicz and Shah (2015).

The procedure described above is based on the concept of *supervised learning*, in which it is assumed that in the training data set all labels attributed to observed data vectors are exactly known. However, in many practical situations, a process of assigning labels (labeling) to predictor vectors may be costly, and thus practically impossible. In such a case, we may have a large number of predictor vectors, but only a part of them (sometimes even small) has assigned respective labels. A real-life example of this situation is considered in the fourth section of this paper. To cope with this problem, one can use a *semi-supervised learning* approach. There exist many semi-supervised learning algorithms used for classification purposes. In this paper, we use a simple approach, described in Witten et al. (2011). This approach is a kind of the EM (expectation-maximization) procedure. At the first step of this procedure, a classifier is built using only labeled data vectors. Then, the obtained classifier is used to assign labels to the remaining unlabeled data vectors. Now, this extended labeled data set (with original and estimated labels) is used for the construction of a new classifier. Then, this new classifier is used for assigning new labels for vectors with originally unknown labels. The whole procedure is repeated iteratively several times until convergence. Unfortunately, this procedure does not guarantee the improvement of the classification accuracy, but in many cases, such improvement can be achieved. In Sect. 4, we present an example showing some benefits (but also some dangers) related to the application of a semi-supervised learning methodology.

2.1 Classifiers for Binary Classification—Logistic Regression

Binary classification is historically the oldest problem of machine learning. Since the works of R. Fisher on discriminant analysis, many different approaches have been proposed. The most popular groups of binary classification algorithms are decision trees and regression methods. Both approaches are implemented in machine learning software packages. In this paper, we use a classifier based on logistic regression, as it is easier for practical implementation. Moreover, according to Hastie et al. (2008), it is probably the best regression tool for the analysis of discrete data. In regression-based classifiers, in order to classify an object into one of two classes, 0 or 1, one has to calculate probabilities $P(Y = 1|X_1 = x_1, \ldots, X_m = x_m)$, and $P(Y = 0|X_1 = x_1, \ldots, X_m = x_m) = 1 - P(Y = 1|X_1 = x_1, \ldots, X_m = x_m)$. A new object described by a vector $(x_1^*, \ldots, x_m^*)$ of attributes (predictors) is classified to the class for which the respective probability is greater. In the linear logistic model, these probabilities are estimated through applying non-linear transformation to the linear function g (the so-called link function) of observed data

$$g(x_1, \ldots, x_m) = \beta_0 + \beta_1 x_1 + \cdots + \beta_m x_m. \tag{1}$$

Thus, it belongs to a class of generalized linear models (GLM). The probability of class 1 is calculated using the logistic function according to the formula

$$P(Y = 1|X_1 = x_1, \ldots, X_m = x_m) = \frac{1}{1 + \exp(g(x_1, \ldots, x_m))}. \tag{2}$$

Parameters $(\beta_0, \beta_1, \ldots, \beta_m)$ of this model have to be estimated from the training data consisting of n observations. This can be done by the maximization, with respect to $\beta_0, \beta_1, \ldots, \beta_m$, of the log-likelihood function (Witten et al. 2011):

$$\begin{aligned} L(\beta_0, \beta_1, \ldots, \beta_m) = \textstyle\sum_{i=1}^{n} & [(1 - y_i) \log(1 - P(Y = 1|x_1, \ldots, x_m)) \\ & + y_i \log(P(Y = 1|x_1, \ldots, x_m))]. \end{aligned} \tag{3}$$

To solve this problem one has either to solve the log-likelihood equations using the iteratively reweighted least squares method (see Hastie et al. 2008, for details) or to use any optimization tool (such as that available in MS Excel) for the direct maximization of (3). When we use available software packages we have to note that in some of them in the formula (2) the function $g(x_1, \ldots, x_m)$ is replaced by $-g(x_1, \ldots, x_m)$. If this takes place, the estimated values of the parameters $(\beta_0, \beta_1, \ldots, \beta_m)$ have opposite signs.

In the considered case of the two-state process, the training set used for building a classifier must consist of at least one sequence of instances when the process remains in the state 0, and at least one sequence of instances when the process remains in the state 1. When the process is in the state 0, we observe the results of classification equal to 0 with probability p_{00} equal to the estimated *Specificity* of the classifier, and the results of classification equal to 1 with probability $p_{01} = 1 - p_{00}$. On the other hand, when the process is in the state 1, we observe the results of classification equal to 1 with probability p_{11} equal to the estimated *Sensitivity* of the classifier. The probability of observing 0 in this state of the process is then equal to $p_{10} = 1 - p_{11}$.

2.2 *Combined Classifiers*

One of the most difficult practical problems with classification is missing data. Most classifiers need a sufficiently large training data set with all values of attributes known. However, in many practical cases, especially for medical data, it is hardly attainable. When the number of missing attribute data is small, one can use the so-called imputation methods to replace missing data with their appropriate predictions. In some situations, however, this cannot be done. Consider, for example, two subsets of attributes observed in periods of different lengths. If we want to build a classifier using the whole set of attributes, we would have to rely only on these time points when the observations from both subsets are available. In such case, we may lose valuable information from the time moments when the information from only one subset of attributes is available.

In this paper, we propose to consider, as an alternative to a usual classifier that uses all attribute data, a group of classifiers that are built using separate subsets of attributes. When the results of classification using these classifiers are fully concordant, there is no problem with the determination of the final classification result. When the accuracies of the classifiers are similar, one can use majority voting (with a supplementary rule for the case when the number of classifiers is even) for establishing a designated class. The problem begins when these results are not concordant and the accuracies of individual classifiers are significantly different. To avoid such problems, in this paper, we consider only two decision rules: the class is designated either if decisions of all classifiers are the same (strong confirmation) or if it is not possible to confirm strongly the alternative (i.e., there is at least one indication of the considered class). The choice of the applied rule should depend upon the possible consequences of an undertaken decision.

The calculation of the statistical properties of such combined classifiers is not straightforward, predictions are typically correlated. In general, the separate classifiers are jointly distributed according to the multivariate Bernoulli distribution (Dai et al. 2013). Mathematical representation of this distribution is complicated and involves many parameters to be estimated (see Teugels 1990). Thus, for a limited amount of available data general models are difficult for implementation. In this paper, we propose to use its simplification by assuming that the results of classification obtained by different separate classifiers are *equicorrelated*. By the "equicorrelation" we mean that for all pairs of classifiers the correlation coefficient ρ takes the same value. This model was considered by Gupta and Tao (2010) for solving multiple testing problems. Using their model we assume that we have the results of l classifiers $Y^1, \ldots, Y^l$, and that the probabilities of observing 1 are equal to $p_1, \ldots, p_l$, respectively.

Consider first the simplest case when we combine the results of only two classifiers. Let $Y^{(2)} = Y^1 + Y^2$, and $p_1 = P(Y^1 = 1)$, $p_2 = P(Y^2 = 1)$. Then (Gupta and Tao 2010):

$$P(Y^{(2)} = 0) = (1 - p_1)(1 - p_2) + \rho\sqrt{p_1 p_2 (1 - p_1)(1 - p_2)}, \tag{4}$$

$$P(Y^{(2)} = 2) = p_1 p_2 + \rho\sqrt{p_1 p_2 (1 - p_1)(1 - p_2)}, \tag{5}$$

and

$$P(Y^{(2)} > 0) = 1 - P(Y^{(2)} = 0), \tag{6}$$

where ρ is the correlation coefficient between random variables Y^1, Y^2.

When the results of $l = 3$ classifiers are combined, we have $Y^{(3)} = Y^1 + Y^2 + Y^3$, and $p_1 = P(Y^1 = 1)$, $p_2 = P(Y^2 = 1)$, $p_3 = P(Y^3 = 1)$. Then (Gupta and Tao 2010)

$$\begin{aligned} P(Y^{(3)} = 0) = (1 - p_3)P(Y^{(2)} = 0) + \rho[&\sqrt{p_1 p_3 (1 - p_1)(1 - p_3)}(1 - p_2) \\ &+ \sqrt{p_2 p_3 (1 - p_2)(1 - p_3)}(1 - p_1)], \end{aligned} \tag{7}$$

$$\begin{aligned} P(Y^{(3)} = 3) = p_3 P(Y^{(2)} = 2) + \rho[&\sqrt{p_1 p_3 (1 - p_1)(1 - p_3)}p_2 \\ &+ \sqrt{p_2 p_3 (1 - p_2)(1 - p_3)}p_1] \end{aligned}, \tag{8}$$

and

$$P(Y^{(3)} > 0) = 1 - P(Y^{(3)} = 0), \tag{9}$$

where in this case ρ is the correlation coefficient between random variables $Y^{(l)}$ and $Y^{(k)}$ for $\forall l \neq k$. Similar results can be obtained using the recursive formulae given in Gupta and Tao (2010) if a number of classifiers is greater than 3.

In the case of two classifiers, the estimation of ρ is straightforward. In the case of three (or more) classifiers, the situation is slightly more complicated. In this case, we use a basic result of probability, that for random variables $X_1, \ldots, X_l$ the following equality holds:

$$\mathrm{Var}\left(\sum_{i=1}^{l} X_i\right) = \sum_{i=1}^{l} \mathrm{Var}(X_i) + \sum_{i=1}^{l}\sum_{j\neq i}^{l} \mathrm{Cov}(X_i, X_j) \tag{10}$$

If our observation are standardized in such a way that $\mathrm{Var}(X_i) = 1, i = 1, \ldots, l$, and equicorrelated, than we have the following equation:

$$\mathrm{Var}\left(\sum_{i=1}^{l} X_i\right) = l + l(l-1)\rho \tag{11}$$

The estimation of ρ is now the following. First, for each considered classifier we calculate the standard deviation of the obtained results of classification. Then, we standardize these results by dividing them by the respective standard deviation, and for each element of the training set (with all values of attributes available!) we calculate the sum of these standardized values. Finally, we calculate the variance of these sums, insert it into (11), and solve this equation with respect to ρ.

In this paper, we consider the combination of two classifiers. The combination of a greater number of classifiers is, of course, possible, but in such case, the assumption of equicorrelation often becomes less adequate.

3 Monitoring the State-Related Observational Data

The results of the classification described in the previous section form a series of zero-one observations with probabilities of observing "ones" depending upon the actual state of a monitored process. From the perspective of statistical process control (SPC), we can consider the current state of a process as the in-control state, and the transition from this state to another one we can consider as the transition to the out-of-control state. Therefore, we can consider the monitoring of our process as the monitoring of a process with binary observations $Y_i, i = 1, 2, \ldots,$ where $Y_i \in 0, 1$.

SPC procedures for binary observations, also known as attribute data, are well-known. Basic procedures are described, e.g., in the book by Montgomery (2011), and references to other procedures can be found in Woodall (1997). For monitoring individual observations, and this is our case, Montgomery (2011) recommends CUSUM and EWMA control charts. Such control charts for attribute data are well-known, but they are designed under the assumption of independent (serially non-correlated) observations. Unfortunately, for processes considered in this paper, this assumption may not be satisfied. Therefore, we should look for a procedure applicable to serially dependent data.

In the case of serially dependent real-valued (variables) data, different control charts have been already proposed. They are based on mathematical models of real-valued time series, such as, e.g., the Box-Jenkins ARMA models. Unfortunately, in the case of discrete data, a direct application of this approach is not, as it was proved by Steutel and Van Haarn (1979), possible. To cope with this problem, several approaches have been proposed. They are described, e.g., in an overview paper by McKenzie (2003).

The most popular approach is based on the concept of "thinning" already introduced by Steutel and Van Haarn (1979). This approach is especially useful when the discrete time series has marginals defined on countable and infinite sets and described by such distributions like Poisson or geometric. When the marginal distribution is defined on a finite set, some additional assumptions must be made. For example, in the case of a binomial marginal distribution, an autoregression-type model was proposed by McKenzie (1985), and further generalized by Weiß (2009a). This model, which belongs to a more general class of INAR models, in the case of binary data is very simple.

Let $Y_t, t = 0, 1, \ldots$ be sequence of binary 0, 1 observations. Then, according to McKenzie's INAR(1) binary autoregression model we have

$$Y_t = A_t Y_{t-1} + B_t(1 - Y_{t-1}), \tag{12}$$

where A_t and B_t are sequences of i.i.d. random variables with $P(A_t = 1) = \alpha$ and $P(B_t = 1) = \beta$. The INAR(1) model (12) describes a two-state Markov chain with the autoregressive AR(1) structure. For random variables generated by (12) both conditional mean and variance are linear functions of Y_{t-1}. Let $\rho = \mathrm{Cor}(Y_t, Y_{t-1})$ and $p = P(Z = 1)$, where the random variable Z is distributed according to the Bernoulli distribution with parameter p. If $\beta = p(1 - \rho)$ and $\alpha = \beta + \rho$, then the marginal distribution of Y_t is the same as that of Z. Moreover, the autocorrelation function is equal to $\rho_X(k) = \rho^k, k = 0, 1, \ldots$.

An alternative approach for modeling integer-valued time series, historically an older one, was proposed in a series of papers written by Jacobs and Lewis. The models introduced by these authors are referred to as DARMA models. In this paper, we consider the simplest one, known as the DAR(1), proposed in Jacobs and Lewis (1978). In the DAR(1) model, consecutive observations are generated by the following mechanism:

$$Y_t = V_t Y_{t-1} + (1 - V_t) Z_t, \tag{13}$$

where V_t are i.i.d. random variables with $P(V_t = 1) = \rho$, and Z_t are i.i.d. random variables described by a certain (discrete) probability distribution π. If Y_0 is generated by π, then (13) describes a stationary process whose marginal distribution is π. Moreover, the expectation of $Y_t|Y_{t-1}$ is a linear function of Y_{t-1}, and the variance of $Y_t|Y_{t-1}$ is a quadratic function of Y_{t-1} (McKenzie 2003). In the case considered in this paper, random variables Z_t are distributed according to the Bernoulli distribution with $P(Z_t = 1) = p$. It was proved that the autocorrelation of Y_t defined by (13) is the same as in the case of the INAR(1) model. Moreover, the process described by the DAR(1) model is a two-state Markov chain, which is described by the following transition matrix (Jacobs and Lewis 1978):

$$T_{ij} = \begin{bmatrix} \rho + (1-\rho)(1-p) & (1-\rho)p \\ (1-\rho)(1-p) & \rho + (1-\rho)p \end{bmatrix}. \tag{14}$$

The same transition matrix describes the INAR(1) process for binary data. Therefore, although in general discrete processes described by the INAR(1) and the DAR(1) models have different correlation structures, in the case of binary data these models have the same properties. Thus, monitoring procedures designed under the assumption of any of these two models should be the same.

Control charts for correlated binary data have been proposed in several papers by Weiß. For example, he considered in Weiß (2009b) monitoring the INAR process with binomial marginals using three control schemes: moving average (MAV) chart, conditional control chart, and a chart based on runs. Some additional monitoring schemes based on less popular SPC statistics were also proposed in Weiß (2012). Note, that binary data considered in this paper are a special case of binomial data. Therefore, one can use the charts proposed in Weiß (2009b) to monitor such kind of data. Consider, for example, a moving average chart based on a data window of length m. Let $p = P(Y_t = 1)$, and ρ be the coefficient of autocorrelation of the monitored process. On the MAV chart we plot averages $\bar{Y}_t^{(m)} = (Y_{t-m+1} + \cdots + Y_t)/m$. The expected value of $\bar{Y}_t$ is equal to p, and the variance is given by the following formula (Weiß 2009b):

$$\mathrm{Var}\left(\bar{Y}_t^{(m)}\right) = \frac{p(1-p)}{m}\frac{1+\rho}{1-\rho}\left(1 - \frac{2}{m}\frac{\rho}{1-\rho^2}(1-\rho^m)\right). \tag{15}$$

A control chart based on k-sigma control limits has the central line $CL = p$, and control limits $LCL = \max\left(0, p - k\sqrt{\mathrm{Var}(\bar{Y}_t^{(m)})}\right)$, $LCL = \min\left(1, p + k\sqrt{\mathrm{Var}(\bar{Y}_t^{(m)})}\right)$. Note, that for the first $m-1$ observations the control limits should be appropriately recalculated.

Classical Shewhart control charts (with symmetric control limits) for binary data have poor properties when sample sizes are small. This is caused by the asymmetry of the probability distribution of the average of binary random variables. Simulation experiments reveal that this problem also exists in the case of the moving average chart proposed by Weiß (2009b) when the length of a data window is less than 10. In certain applications, it is not advisable to use a larger data window, as by increasing its length we decrease the rate of false alarms, but on the other hand, we increase the expected time to a real alarm. This happens, e.g., in a real-life problem considered in the next section of this paper. To avoid problems related to the asymmetry of data (Weiß 2009c), instead of original observations, considered their "jumps" defined as $J_t = Y_t - Y_{t-1}$, where Y_t is distributed according to the binomial distribution. He proved that the marginal distribution of J_t has all odd moments, including the mean and skewness, equal to 0. In the case of Bernoulli data, which is the special case of the problem considered in Weiß (2009c), the variance of J_t is equal to

$$V_J = 2p(1-p)(1-\rho). \tag{16}$$

Moreover, the autocorrelation function for the process J_t equals (Weiß 2009c).

$$\rho_J(k) = -\frac{1-\rho}{2}\rho^{k-1}, k = 1, 2, \ldots. \tag{17}$$

These properties were used by Weiß (2009c) in his proposal of a Shewhart-type control chart for dependent binomial data.

When the data are binary, the probability distribution of "jumps" J_t is symmetric, and the probability distribution of moving averages is also symmetric. Using simple, but tedious algebra, we have found that the variance of the moving average statistic based on J_t can be calculated from the following formula:

$$\mathrm{Var}(\bar{J}_t^{(m)}) = \frac{2p(1-p)(1-\rho)}{m^2}\left[m - (1-\rho)\rho^{m-1}\left(1-\frac{1}{\rho}\right)^{-2}\left(\frac{1}{\rho} + \left((m-1)\frac{1}{\rho} - m\right)\frac{1}{\rho^m}\right)\right]. \tag{18}$$

Unfortunately, for binary data, the usage of a control chart based on "jumps" is of very limited value. It is easy to show that in the case of a MAV chart only three distinct values of the plotted statistic are possible: $-1/m, 0, 1/m$. So if we set a standard deviation multiplier to a certain value only one of two solutions is possible: either to accept all data points or to react to nearly all changes in the results of classification. In the first case, we are not able to generate alarms, and in the second case, the rate of false alarms is usually not acceptable. In the case of observed "jumps", one can think about a control chart based on a moving variance. This solution might be useful in practice, but its theoretical justification requires future investigations.

Another control scheme proposed in the literature for binary data is based on the concept of runs. In processes described by Markov chains of the first order, such as the INAR(1), the conditional probability distribution of the run length is given by Weiß (2009b)

$$r(k) = p_{1|1}^{k-1}, k = 1, 2, \ldots, \tag{19}$$

where $p_{1|1} = P(Y_t = 1|Y_{t-1} = 1)$. Weiß (2009b) proposed a control chart based on this distribution. In practical applications, however, a simpler procedure is used. An alarm signal is generated when the maximal observed run exceeds a certain critical value. The probability of such an alarm can be calculated from (19). Jacobs and Lewis (1978) derived the probability generating function for the random variable T that describes the lengths of runs. This function may be transformed into the moment generating function which allows for calculating the moments of runs of a given order.

4 Application—The Case of Bipolar Disorder

Bipolar disorder (BD) is a mental illness affecting over 2% of the world's population (Grande et al. 2016). BD is characterized by episodic fluctuations between mood phases ranging from depression, through mixed and euthymic, to manic episodes. It is a chronic and recurrent disease having a serious impact on psychosocial functioning, cognition, and quality of life (Catala-Lopez et al. 2013). Also, the suicide rate of psychiatric patients suffering from BD is the highest among all mental disorders (Chen and Dilsaver 1996). Therefore, the detection of early symptoms of illness episodes, the so-called affective states, is crucial. Also, early treatment significantly decreases the severity of the symptoms leading to the improvement of the patient's well-being and reduction of the treatment costs.

The dynamic growth and spread of smartphone technology allow real-time monitoring of a patient in naturalistic settings through self-monitoring, as well as through the monitoring of automatically collected objective data, such as speech activities and behavioral activities (Faurholt-Jepsen et al. 2014). Consequently, smartphone apps enable monitoring of patient's behavior, social interactions, patient's voice signal, movements or changes of localization, and more. Moreover, the objective data collected via smartphones are correlated with scores on the depression (e.g., HDRS) and mania (e.g., YMRS) scales (Gruenerbl et al. 2015), and maybe used as a state marker for monitoring of illness activity in patients with BD. BD patients are generally open to the use of smartphones and wearables to help them monitor and assess their mental state (Schwartz et al. 2016). Recent research confirms that smartphones become an increasingly effective tool for the assessment of BD patients' affective state and early detection of a phase change (Faurholt-Jepsen et al. 2014; Gruenerbl et al. 2015), and changes in patient's behavioral activities captured with smartphone usage are regarded as potential sensitive measures of changing course of affective states (depression, mania, mixed state) in BD. However, the detection of the change of a patient state is a challenging task because the available data is collected from the smartphone and hence indirect.

In this paper, we pursue an alternative approach. We monitor the two-state process and try to detect a change of the process (patient's) state by such a semi-supervised learning approach (statistical process control). In this spirit, we follow the idea proposed by Kaczmarek-Majer et al. (2018) and by Hryniewicz et al. (2019a) and use statistical process control as an effective methodology to build patient-dependent models and generate alarms when the patient's behavior related to smartphone usage changes.

From a preliminary analysis of available data, performed using methods of data mining and the expertise of psychiatrists, we have chosen six predictors of the patient's state. Three of them (X_1—average daily number of phone calls, X_2—average daily length of phone calls, X_3—standard deviation of daily phone calls) were related to the phone activity of a patient. The remaining three (X_4—average daily number of SMSes, X_5—average daily SMS length, X_6—standard deviation of daily SMS length) were related to the patient's activity in writing SMSes. The choice of these predictors was further confirmed by preliminary statistical analysis.

At the first stage, we looked for an appropriate classifier. As the training set, we used observations from a sequence of 28 days when the monitored patient was in the state of mania (State 1), followed by observations from a sequence of 41 days when the monitored patient was in the state of euthymia (State 0). For these data, we

Table 2 Comparison of classifiers—sums of ranks

Classifier	10-fold CV	Test set	Total
LogReg	26	10	36
RandForest	23	29.5	52.5
RandTree	14	29.5	43.5
OneR	41	15	56
Jrip	30	24	54
AdaBoost	12	40	52

built several classifiers (logistic regression, random tree, random forest, OneR, Jrip, AdaBoost) using publicly available software WEKA. The quality of these classifiers was tested on the training set using the stratified ten-fold cross-validation, and on the test set consisting of data collected from a sequence of 37 days when the monitored patient was in the state of euthymia (State 0), followed by observations from a sequence of 12 days when the monitored patient was in the state of mania (State 1). Both data sets are available on request from the authors. For evaluation purposes, we used 6 quality measures available in WEKA. For each considered measure we ranked the evaluated classifiers. The sums of respective ranks are presented in Table 2.

The results of evaluation presented in Table 2 confirm the opinion of Hastie et al. (2008) that Logistic Regression is probably the best regression tool for the analysis of discrete data. Thus, we have chosen this classifier, described in details in Sect. 2.1, for further analyses.

From the analysis of all data contained in the training test, we have estimated the following link function:

$$g_f(x_1,\ldots,x_6) = 1.3929 - 0.0088x_1 - 0.0161x_2 + 0.0063x_3 + 0.0043x_4 - 0.1049x_5 + 0.0838x_6. \tag{20}$$

When we considered separately data related to phone calls and sent SMSes, where in both cases the number of days with valid data was greater than the number of days with valid data available for both subsets of attributes (predictors), the respective link functions are the following:

$$g_c(x_1,\ldots,x_3) = 0.2974 - 0.0036x_1 - 0.0185x_2 + 0.0078x_3 \tag{21}$$

$$g_s(x_4,\ldots,x_6) = 0.5547 + 0.0027x_4 - 0.1114x_5 + 0.0903x_6 \tag{22}$$

Then, we have inserted, respectively, (20)–(22) into (2), and obtained formulae for the calculation of the probability that a patient is in the state of mania. When this probability is greater or equal 0.5 we classify the patient's state as manic (State1). Otherwise, the state is classified as euthymic (State0).

In the first stage of building a control chart, we consider data collected from a patient being in the state of euthymia (State 0). By applying the already trained classifier we obtain a series of binary values. From these data, we calculate the estimated probability of obtaining the result of classification equal to 1, denoted by $p^\star$, and the estimated coefficient of autocorrelation of the observed results of classification, denoted by $\rho^\star$. These values are now used for the construction of a moving average chart with the window of length m. The central line of this chart is equal to $p^\star$. The standard deviation needed for the calculation of control limits is calculated as the square root of the variance calculated according to (15) with p and ρ replaced by their respective estimated values. Finally, we have to choose the value of the standard deviation multiplier.

When we considered the results of classification using the classifier based on all of the attributes, the estimated values of p and ρ were equal to 0.244 and 0.427, respectively. Then, these values have been used for the calculation of the standard deviation of the moving average statistic for $m = 3$, which in this case takes the value of 0.322. When we take a standard deviation multiplier equal to 3, as we usually do for Shewhart control charts, both control limits are beyond the interval [0, 1], and such a control chart is useless. Therefore, we have set $k = 2$, and obtained a valid upper control limit equal to 0.888. Taking k smaller than the usual value of 3 can have other justification. In the considered case, the empirical distribution of MAV statistic is strongly leptokurtic (the excess kurtosis is positive, and equal to 0.46). Thus, the probability mass of this distribution is more concentrated around the mean value than in the case of the normal distribution.

When we use the control chart designed this way for the analysis of classification data for the test period, we arrive at the results presented in Fig. 1. We can see that when the monitored patient is in the euthymic state (first 37 data points) there is no alarm. However, when the affective state changed to manic, this change was detected on the third day (i.e., when all data from days when the patient was in State 1 were used for the calculation of the MAV statistic). One can also notice the sequence of small values of the MAV statistic in the final sequence of observations. This is probably due to the effect of medical treatment, and indicates that the patient has returned to the stable euthymic state.

When combined classifiers are used for monitoring purposes the procedure is slightly different. First, the sequences of classification results are calculated on their respective data sets from the training period. For the classification obtained by the sub-classifier based on phone call data, the estimated probability of observing positive results of classification was equal to 0.22. For the classification obtained by the sub-classifier based on SMSes data, the analogous estimated probability was equal to 0.17. The correlation between the results of both classifications, estimated using data obtained at data points when both results of classification were available, was equal to 0.34. These values have been used for the calculation of positive results of classifications for two types of combined classifiers. The first combined classifier, coined as the CB1, yields the positive result of classification when at least one of two considered sub-classifiers gives a positive result. The second combined classifier, coined as the CB2, yields the positive result of classification when both considered sub-classifiers give a positive result. The probabilities of observing positive results of classification for the CB1 and CB2 classifiers were equal to 0.34 and 0.05, respectively. The remaining steps leading to the construction of MAV control charts for the considered combined classifiers are the same as in the case of the classifier based on all observed attributes.

In the case of the CB1 classifier, the autocorrelation of the results of classification (on training data) was equal to −0.06, and the standard deviation for the MAV statistic (for $m = 3$) was equal to 0.263. Hence, the upper control limit for the MAV chart, and $k = 2$, was equal to 0.858. The results of the application of this chart to the test classification data obtained using the combined CB1 classifier are presented in Fig. 2. We can see that the monitored process plotted on the MAV chart is nearly the same as in the case of the classifier based on all attributes. The only difference is for one time point where a false alarm in the state of euthymia was observed.

The performance of the MAV chart based on the CB2 classifier is significantly different. The autocorrelation of the results of classification (on training data) was equal to −0.05, and the standard deviation for the MAV statistic (for $m = 3$) was equal to 0.12. This standard deviation is significantly smaller than in the previous

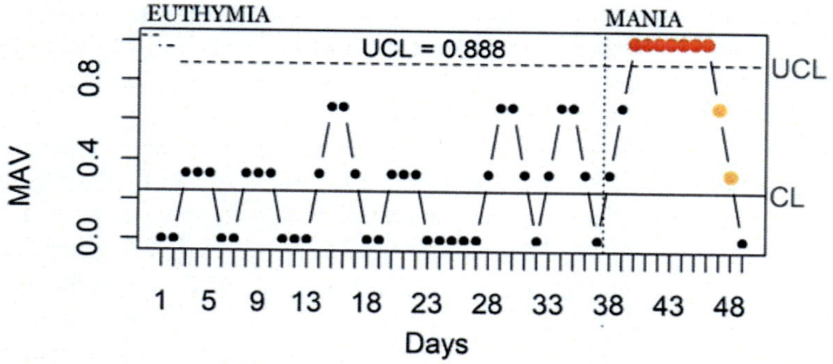

Fig. 1 MAV-3 control chart—classification using all 6 attributes, two-sigma control limits

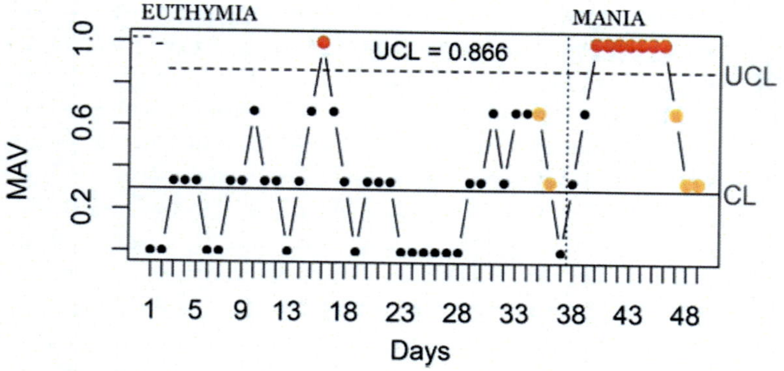

Fig. 2 MAV-3 control chart—classification using the CB1 classifier, two-sigma control limits

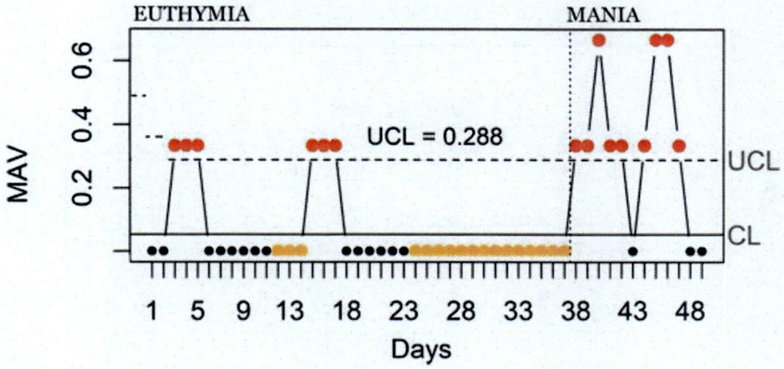

Fig. 3 MAV-3 control chart—classification using the CB2 classifier, two-sigma control limits

cases. Therefore, the upper control limit for the MAV chart, for the same value of m, is significantly smaller than for the previously considered MAV charts. Thus, for $k = 2$ one can observe 6 false alarms (in 2 sequences of 3 signals each), and real alarms starting on the very first day of the change of state. This means that in the case when false alarms are less costly than the delay in the revealing of the real change of state this chart outperforms the previous charts. The results of the application of this chart to the test classification data obtained using the combined CB2 with $k = 2$ classifier are presented in Fig. 3. We can see that the monitored process plotted on the MAV chart is nearly the same as in the case of the classifier based on all attributes. The only difference is for a one time point where a false alarm in the state of euthymia was observed.

When we set $k = 3$ the situation is different. We observe no false alarms, but also few real alarms, as presented in Fig. 4. Control charts with three-sigma control limits are preferred when costs of false alarms are high, and delay times of true alarms are not critical. In the problem of monitoring BD patients, considered in this paper, this is not the case. Therefore, the control chart with two-sigma limits based on the CB2 classifier is a valuable alternative to the control chart based on the classifier that uses all observed predictors.

The main problem in building classifiers used for the determination of the current state of BD patients is related to the labeling of data from a training set. Such labels can be assigned only by psychiatrists. One can assume that the diagnose is with great probability correct for a few days around the day in which a personal contact between a patient and a physician took place. For other days the labels that are assigned to sets of context variables are less probable. Therefore, in a period in which we collect training data, some labels may be considered as known, but the remaining labels can be considered as unknown. This is exactly the problem the semi-supervised learning is dealing with.

In the considered case of a BD patient, some labels assigned by psychiatrists are less credible. Therefore, in building our classifier, we can consider them as unknown. In this particular example we have nearly 60% of unknown labels. When we take

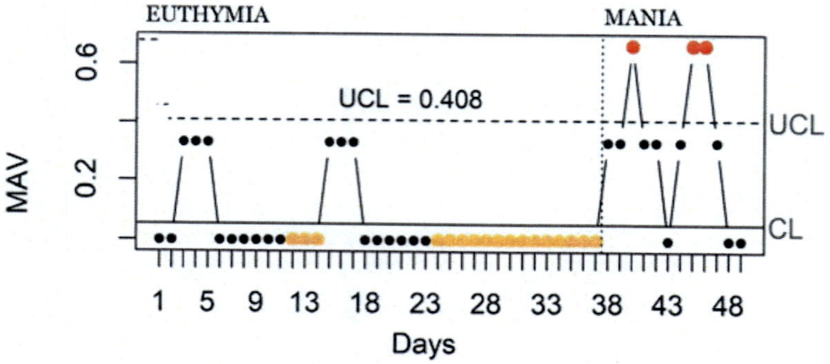

Fig. 4 MAV-3 control chart—classification using the CB2 classifier, three-sigma control limits

into account only the labeled data, the logistic regression model is described by the following link function:

$$g_f(x_1, \ldots, x_6) = -1.5985 + 0.0060x_1 - 0.0135x_2 + 0.0031x_3 - 0.0090x_4 + 0.1274x_5 \\ -0.0077x_6. \tag{23}$$

This model is used as the starting point of the iterative process described in Sect. 2. After 5 steps the iteration process stabilizes, and the final logistic regression model, obtained using the semi-supervised learning methodology, is given by the following link function:

$$g_f(x_1, \ldots, x_6) = 2.3422 - 0.0298x_1 - 0.0030x_2 - 0.0018x_3 + 0.0010x_4 - 0.1359x_5 \\ +0.1064x_6. \tag{24}$$

Now, let us compare both classifiers, i.e., one based on a subset of the whole training set for which we have fully known data (denoted by SL), and the other, with the partially known data, where unknown labels in the training set are estimated using the semi-supervised learning approach (denoted by SSL). For the evaluation of the SSL classifier on the full training set (SSL-F), we assumed that all the labels in this set are known. For the SL classifier, we consider two cases. In the first case, the classifier is evaluated on the subset of the training test for which all labels are known (SL-K), and in the second case, the classifier is evaluated on the full training data set (SL-F). Then, we evaluated both these classifiers on the test set for which we assume that all labels are known. The results of the comparison are presented in Table 3. The results of the comparison presented in Table 3 show that for the training set with records labeled by known labels the classifier built on this particular data (SL-K) has better quality than the classifier built on the extended training data set with known and unknown labels (SSL-F), applied to the set with known labels. The quality of classification significantly deteriorates when we apply this classifier to the full training data set (SL-F). When applied to the test data, the SSL classifier performs much better. A possible explanation of this is the following: the training

Table 3 Comparison of SL and SSL classifiers

	Training data			Test data	
	SSL-F	SL-K	SL-F	SSL	SL
Accuracy	0.6232	0.6333	0.4493	0.7959	0.4286
Precision	0.5714	0.6522	0.4138	0.5625	0.2778
Sensitivity	0.2857	0.8333	0.8571	0.7500	0.8333
Specificity	0.8537	0.3333	0.1707	0.8108	0.2973
F1	0.3810	0.7317	0.5581	0.6429	0.4167

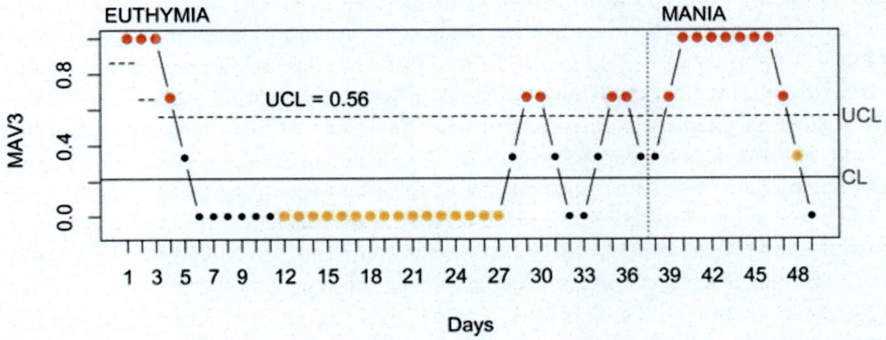

Fig. 5 MAV-3 control chart—classification using the SSL classifier, two-sigma control limits

set with known labels is not very similar neither to the full training set nor to the test set. However, the extended training set is much more similar to the test set, and thus the results of classification using the SSL classifier built using all training data are better.

Good properties of a classifier built on the extended training data set using an SSL classifier do not mean, unfortunately, that a control chart based on predictions made by these classifiers performs better than a control chart based on complete data. In Fig. 5, we can see an excessive rate of false alarms. The alarm signals at the beginning of the Euthymia (In-control) period can be explained as artifacts of the previous health state of the patient. The remaining false alarms are due to the low value of a control limit. This value is low because in the training period the results of a classification in the state of Euthymia are less correlated than those observed in the test period. Moreover, the percentage of false alarms is also slightly lower. This results in the underestimation of the variability of charted data. One can also use these findings as a kind of warning, that control limits of a chart should be calculated from a period that is not used for building a classifier. Unfortunately, in the considered case the amount of available data was too small to follow this recommendation.

5 Conclusions and Future Research

This is one of the first studies with application of the statistical process control methodology in the context of supporting the smartphone-based monitoring of bipolar disorder patients and illustrates monitoring of the patient's (indirect) state using (directly observed) smartphone-based daily aggregates about calling and texting (sending SMSes). The proposed approach is illustrated with real-life data captured from the observational study. The practical motivation of the proposed approach is to detect the state change of a BD patient based on the newly collected indirect data for each patient. We have proposed a monitoring procedure which is based on the combination of different approaches from such areas as data mining and statistical process control. In particular, the transition from observable data to the predictions of the interesting, latent characteristics was based on supervised learning (classification) or semi-supervised learning (classification for which a share of labels are missing). The proposed procedure appears to be effective in the considered practical case. During its implementation, several questions have arisen that need future investigations.

The first group of such questions is related to the classification methodology. It is well-known that some types of classifiers, commonly considered as very good (such as, e.g., Random Forests), for small training data sets are characterized by high generalization errors. This phenomenon is also known as "overfitting", as such classifiers yield perfect classification of training data, but fail when applied for other data sets of the same type. The results presented in Table 2 clearly describe this problem. This is exactly the case when we applied such classifiers for the data from our training data set. For such "perfect" classification, it is not possible to propose a control chart, as the standard deviation of the results of classification is equal to zero. As such situations may happen in practice, it is advisable to consider the application of more sophisticated classifiers obtained using the method of regularization (for regression-type classifiers) or tree pruning (for decision-tree classifiers). The application of such classifiers for small data sets also requires future investigation. Another problem is related to the time structure of classified data. For the most popular classifiers, the possible serial correlation of consecutive results of classification is not taken into account. Considering such possible autocorrelation structure may lead to better results of classification. One can also think about the application of the so-called "one-class" learning. These methods are used for the detection of outliers or other anomalies. Therefore, one can think about their sequential application for process data. This approach is under investigation in our current works.

The second group of questions is related to problems of monitoring. In this research, we have used moving average (MAV) charts. The main reason for doing this is the availability of closed-form formulae that can be used for the design of charts. However, the application of other charts, like EWMA or CUSUM, should also be considered in future investigations. One should also consider the application of charts based on runs, as these charts seem to be very natural for the considered type of data. Another problem, which cannot be forgotten is related to the size of the training data. Our predicted charted data are binary. It is a well-known fact, that for this type of data the sample used for the design of a control chart cannot be small, for example, in the real-life example considered in this paper.

In this paper, we have noticed that MAV control charts based on "jumps' are rather not applicable for the monitoring of binary data. However, one can think about a monitoring procedure based on the moving variance. This approach may be useful in the case of Bipolar Disorder when we take into consideration the so-called *mixed* state of health. In this state, we observe a mix of symptoms of depression and mania. Preliminary investigations show that this state is difficult for identification when we use only average values of observed random variables, and the consideration of data variability may lead to more efficient monitoring.

From a medical point of view, the results presented in this paper may be viewed upon as "proof of concept". More generally applicable results may be expected if we use a more diversified set of observations collected from different patients.

Ethical Issues

The study obtained the consent of the Bioethical Commission at the District Medical Chamber in Warsaw (agreement no. KB/1094/17).

Acknowledgements The study was submitted to the Office for Registration of Medicinal Products, Medical Devices and Biocidal Products in accordance with Polish law. This work was partially financed from EU funds (Regional Operational Program for Mazovia)—a project entitled "Smartphone-based diagnostics of phase changes in the course of bipolar disorder" (RPMA.01.02.00-14-5706/16-00). The authors thank psychiatrists that participated in the observational trial for their commitment and advice. The authors thank the researchers Weronika Radziszewska and Anna Olwert from Systems Research Institute, Polish Academy of Sciences for their support as well as Monika Dominiak and Professor Łukasz Święcicki for their support during the clinical assessment and for their advice about bipolar disorder.

References

Catala-Lopez, F., Genova-Maleras, R., Vieta, E., & Tabares-Seisdedos, R. (2013). The increasing burden of mental and neurological disorders. *European Neuropsychopharmacology*, *23*, 1337–9.

Chen, Y. W., & Dilsaver, S. C. (1996). Lifetime rates of suicide attempts among subjects with bipolar and unipolar disorders relative to subjects with other axis I disorders. *Biological Psychiatry*, *39*, 896–899.

Dai, B., Ding, S., & Wahba, G. (2013). Multivariate Bernoulli distribution. *Bernoulli*, *19*, 1465–1483.

Duda, R., Hart, P., & Stork, D. (2000). *Pattern classification* (2nd ed.). New York: Wiley-Interscience.

Faurholt-Jepsen, M., Frost, M., Vinberg, M., Christensen, E. M., Bardram, J. E., & Kessing, L. V. (2014). Smartphone data as objective measures of bipolar disorder symptoms. *Psychiatry Research*, *217*, 124–127.

Grande, I., Berk, M., Birmaher, B., & Vieta, E. (2016). Bipolar disorder. *The Lancet*, *387*(10027), 1561–1572.

Grünerbl, A., Muaremi, A., Osmani, V., Bahle, G., Oehler, S., Tröster, G., et al. (2015). Smartphone-based recognition of states and state changes in bipolar disorder patients. *IEEE Journal of Biomedical and Health Informatics*, *19*(1), 140–148.

Gupta, R. C., & Tao, H. (2010). A generalized correlated binomial distribution with application in multiple testing problems. *Metrika*, *71*, 59–77.

Hastie, T., Tibshirani, R., & Friedman, J. (2008). *The elements of statistical learning: Data mining, inference, and prediction* (2nd ed.). New York: Springer.

Hryniewicz, O. (2015). SPC of processes with predicted data: Application of the data mining methodology. In Knoth, S., & Schmid, W. (Eds.) *Frontiers in statistical quality control* (Vol. 11, pp. 219–235). Heidelberg: Springer.

Hryniewicz, O., Kaczmarek-Majer, K., & Opara, K. R. (2019). Control charts based on fuzzy costs for monitoring short autocorrelated time series. *International Journal of Approximate Reasoning*, *114*, 166–181.

Hryniewicz, O., Kaczmarek-Majer, K., & Opara, K. R. (2019). Monitoring two-state processes using indirectly observed data. In *Proceedings of 12th International Workshop on Intelligent Quality Control (ISQC)*. Hong Kong, August 2019, City University of Hong Kong.

Jacobs, P. A., & Lewis, P. A. W. (1978). Discrete time series generated by mixtures I: Correlational and runs properties. *Journal of the Royal Statistical Society, Series B 40*, 94–105.

Japkowicz, N., & Shah, M. (2015). *Evaluating learning algorithms*. Classification perspective. New York: Cambridge University Press.

Kaczmarek-Majer, K., Hryniewicz, O., Opara, K., Radziszewska, W., Owsinski, J., & Zadrozny, S. (2018). Control charts designed using model averaging approach for phase change detection in bipolar disorder. In: Destercke, S., Denoeux, T., Gil, M., Grzegorzewski, P., & Hryniewicz, O. (Eds.) *Uncertainy modelling in data science*. Advances in Intelligent Systems and Computing (Vol. 832, pp. 115–123). Springer International.

McKenzie, E. (1985). Some simple models for discrete variate time series. *Water Resources Bulletin*, *21*, 645–650.

McKenzie, E. (2003). Discrete variates time series. In Shanbhag, D. N., & Rao, C. R. (Eds.) *Handbook of statistics* (Vol. 21, pp. 573–605). Amsterdam: Elsevier Science B.V.

Montgomery, D. C. (2011). *Introduction to statistical quality control* (6th ed.). New York: Wiley.

Powers, D. M. W. (2011). Evaluation: From precision, recall and F-measure to ROC, informedness, markedness & correlation. *Journal of Machine Learning Technologies*, *2*, 37–63.

Schwartz, S., Schultz, S., Reider, A., & Saunders, E. F. H. (2016). Daily mood monitoring of symptoms using smartphones in bipolar disorder: A pilot study assessing the feasibility of ecological momentary assessment. *Journal of Affective Disorders*, *191*, 88–93.

Steutel, F. W., & Van Haarn, K. (1979). Discrete analogues of self-decomposability and stability. *The Annals of Probability*, *7*, 893–899.

Teugels, J. L. (1990). Some representations of the multivariate Bernoulli and binomial distributions. *Journal of Multivariate Analysis*, *32*, 256–268.

Weiß, C. H. (2009). A new class of autoregressive models for time series of binomial counts. *Communications in Statistics - Theory and Methods 38*, 447–460.

Weiß, C. H. (2009). Monitoring correlated processes with binomial marginals. *Journal of Applied Statistics*, *36*, 399–414.

Weiß, C. H. (2009). Jumps in binomial AR(1) processes. *Statistics and Probability Letters*, *79*, 2012–2019.

Weiß, C. H. (2012). Continuously monitoring categorical processes. *Quality Technology and Quantitative Management*, *9*, 171–188.

Witten, I. H., Frank, E., & Hall, M. A. (2011). *Data mining: Practical machine learning tools and techniques* (3rd ed.). Amsterdam: Elsevier.

Woodall, W. H. (1997). Control charts based on attribute data: Bibliography and review. *Journal of Quality Technology*, *29*, 172–183.

Monitoring Image Processes: Overview and Comparison Study

Yarema Okhrin, Wolfgang Schmid, and Ivan Semeniuk

Abstract In this paper, an overview of recent developments on monitoring image processes is presented. We consider a relatively general model wherein the in-control state spatially correlated pixels are monitored. The control charts described are based on non-overlapping regions of interest. This leads to a dimension reduction but, nevertheless, we still face a high-dimensional data set. We consider residual charts and charts based on the generalized likelihood ratio (GLR) approach. For the calculation of the control statistic of the latter chart, the inverse of the covariance matrix of the process must be determined. However, in a high-dimensional setting, this is time consuming and moreover, the empirical covariance matrix does not behave well in such a case. This is the reason why two further control charts are considered which can be regarded as modifications of the GLR statistic. Within an extensive simulation study, the presented control charts are compared with each other using the median run length as a performance criterion.

Keywords Statistical process control · Statistical image analysis · Image monitoring · High-dimensional data

1 Introduction

Image analysis deals with the extraction of meaningful information from images (mainly, from digital images) by means of digital image processing techniques. The main aim is the improvement of pictorial information for human interpretation and

Y. Okhrin
Department of Statistics, University of Augsburg, Augsburg, Germany
e-mail: yarema.okhrin@wiwi.uni-augsburg.de

W. Schmid (✉) · I. Semeniuk
Department of Statistics, European University Viadrina, Frankfurt(Oder), Germany
e-mail: schmid@europa-uni.de

I. Semeniuk
e-mail: semeniuk@europa-uni.de

S. Knoth and W. Schmid (eds.), *Frontiers in Statistical Quality Control 13*,
Frontiers in Statistical Quality Control,
https://doi.org/10.1007/978-3-030-67856-2_9

processing of image data for tasks such as storage, transmission, and extraction of pictorial information (cf. Gonzalez and Woods 2018). Since the resolution of digital cameras has dramatically increased in recent years and the number of possible applications is diverse, the analysis of image data has become a popular field of research.

A digital image consists of a number of pixels, which can be considered as a realization of a stochastic process. Several statistical approaches have been proposed to model image processes as, e.g., Kalman filtering, Markov random fields, hidden Markov processes, and Bayesian approaches. An overview is given in, e.g., Fieguth (2010). These methods play an important role in feature extraction, image pattern classification, etc. Since the number of pixels in a digital image is usually huge, we are faced with a high-dimensional data set.

The objective of this paper is to monitor an image process over time. Such problems can be found in many fields of application. Within Industry 4.0 completely new measurement methods have been introduced as, e.g., sensors that regularly take photos from the production process. This method is frequently applied within a 3D printing process (cf. Colosimo 2018). A similar problem can be observed within the printing process of a journal, where the brightness of the cover should be checked for changes. We can also find many examples in medicine as, e.g., the early detection of tumors and vascular changes. In all of these applications, the aim is of course to detect any deviations as soon as possible after their occurrence. Consequently, the number of possible applications is huge but, surprisingly, there are only a few papers dealing with image monitoring.

Among the first to apply control charts to image data were Horst and Negin (1992). Their purpose was to improve the productivity of web process applications. Armingol et al. (2003) took into account illumination changes through a transformation of the pixel values of the image. They constructed individual moving-range control charts for each pixel. A disadvantage of their approach is that the correlation structure of neighboring pixels is not taken into account. Nembhard et al. (2003) combined control charts for variable data with the EWMA control chart. Hotelling's T^2 control chart was widely applied in image analysis, e.g., by Mason et al. (1997), Tong et al. (2005), Liu and MacGregor (2006). Lin (2007a, b) combined multivariate control charts and wavelets to detect defects in electronic components. Lin et al. (2008) compared a wavelet and Hotelling's T^2 control chart with a wavelet and a principal component approach in detecting defects in LED chips. Jiang et al. (2005) used a spatial exponentially weighted moving average chart to find defects in LCD monitors and Lu and Tsai (2005) used a spatial $\bar{x}$ chart for the same application.

An overview on control charting with images is given in Megahed et al. (2011). In Koosha et al. (2017), a nonparametric regression method using wavelet basis functions is developed to extract features from grayscale image data. The extracted features are monitored over time to detect out-of-control situations using a generalized likelihood ratio control chart. Methods of machine learning have been used to monitor image processes as well. For example, Rafajłowicz and Rafajłowicz (2017) apply the K-medoids clustering algorithm for colored RGB images.

Otto and Seckmeyer (2019) consider overlapping regions of interest and introduce a multivariate EWMA chart taking into account correlated pixels.

Recently, Okhrin et al. (2019) derived several new control charts based on the generalized likelihood ratio approach. In their paper, they take the spatial correlation of the pixels into account. Their approach uses non-overlapping regions of interest.

The main purpose of the present paper is to give an overview of the topic and to compare the charts proposed by Okhrin et al. (2019) within an extensive simulation study.

The paper is structured as follows. In Sect. 2, we give a brief introduction into image analysis and statistical image analysis. Here the statistical model is explained which will be used in the rest of the paper. In Sect. 3, we present some control charts, the residual approach and the charts introduced in Okhrin et al. (2019). In particular, the high-dimensional setting of the underlying problem is discussed in detail. Section 4 provides a comparison study of the procedures discussed in Sect. 3. Several out-of-control situations are treated. As a measure of performance, the median run length is used. It turns out that there is no chart that dominates the others in all of the considered situations. However, an overall good performance can be observed for the chart, which is based on the generalized likelihood ratio approach.

2 Image Analysis

In this section we briefly describe some basic concepts of image processing with a focus on statistical image processing. More details can be found in, e.g., Fieguth (2010), Gonzalez and Woods (2018), Réfrégier and Goudail (2013), Sonka et al. (2014).

2.1 Digital Image Fundamentals

From the mathematical point of view, an image can be seen as a function $f : D \to W$ with $D \subset I\!R^2$ for a 2-dimensional (2D) image and $W \subset I\!R^k$. Frequently, D is a rectangle. For a black-white image, W consists only of two values (usually 0 and 1, 0 stands for black color and 1 for white color). For an 8-bit color image, $f(x, y)$ is a vector of 3 individual components RGB (red, green, blue) with values between 0 and $2^8 - 1 = 255$ for each component (cf. Sonka et al. 2014). If all elements are equal to zero, we obtain a black image, and if all are equal to 255, the resulting image is white. Today, it is common to work with 24-bit or 32-bit images, which provide a much wider variety of colors. To standardize the operations on images, the values are typically rescaled to $[0, 1]$. If all components are set equal, which is equivalent to $k = 1$, we obtain a grayscale image. In the following, we will exclusively deal with grayscale images.

In order to process an image, it must be represented by a discrete data structure. A digital image can be obtained from an image by sampling and quantization (e.g., Sonka et al. 2014; Gonzalez and Woods 2018). It is given by $f(i\Delta x, j\Delta y)$ for $i = 1, ..., l,\ j = 1, ..., m$, where Δx and Δy are geometric length and width of the area of interest. f is called intensity function and the values of f are the intensities of the corresponding pixels. In the following we will use the shorter notation $f(i, j)$ instead of $f(i\Delta x, j\Delta y)$. We consider a digital image, further briefly image, that has l rows and m columns of pixels. Thus, the image can be written as an array

$$\begin{bmatrix} f(1,1) & \dots & f(1,m) \\ f(2,1) & \dots & f(2,m) \\ \vdots & \ddots & \vdots \\ f(l,1) & \dots & f(l,m) \end{bmatrix}.$$

Nowadays, the resolution of 4K HD television is up to 4096×2160 pixels and the resolution of images taken by high-end smartphones up to 4032×3024 pixels.

In order to sharpen and smooth an image, it is usually pre-processed. There are several tools available for doing this. Using an intensity transformation, the whole figure can be transformed as a unity, while spatial filtering makes use of local smoothing. Many approaches provide a transformation of the image into its frequency domain, e.g., the Fourier-related transforms, the Walsh–Hadamard transform, wavelet transforms, discrete cosine transform, etc. A detailed overview on various approaches are given in Gonzalez and Woods (2018), Sonka et al. (2014).

2.2 *Statistical Image Analysis*

Real images are influenced by random errors. The errors may have different sources such as technical issues, lighting, particles in the air, instability of the object, etc. Thus, an image could be treated as a realization of a stochastic process. Several proposals have been made in the literature to model image processes like, e.g., Kalman filtering, Markov random fields, hidden Markov processes, Bayesian approaches, etc. An overview can be found in, e.g., Fieguth (2010).

In the following, we want to make use of the linear error model

$$\tilde{Y}(i, j) = f(i, j) + \varepsilon(i, j)\ ,\ i = 1, ..., l,\ \ j = 1, ..., m. \tag{1}$$

This means that there is only an additive noise influencing the pixel intensities $f(i, j)$ in the image. We will assume that the intensities $\{f(i, j)\}$ and the error variables $\{\varepsilon(i, j)\}$ are orthogonal.

Frequently, the random variables $\varepsilon(i, j),\ i = 1, ..., l,\ \ j = 1, ..., m$ are assumed to be spatially independent and normally distributed with mean 0 and variance σ^2. In that case, the image has usually been pre-processed and the error quantities explain the deviation between the observed pixel intensities and the smoothed values, which are also described as the nominal image. This is of course a great restriction and it is not fulfilled in many applications. Gonzalez and Woods (2018) describe several cases where this approach does not work (quantum-limited imaging, such as in X-ray, nuclear-medicine imaging, etc.).

In this paper, we consider a more general model and we do not make use of the independence assumption for the noise process, since in our opinion it is too restrictive. We assume that the error quantities follow a matrix-valued distribution (Gupta and Nagar 2018). Such an approach was also made in Rafajłowicz (2018). Here we make use of the matrix-valued normal distribution assuming certain types of covariance matrices as it is done in spatial statistics (cf. Cressie 1992; Cressie and Wikle 2015). Consequently, the pixel intensity process is spatially correlated.

3 Monitoring Procedures for the Pixel Process in the Time Domain

Subject of this paper is to monitor an image process over time. Such problems arise in many fields of application. In environmetrics, satellite data provide a good possibility to monitor a large area with the aim to detect forest fires, changes in glaciers, floods, etc. Within Industry 4.0, completely new measurement methods have been introduced, e.g., sensors. There are sensors taking photos from the production process and this method is frequently applied within a 3D printing process (cf. Colosimo 2018). This problem is also related to the printing process of a journal where the brightness of the cover should be checked for changes. We can find many examples in medicine, such as the early detection of tumors, vascular changes, etc. Further, a military unit is interested in detecting foreign aircraft in its own airspace. Thus, the number of possible applications is very diverse. However, the field is still at an early stage.

In principle, it is possible to monitor an image in time and in frequency domains. Here we want to focus on monitoring procedures in the time domain only. For analysis in the frequency domain, the image would be firstly transformed by a suitable transformation as, e.g., wavelets, etc. But then the question arises how a change in the original process will be influenced by the transformation. We will not discuss this topic here.

The aim of a monitoring procedure is to detect a significant change as soon as possible after its occurrence. Such problems are subject of statistical process control (SPC, cf. Montgomery 2009). The most important tools in SPC are control charts. Control charts have been widely used in engineering. In that context, the process is mostly assumed to be univariate and independent over time. In the present case, we have a multivariate process. Control charts for multivariate independent processes have been studied by various authors. The first control chart for independent and multivariate normally distributed random vectors was derived by Hotelling (1947).

It is based on the Mahalanobis distance between the observations and the target mean vector. Lowry et al. (1992), Sparks (1992) extended the EWMA chart to multivariate data by using a multivariate EWMA recursion. Further generalizations of the EWMA chart were given by, e.g., Fassó (1999), Hawkins et al. (2007). There are several extensions of the univariate CUSUM scheme to the multivariate case. Because the direct application of the sequential probability ratio test (SPRT) of Wald to independent multivariate normally distributed variables leads to a control chart which is not directionally invariant (cf. Healy 1987), i.e., the distribution of its run length in the out-of-control state depends as well on the direction as on the magnitude of the change, several authors proposed control schemes having this desirable property like, e.g., Crosier (1988), Pignatiello Jr and Runger (1990), Ngai and Zhang (2001).

Unfortunately, these approaches cannot be directly applied to monitor an image process since an image is a high-dimensional data set and these methods have been introduced for small and medium dimensions. Thus, it is necessary either to modify these approaches or to introduce new ones.

Nowadays, all monitoring procedures of image processes are based on aggregated characteristics of an image, such as, e.g., entropy, spatial entropy, means, etc. The reason for considering these characteristics is that a pixel to pixel analysis leads to dramatic theoretical and computational problems. Particularly, since we assume correlated pixels within an image, such an analysis can only be successful if the underlying covariance matrix of the pixels has a specific structure. Else, the number of parameters with respect to the available number of data is huge and the analysis suffers from the curse of dimensionality.

In order to simplify the problem, sub-images, the so-called regions of interest (ROIs), are usually considered (e.g., in Megahed et al. 2012; Koosha et al. 2017). They are obtained by splitting the whole image into smaller sub-images. The local characteristics for ROIs are usually defined as local measures of location and local measures of variation. Our aim is to monitor these local characteristics. The advantage of introducing sub-images is of course that the dimension behind the problem is reduced. Nevertheless, we are faced with a high-dimensional problem and the classical control charts for multivariate processes cannot be applied. In order to get efficient control charts, the methods of multivariate process control have to be combined with the newest results on high-dimensional data analysis and on spatiotemporal statistics.

Monitoring the ROIs offers two further advantages compared to monitoring classical scalar or multivariate data. First, if a control chart signals, we can use the information for the individual ROIs to identity the approximate location of the change. Thus, we have not only a hint that a change occurred, but also where exactly it has happened. This is a crucial information in practice. Second, if we know the position of the change, we can also determine its magnitude. In the next step, this can be used to identify the causes of the shift and later to adjust the production line appropriately. The practitioner gets useful insights into the causes and the extent of the faults.

3.1 Model

Our aim is to detect whether there has been a change within an image process. Since the image process may be quite complicated, we focus in this paper on the situation that the main subject of the image is fixed and not changing over time. For instance, the image could show the cover of a journal taken by a static camera within the printing process and our aim is to detect changes in its brightness.

Next, we consider images over time. Let

$$\tilde{Y}_t(i,j) = f(i,j) + \varepsilon_t(i,j), \quad i = 1, ..., l, \quad j = 1, ..., m \tag{2}$$

be the image process at time $t \geq 1$, $\tilde{Y}_t(i,j)$ are the elements of $l \times m$ matrix $\tilde{\mathbf{Y}}_t$. Since the image is static, it follows that $f(i,j)$ does not change over time and does not depend on t. The quantity $\varepsilon_t(i,j)$ denotes a noise process, which is assumed to be independent over time but not over space. Thus, we are working under much weaker assumptions than it is usually done in literature, where mostly independence over space is assumed too. Our intention is to derive suitable control procedures for such a situation. A more general case of a non-static subject will be treated in a forthcoming paper.

The change point model is defined as

$$\tilde{\mathbf{X}}_t = \begin{cases} \tilde{\mathbf{Y}}_t & \text{for } t < \tau \\ \tilde{\mathbf{Y}}_t + \mathbf{A} & \text{for } t \geq \tau \end{cases} \tag{3}$$

for $t \geq 1$ with $\mathbf{A} \neq \mathbf{0}$ and $\tau \in \mathbb{N} \cup \{\infty\}$. If $\tau = \infty$ we say that the image process is in control and no change has happened. If $\tau < \infty$ then the image process is out of control starting from time point τ. Sometimes $\{\tilde{\mathbf{X}}_t\}$ is also called the observed process and $\{\tilde{\mathbf{Y}}_t\}$ the target process.

In practice, it is usually assumed that a pre-run of the target process is given which is used to estimate f. We assume that f is known. We will not discuss the influence of parameter estimation within this paper. This will be done in a forthcoming contribution.

3.2 Residual Charts

Taking into account all pixels for the analysis is a very challenging task, since a simple cell phone image nowadays already consists of around 4 million pixels and thus, we have to monitor a process with 4 million components over time.

In order to reduce the complexity of the problem, the image is partitioned into sub-images (regions, ROIs) I_{ij} of size $h_1 \times h_2$, $i = 1, ..., r_1$, $j = 1, ..., r_2$ assuming that $l = h_1 r_1$ and $m = h_2 r_2$, $r_1, r_2 \in \mathbb{N}$. Figure 1 describes this partitioning schematically. It is often assumed that $h_1 = h_2$. Most of the proposed charts are based on

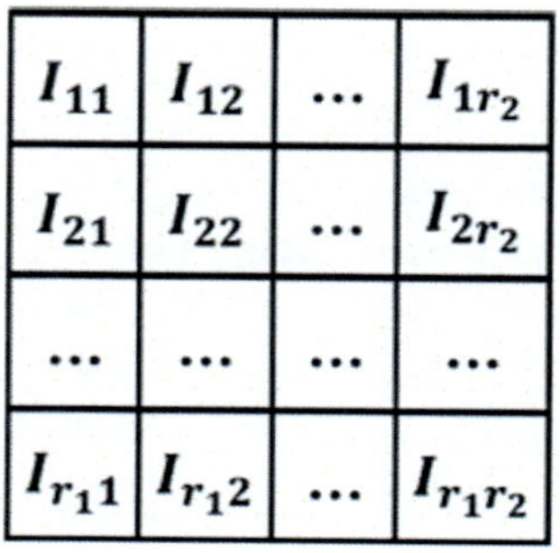

Fig. 1 Partitioning of an image in $r_1 \cdot r_2$ regions of interest (ROI)

local characteristics of I_{ij}, like the mean, weighted mean, standard deviation, etc., of the pixel intensities within a given ROI.

It is important to stress that we consider non-overlapping ROIs. Overlapping ROIs allow for more flexibility in the dimension reduction, but lead to an extreme complexity in the dependence structure. For this reason, we consider disjoint ROIs, although the suggested technique can be directly extended to the overlapping case as well.

In Megahed et al. (2012), the authors applied a control chart to the residual process. They assumed overlapping ROIs and considered the residual process $\hat{\boldsymbol{\varepsilon}}_t$. Let $\tilde{T}_{t,ij}$ be the mean of $\hat{\varepsilon}_{t,vu}$ on I_{ij} at time point t, i.e., $\tilde{T}_{t,ij} = \frac{1}{|I_{ij}|}\sum_{(v,u)\in I_{ij}} \hat{\varepsilon}_{t,vu}$. Assuming that $\{\tilde{T}_{t,ij}\}$ are independent over t, i, j, and that $\tilde{T}_{t,ij} \sim N(\mu_{ij}, \sigma_{ij}^2)$ in the in-control case, they derived a generalized likelihood ratio chart for a mean shift model as described in (2). Here the quantities μ_{ij} and σ_{ij}^2 are assumed to be known. In practice, they can be estimated via a pre-run but the influence of parameter estimation on the chart is usually not addressed.

Moreover, this approach is based on the residual process assuming its independence. Note that even in the case of a multiple linear regression the residual process is not independent. Therefore, this assumption is questionable. Further, a detailed comparison of residual charts and so-called modified charts are given in Knoth and Schmid (2004). Recently, it has been shown in Rabyk and Schmid (2016) that modified charts behave much better than residual charts if the data is highly correlated.

3.3 Control Charts Based on the GLR Approach

Here we briefly want to describe the control charts recently introduced in Okhrin et al. (2019). The ROIs are built as in the previous subsection, thus, in total there are $r_1 \cdot r_2$ ROIs. We do not consider the residual process but the original process, i.e., the images directly. Let $\tilde{T}_{t,ij} = \tilde{X}_{t,ij} = \frac{1}{|I_{ij}|}\sum_{(v,u)\in I_{ij}} \tilde{X}_{t,vu}$. Suppose that in the in-control state $\mathbf{T}_t$=$(\tilde{T}_{t,11}, \tilde{T}_{t,21}, \ldots, \tilde{T}_{t,r_11}, \ldots, \tilde{T}_{t,r_1r_2})'$ is multivariate normally distributed with mean $\boldsymbol{\mu}$ and covariance matrix $\mathbf{G}$. Note that $\boldsymbol{\mu}$ is a $r_1 \cdot r_2$-dimensional vector and $\mathbf{G}$ is a $(r_1 \cdot r_2) \times (r_1 \cdot r_2)$-dimensional matrix. If the size of the original image is 2000×2000 pixels, then taking, e.g., $h_1 = h_2 = 100$ the size reduces to 400 sub-images and then, as a result, $\mathbf{T}_t$ is a 400-dimensional vector.

Note that in this paper we do not assume that the variables $\tilde{T}_{t,ij}$, $i = 1, ..., r_1$, $j = 1, ..., r_2$ for a fixed time point t are independent as it has been done in the previously mentioned papers. Our aim is to monitor the mean behavior of a sequence of independent multivariate normally distributed random vectors. Several methods have been proposed in the literature to deal with this problem (cf. Sect. 3). These methods have been studied in the multivariate case assuming small dimensions as, e.g., values between 2 and 10. Here the dimension is much higher! Moreover, these approaches depend on a certain design parameter (smoothing matrix, reference matrix) which has to be chosen earlier and the choice of these quantities is difficult in the high-dimensional case.

Assuming the change point model (3), we want to derive a generalized likelihood ratio chart for the model under distributional assumptions above. This chart does not depend on any additional design parameters. We denote $r = r_1 \cdot r_2$ and the mean values of the regions of interest $I_{11}, I_{21}, ..., I_{r_1 1}, ... I_{r_1 r_2}$ at time point t by $T_{t,1}, ..., T_{t,r}$. We assume that

$$\mathbf{T}_t \sim \begin{cases} \mathcal{N}_r(\boldsymbol{\mu}, \mathbf{G}), & t < \tau, \\ \mathcal{N}_r(\boldsymbol{\mu} + \boldsymbol{\Delta}, \mathbf{G}), & t \geq \tau, \end{cases} \quad t \geq 1, \tag{4}$$

with $\boldsymbol{\Delta} \neq \mathbf{0}$ and $\tau \in I\!N \cup \{\infty\}$. If $\tau = \infty$ we say that the image process is in control and no change has happened. If $\tau < \infty$ then the image process is out of control starting from time point τ.

Note that this assumption is not fulfilled if the subject changes or moves over time since the assumption of identically distributed random vectors is no longer fulfilled.

Of course, $\mathbf{T}_t$ can also be considered as a random matrix. If $g = (j-1)r_1 + i \in \{1, ..., r_1 r_2\}$ with $1 \leq i \leq r_1$ and $1 \leq j \leq r_2$, then let $\tilde{T}_{t,ij} = T_{t,g}$. Denoting $\tilde{\mathbf{T}}_t = (\tilde{T}_{t,ij})$ it holds that $\text{vec}(\tilde{\mathbf{T}}_t) = \mathbf{T}_t$.

For that reason, it does not matter whether we consider the mean values of the regions of interest as a matrix or a vector.

Further, we assume that $\mathbf{T}_1, \mathbf{T}_2, ...$ are independent. Applying the generalized likelihood ratio approach (e.g., Reynolds Jr and Lou 2010; Bodnar and Schmid 2011), Okhrin et al. (2019) derived a control procedure which does not depend on any reference values or smoothing parameters. The control statistic at time point $n \geq 1$ is given by

$$R_n = \max_{1 \leq \eta \leq n} (n - \eta + 1) \hat{\boldsymbol{\Delta}}'_{\eta,n} \mathbf{G}^{-1} \hat{\boldsymbol{\Delta}}_{\eta,n}$$

with

$$\hat{\boldsymbol{\Delta}}_{\eta,n} = \frac{1}{n - \eta + 1} \sum_{t=\eta}^{n} \mathbf{T}_t - \boldsymbol{\mu}.$$

A signal is given at time $n \geq 1$ if $R_n > C$ with a suitable constant C, which is called the control limit.

In order to calculate R_n, the following recursive presentation of $\hat{\boldsymbol{\Delta}}_{\eta,n}$ can be used

$$\hat{\boldsymbol{\Delta}}_{\eta,n} = (1 - \frac{1}{n-\eta+1})\hat{\boldsymbol{\Delta}}_{\eta,n-1} + \mathbf{T}_n - \boldsymbol{\mu}.$$

For the determination of R_n, it is necessary to invert the $(r_1 \cdot r_2) \times (r_1 \cdot r_2)$ matrix $\mathbf{G}$. It is usually a high-dimensional matrix. In our example with a cell phone image, it could be a 400×400 matrix. Since in practice $\boldsymbol{\mu}$ and $\mathbf{G}$ are both unknown, they have to be estimated by a pre-run. Let us assume that a pre-run of the target process is provided, say $\mathbf{y}_{1-p}, \ldots, \mathbf{y}_0$. Then we can estimate $\boldsymbol{\mu} = (\mu_1, \ldots, \mu_{r_1}, \ldots, \mu_r)'$ by

$$\hat{\mu}_{(j-1)r_1+i} = \frac{1}{p}\sum_{t=1-p}^{0} \bar{y}_{t,(j-1)r_1+i}\,,\ i = 1, \ldots, r_1\,,\ j = 1, \ldots, r_2\,,$$

$$\bar{y}_{t,(j-1)r_1+i} = \frac{1}{|I_{ij}|}\sum_{(v,u)\in I_{ij}} y_{t,vu}\,,\ i = 1, \ldots, r_1\,,\ j = 1, \ldots, r_2\,,$$

then

$$\mathbf{G} = \big(Cov(T_{t,e}, T_{t,e'})\big)_{e,e'=1,\ldots,r} = (g_{ee'})_{e,e'=1,\ldots,r}\,,$$

and an estimator is given by

$$\hat{g}_{ee'} = \frac{1}{p}\sum_{t=1-p}^{0}(\bar{y}_{t,e} - \hat{\mu}_e)(\bar{y}_{t,e'} - \hat{\mu}_{e'}).$$

However, these estimators will only provide suitable results if p is large compared to $r_1 r_2$. In practice, this is usually not the case and the classical sample estimators fail in a high-dimensional context (e.g., Bai and Saranadasa 1996).

The shrinkage approach (see Ledoit and Wolf 2004) provides another possibility to estimate the covariance matrix. It is a nonparametric estimator which works well even in the high-dimensional case. Thus, it can be applied in situations where the dimension $r_1 r_2$ is of moderate size with respect to p. However, this approach will also fail if p is small. Then a parametric or a semiparametric method seems to be more successful and it is necessary to impose some assumptions on the structure of $\mathbf{G}$. In practice, it is reasonable to assume that more distant observations exhibit a weaker correlation than observations lying closer. We might even consider independence starting from a certain distance. It is also possible to make use of an isotropic covariance matrix with an exponential or a Matern covariance function (Cressie 1992). In this case, the estimation of $\mathbf{G}$ is much more robust and easier since we estimate only a few parameters. A further possibility, as it was mentioned already, is to assume a matrix-variate normal distribution in the form $\tilde{\mathbf{T}}_t \sim \mathcal{N}_{r_1,r_2}(\tilde{\boldsymbol{\mu}}, \mathbf{A}, \mathbf{B})$, where $\tilde{\boldsymbol{\mu}}$ is the matrix of mean values, $\mathbf{A}$ is the matrix which describes the covariances between the rows and $\mathbf{B}$ is the matrix that describes the covariances between the columns of $\tilde{\mathbf{T}}_t$ (Rafajłowicz 2018).

Efficient approximations of the covariance matrices of large data sets have been discussed by several authors. The following approaches have been recently developed to tackle the large-matrix-problem by applying a low-rank approximation of the spatial process (e.g., Cressie and Johannesson 2006; Shi and Cressie 2007; Cressie and Johannesson 2008, Katzfuss and Cressie 2009), by introducing sparseness to $\mathbf{G}$ (Furrer et al. 2006) and a combination of both approaches (Zhang et al. 2015). A comparison of these attempts was given in Vetter et al. (2016).

A disadvantage of the control statistic obtained by the GLR approach is that one needs the inverse of $\mathbf{G}$ what may be a computationally demanding task. In fact, the control statistic R_n consists of a Mahalanobis distance. Bai and Saranadasa (1996), Chen and Qin (2010) consider a similar problem. Following Bai and Saranadasa (1996), the quantity $(n-\eta+1)\hat{\boldsymbol{\Delta}}'_{\eta,n}\mathbf{G}^{-1}\hat{\boldsymbol{\Delta}}_{\eta,n}$ is replaced by the quantity $(n-\eta+1)\hat{\boldsymbol{\Delta}}'_{\eta,n}\hat{\boldsymbol{\Delta}}_{\eta,n} - tr(\mathbf{G})$, which has to be normalized suitably. Okhrin et al. (2019) determined the variance of this quantity. Using this quantity their control statistic is given by

$$M_n = \frac{\max_{1\le\eta\le n}(n-\eta+1)\hat{\boldsymbol{\Delta}}'_{\eta,n}\hat{\boldsymbol{\Delta}}_{\eta,n} - tr(\mathbf{G})}{\sqrt{2\,tr(\mathbf{G}^2)}}.$$

Chen and Qin (2010) provided an improvement of the approach of Bai and Saranadasa (1996). Applying their procedure in the present case, the control statistic is based on

$$\sum_{t,t'=\eta,t\neq t'}^{n} (\mathbf{T}_t - \boldsymbol{\mu})'(\mathbf{T}_{t'} - \boldsymbol{\mu}).$$

Note that for $\eta = n$ the value of this statistic equals zero.

Okhrin et al. (2019) determined the first two moments of this statistic and following the attempt of Chen and Qin (2010) they introduce the control statistic

$$U_n = \frac{1}{\sqrt{tr(\mathbf{G}^2)}} \max\left\{0, \max_{1\le\eta\le n-1} \frac{\sum_{t,t'=\eta,t\neq t'}^{n}(\mathbf{T}_t - \boldsymbol{\mu})'(\mathbf{T}_{t'} - \boldsymbol{\mu})}{\sqrt{2(n-\eta+1)(n-\eta)}}\right\}.$$

Note that in Bai and Saranadasa (1996), Chen and Qin (2010) the underlying statistics do not contain a maximum as in our case. Thus, the results of these authors cannot be used to characterize the asymptotic distribution of our control statistics.

For calculating the control statistics M_n and U_n only $O(r_1^2 r_2^2)$ operations are necessary while for the determination of R_n an inverse matrix has to be calculated which is more time-intensive. The Gauss–Jordan elimination method needs $O(r_1^3 r_2^3)$ operations. Thus, these quantities can be determined much faster and consequently they can be applied to smaller ROIs.

A control chart gives a signal if the corresponding control statistics exceeds a control limit, for example, $U_n > C_U$. The run length of the chart is then defined as

$$RL_U = min\{n \in I\!N \mid U_n > C_U\}.$$

The control limits are determined such that in the in-control state the expectation or the median of the run length (ARL and MRL, respectively) attains a prespecified value ξ. However, the ARL is not a suitable choice if the distribution of the run length is skewed or heavy-tailed. For this reason, we consider the MRL which is robust against these artifacts, and the control limit C_U solves the equation

$$\text{Median}(RL_U) = \xi.$$

Similarly, we determine the control limits for the remaining two charts.

4 Comparison Study

In order to compare the introduced charts, we consider as the in-control process a static image with a chessboard pattern. The in-control images were simulated as follows. The size of every image is 100×100 pixels and it is partitioned into dark and bright squares of size 20×20 pixels ordered as on a chessboard. This results in 12 bright and 13 dark squares. The intensity $\tilde{Y}_{ij}, i, j = 1, \ldots, 100$ of every pixel is assumed to follow normal distribution with parameters f_{ij} and σ^2_{ij}. The mean value $f_{ij}, i, j = 1, \ldots, 100$ of the bright pixel is set equal to 0.8 and of the dark pixel 0.2. Note, in the RGB coding we would set all three components of the intensity vector equal to these values. The variance $\sigma^2_{ij}, i, j = 1, \ldots, 100$ equals 0.03^2. We introduce the spatial correlation of the pixel intensities by exploiting the Euclidean distance between the pixels and the exponential transformation. More precisely, $Corr(\tilde{Y}_{ij}, \tilde{Y}_{i'j'}) = 0.9^{\sqrt{(i'-i)^2+(j'-j)^2}}$ for $i, j, i', j' = 1, \ldots, 100$. Recall that the intensities must belong to the unit interval. If it is not the case, then we resample the value until this restriction is fulfilled. This happens rarely since the standard deviation for every pixel is 0.03.

Figure 2 shows two images simulated under the in-control conditions in a zoom-in mode. Despite looking identical, a closer look reveals variation in the brightness of the dark and bright squares due to the noise in the underlying data generating process. The changes are small enough not to distort the subject of the image. We assume that it is acceptable for images under the in-control conditions to vary in such a way.

The size of the ROIs is set to 10×10 pixels implying a total of 100 non-overlapping ROIs which cover the whole area of the image. This is schematically shown in Fig. 3. As mentioned above, it is possible to consider overlapping ROIs of different sizes. This increases the flexibility of the charts, but makes the implementation and the theoretical results more demanding.

Fig. 2 Two zoomed-in images with the chessboard subject (the actual size is 100 × 100 pixels) simulated under the in-control conditions

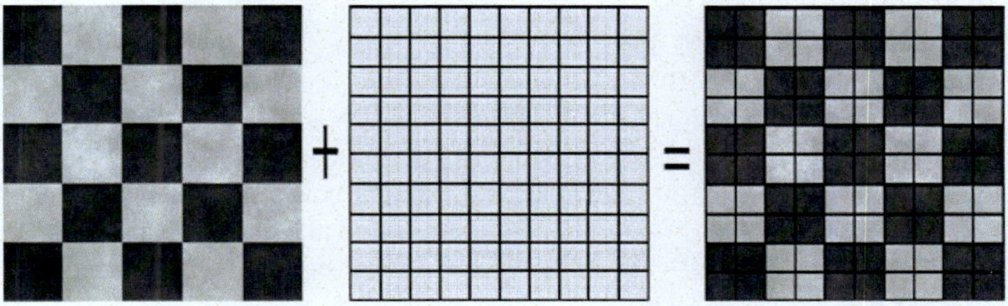

Fig. 3 Partitioning of 100 × 100 pixels image in 10 × 10 ROIs

For the fixed ROIs we derive the mean and the covariance matrix relying on the distributional assumptions of the pixel intensities. Note that the exact distribution of the control statistics R_n, M_n and U_n introduced in the previous chapters is not a standard distribution. For that reason, the control limits of the control charts were calculated using simulations. As a calibration criterion, we select the in-control median run length (MRL) to be equal to $\xi = 100$. In order to calculate the control limits, the regula falsi method is applied and in each step the MRL is estimated using 10^3 independent repetitions. This leads to $C_R = 147.6708$, $C_M = 4.0073$, and $C_U = 3.6169$.

To check the ability of the charts to signal in the out-of-control state, we consider four scenarios for a change in the mean value of the image areas. In each scenario, we change the intensity of a particular part of the figure. The heterogeneity of forms and shadings allows us to verify the efficiency of the suggested control schemes for different types of changes. There must be obviously a link between the size of the changed area and the size of ROIs. For example, small changes are easier to detect with small ROIs. Megahed et al. (2012) suggest to use the Dice similarity coefficient (DSC) to investigate how good the chart estimates the size of the change. Since in our case the ROIs are of fixed size, we use DSC as an assessment tool to relate ROIs and the area of change. Let F denote the area of the change. Then DSC for the ith ROI is defined as

$$DSC_i = \frac{2|F \cap I_i|}{|F| + |I_i|}.$$

In our setting, we look at the maximum of DSC_i's, i.e., $DSC_{max} = \max_i \; DSC_i$, and thus, the largest fraction of the fault area covered by a single ROI. DSC equal one implies that the fault area is completely covered by a single ROI. Small values of ROI indicate that every ROI covers only a small fraction of the faulty area.

Type of changes:

(a) The square of the chessboard in the third row and second column becomes brighter. All pixels of the bright square of the size 20×20 pixels, which is fully covered by 4 ROIs, change its mean value from 0.8 to $0.8 + \delta$ with $\delta \in \{0.005, 0.01, \ldots, 0.05\}$. Figure 4 contains the out-of-control images with $\delta = 0.005$, 0.05 and 0.09. By a pure visual inspection, only the shift on the right-hand side can be identified, but this is hardly possible for smaller shifts.
In Fig. 5, the histogram of the distribution of the out-of-control MRL for the shift $\delta = 0.02$ is plotted to give the reader a feeling of how long it can take before the chart detects this change. The histogram shows that one needs less than 35 observations. DSC_{max} is 0.4 here and indicates that a single ROI covers less than a half of the shifted area.

(b) The part of the square of the chessboard in the third row and second column of the size 10×10 pixels (the part in the left upper corner) becomes brighter. The intensity shifts from 0.8 to $0.8 + \delta$ with $\delta \in \{0.005, 0.01, \ldots, 0.05\}$. The shifted area is covered by a single ROI and thus, $DSC_{max} = 1$.

(c) The intersection of four squares of size 5×5 pixels on the chessboard in the second row and second column, second row and third column, third row and second column, third row and third column of the total size 10×10 pixels becomes brighter. The intensity shifts from 0.8 and 0.2 to $0.8 + \delta$ and $0.2 + \delta$ with $\delta \in \{0.005, 0.01, \ldots, 0.05\}$, respectively. This shift represents the changes

Fig. 4 Two zoomed-in images of the chessboard simulated for the out-of-control situation described in **a**. The shifts are $\delta = 0.005$, 0.05, and 0.09 (from left to right)

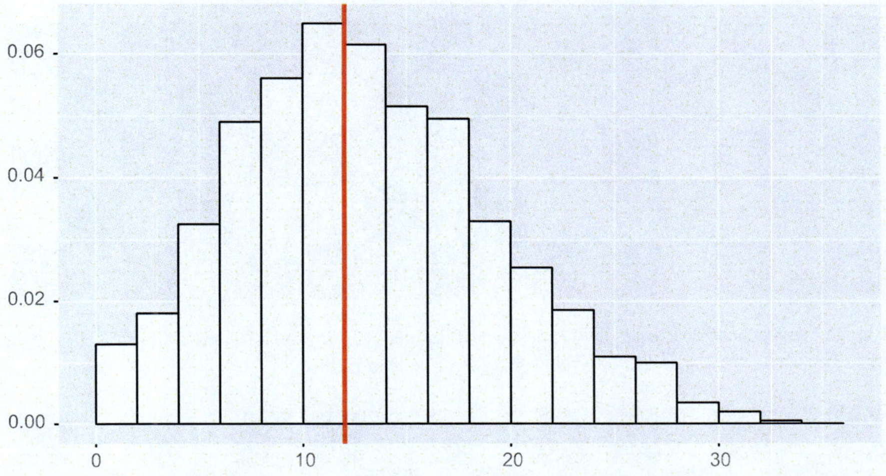

Fig. 5 Histogram of the out-of-control RL for the R_n chart. Case **a**, $\delta = 0.02$, 10^3 repetitions. The red line indicates the out-of-control MRL

Fig. 6 Two zoomed-in images of the chessboard simulated for the out-of-control situation described in **b**. The shifts are $\delta = 0.005$, 0.05, and 0.09 (from left to right)

in 4 neighboring ROIs. 25% of pixels in each of these ROIs change their intensities and $DSC_{max} = 0.25$.

(d) The whole image becomes brighter. The values change from 0.8 and 0.2 to $0.8 + \delta$ and $0.2 + \delta$ with $\delta \in \{0.005, 0.01, \ldots, 0.05\}$.

Figures 6, 7, and 8 explain the changes in scenarios (b)–(d). Figure 9 contains the out-of-control MRLs for the described scenarios (a)–(d). All MRLs start at the target value of 100 and monotonically decrease to one. The spread of convergence heavily depends on the scenario of shifts and on the type of the chart. For small shifts, the charts need more iteration to identify it, whereas large changes (scenarios (a) and (d)) are detected very quickly. In order to better understand the distribution of the run lengths, we additionally consider their means and standard deviations in the Table 1. We restrict the discussion to scenario (a) only. The values for other scenarios look very similar and we drop them for space reasons. We observe that for larger shifts

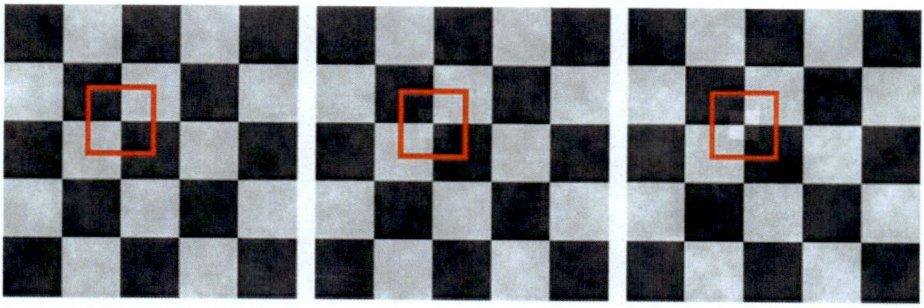

Fig. 7 Two zoomed-in images of the chessboard simulated for the out-of-control situation described in **c**. The shifts are $\delta = 0.005$, 0.05, and 0.09 (from left to right)

Fig. 8 Two zoomed-in images for the chessboard simulated for the out-of-control situation in **d**. The shifts are $\delta = 0.005$, 0.05, and 0.09 (from left to right)

the mean and the median are very close. Only for small changes, the mean becomes larger indicating that there are some very large run lengths in the sample. The values of the standard deviation are large for small shifts, but quickly decrease with larger δ. Also, to estimate the size and the shift in the mean intensities of ROIs, one can plot a diagram as in Fig. 10. We put the ordered ROIs (from top to bottom, left to right) using the partitioning in Fig. 3 on the horizontal axis in Fig. 10 whereas the mean intensities are plotted on the vertical axis. Black circles indicate the in-control mean intensities; blue circles are estimated mean intensities after the chart signaled. In scenario (a) the faulty area is completely covered by the ROIs 25, 26, 35, and 36. This is clearly indicated by the spikes in the average intensities.

For the considered out-of-control scenarios, the chart based on the control statistic R_n shows the best overall performance. This is not surprising since it is obtained using the GLR approach and the other charts are only approximations to this quantity. The disadvantage of R_n is that in contrary to M_n and U_n it depends on the inverse of the covariance matrix. This quantity is difficult to determine in a high-dimensional situation. Thus, the main advantages of M_n and U_n can be observed in a high-dimensional situation since these statistics can be easier evaluated than R_n. Moreover, here we do not discuss the influence of parameter estimation on the charts. The determination of the inverse is even more complicated taking into account the

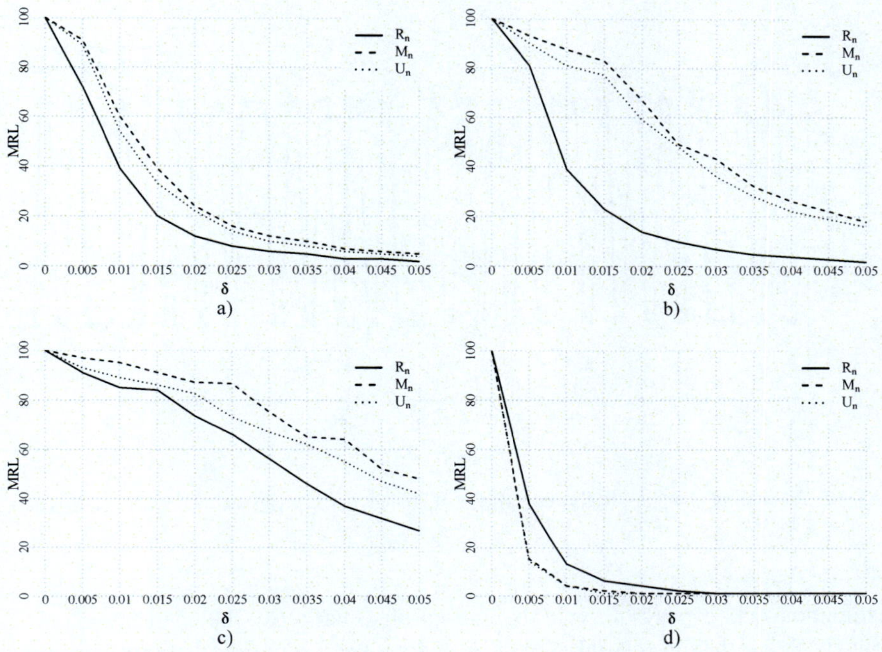

Fig. 9 Out-of-control MRLs of the control charts R_n, M_n, and U_n for the scenarios described in **a–d**

Table 1 Out-of-control MRLs, means, and standard deviations in scenario (a)

Shift, δ	0.005	0.01	0.015	0.02	0.025	0.03	0.035	0.04	0.045	0.05
R_n										
MRL	72	39	20	12	8	6	5	3	3	2
$Mean$	90.004	41.492	21.042	13.085	8.599	6.128	4.636	3.567	2.808	2.140
$Std.Dev.$	70.953	23.039	11.077	6.435	3.905	2.749	2.038	1.567	1.201	0.947
M_n										
MRL	91	60	39	24	16	12	10	7	6	5
$Mean$	118.673	70.111	40.456	25.535	16.919	12.495	9.632	7.449	5.910	4.881
$Std.Dev.$	107.124	47.673	23.533	13.407	8.289	5.550	4.316	3.084	2.572	2.103
U_n										
MRL	89	54	33	22	14	10	8	6	5	4
$Mean$	115.175	62.479	36.540	22.537	15.185	10.839	8.158	6.183	4.741	3.715
$Std.Dev.$	98.064	43.344	21.450	12.102	7.854	5.200	3.826	2.899	2.278	1.771

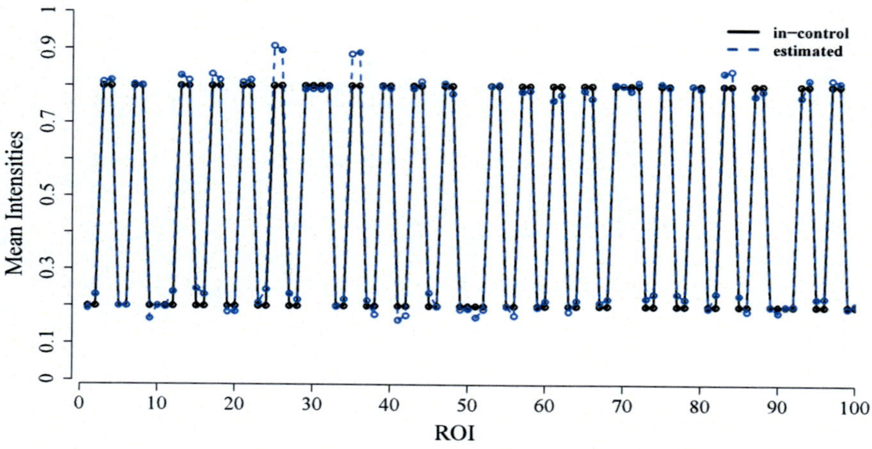

Fig. 10 Estimating changing ROIs and its sizes of change in the mean intensities for the R_n chart. Case **a**, $\delta = 0.05$

estimation risk. Nevertheless, in the in-control scenario (d), the chart based on U_n signals earlier than the chart based on R_n. Moreover, U_n uniformly outperforms the chart with M_n in all considered cases. Obviously, if smaller parts of the image are defected, the charts react slower. If the whole image is effected as in (d), the charts signal faster.

Intuitively, the ability of the charts to signal depends on the choice of the size of the minimum ROI. As it is analyzed in Megahed et al. (2012) the size of the ROI should be chosen in relation to the size of the area of the expected defect. If the size of the ROI is smaller than the expected area of the defect, then the chart is supposed to signal faster than in the case where the size of the ROI is larger.

5 Conclusions

In this paper, we discuss the problem of monitoring an image process over time. We give an overview of the existing literature with a focus on the recent approach of Okhrin et al. (2019). Since the number of pixels is huge, we face a high-dimensional problem and for that reason methods for high-dimensional data should be used in this context. While many authors assume an independent residual process, we take the spatial correlation structure of the pixels into account. In order to reduce the dimensionality of the data, we build non-overlapping ROIs for every image and use local statistical characteristics of those ROIs. We consider three possible control charts and motivate them. In an extensive simulation study, we compare these three control designs with each other. This is done for various out-of-control situations.

The chart based on the generalized likelihood ratio approach shows the best overall performance. Two other considered control charts make use of a simpler control statistics which are much easier to handle in a high-dimensional setting. However, they do not seem to have such a good performance as the chart based on the GLR.

References

Armingol, J. M., Otamendi, J., De La Escalera, A., Pastor, J. M., & Rodriguez, F. J. (2003). Statistical pattern modeling in vision-based quality control systems. *Journal of Intelligent and Robotic Systems*, *37*(3), 321–336.

Bai, Z., & Saranadasa, H. (1996). Effect of high dimension: By an example of a two sample problem. *Statistica Sinica*, *6*(2), 311–329.

Bodnar, O., & Schmid, W. (2011). CUSUM charts for monitoring the mean of a multivariate Gaussian process. *Journal of Statistical Planning and Inference*, *141*(6), 2055–2070.

Chen, S. X., & Qin, Y. L. (2010). A two-sample test for high-dimensional data with applications to gene-set testing. *The Annals of Statistics*, *38*(2), 808–835.

Colosimo, B. M. (2018). Modeling and monitoring methods for spatial and image data. *Quality Engineering*, *30*(1), 94–111.

Cressie, N. (1992). Statistics for spatial data. *Terra Nova*, *4*(5), 613–617.

Cressie, N., & Johannesson, G. (2006). Spatial prediction for massive datasets. Mastering the Data Explosion in the Earth and Environmental Sciences: Australian Academy of Science Elizabeth and Frederick White Conference.

Cressie, N., & Johannesson, G. (2008). Fixed rank kriging for very large spatial data sets. *Journal of the Royal Statistical Society: Series B (Statistical Methodology)*, *70*(1), 209–226.

Cressie, N., & Wikle, C. K. (2015). *Statistics for spatio-temporal data*. Hoboken: Wiley.

Crosier, R. B. (1988). Multivariate generalizations of cumulative sum quality-control schemes. *Technometrics*, *30*(3), 291–303.

Fassó, A. (1999). One-sided MEWMA control charts. *Communications in Statistics-Simulation and Computation*, *28*(2), 381–401.

Fieguth, P. (2010). *Statistical image processing and multidimensional modeling*. Berlin: Springer Science & Business Media

Furrer, R., Genton, M. G., & Nychka, D. (2006). Covariance tapering for interpolation of large spatial datasets. *Journal of Computational and Graphical Statistics*, *15*(3), 502–523.

Gonzalez, R. C., & Woods, R. E. (2018). *Digital image processing*. Upper Saddle River: Prentice Hall.

Gupta, A. K., & Nagar, D. K. (2018). *Matrix variate distributions*. Boca Raton: Chapman and Hall/CRC

Hawkins, D. M., Choi, S., & Lee, S. (2007). A general multivariate exponentially weighted moving-average control chart. *Journal of Quality Technology*, *39*(2), 118–125.

Healy, J. D. (1987). A note on multivariate CUSUM procedures. *Technometrics*, *29*(4), 409–412.

Horst, R., & Negin, M. (1992). Vision system for high-resolution dimensional measurements and on-line SPC: Web process application. *IEEE Transactions on Industry Applications*, *28*(4), 993–997.

Hotelling, H. (1947). Multivariate quality control illustrated by the air testing of sample bombsights. In *Techniques of statistical analysis*. New York: McGraw Hill.

Jiang, B., Wang, C., & Liu, H. (2005). Liquid crystal display surface uniformity defect inspection using analysis of variance and exponentially weighted moving average techniques. *International Journal of Production Research*, *43*(1), 67–80.

Katzfuss, M., & Cressie, N. (2009). Maximum likelihood estimation of covariance parameters in the spatial-random-effects model. In *Proceedings of the Joint Statistical Meetings*.

Knoth, S., & Schmid, W. (2004). Control charts for time series: A review. In *Frontiers in statistical quality control* (Vol. 7). Berlin: Springer.

Koosha, M., Noorossana, R., & Megahed, F. (2017). Statistical process monitoring via image data using wavelets. *Quality and Reliability Engineering International*, *33*(8), 2059–2073.

Ledoit, O., & Wolf, M. (2004). A well-conditioned estimator for large-dimensional covariance matrices. *Journal of Multivariate Analysis*, *88*(2), 365–411.

Lin, H. D. (2007a). Automated visual inspection of ripple defects using wavelet characteristic based multivariate statistical approach. *Image and Vision Computing*, *25*(11), 1785–1801.

Lin, H. D. (2007b). Computer-aided visual inspection of surface defects in ceramic capacitor chips. *Journal of Materials Processing Technology*, *189*(1–3), 19–25.

Lin, H. D., Chung, C. Y., & Lin, W. T. (2008). Principal component analysis based on wavelet characteristics applied to automated surface defect inspection. *WSEAS Transactions on Computer Research*, *3*(4), 193–202.

Liu, J. J., & MacGregor, J. F. (2006). Estimation and monitoring of product aesthetics: Application to manufacturing of "engineered stone" countertops. *Machine Vision and Applications*, *16*(6), 374.

Lowry, C. A., Woodall, W. H., Champ, C. W., & Rigdon, S. E. (1992). A multivariate exponentially weighted moving average control chart. *Technometrics*, *34*(1), 46–53.

Lu, C. J., & Tsai, D. M. (2005). Automatic defect inspection for LCDs using singular value decomposition. *The International Journal of Advanced Manufacturing Technology*, *25*(1–2), 53–61.

Mason, R. L., Tracy, N. D., & Young, J. C. (1997). A practical approach for interpreting multivariate T^2 control chart signals. *Journal of Quality Technology*, *29*(4), 396–406.

Megahed, F. M., Woodall, W. H., & Camelio, J. A. (2011). A review and perspective on control charting with image data. *Journal of Quality Technology*, *43*(2), 83–98.

Megahed, F. M., Wells, L. J., Camelio, J. A., & Woodall, W. H. (2012). A spatiotemporal method for the monitoring of image data. *Quality and Reliability Engineering International*, *28*(8), 967–980.

Montgomery, D. C. (2009). *Statistical quality control*. New York: Wiley.

Nembhard, H. B., Ferrier, N. J., Osswald, T. A., & Sanz-Uribe, J. R. (2003). An integrated model for statistical and vision monitoring in manufacturing transitions. *Quality and Reliability Engineering International*, *19*(6), 461–476.

Ngai, H. M., & Zhang, J. (2001). Multivariate cumulative sum control charts based on projection pursuit. *Statistica Sinica*, *11*(3), 747–766.

Okhrin, Y., Schmid, W., & Semeniuk, I. (2019). *New approaches to monitor image data with spatially correlated intensities*. Technical report.

Otto, P., & Seckmeyer, G. (2019). *Parallelized monitoring of dependent spatiotemporal processes*. Technical report.

Pignatiello, J. J, Jr., & Runger, G. C. (1990). Comparisons of multivariate CUSUM charts. *Journal of Quality Technology*, *22*(3), 173–186.

Rabyk, L., & Schmid, W. (2016). EWMA control charts for detecting changes in the mean of a long-memory process. *Metrika*, *79*(3), 267–301.

Rafajłowicz, E. (2018). Classifiers for matrix normal images: Derivation and testing. In *International Conference on Artificial Intelligence and Soft Computing*. Springer

Rafajłowicz, E., & Rafajłowicz, W. (2017). Image-driven decision making with application to control gas burners. In *IFIP International Conference on Computer Information Systems and Industrial Management*. Springer.

Réfrégier, P., & Goudail, F. (2013). *Statistical image processing techniques for noisy images: An application-oriented approach*. Berlin: Springer Science & Business Media.

Reynolds, M. R, Jr., & Lou, J. (2010). An evaluation of a GLR control chart for monitoring the process mean. *Journal of Quality Technology*, *42*(3), 287–310.

Shi, T., & Cressie, N. (2007). Global statistical analysis of MISR aerosol data: A massive data product from NASA's terra satellite. *Environmetrics: The Official Journal of the International Environmetrics Society*, *18*(7), 665–680.

Sonka, M., Hlavac, V., & Boyle, R. (2014). *Image processing, analysis, and machine vision*. Boston: Cengage Learning.

Sparks, R. S. (1992). Quality control with multivariate data. *Australian Journal of Statistics*, *34*(3), 375–390.

Tong, L. I., Wang, C. H., & Huang, C. L. (2005). Monitoring defects in IC fabrication using a Hotelling T^2 control chart. *IEEE Transactions on Semiconductor Manufacturing*, *18*(1), 140–147.

Vetter, P., Schmid, W., & Schwarze, R. (2016). Spatio-temporal statistical analysis of the carbon budget of the terrestrial ecosystem. *Statistical Methods & Applications*, *25*(1), 143–161.

Zhang, B., Sang, H., & Huang, J. Z. (2015). Full-scale approximations of spatio-temporal covariance models for large datasets. *Statistica Sinica*, *1*, 99–114.

Parallelized Monitoring of Dependent Spatiotemporal Processes

Philipp Otto

Abstract With the growing availability of high-resolution spatial data, such as high-definition images, three-dimensional point clouds of light detection and ranging (LIDAR) scanners, or communication and sensor networks, it might become challenging to detect changes and simultaneously account for spatial interactions in a timely manner. To detect local changes in the mean of isotropic spatiotemporal processes with locally constrained dependence structures, we have proposed a monitoring procedure that can be completely run on parallel processors. This allows for fast detection of local changes (i.e., in the case that only a few spatial locations are affected by the change). Due to parallel computation, high-frequency data could also be monitored. Hence, we additionally focused on the processing time required to compute the control statistics. Finally, the performance of the charts has been analyzed using a series of Monte Carlo simulation studies.

Keywords Computational statistics · Covariance tapering · Distributed computing · Spatiotemporal monitoring

1 Introduction

In the era of big data, the amount of available data that could also be used for process monitoring is rapidly growing. For instance, the resolution and size of images have been massively growing over recent years. Although it is debatable whether the information content is increasing with an increasing size/resolution of the images, a high amount of data must be handled. Another example can be seen in remotely sensed data from satellites, such as the concentration of atmospheric pollutants, or data resulting from three-dimensional (3D) point clouds of light detection and ranging (LIDAR) scanners. These scanners could measure millions of data points per second. In particular, for autonomous driving, data from such scanners could be of interest

P. Otto (✉)
Institute of Cartography and Geoinformatics, Leibniz University Hannover, Hanover, Germany
e-mail: otto@ikg.uni-hannover.de

S. Knoth and W. Schmid (eds.), *Frontiers in Statistical Quality Control 13*,
Frontiers in Statistical Quality Control,
https://doi.org/10.1007/978-3-030-67856-2_10

and must be processed within very short time frames. All these examples have in common that they have some natural ordering in space, inducing dependence on the spatial dimension. As Tobler (1970) stressed, adjacent observations are more related than observations that are more distant from each other. This paradigm is known as the first law of geography. It is important to stress at this point that this natural spatial ordering could also be interpreted as a network structure. Basically, networks could be considered to have a specific order in a certain, probably non-Euclidean, space. Thus, data coming from communication or sensor networks are further examples that are of interest for this paper (for example, see Fu and Jeske 2014; Wilson et al. 2016; Woodall et al. 2017; Jeske et al. 2018 for monitoring procedures of networks).

More precisely, the focus of this paper is on monitoring dependent spatiotemporal processes with a large number of spatial locations. From a computational point of view, the proposed monitoring procedure should be fully scalable for growing spatial dimensions. That is, if the number of spatial locations is growing, it should be feasible to monitor the process within the same amount of time using more processing units. Simultaneously, we account for the natural spatial dependence of the process. To date, monitoring procedures for spatially dependent data or image data are rarely discussed in the literature. For instance, Jiang et al. (2011) discussed surveillance methods for data that are correlated in space. Along with modeling approaches, monitoring procedures for complex spatial data structures have also been discussed by Colosimo (2018). In contrast, Garthoff and Otto (2017) considered methods of statistical process control to detect purely spatial changes. Furthermore, Megahed et al. (2011, 2012) focused on images and proposed to define regions of interest for which a characteristic quantity based on the sample mean and covariance can be constructed. Another approach uses a spatial scan statistic for cluster detection to monitor spatial processes, which was proposed by Sparks and Patrick (2014).

To motivate the procedure and the need for a fully scalable approach, we additionally introduce an empirical example from the field of meteorology. In Fig. 1, a rooftop camera located in Hanover is shown, which constantly takes pictures of the daytime sky, which are the so-called all-sky images (for more details and applica-

Fig. 1 All-sky images. Left: camera on the rooftop making constant all-sky images (© Philipp Otto); right: one instance of the resulting sequence of images (© Institute for Meteorology and Climatology, Leibniz Universität Hannover)

tions, we refer to Crisosto et al. 2018; Hofmann and Seckmeyer 2017; Tohsing et al. 2013). One instance of this sequence of images is shown on the right-hand side. From a meteorological point of view, clouds building an ergodic system indicate stable weather conditions. That is, the weather will not change in this case. On the contrary, if the clouds result from a non-ergodic system, the weather will change. Thus, clouds, and in particular, all-sky images could be used for weather-change detection. This is relevant for all energy producers because prices of electricity are highly volatile and depend to a large extent on the overall energy supply. Furthermore, prices on the electricity market could even be negative (i.e., producers must pay if they produce electricity). Hence, timely detection of weather changes is of high interest, especially in times of growing production capacities using renewable energy resources.

However, monitoring such processes leads to new challenges from a statistical and computational point of view. More precisely, these images typically have a high resolution/definition, and they could be taken at a high frequency. As these images are coming from a natural process, we assume that the process cannot be stopped (i.e., it is continuously running regardless of whether the control chart signaled a change or not). Furthermore, the process is spatially dependent due to the natural spatial ordering of the locations.

The remainder of the paper is structured as follows. In the next section, we discuss a Gaussian spatial dependence model and several potential fields of applications. Further, a monitoring procedure for such processes is introduced, which allows for full parallelization of the computation of the control statistics. More precisely, an exponentially weighted moving average (EWMA) chart is considered. In the ensuing section, the results from a Monte Carlo simulation are reported to evaluate the performance of the chart and to show some limitations of the approach. Eventually, Sect. 5 concludes the paper.

2 Spatial Dependence Models

The main objective of the paper is to monitor spatiotemporal processes lying in a multidimensional space. That is, deviations of the observed process $\{X_t(\boldsymbol{s}) : t \in \mathbb{Z}, \boldsymbol{s} \in D\}$ from a so-called target process $\{Y_t(\boldsymbol{s}) : t \in \mathbb{Z}, \boldsymbol{s} \in D\}$ should be detected as soon as possible. More precisely, the process is observed at several locations $\boldsymbol{s}$ in a finite q-dimensional space D (i.e., $D \subset \mathbb{R}^q$). For simplicity, we assume that we observe the process at the same set of locations $\{\boldsymbol{s}_1, \ldots, \boldsymbol{s}_n\}$ for all time points t. Furthermore, we assume that the target process could be dependent in space but is independent over time.

This allows for a wide range of applications in various fields of research. For instance, if the set of locations $\{\boldsymbol{s}_1, \ldots, \boldsymbol{s}_n\}$ lies in a two-dimensional (2D) unit grid $\mathbb{Z}^2$, sequences of images could be monitored. In particular, this could be relevant in production engineering, where the production outcome could be assessed using images. It is important to note that the monitoring method is also applicable to high-resolution images that have been taken at a high frequency (e.g., the all-sky images discussed in Sect. 1). In such settings, the process is called a spatial lattice process.

Another example can be found in environmetrics, where it could be of interest to monitor air pollutants in the atmosphere (i.e., D is a 3D continuous space, or at ground measurement stations, D would be a 2D space in this case). These settings cover all spatial point processes and marked point processes. For instance, data measured by LIDAR sensors or satellite remote sensing represent such kind of processes. Eventually, data could also be observed at irregular polygons, such as municipalities or counties. For instance, certain health indicators or incidence rates could be monitored for disease surveillance.

The spatial dependence is characterized by

$$Cov(Y_t(\boldsymbol{s}_i), Y_t(\boldsymbol{s}_j)) = C(\boldsymbol{s}_i - \boldsymbol{s}_j) \tag{1}$$

for $i \neq j$, where $C : \mathbb{R}^+ \to \mathbb{R}^+$ is a function of the difference $\boldsymbol{s}_i - \boldsymbol{s}_j$. If C is only a function of the distance $||\boldsymbol{s}_i - \boldsymbol{s}_j||$, the process is called isotropic, but this restriction is not needed for monitoring. That is, our approach is suitable for both isotropic and anisotropic processes. Assuming additionally that the expectation is constant over space and time, that is,

$$E(Y_t(\boldsymbol{s}_i)) = \mu \qquad \text{for all } i \text{ and } t, \tag{2}$$

the process is weakly stationary.

Furthermore, the covariance function C defines a covariance matrix $\boldsymbol{\Sigma} = (\sigma_{ij})_{i,j=1,\ldots,n}$, where

$$\sigma_{ij} = \begin{cases} C(\boldsymbol{s}_i - \boldsymbol{s}_j) & \text{for } i \neq j \\ Var(Y(\boldsymbol{s}_i)) & \text{for } i = j. \end{cases} \tag{3}$$

Obviously, this covariance matrix has a dimension $n \times n$, which is usually very large for empirical applications. For example, high-definition images have a resolution of 1920×1080 pixels, which leads to a $2,073,600$-dimensional covariance matrix, which might be infeasible to compute.

Subsequently, we assumed that the target process $\boldsymbol{Y}_t = (Y_t(\boldsymbol{s}_1), \ldots, Y_t(\boldsymbol{s}_n))'$ is a multivariate Gaussian process

$$\boldsymbol{Y}_t \sim N_n(\boldsymbol{\mu}, \boldsymbol{\Sigma}) \qquad \text{for all } t, \tag{4}$$

where $\boldsymbol{\mu}$ is the constant mean vector. Note that this implies that the process is independent over time. Moreover, this covers spatial autoregressive models, if $\boldsymbol{\mu} = (I - \rho W)^{-1}\boldsymbol{\mu}^*$ and $\boldsymbol{\Sigma} = (I - \rho W)^{-1} D^* (I - \rho W)^{-1}$ with a known weighting matrix W. Thus, these autoregressive models are special cases of the considered setting with a known function C.

3 Monitoring Procedure for High-Resolution Images

In the following section, the focus is on the parallel monitoring procedure, which accounts for the spatial dependence and is simultaneously applicable for large spatial dimensions (i.e., n is large). Thus, we initially define an out-of-control model for changes in the mean of the process. Then, the idea of parallel monitoring is described. The proposed procedure can be run fully in parallel without the need to combine the results of the parallel processes at each time point. From a computational perspective, the procedure is, therefore, highly scalable. Eventually, a suitable control characteristic and a multivariate EWMA chart are proposed to monitor such processes.

3.1 *Out-of-Control Model*

In this paper, we focuse on mean changes denoted by $\boldsymbol{a} \in \mathbb{R}^n \setminus \{\mathbf{0}\}$ only. However, control procedures for monitoring the covariance or simultaneous procedures for mean and covariance monitoring could be constructed in an analogous manner. For changes in the mean, the observed process $\boldsymbol{X}_t = (X_t(\boldsymbol{s}_1), \ldots, X_t(\boldsymbol{s}_n))'$ can be specified as

$$\boldsymbol{X}_t = \begin{cases} \boldsymbol{Y}_t & \text{for } t < \tau \\ \boldsymbol{a} + \boldsymbol{Y}_t & \text{for } t \geq \tau \, . \end{cases} \tag{5}$$

If the change point $\tau = \infty$, the observed process always coincides with the target process, and it would be called in control. Apparently, both processes would also be equal if $\boldsymbol{a} = \mathbf{0}$. It is worth noting that the mean change does not necessarily affect all locations $\boldsymbol{s}_1, \ldots, \boldsymbol{s}_n$, but it could also affect only a few locations, meaning at least one.

For instance, we considered a random field of size 3×3 (i.e., $n = 9$), and a change in the mean of the first location (i.e., $\boldsymbol{a} = (2, 0, 0, 0, 0, 0, 0, 0, 0)'$). In Fig. 2, a simulated random field of this size is depicted as single time series with $\boldsymbol{a}$ plotted as a solid red line. Note that the time series plots are shown in the correct spatial ordering (i.e., the distance between each plot corresponds to the true distance of the locations). To visualize the distance to the first location, where the mean change occurs, the background of the plots is colored according to the Euclidean distance from $\boldsymbol{s}_1$. The covariance matrix $\boldsymbol{\Sigma}$ results from an exponential covariance function shown in Fig. 3. The respective entries of $\boldsymbol{\Sigma}$ are highlighted by circles and the background colors correspond to the colors in Fig. 2.

It is essential to account for spatial dependence when monitoring the process. This is important not only because the mean change could be diminished due to spatial interactions (especially in the case of spatial autoregressive covariance structures), but one could also be interested in identifying the origin of the change. In particular, this is of interest if the mean change does not affect $\boldsymbol{Y}_t$ directly but affects another

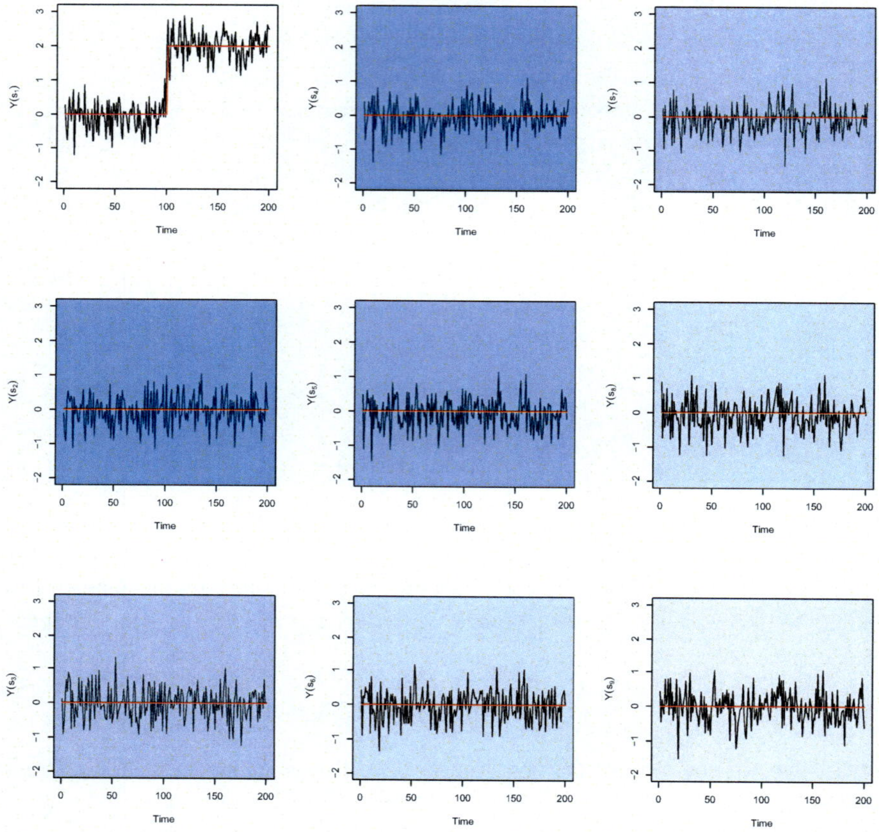

Fig. 2 Simulated spatiotemporal process depicted as a single time series where the natural spatial ordering was kept. The background colors indicate the distance from the first location $\boldsymbol{s}_1$, where a change in the mean occurs at $\tau = 100$. The vector $\boldsymbol{a}$ is shown as a red line

latent process. In this case, the mean shift in location $\boldsymbol{s}_1$ would have an influence on all other locations due to the spatial dependence. That is, if we do not account for spatial interactions, the source cannot be correctly identified.

For monitoring and detecting possible mean changes, we sequentially tested whether the mean of the observed process coincides with the mean of the target process at each time point (i.e., whether there is mean change or not). Hence, at time t, the decision problem is given by

$$H_{0,t} : E(\boldsymbol{X}_t) = \boldsymbol{\mu} \quad \text{versus} \quad H_{1,t} : E(\boldsymbol{X}_t) \neq \boldsymbol{\mu}\,, \tag{6}$$

(i.e., under the null hypothesis, the expectation of the observed process equals the true target mean).

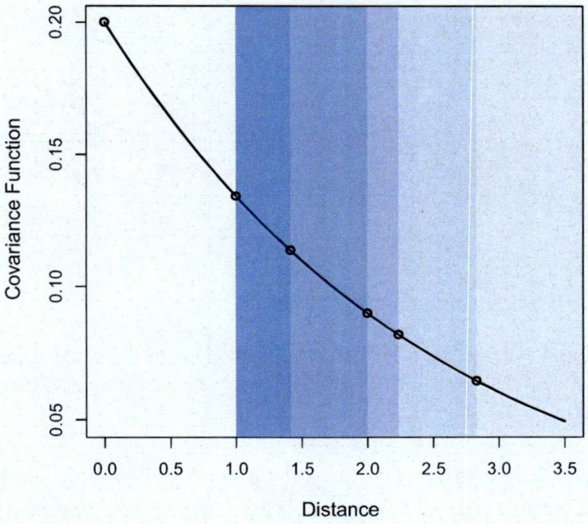

Fig. 3 Spatial covariance function for the simulated random field in Fig. 2

3.2 Parallel Monitoring

To account for spatial interactions, one could easily construct a characteristic quantity resulting from a transformation of the current observation with the target mean and covariance matrix, that is,

$$\boldsymbol{T}_t = \boldsymbol{\Sigma}^{-1/2} (\boldsymbol{X}_t - \boldsymbol{\mu}) \ . \tag{7}$$

As the target process is a Gaussian process, the residuals would be normal, and the derivation of the moments of the characteristics is based on Isserlis' theorem (see Brillinger 1981).

However, the computation of the characteristic $\boldsymbol{T}_t$ requires the Cholesky decomposition of the inverse of $\boldsymbol{\Sigma}$. Common algorithms have a computational complexity of $O(n^3)$. Although this matrix only has to be computed once, it might not be feasible if the number of locations is large. Nevertheless, the computation of the characteristic requires the multiplication of a matrix and a column vector at each time point t, which has a complexity of $O(n^2)$ for the straightforward implementation.

Thus, we propose to split the problem into several chunks, which can be run fully in parallel processes. More precisely, the set of locations is split into several subsets of equal size and a characteristic quantity is computed for each subset. For instance, an image of size 100×100 could be split into smaller windows of size 10×10 (i.e., 100 of these windows cover the whole image). For the empirical motivation given in Sect. 1, we illustrate one possible way of sectioning the image in Fig. 4. If the size of the subsets is denoted by p, a characteristic quantity of the ith window,

$$\boldsymbol{T}_{t,i}^{(p)} = \boldsymbol{\Sigma}_p^{-1/2} \left(\boldsymbol{X}_{t,i}^{(p)} - \boldsymbol{\mu}_p \right) , \tag{8}$$

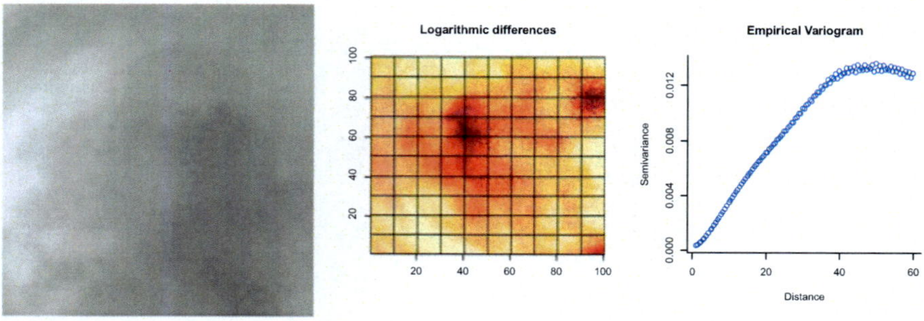

Fig. 4 Illustration of the window selection for the empirical example. Left: all-sky image (© Philipp Otto); center: first logarithmic differences between two images; right: estimated spatial variogram

has only a dimension of p instead of n. The $p \times p$-dimensional covariance matrix $\boldsymbol{\Sigma}_p$ can easily be constructed from the covariance function C as in (3). Similarly, it can be computed for spatial autoregressive models. If the in-control C is unknown, it could also be estimated in a pre-estimation step, e.g., by variogram estimators (see Fig. 4 right-hand plot).

In general, $\boldsymbol{T}_t^{(p)}$ is not independent and normally distributed even if the process is in control because positive entries of the covariance matrix $\boldsymbol{\Sigma}$ are neglected or, in other words, are set to zero. However, if these entries that are set to zero are small, the characteristic is very close to an independent and normally distributed random vector if the process is in control. In spatial statistics, this approach is commonly known as covariance tapering. A tapered covariance function C_{tap} results from multiplying the underlying covariance function C with a tapering matrix C_θ, which is assumed to be a valid covariance matrix with zero entries for all distances larger than θ. Note that the direct product of two positive definite matrices is again positive definite according to the Schur product theorem; thus, C_{tap} is a valid covariance function.

On one hand, a small window size reduces the computational complexity in terms of computation time and required memory, which is needed for each chunk of data (i.e., for computing the characteristic quantity and control statistics of the multivariate EWMA chart). In the next section, we explain the suggested control procedure in more detail. In contrast, a small window size also means that more parallel processes are needed to cover the entire set of locations. As the costs for hardware are typically lower than the opportunity costs incurred by longer computation times, this drawback is not critical. More severe drawbacks must be seen in the lower accuracy if the range of the spatial dependence exceeds the window size. Alternatively, larger windows improve the accuracy of the chart while increasing the computational complexity. In Fig. 5, we illustrate this trade-off for a 2D image and selected window sizes.

Moreover, we implicitly assume that the observations $\boldsymbol{X}_{t,i}^{(p)}$ are available at all locations of each of the p windows. However, in spatial settings, where the data are measured by remote sensing (e.g., Katzfuss and Cressie 2011) or using wearable sensors (e.g., Baghdadi et al. 2019), we are often faced with incomplete data. First, missing data could be imputed for each window. This can be done in parallel, but

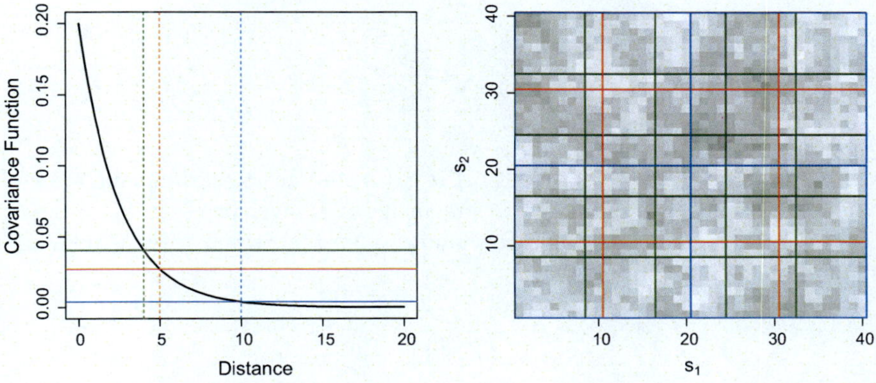

Fig. 5 Illustration of the trade-off between the accuracy improvements (due to less covariance tapering) and the window size

the proportion of missing data per window should not be too large. To not induce the spatial dependence structure of the target process, the covariance function C should not be used for imputation. Instead, simple kriging techniques can be applied, like ordinary kriging or inverse-distance kriging. Second, it could be reasonable to aggregate the data in advance (see, e.g., Zwetsloot and Woodall 2019), such that the monitoring procedure would be based on sample means of the observation across several locations or time points.

3.3 Control Characteristic and Multivariate EWMA Chart

In this paper, we consider a multivariate EWMA control chart to illustrate the procedure of parallelized monitoring. The chart is based on the approaches in Śliwa and Schmid (2005a, b), including a smoothing parameter $\lambda \in (0, 1]$. More precisely, a multivariate EWMA recursion is applied to the characteristic quantity $\boldsymbol{T}_{t,i}^{(p)}$ of each window, that is

$$\begin{aligned} \mathbf{Z}_{t,i}^{(p)} &= (1-\lambda)\,\mathbf{Z}_{t-1,i}^{(p)} + \lambda\,\boldsymbol{T}_{t,i}^{(p)} \\ &= (1-\lambda)^{d}\,\mathbf{Z}_{0}^{(p)} + \lambda \sum_{k=0}^{t-1} (1-\lambda)^{i}\,\boldsymbol{T}_{t-k,i}^{(p)}\,, \end{aligned} \tag{9}$$

where the starting value is equal to the target value, i.e., $\mathbf{Z}_{0}^{(p)} = E_{\tau=\infty}\left(\boldsymbol{T}_{t,i}^{(p)}\right)$.

For the in-control state, we can derive the moments of $\mathbf{Z}_{t,i}^{(p)}$. To be precise

$$E_{\tau=\infty}\left(\mathbf{Z}_{t,i}^{(p)}\right) = E_{\tau=\infty}\left(\boldsymbol{T}_{t,i}^{(p)}\right) \tag{10}$$

and

$$Cov_{\tau=\infty}\left(\mathbf{Z}_{t,i}^{(p)}\right) = \lambda^2 \sum_{k=0}^{t-1}\sum_{l=0}^{t-1}(1-\lambda)^{k+l}\Gamma(l-k)\,, \tag{11}$$

where $\Gamma(i-j) = Cov_{\tau=\infty}(\boldsymbol{T}(t-i), \boldsymbol{T}(t-j))$. For an increasing window size, the moments converge to the in-control moments of T_t given by (7).

We can construct a statistic for each window i as the Mahalanobis distance between $\mathbf{Z}_{t,i}^{(p)}$ and its in-control mean, that is

$$\psi_{t,i} = \left(\mathbf{Z}_{t,i}^{(p)} - E_{\tau=\infty}\left(\mathbf{Z}_{t,i}^{(p)}\right)\right)' \left(Cov_{\tau=\infty}\left(\mathbf{Z}_{t,i}^{(p)}\right)\right)^{-1}\left(\mathbf{Z}_{t,i}^{(p)} - E_{\tau=\infty}\left(\mathbf{Z}_{t,i}^{(p)}\right)\right). \tag{12}$$

Eventually, the control statistic of the multivariate EWMA chart is given by

$$T_t = \max_{i=1,\ldots,K}\{\psi_{t,i}\}\,, \tag{13}$$

where K is the total number of windows, which results from the overall sample size and the size of the windows. Similarly, Woodall and Ncube (1985) proposed running univariate charts in parallel to monitor multivariate processes.

It is worth noting that we assume that the process cannot be stopped after a signal occurred, like it is the case for the motivating example considered in Sect. 1. Thus, the control chart can be run completely in parallel without the need to check whether there was a signal in one of the windows. More precisely, the chart signals if the control statistic $\psi_{t,i}$ exceeds a certain threshold, the upper control limit (UCL), in one of the windows. However, the control statistics of the remaining windows where the statistics did not exceed the UCL are not set to zero. This would require an additional combining step, which is undesirable from a computational point of view. Nevertheless, a combining step could be important in practice. Thus, several adaptations of the proposed parallelized charts are discussed in the following paragraphs. All of them reduce the computation efficiency in terms of time and/or memory but can get important for practical applications.

For instance, the set $\{\psi_{t,i} : i = 1, \ldots, K\}$ can be used to gain further insights on the observed process. With a certain temporal delay, the results from all parallel processes can be combined, while the charts are continuously running in parallel. In the easiest case, the control statistics could be combined, only if the chart signaled at t. In this case, $\{\psi_{t,i}\}$ may provide further information on the detected fault. For instance, these statistics show which windows are affected by the structural break. That is, only these windows, for which $\psi_{t,i}$ exceeds the upper control limit, must be inspected to find reasons for the change. This reduces the inspection's costs.

If one decides for combining $\psi_{t,i}$ at all time points t, variable sampling intervals (VSI) could also be implemented instead of fixed sampling intervals (e.g., Reynolds JR 1996; Reynolds et al. 1990; Reynolds Jr and Arnold 2001). In the

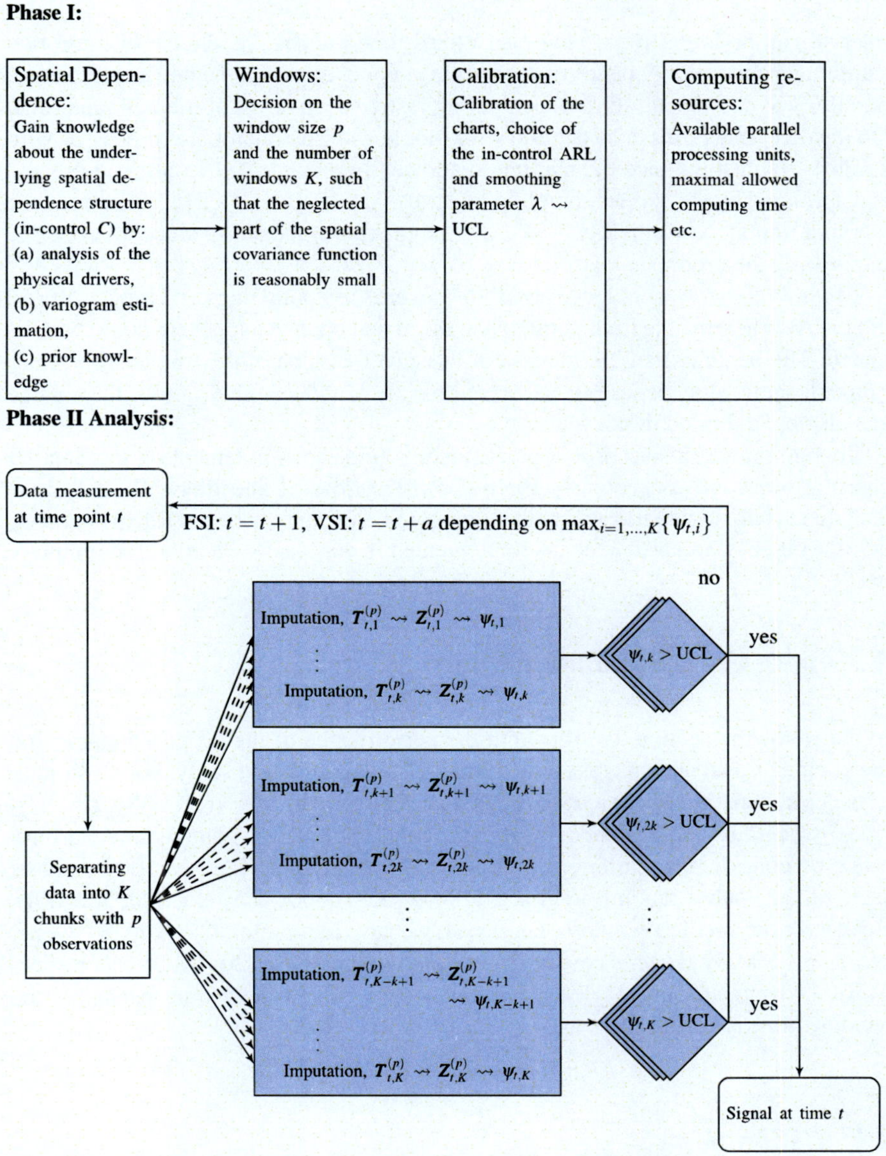

Fig. 6 Guide for implementing the parallel monitoring procedure (above: Phase I analysis, below: Phase II implementation). In case of an FSI chart, all steps colored in blue can be run fully in parallel without the need for a combining step. For VSI charts, the results from all parallel processes have to be combined in order to compute the maximum of all statistics $\psi_{t,i}$

case of VSI charts, a second warning limit would be implemented, which is between zero and the upper control limit. If $\max_{i=1,\ldots,K}\{\psi_{t,i}\}$ exceeds this limit, the process is sampled more frequently (e.g., at every time point) and otherwise the sampling interval can be larger (e.g., only every third time point). The choice of these two sampling frequencies depends on the specific applications. On the one hand, running parallel VSI charts requires less computing resources in terms of memory and time. On the other hand, this does not allow for monitoring spatiotemporal processes with a higher frequency, since the maximum allowed frequency of the observed process is determined by the shorter sampling intervals (i.e., if $\max_{i=1,\ldots,K}\{\psi_{t,i}\}$ exceeds the warning limits). Nevertheless, VSI charts might be beneficial to reduce the costs for monitoring the process.

Moreover, $\{\psi_{t,i}\}$ could be saved for several temporal lags, i.e., $\{\psi_{t-l,i} : l = 0, \ldots, L\}$. Thereby, the time point, when the actual change happened, could be estimated. The beginning of an increase of $\psi_{t,i}$ gives an idea of the true change point. Since the control statistics are also saved for all windows, the origin of the change can also be identified in this way.

To summarize this section, a practitioner's guide for implementing the parallel charting procedure is shown in Fig. 6. Both the Phase I and Phase II analysis is depicted in two flow charts, where the steps colored in blue can be run fully in parallel, while white boxes represent steps implemented in classical sequential programs.

4 Monte Carlo Simulation Study

In the following section, we illustrate the performance of the proposed charts and the computational advantages using a Monte Carlo simulation study. For all simulations, we assume the spatial process lies on a 2D unit grid $D = \{(s_1, s_2)' \in \mathbb{Z}^2 : 1 \leq s_1, s_2 \leq d\}$. Thus, the number of locations is equal to d^2 at each point of time. Furthermore, we consider three different window sizes, namely $p = 4, 25$, or 100, and three sizes of the spatial random field $n \in \{400, 2500, 10000\}$ (i.e., $d \in \{20, 50, 100\}$). Hence, for the smallest window size $p = 4$, we have to monitor $K = 100$, $K = 625$, and $K = 2500$ windows to cover to the whole sample of $d = 20, 50$, or 100, respectively. In contrast, only $K = 4$, $K = 25$, or $K = 100$ windows are needed if the windows size is 100.

4.1 *Calibration*

In Table 1, the UCLs are reported for all considered settings. The control limits were obtained using a simulation study with 10000 replications. The control charts have been calibrated for an in-control average run length (ARL) of 100 assuming independent normally distributed random variables within each window. Not surprisingly, the UCLs are increasing with the increasing window size p, increasing number of windows K, and increasing smoothing parameter λ.

It is important to stress that we intentionally calibrated the charts in this way to illustrate the effect of neglecting important parts of the spatial dependence. That is,

Table 1 Upper control limits for $p \in \{4, 25, 100\}$ and $n \in \{400, 2500, 10000\}$ observations per point of time. The smoothing parameter λ ranges from 0.1 to 1.0

λ	$p = 4$			$p = 25$			$p = 100$		
	$K = 100$	$K = 625$	$K = 2500$	$K = 16$	$K = 100$	$K = 400$	$K = 4$	$K = 25$	$K = 100$
0.1	22.5371	26.6747	29.7441	52.5296	59.0027	63.4895	140.832	152.140	159.683
0.2	23.1440	27.2215	30.2197	53.5366	59.7057	64.1167	142.824	153.603	160.805
0.3	23.3506	27.3777	30.3415	53.8971	59.9981	64.3200	143.683	154.052	161.128
0.4	23.4186	27.4287	30.3716	54.0847	60.0456	64.3897	144.041	154.140	161.228
0.5	23.4829	27.4382	30.4025	54.1080	60.0647	64.4045	144.134	154.290	161.294
0.6	23.4933	27.4806	30.4024	54.1650	60.1426	64.4567	144.178	154.296	161.274
0.7	23.5134	27.4443	30.4193	54.1899	60.1342	64.4004	144.223	154.357	161.308
0.8	23.4733	27.4735	30.4298	54.1902	60.1113	64.4300	144.342	154.319	161.324
0.9	23.5055	27.4644	30.4147	54.2088	60.1368	64.4045	144.262	154.293	161.285
1.0	23.4939	27.4524	30.4351	54.1803	60.1480	64.3888	144.275	154.318	161.314

adjacent observations lying in neighboring windows, even though they are close, are completely ignored for the computation of $\boldsymbol{T}_{t,i}^{(p)}$. Thus, it was not adjusted for an important share of spatial dependence, especially at the edges of the windows. Of course, this has more influence on the smaller considered windows. Basically, this is the trade-off to ensure that the monitoring procedure is fully scalable from a computational perspective and can, therefore, also be applied in the presence of big data. Because the charts are calibrated for independent normal variables (i.e., the ideal situation without neglecting important entries of the spatial covariance matrix), we could illustrate the influence of this effect regarding the in-control and out-of-control performance in the ensuing Monte Carlo simulation study.

4.2 *Performance of the Proposed Parallelized Chart*

Initially, we illustrate the proposed parallelized EWMA chart by a simulated random process with $n = 100 \times 100 = 10000$ spatial locations and $T = 200$ time points. The spatial dependence is exponentially decaying. Moreover, the mean changes at a few locations after $t = 100$ (see Fig. 7, top row). To be precise, the area of the fault is 6×6 pixels and, thus, lies within several windows, which have been chosen as 5×5 windows over full unit grid. That is, all windows are differently affected by the mean change. One possible way to quantify the affected area per window is the Dice similarity coefficient (DSC, Sørensen et al. 1948; Dice 1945), which is proportional to the ratio between the number of locations with a mean change and the total number of locations. To be precise, it is given by

$$\text{DSC} = \frac{2 \times \text{number of affected locations}}{\text{number of affected locations} + \text{number of locations per window}} .$$

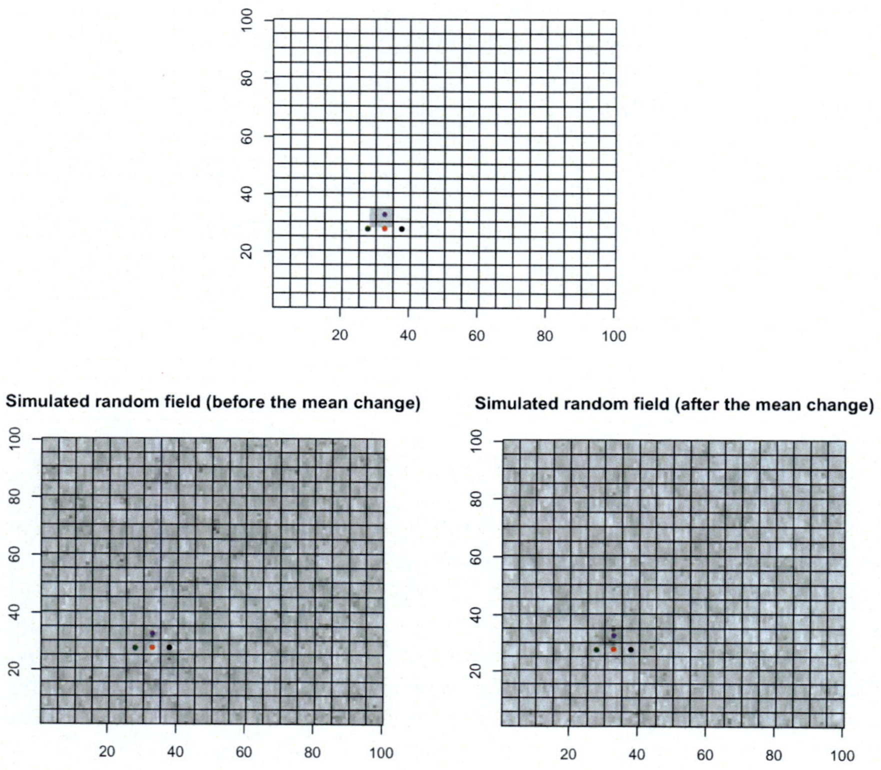

Fig. 7 Example of a parallelized EWMA chart. Top row: simulated mean change from zero to one in all gray colored locations, where the squares represent the 5×5 windows (left); bottom row: simulated random fields before and after the mean change

This leads to four different DSC values, which could be attained in this setting. We exemplarily pick four locations with different DSC values and depict the average control statistics of 100 replications in Fig. 8. Regarding the first location (colored in black), there is no change in the mean, but it is located close to the fault, which spills over to this location. However, we see that $\psi_{t,i}$ is almost not influenced, since the parallelized control charts adjust for these spatial spillover effects. The slight increase of the control statistic after $t = 100$ can be explained by the neglected part of the spatial covariance function. In the spatial representation of the average control statistics (bottom row of Fig. 8), one can see that the fault's location is clearly separated. Thus, $\psi_{t,i}$ can be used to find the origin of the mean change.

Below, the focus is on the performance of the parallelized charts. To evaluate the effect of parallelizing the monitoring procedure, 2D random fields with $n = 10000$ observations are simulated. The spatial covariance function coincides with

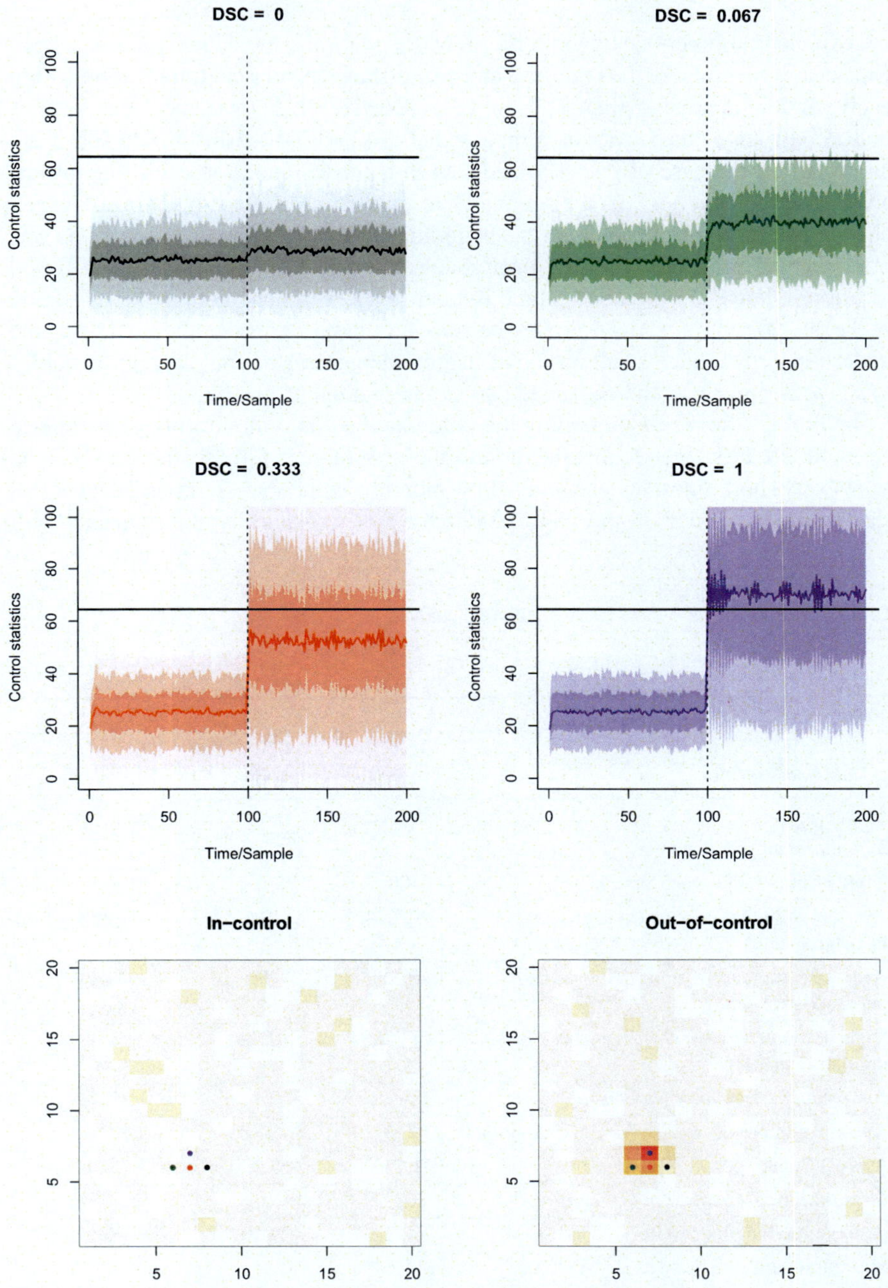

Fig. 8 Example of a parallelized EWMA chart. Average control statistics of selected locations (indicated by the dots of corresponding colors in Fig. 7) and their ±(1, 2, 3)-standard deviation bands (100 replications). The bold horizontal line corresponds to the upper control limit. Bottom row: Average control statistics in a spatial representation before and after the mean change

the exponential function shown in Fig. 3. More precisely, the window size has been chosen as $p = 4$, 25, or 100 (i.e., the minimal distance from a central location within each window is smaller than 1, 2, or 10, respectively). For instance, this implies that all entries of the covariance matrix that are highlighted in blue in Fig. 3 are neglected for the case of $p = 4$. Beside, an in-control simulation study, the out-of-control performance has also been analyzed. In the out-of-control settings, a mean change occurring immediately at the beginning, $\tau = 0$, at only a few locations was implemented, more precisely, the mean changes from 0 to 2 at the first 10 out of 10000 locations. Hence, the area of the fault is small. However, note that the interpretation of the out-of-control ARLs is limited because the charts were not calibrated using the underlying covariance function C but using independent normal random variables to show the impact of the neglected part of the spatial correlation.

In Table 2, the results regarding the computational and monitoring performance are reported. First, we see that the monitoring procedure is fully scalable and can be run in very short times, which would allow monitoring the spatiotemporal data of this size with a frequency of up to 343.29 kHz for $p = 4$. Naturally, the frequency will

Table 2 Results of the Monte Carlo simulation study

	Window size		
	$p = 4$	$p = 25$	$p = 100$
Parameters			
λ	0.5	0.5	0.5
n	10000	10000	10000
UCL	30.4025	64.4045	161.294
Computational performance			
Number of parallel processes K	2500	400	100
Average computation time (in log seconds, in-control state)	−12.7463	−9.4351	−9.3803
Average computation time (in log seconds, out-of-control state)	−12.8921	−10.9163	−9.0527
Highest frequency for monitoring (in kHz, in-control state)	343.2884	12.5205	11.8521
Monitoring performance (in-control)			
Average number of signals per 100 time points	3.3135	1.5385	1.2270
Average run length	31.782	68.329	84.844
Monitoring performance (out-of-control)			
Average run length ($\tau = 0$)	2.423	2.810	3.184
Minimum fault size per window (DSC)	0	0	0
Average fault size per window (DSC)	0.0013	0.0017	0.0018
Maximum fault size per window (DSC)	0.667	0.333	0.182
Number (percentage) of windows with DSC > 0	5 (0.002)	2 (0.005)	1 (0.01)

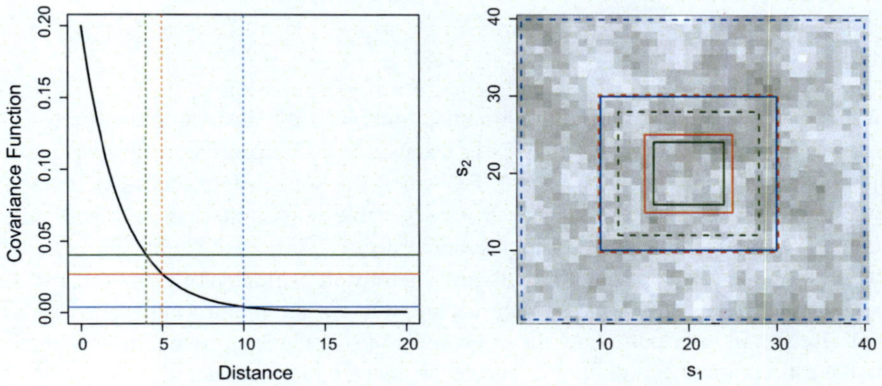

Fig. 9 Illustration of framed windows of different size, namely $p = 64$ (green), $p = 100$ (red), and $p = 400$ (blue)

approach some limit if the window size is increasing. Note that even parallelized VSI charts cannot increase this maximal frequency as the bottleneck is determined by the shorter sampling interval. Furthermore, the computation times are similar regardless of whether the process is in control or not.

Second, the above-mentioned effect can be observed by examining the number of false signals in the in-control state. Because of the window selection, especially for small windows of size 4 or 25, too many in-control signals exist, and the in-control ARL deviates from the target ARL of 100. Certainly, for practical applications, the chart should be calibrated using the true covariance function C to retain the target ARL. Because the signal-to-noise ratio is rather large, the change can be rapidly detected in all cases. For $p = 4$, there are five windows, in which two of four locations are affected by the mean change. This ratio decreases with an increased window size (i.e., 5 of 25 locations (in two windows) for $p = 25$ and 10 of 100 locations (in one window) for $p = 100$). This leads to a maximum DSC of $2/3$ ($p = 4$), $1/3$ ($p = 5$), and $2/11$ regarding $p = 10$, while the average DSC is roughly 0.001 in all settings.

To improve the accuracy of the charts while still being fully scalable, we suggest using overlapping framed windows instead of the simple window approach illustrated above. In particular, each window should be framed to a certain extent, as depicted in Fig. 9 for three example windows of different sizes. When computing the quantity $\boldsymbol{T}_{t,i}^{(p)}$, all observations within the window and its frame should be included. Further, only the values of $\boldsymbol{T}_{t,i}^{(p)}$ lying within the window (but not the frame) are used for monitoring. Although this approach increases the computational complexity, the above-described edge effects of the windows can be avoided.

5 Conclusion

In this paper, the application of parallelized multivariate EWMA control charts for monitoring spatiotemporal processes is discussed. To be precise, a fairly general spatiotemporal setting is considered, for which the spatial dependence is defined by a known covariance function C. It is reasonable to assume that the covariance function is known if the underlying physical process is fully understood. However, if the spatial dependence structure is unknown, it must be estimated (e.g., by variogram or semivariogram estimators), which has an effect on the monitoring procedure as well. Hence, in the future, how the proposed control charting procedure is affected by the estimation uncertainty of C should be analyzed in more detail.

The motivation to consider parallel multivariate control charts relates to the fact that spatial dependence decreases with an increasing distance between two locations. Thus, if the distance between two such locations is large enough, the covariance could be set to zero. In spatial statistics, this approach is known as covariance tapering. However, it still requires processing the full dataset and multiplying a large dimensional, but sparse matrix and a vector. Thus, the computational complexity increases with an increasing distance. In contrast, the proposed monitoring procedure is scalable to any number of spatial locations by an increase in the number of parallel processes. However, important entries of the covariance matrix could be lost if the windows for the parallel monitoring are not carefully selected. We illustrated this effect using a Monte Carlo simulation study. Furthermore, a framed window approach is briefly discussed. However, this alternative window definition must be analyzed in further detail in future research.

References

Baghdadi, A., Cavuoto, L. A., Jones-Farmer, A., Rigdon, S. E., Esfahani, E. T., & Megahed, F. M. (2019). Monitoring worker fatigue using wearable devices: A case study to detect changes in gait parameters. *Journal of Quality Technology*, 1–25.

Brillinger, D. (1981). Some aspects of the analysis of the evoked response experience. In A. Saleh, M. Csorgo, D. Dawson, & J. Rao (Eds.), *Proceedings of the International Symposium on Statistics and Related Topics*. Amsterdam: North-Holland Pub. Co.

Colosimo, B. M. (2018). Modeling and monitoring methods for spatial and image data. *Quality Engineering*, *30*(1), 94–111.

Crisosto, C., Hofmann, M., Mubarak, R., & Seckmeyer, G. (2018). One-hour prediction of the global solar irradiance from all-sky images using artificial neural networks. *Energies*, *11*(11), 2906.

Dice, L. R. (1945). Measures of the amount of ecologic association between species. *Ecology*, *26*(3), 297–302.

Fu, Y., & Jeske, D. R. (2014). SPC methods for nonstationary correlated count data with application to network surveillance. *Applied Stochastic Models in Business and Industry*, *30*(6), 708–722.

Garthoff, R., & Otto, P. (2017). Control charts for multivariate spatial autoregressive models. *AStA Advances in Statistical Analysis*, *101*(1), 67–94.

Hofmann, M., & Seckmeyer, G. (2017). A new model for estimating the diffuse fraction of solar irradiance for photovoltaic system simulations. *Energies*, *10*(2), 248.

Jeske, D. R., Stevens, N. T., Tartakovsky, A. G., & Wilson, J. D. (2018). Statistical methods for network surveillance. *Applied Stochastic Models in Business and Industry*, *34*(4), 425–445.

Jiang, W., Han, S. W., Tsui, K.-L., & Woodall, W. H. (2011). Spatiotemporal surveillance methods in the presence of spatial correlation. *Statistics in Medicine*, *30*(5), 569–583.

Katzfuss, M., & Cressie, N. (2011). Spatio-temporal smoothing and EM estimation for massive remote-sensing data sets. *Journal of Time Series Analysis*, *32*(4), 430–446.

Megahed, F. M., Wells, L. J., Camelio, J. A., & Woodall, W. H. (2012). A spatiotemporal method for the monitoring of image data. *Quality and Reliability Engineering International*, *28*(8), 967–980.

Megahed, F. M., Woodall, W. H., & Camelio, J. A. (2011). A review and perspective on control charting with image data. *Journal of Quality Technology*, *43*(2), 83–98.

Reynolds, M. R., Amin, R. W., & Arnold, J. C. (1990). CUSUM charts with variable sampling intervals. *Technometrics*, *32*(4), 371–384.

Reynolds, J. R., & M. R., (1996). Variable-sampling-interval control charts with sampling at fixed times. *IIE Transactions*, *28*(6), 497–510.

Reynolds, M. R, Jr., & Arnold, J. C. (2001). EWMA control charts with variable sample sizes and variable sampling intervals. *IIE Transactions*, *33*(6), 511–530.

Śliwa, P., & Schmid, W. (2005a). Monitoring the cross-covariances of a multivariate time series. *Metrika*, *61*, 89–115.

Śliwa, P., & Schmid, W. (2005b). Surveillance of the covariance matrix of multivariate nonlinear time series. *Statistics*, *39*, 221–246.

Sørensen, T. (1948). A method of establishing groups of equal amplitude in plant sociology based on similarity of species content and its application to analyses of the vegetation on Danish commons.

Sparks, R., & Patrick, E. (2014). Detection of multiple outbreaks using spatio-temporal EWMA-ordered statistics. *Communications in Statistics-Simulation and Computation*, *43*(10), 2678–2701.

Tobler, W. R. (1970). A computer movie simulating urban growth in the Detroit region. *Economic Geography*, *46*(sup1), 234–240.

Tohsing, K., Schrempf, M., Riechelmann, S., Schilke, H., & Seckmeyer, G. (2013). Measuring high-resolution sky luminance distributions with a CCD camera. *Applied Optics*, *52*(8), 1564–1573.

Wilson, J. D., Stevens, N. T., & Woodall, W. H. (2016). Modeling and detecting change in temporal networks via a dynamic degree corrected stochastic block model. arXiv preprint arXiv:1605.04049.

Woodall, W. H., & Ncube, M. M. (1985). Multivariate CUSUM quality-control procedures. *Technometrics*, *27*(3), 285–292.

Woodall, W. H., Zhao, M. J., Paynabar, K., Sparks, R., & Wilson, J. D. (2017). An overview and perspective on social network monitoring. *IISE Transactions*, *49*(3), 354–365.

Zwetsloot, I. M., & Woodall, W. H. (2019). A review of some sampling and aggregation strategies for basic statistical process monitoring. *Journal of Quality Technology*, 1–16.

Product's Warranty Claim Monitoring Under Variable Intensity Rates

Wenpo Huang, Wei Jiang, and Chengyou Shi

Abstract Product manufacturers have paid great attentions to monitoring number of warranty claims for sold product as high claims trigger improvement opportunities and/or incur excessive operational costs. Poisson distribution has been widely used to model the claim number with the pooled Poisson intensity rate being referred as the nominal failure intensity rate. Since products used by different customers are heterogeneous, failure intensity rates vary from product to product. The counts of warranty claims are often skewed and over-dispersed. Negative binomial (NB) distribution which is the compound of the Poisson-gamma mixture distribution has been widely used to model the over-dispersed count data. However the use of the NB distribution may trigger signals more than expected when the intensity rates are not randomized from time to time. In this paper, the impact of time-varying intensity rates is investigated. We show that conventional control limits based on the NB distribution-based Shewhart chart should be lowered to accommodate the reduced variation of counts when products intensity rates become constant from time to time.

Keywords Warranty counts · Hierarchical model · Over-dispersion · Weighted log-likelihood

1 Introduction

For long time, most manufacturers provide warranty coverage on sold products as an obligatory service to consumers and product warranty service becomes prominent for boosting revenue. Manufacturers collect warranty claims in order to monitor and

W. Huang (✉)
Hangzhou Dianzi University, Hangzhou, China
e-mail: bobhuang09@hotmail.com

W. Jiang · C. Shi
Shanghai Jiao Tong University, Shanghai, China
e-mail: jiangwei@sjtu.edu.cn

C. Shi
e-mail: scy320681@163.com

S. Knoth and W. Schmid (eds.), *Frontiers in Statistical Quality Control 13*,
Frontiers in Statistical Quality Control,
https://doi.org/10.1007/978-3-030-67856-2_11

analyze these claims for quality or reliability related problems. Therefore appropriate modelling of warranty claims is essential to help manufacturers examine production processes and prepare accurate inventory stockings for upcoming warranty claims.

Poisson distribution is one of the most popular distributions to model various types of count data including the number of defective items, transportation volumes, and customer arrivals. From the Poisson limiting theorem, counts generated from a large number of population which follow the binominal distribution are known to converge to the Poisson distribution. In literature, warranty claim is often modeled as the Poisson distribution (Kalbfleisch et al. 1991; Wu and Meeker 2002). Kalbfleisch et al. (1991) used the Poisson model to analyze automobile warranty data. They found that the repair counts obtained from a large fleet of cars are expected to be close to Poisson when the repair rates are small. Wu and Meeker (2002) also claimed that individual product units generally display heterogeneity in claim rates due to variations in usage and environment, but total claims across large numbers of units tend to be well approximated by Poisson distributions.

The homogeneous Poisson process (HPP) can be extended to the nonhomogeneous Poisson process (NHPP) when the Poisson intensity (or the rate) is time-varying. Majeske (2007) proposed an NHPP model with a parametric intensity rate function to analyze automobile warranty data. Three types of intensity rates, Weibull-uniform, power law, and linear hazard functions, are considered. Fredette and Lawless (2007) focused on the prediction of the total number of warranty claims over a specified time period. The Poisson intensity is given by the product of an unit-wise factor and a common shape function. Lawless et al. (2012) used the Nelson–Aalen estimator to estimate the intensity in nonparametric way.

There are two assumptions related to the use of Poisson distribution: equi-dispersion and independence. The equi-dispersion assumption requires the mean and variance of count data to be equal, which can often be violated for many real count processes. Extensions of Poisson distribution provide solutions for these data. Consul and Jain (1973) brought forth a distribution with two parameters that generalizes the usual Poisson distribution in a flexible way. The negative binomial (NB) distribution is used as an adequate model for over-dispersed data. Sparks et al. (2010) used the NB distribution to model the over-dispersed daily disease counts with a non-homogeneous mean. Urbieta et al. (2017) used the NB distribution to model the daily number of hospitalizations and compared the performance of the EWMA and CUSUM chart for the NB distribution. The COM-Poisson distribution which is another two-parameter generalization of Poisson distribution has been used to model a wide range of over-dispersion and under-dispersion. Sellers et al. (2012) surveyed different COM-Poisson models and their applications in marketing, transportation and biology. Beside of parametric distributions, Qiu et al. (2019) proposed a nonparametric method to model multivariate count data.

To account for over-dispersion, Christensen et al. (2003) proposed a hierarchical model for count data with over-dispersion and excessive zeros compared with Poisson distribution. They assumed that the Poisson intensity of the population is randomly distributed other than constant. They further showed that the hierarchical Poisson-gamma model actually leads to the NB distribution. We can see that all of the above approaches model the population count but ignore the individual heterogeneity. As discussed by Akbarov and Wu (2012), over-dispersion in the warranty claims arises mainly due to two reasons. The first is the heterogeneity of units, that is the differences in the intrinsic reliability of individual units. The second is the heterogeneity of users as products are used in different operating environments and in different usage rates. It is necessary to extend count models by including unit-specific random effects in populations with heterogeneous units (Fredette and Lawless 2007).

The remainder of this article is as follows. Section 2 reviews existing approaches of warranty claims modelling and proposes a hierarchical model to take the unit heterogeneity into consideration. Section 3 builds the log-likelihood-based charting statistic to monitor the warranty counts. The performance of proposed control chart is evaluated in Sect. 4 and illustrated by a real data example in Sect. 5. Section 6 provides concluding remarks.

2 Modelling the Warranty Claims

Let n be the number of units under warranty and $X_{i,t}$ be the number of warranty claims between time $t-1$ and time t from unit i. We denote $X_t = \sum_{i=1}^{N} X_{i,t}$ as the total claims from all units. When $X_{i,t}$ is Poisson distributed with incidence rate or intensity λ, i.e. $X_{i,t} \sim \text{Poiss}(\lambda)$, we have $X_t \sim \text{Poiss}(n\lambda)$. The sample size n can be constant or vary with time t. Various control schemes have been proposed to monitor the change in Poisson intensity λ when the sample size n is time-varying. Readers may refer to (Jiang et al. 2011; Richards et al. 2015) for detailed discussion. Richards et al. (2015) referred to this process as nonhomogeneous Poisson process (NHPP), since the intensity rate $n_t\lambda$ of the total claim X_t is also time-varying. In literature, the nonhomogeneous Poisson process has been widely used to model the number of warranty claims. It is assumed that the claims X_t are independent and follow the Poisson distribution with mean (or the intensity rate) λ_t. Different types of λ_t are discussed in Majeske (2007) when they are widely used to model the number of automobile warranty claims.

The grouped counts X_t are often shown to be over-dispersed, that is the variance in the marginal distribution of grouped count exceeds the mean. To tackle the issue of over-dispersion, Christensen et al. (2003) introduced the random effect into the Poisson rate model and proposed a hierarchical Poisson-Gamma mixed model to describe the over-dispersion phenomenon. They assumed that the Poisson intensity rate λ follows the Gamma distribution with mean μ and variance $\gamma\mu^2$, thus the probability mass function (PMF) of the count X_t is

$$h(x) = \int_0^\infty f(x; n\lambda) g(\lambda; \mu, \gamma) d\lambda$$

where $f(x; n\lambda) = \frac{(n\lambda)^x}{x!}\exp(-n\lambda)$ is the PMF of the Poisson distribution with mean $n\lambda$ and $g(\lambda; \mu, \gamma)$ is the probability density function (PDF) of the Gamma distribution with shape parameter $1/\gamma$ and scale parameter $\mu\gamma$. It is well known that above integration reduces to the negative binomial distribution with mean $n\mu$ and variance $n\mu(1 + n\mu\gamma)$ which shows the over-dispersion is related to the parameter γ.

It is worth noting that all above approaches assume the warranty claims from units all follow the Poisson distribution with the same intensity rate λ, thus all units are homogeneous. Due to the randomness of each unit, it is more reasonable to assume that the warranty claims of units are heterogeneous distributed, that is the Poisson intensity rates $\{\lambda_i\}_{i=1}^n$ of all units are different (Akbarov and Wu 2012).

To cope with the heterogeneity among units, we model the mixed nonhomogeneous Poisson processes by including unit-specified random effects in the Poisson process model. The Poisson intensity $\lambda_{i,t}$ of the claim $X_{i,t}$, modelled as the multiplication of a scaler and a function of time t, has the following parametric form

$$\lambda_{i,t} = \alpha_i h(t) \tag{1}$$

where $h(t)$ describes the shape of the rate function and α_i represents overall random variable which depends on the usage of each unit and is randomly distributed, see also in (Fredette and Lawless 2007; Lawless et al. 2009; Lawless and Crowder 2010; Akbarov and Wu 2012). In the next section, a likelihood ratio based approach is proposed to monitor the counts generated from model (1).

3 Control Chart for Monitoring Claims with Heterogeneity Poisson Intensities

As discussed in Sect. 2, the scale parameters α_i of the claims should be different across unit due to the heterogeneity in units and users. In literature, it is often assumed that α_i is randomly drawn from a gamma distribution. Simulation suggests predictions based on such assumption are robust to some types of misspecification (Fredette and Lawless 2007). The gamma assumption provides adequate predictions of automobile warranty claims if the α_i's are not actually random or if they are random but their actual distribution is not gamma.

For sake of simplicity, we only focus on the case that the shape of the rate function is constant, i.e. $h(t) \equiv h$. In such case $\lambda_{i,t}$ becomes $\alpha_i h$ which is denoted as λ_i. We assume that λ_i is randomly drawn from the gamma distribution with shape parameter α and scale parameter Λ/α. Based on Sect. 2, we have the following hierarchical model:

$$\lambda_i \sim Gamma(\alpha, \Lambda/\alpha) \tag{2a}$$
$$X_{i,t}|\lambda_i \sim \text{Poiss}(\lambda_i). \tag{2b}$$

From $\mathrm{Var}(\lambda_i) = \Lambda^2/\alpha$, it can be observed that the shape parameter α reflects units' heterogeneity (or aggregation) level. Small/large α leads to high/low variation among units.

It is worth noting that although the conditional counts $X_{i,t}|\lambda_i$ when given the Poisson intensity in Eq. (2) are independent, the marginal counts $X_{i,t}$, $t = 1, 2, \cdots$ are not independent for the randomness of λ_i. In fact the covariance between $X_{i,t}$ and $X_{i,t-1}$ can be obtained by

$$\begin{aligned}
\mathrm{Cov}[X_{i,t}, X_{i,t-1}] &= \mathbb{E}[X_{i,t}X_{i,t-1}] - \mathbb{E}[X_{i,t}]\mathbb{E}[X_{i,t-1}] \\
&= \mathbb{E}_{\lambda_i}[\mathbb{E}[X_{i,t}X_{i,t-1}|\lambda_i]] - \Lambda^2 \\
&= \mathbb{E}_{\lambda_i}[\lambda_i^2] - \Lambda^2 \\
&= \mathrm{Var}(\lambda_i) + (\mathbb{E}_{\lambda_i}[\lambda_i])^2 - \Lambda^2 \\
&= \Lambda^2/\alpha.
\end{aligned}$$

From the decomposition of variance $\mathrm{Var}[X_{i,t}] = \mathbb{E}[\mathrm{Var}(X_{i,t}|\lambda_t)] + \mathrm{Var}[\mathbb{E}(X_{i,t}|\lambda_t)] = \Lambda + \Lambda^2/\alpha$, we can further get $\mathrm{Corr}[X_{i,t}, X_{i,t-1}] = \Lambda/(\alpha + \Lambda)$. This verifies the argument that the time series of $X_{i,t}$ are correlated when the usage rates are heterogeneous.

Suppose we are simultaneously monitoring n units whose warranty claims are $X_{i,t}$. We aim to detect the possible change in the scale parameter of gamma distribution which arises from the quality issue in the manufacturing or the environment change in use. It is equivalent to consider the following hypothesis testing:

$$H_0 : \Lambda = \Lambda_0 \quad H_1 : \Lambda > \Lambda_0.$$

By integrating the conditional Poisson distribution and the gamma marginal distribution, we can obtain the log-likelihood of unit i at time t from the negative binomial distribution. The log-likelihood of the count $X_{i,t}$ can be written as

$$\ell_{i,t}(\Lambda) = \alpha \log \alpha + X_{i,t} \log \Lambda - (\alpha + X_{i,t}) \log(\alpha + \Lambda) + \sum_{z=0}^{X_{i,t}-1} \log(\alpha + z).$$

By summing up of all units, the log-likelihood of all units at time t should be

$$\ell_t(\Lambda) = n\alpha \log \alpha + X_t \log \Lambda - (n\alpha + X_t) \log(\alpha + \Lambda) + \sum_{i=1}^{n} \sum_{z=0}^{X_{i,t}-1} \log(\alpha + z).$$

The maximum likelihood estimator of Λ is $\hat{\Lambda}_t = \max(X_t/n, \Lambda_0)$. Based on this, we can obtain the likelihood ratio-based charting statistic at time t

$$T_t = \ell_t(\hat{\Lambda}_t) - \ell_t(\Lambda_0) = n\left(\hat{\Lambda}_t \log\frac{\hat{\Lambda}_t}{\Lambda_0} - (\alpha + \hat{\Lambda}_t)\log\frac{\alpha + \hat{\Lambda}_t}{\alpha + \Lambda_0}\right). \tag{3}$$

We refer to this chart as the T_{LR} chart. The log-likelihood ratio in Eq. (3) can then be used to construct the CUSUM charting statistic. The CUSUM chart and EWMA chart based on the log-likelihood ratio will be explored in our further work. In this paper, because we focus on the impact of randomized Poisson intensities on the chart performance, only the T_{LR} chart is studied. As the hierarchical Poisson-gamma model actually leads to the negative binomial distribution, various control charts for the negative binomial distribution have been proposed such as (Radaelli 1994). The difference is that control charts for the negative binomial distribution assume that all observations are independent across time while the counts in the hierarchical model in (2) are time correlated.

The T_{LR} chart signals when the charting statistic T_t first exceeds the control limit L. The run length is defined by

$$RL = \min\{t : T_t > L, t \geq 1\}$$

where the control limit is selected to provide the pre-specified in-control average run length (ARL). Discussion of the T_{LR} chart and the simulation procedure are provided in the next section.

4 Performance Evaluation

Because $X_{t,i}$s are marginally negative binomial distributed, a straightforward way is to use the quantiles of the negative binomial distribution as the control limit. For example, to obtain an in-control ARL of 100, the control limit can be selected as the 99th percentile. This approach is referred as Approach I. We should be aware that this approach assumes that the charting statistics are independent across time. That implies that the Poisson intensities λ_i are resampled from the gamma distribution each time.

More reasonable approach is to first generate λ_i from the gamma distribution and then keep λ_i fixed when generating the Poisson counts. This approach is referred as Approach II. As discussed in Sect. 3, the charting statistics become time-correlated. The main difference between two approaches lies in the way of generating λ_i. The λ_is of Approach I are time-varying and independent distributed, thus more volatile while the λ_is of Approach II are constant once they are generated. Therefore the variance of charting statistics in Approach I is greater than the variance of Approach II. It can also be explained by the conditional variance identity

$$\text{Var}[T_t] = \mathbb{E}[\text{Var}(T_t|\lambda_t)] + \text{Var}[\mathbb{E}(T_t|\lambda_t)].$$

Table 1 Control limits and ARL values of Approach I when $\text{ARL}_0 = 100$, $\Lambda_0 = 1.0$ and $n = 100$

Λ	$\alpha = 0.6$	0.8	1.0	1.2	1.4	1.6	1.8	2.0
	L =3.31	3.28	3.26	3.27	3.34	3.38	3.33	3.42
1.05	75.76	69.06	65.83	63.41	62.26	61.32	60.49	59.8
1.10	41.47	34.68	31.11	28.97	28.97	27.99	27.12	24.73
1.15	21.88	17.72	15.67	13.97	13.38	12.96	12.33	11.17
1.20	12.57	9.86	8.6	7.62	7.42	7.03	6.49	6.03
1.30	5.26	4.12	3.51	3.17	3.02	2.84	2.69	2.54
1.40	2.86	2.31	2.03	1.84	1.75	1.67	1.58	1.53
1.50	1.89	1.59	1.44	1.33	1.29	1.25	1.21	1.18

Thus the control limit of Approach I should be wider than Approach II. Using the control limit of Approach I to monitor process of Approach II can be misleading when monitoring counts from the hierarchical model (2).

Table 1 provides the control limits of ARL values of Approach I when $n = 100$. The in-control Λ_0 is set as 1.0 while the in-control ARL is set as 100. We first use simulation to find the 99th percentile of the charting statistic, then set it as the control limit, and finally use this limit to obtain the probability of exceeding the limit and the out-of-control ARL.

However the ignorance of correlation in warranty claims can lead to unexpected charting performance when control limits from Approach I are directly applied to counts from the hierarchical model (2). Due to the correlation between $X_{i,t}$ and $X_{i,t-1}$ from model (2), control limits should be adjusted. Simulation procedure is listed below to carry out the approximation of control limits and the evaluation of the performance of Approach II:

(1) set the number of units n, the shape parameter α and overall intensity rate Λ;
(2) generate $\lambda_i, i = 1, \cdots, n$ from $Gamma(\alpha, \Lambda/\alpha)$;
(3) generate $X_{i,t}$ from Poiss(λ_i) and calculate the charting statistic T_t;
(4) repeat Step (3) until $T_t > L$, record the run length;
(5) repeat Step (2)–(4) for 10,000 times, then return the ARL value.

Table 2 provides the control limits and ARL values of Approach II. The scale parameter α ranges from 0.6 to 2.0. Both Tables 1 and 2 show good detection ability of process change in Λ. By comparing with Table 1, one can also find that the control limits in Table 2 are significantly lowered. It should be noted that smaller out-of-control ARL values in Table 1 do not necessarily mean Approach I outperforms Approach II since two simulation approaches target on two different data generation models. It should be noted that the detection of process change can be greatly delayed if using the control limits from Approach I while observations actually come from Approach II, especially when units are more heterogeneous.

It is interesting to find that the chart's performance improves when the shape parameter α increases (or claims are less heterogeneous). In fact this can be explained

Table 2 Control limits and ARL values of Approach II when $\text{ARL}_0 = 100$, $\Lambda_0 = 1.0$ and $n = 100$

Λ	$\alpha = 0.6$	0.8	1.0	1.2	1.4	1.6	1.8	2.0
	L =1.64	1.90	2.10	2.28	2.44	2.50	2.55	2.60
1.05	98.88	96.64	94.41	93.3	91.38	88.36	86.8	84.77
1.10	82.2	78.07	73.59	69.76	66.84	63.42	59.89	56.84
1.15	62.93	58.15	53.06	48.69	42.43	39.36	35.49	31.12
1.20	46.76	40.98	34.52	28.86	23.37	20.84	18.14	15.27
1.30	20.57	15.44	11.34	7.62	6.26	5.46	4.33	3.61
1.40	7.93	5.31	3.79	2.54	2.12	1.88	1.75	1.63
1.50	3.88	2.27	1.64	1.43	1.26	1.21	1.18	1.15

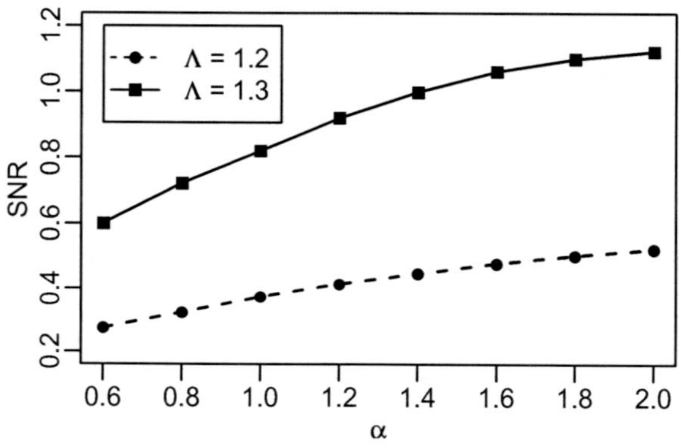

Fig. 1 SNR of T_t for α ranging from 0.6 to 2.0 when $\Lambda_1 = 1.2$ and 1.3

by the signal-to-noise ratio (SNR). For the charting statistic T_t, the SNR is expressed as

$$SNR(T_t) = \frac{\mathbb{E}[T_t|\Lambda_t = \Lambda_1] - \mathbb{E}[T_t|\Lambda_t = \Lambda_0]}{\sigma(T_t|\Lambda_t = \Lambda_0)}. \tag{4}$$

Figure 1 plots the SNRs of T_t when α ranges from 0.6 to 2.0 when Λ_1 is 1.2 and 1.3. It can be observed that the SNR increases with α which explains the proposed T_{LR} chart has better performance under larger α.

5 A Real Data Example

In this section, we use an real data example to illustrate the use of our proposed control scheme. The data contains the repair records of 282 lifts installed at an high speed rail (HSR) station in China. Each record ranging from January, 2013 to December, 2013 includes the device ID, site ID, repair date, failure cause, failure part, repair order, etc. In this study we only focus on the number of warranty claims

Table 3 The monthly repair counts of randomly selected 15 lifts located in an HSR station

ID	Jan	Feb	Mar	Apr	May	Jun	Jul	Aug	Sep	Oct	Nov	Dec	λ_i
FT6-A-32	1	1	1	0	1	0	2	0	0	1	1	3	0.91
FT5-A-9	0	0	2	2	5	1	2	3	2	3	1	2	1.82
FT5-A-8	0	1	1	2	1	1	0	2	3	2	0	2	1.18
FT5-A-7	0	0	1	2	1	0	2	4	5	3	1	0	1.45
FT5-A-5	2	2	5	4	2	3	4	0	1	2	1	2	2.36
FT5-A-2	1	3	0	1	1	0	0	1	0	1	0	2	0.82
FT5-A-17	0	1	0	0	2	0	0	0	2	3	1	1	0.64
FT5-A-14	2	0	1	1	1	3	1	2	2	0	0	1	1.27
EV-02	0	1	0	0	1	1	1	2	1	1	1	0	0.73
EV-01	0	1	0	0	0	1	3	3	1	5	0	2	1.00
B1-FT3-6	4	3	0	1	0	2	1	1	4	2	1	1	1.64
B1-FT2-5	2	2	0	1	0	1	3	0	0	4	4	6	1.64
B1-FT1-41	1	0	1	0	0	0	1	1	0	0	1	2	0.64
B1-FT1-25	0	0	1	0	0	1	2	0	1	1	0	0	0.45
B1-FT1-08	1	1	1	0	0	4	1	1	0	0	1	0	0.82
Mean	0.93	1.07	0.93	0.93	1.00	1.20	1.53	1.33	1.47	1.87	0.73	1.60	

but ignore the claim types. We randomly select 15 lifts as our target and count each lift's number of reported claims in each month. The repair counts are summarized in Table 3. We also summarize the monthly repair numbers which are listed in the last row. It can be found that the average repair number in October is significantly higher than other months. Thus we remove the records of October when estimating the in-control parameters.

The claim intensities of all lifts after removing the October counts are listed in the last column. It can be found that although all lifts are located in the same station, the intensities can be quite different due to the diversities in location, functions and targeted passengers. These differences lead to the heterogeneities of repair rates. To test whether λ_i follows the gamma distribution, we fit them with a gamma distribution shown in Fig. 2. From the cdf fitting and the QQ plot, we can find that the intensities

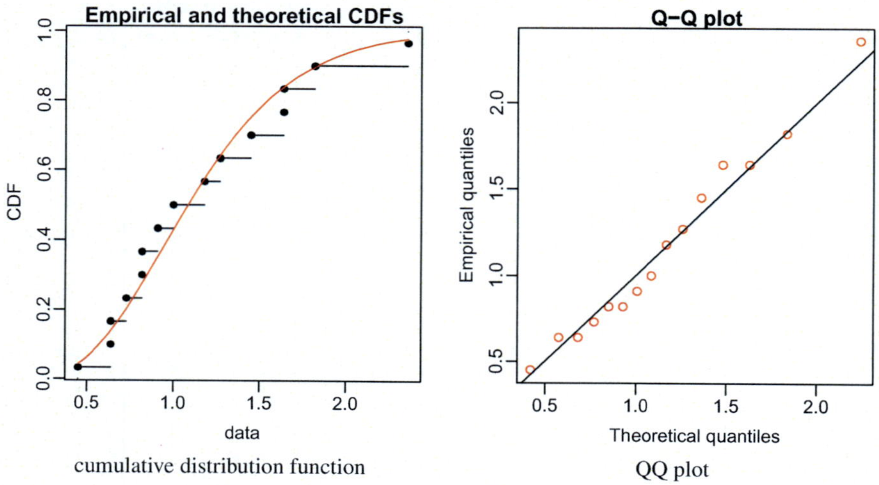

Fig. 2 Fitting λ_i with Gamma distribution

Fig. 3 Monitoring the reported monthly repair counts in Table 3 when the in-control ARL is 30

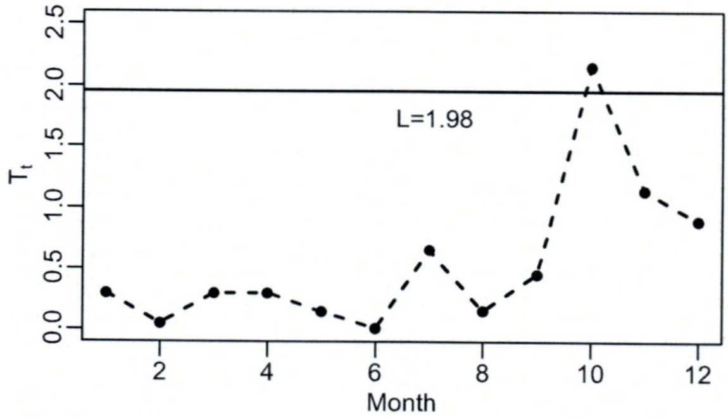

can be well fitted by the gamma distribution. The parameters of the gamma distribution are further estimated by maximizing the log-likelihood function. We find that the MLE of α and Λ are 5.2 and 1.16, respectively. In this study, the in-control ARL is set to 30, that is there is a false alarm every two and a half year. By conducting the simulation procedure provided in Sect. 4, the control limit is found to be 1.98. We use this control limit to monitor the lift repair data provided in Table 3. Figure 3 plots the charting statistics T_t along with the control limit $L = 1.98$. It can be found that the chart signals at the 10th sample which indicates the lifts have higher repair counts in October. This is mainly due to the extremely high usages of all lifts in the HSR station during the National Day holiday which is the first seven days of October.

6 Conclusions

In this study, we first discuss the over-dispersion phenomenon which widely exists in warranty claim data and then model the over-dispersion by a hierarchical Poisson-gamma model. Different from the hierarchical model in Christensen et al. (2003) which assumes that all units have the same Poisson intensities, our proposed model includes unit-specified Poisson intensities which are randomly drawn from the gamma distribution to account for the over-dispersion originated from the unit-specified usage heterogeneity. The control chart based on the log-likelihood is developed to detect possible change in gamma mean. Because the charting statistics are time-correlated due to the random effect of Poisson intensities, it is shown that the control limit should be lowered comparing with the case that the charting statistics are time-independent. We also find that the chart has better detection ability when the randomized Poisson intensity has smaller variation. The performance of the proposed method is evaluated via a lift which contains the repair records of 282 lifts installed at an high speed rail station in China.

In our further studies, we would like to extend this study in several ways. Firstly the constant failure rate would become time-varying. In literature, the failure rate is usually be selected as Weibull, power law, or linear hazard function. It could be challenging when the failure rate becoming time (or age) dependent since units may have different ages under the same calendar day. Secondly, this study only focuses on the Shewhart-type control chart. The CUSUM chart and EWMA chart can be developed based on the log-likelihood ratio. Other approaches including the weighted log-likelihood ratio proposed by Zhou et al. (2012) can also be considered since more recent data are usually more important. Finally, because our hierarchical model assumes that the Poisson intensity rates are random but fixed, another direction in the further study is to use Bayesian approach to update the estimation of unobservable unit-specified random intensities when we have more and more warranty claim data. This shares the similar idea of the self-starting control chart, see also in Shen et al. (2016).

References

Akbarov, A., & Wu, S. (2012). Forecasting warranty claims considering dynamic over-dispersion. *International Journal of Production Economics*, *139*(2), 615–622.

Christensen, A., Melgaard, H., Iwersen, J., & Thyregod, P. (2003). Environmental monitoring based on a hierarchical Poisson-gamma model. *Journal of Quality Technology*, *35*(3), 275–285.

Consul, P. C., & Jain, G. C. (1973). A generalization of the Poisson distribution. *Technometrics*, *15*(4), 791–799.

Fredette, M., & Lawless, J. F. (2007). Finite-horizon prediction of recurrent events, with application to forecasts of warranty claims. *Technometrics*, *49*(1), 66–80.

Jiang, W., Shu, L., & Tsui, K. L. (2011). Weighted CUSUM control charts for monitoring Poisson processes with varying sample sizes. *Journal of Quality Technology*, *43*(4), 346–362.

Kalbfleisch, J., Lawless, J., & Robinson, J. (1991). Methods for the analysis and prediction of warranty claims. *Technometrics*, *33*(3), 273–285.
Lawless, J. F., & Crowder, M. J. (2010). Models and estimation for systems with recurrent events and usage processes. *Lifetime data analysis*, *16*(4), 547–570.
Lawless, J. F., Crowder, M., & Lee, K. A. (2009). Analysis of reliability and warranty claims in products with age and usage scales. *Technometrics*, *51*(1), 14–24.
Lawless, J. F., Crowder, M., & Lee, K. A. (2012). Monitoring warranty claims with CUSUMs. *Technometrics*, *54*(3), 269–278.
Majeske, K. D. (2007). A non-homogeneous Poisson process predictive model for automobile warranty claims. *Reliability Engineering & System Safety*, *92*(2), 243–251.
Qiu, P., He, Z., & Wang, Z. (2019). Nonparametric monitoring of multiple count data. *IISE Transactions*, 1–13.
Radaelli, G. (1994). Poisson and negative binomial dynamics for counted data under CUSUM-type charts. *Journal of Applied Statistics*, *21*(5), 347–356.
Richards, S. C., Woodall, W. H., & Purdy, G. (2015). Surveillance of nonhomogeneous Poisson processes. *Technometrics*, *57*(3), 388–394.
Sellers, K. F., Borle, S., & Shmueli, G. (2012). The COM-Poisson model for count data: A survey of methods and applications. *Applied Stochastic Models in Business and Industry*, *28*(2), 104–116.
Shen, X., Tsui, K. L., Zou, C., & Woodall, W. H. (2016). Self-starting monitoring scheme for Poisson count data with varying population sizes. *Technometrics*, *58*(4), 460–471.
Sparks, R. S., Keighley, T., & Muscatello, D. (2010). Early warning CUSUM plans for surveillance of negative binomial daily disease counts. *Journal of Applied Statistics*, *37*(11), 1911–1929.
Urbieta, P., Ho, L. L., & Alencar, A. (2017). CUSUM and EWMA control charts for negative binomial distribution. *Quality and Reliability Engineering International*, *33*(4), 793–801.
Wu, H., & Meeker, W. Q. (2002). Early detection of reliability problems using information from warranty databases. *Technometrics*, *44*(2), 120–133.
Zhou, Q., Zou, C., Wang, Z., & Jiang, W. (2012). Likelihood-based EWMA charts for monitoring Poisson count data with time-varying sample sizes. *Journal of the American Statistical Association*, *107*(499), 1049–1062.

A Statistical (Process Monitoring) Perspective on Human Performance Modeling in the Age of Cyber-Physical Systems

Fadel M. Megahed, L. Allison Jones-Farmer, Miao Cai, Steven E. Rigdon, and Manar Mohamed

Abstract With the continued technological advancements in mobile computing, sensors, and artificial intelligence methodologies, computer acquisition of human and physical data, often called cyber-physical convergence, is becoming more pervasive. Consequently, personal device data can be used as a proxy for human operators, creating a digital signature of their typical usage. Examples of such data sources include: wearable sensors, motion capture devices, and sensors embedded in work stations. Our motivation behind this paper is to encourage the quality community to investigate relevant research problems that pertain to human operators. To frame our discussion, we examine three application areas (with distinct data sources and characteristics) for human performance modeling: (a) identification of physical human fatigue using wearable sensors/accelerometers; (b) capturing changes in a driver's safety performance based on fusing on-board sensor data with online API data; and (c) human authentication for cybersecurity applications. Through three case studies, we identify opportunities for applying industrial statistics methodologies and present directions for future work. To encourage future examination by the quality community, we host our data, Code, and analysis on an online repository.

F. M. Megahed (✉) · L. A. Jones-Farmer
Miami University, 800 E. High Street, Oxford, OH 45056, USA
e-mail: fmegahed@miamioh.edu

L. A. Jones-Farmer
e-mail: farmerl2@miamioh.edu

M. Cai · S. E. Rigdon
Saint Louis University, 3545 Lafayette Ave., Room 481, St. Louis, MO 63103, USA
e-mail: miao.cai@slu.edu

S. E. Rigdon
e-mail: steve.rigdon@slu.edu

M. Mohamed
University of Alabama at Birmingham, University Hall 4105, 1402 10th Ave. S., Birmingham, AL 35294-1241, USA
e-mail: manarkasem@gmail.com

S. Knoth and W. Schmid (eds.), *Frontiers in Statistical Quality Control 13*,
Frontiers in Statistical Quality Control,
https://doi.org/10.1007/978-3-030-67856-2_12

Keywords Human · Centered data science · Industry 4.0 · Internet of Things · Operator 4.0 · Statistical process control

1 Introduction

With the ever-decreasing costs of wireless networking and continued advancements in mobile computing technologies, we now live in a connected world. The number of internet-connected devices, e.g., sensors, machines/equipment, and medical devices, continues to exponentially increase. The data, networking, and infrastructure supporting these devices are commonly referred to as the *Internet of Things* (IoT) (Gubbi et al. 2013). The *International Data Corporation* (IDC 2019) estimates that there will be 41.6 billion connected devices, generating 79.4 zettabytes ($1\ ZB \approxeq 1$ trillion terabytes) of data in 2025.

There are a large number of opportunities to use these data/devices to transform business operations. It is expected that IoT can lead to new paradigms in:

(a) smart and connected health in health-care operations/management (Leroy et al. 2014; Chen et al. 2018);
(b) Industry 4.0 (Lasi et al. 2014) or the "Industrial Internet of Things" (IIoT) (Jeschke et al. 2017) in manufacturing and supply chain management applications;
(c) smart cities or smart grids, where local governments and/or energy companies use IoT sensors to manage their resources more efficiently;
(d) smart farming, where agricultural decisions are informed by embedded sensors and/or drones (Wolfert et al. 2017); and
(e) autonomous or connected vehicles, which capitalize on a large number of internet-connected sensors.

A common theme among these applications is that the use of wireless technology in these domains is now possible due to the development of engineering systems that capitalize on the seamless integration of computation and physical components. For this reason, Helen Gill of the United States' National Science Foundation (NSF) coined the term *cyber-physical systems (CPS)* around 2006 as a catch-all phrase to capture those technologies (Lee and Seshia 2017). The National Science Foundation (2019) states that:

> CPS will enable capability, adaptability, scalability, resiliency, safety, security, and usability that will expand the horizons of these critical systems. CPS technologies are transforming the way people interact with engineered systems, just as the Internet has transformed the way people interact with information ... Moreover, the integration of artificial intelligence with CPS creates new research opportunities with major societal implications ... While tremendous progress has been made in advancing CPS technologies, the demand for innovation across application domains is driving the need to accelerate fundamental research to keep pace.

There are two main observations that need to be highlighted based on NSF's vision for CPS technologies. First, the expectations regarding the transformational potential of CPS technologies are quite high. We agree with the statement that *fundamental research* is needed to help unlock such potential and allow for innovative applications. Second, it is unfortunate that there is limited discussion of how statistical methodologies can be capitalized upon to be developed to advance the current state of CPS technologies despite the fact that the above synopsis stems from the joint work of several directorates of NSF as well as the research and development (R&D) arms of several U.S. governmental agencies. In our view, the utility of statistical approaches (outside of regression, which non-statisticians often use for baseline comparisons in machine learning applications) is not fully understood by practitioners and researchers engaged with CPS technology. These researchers and practitioners may have had limited exposure to statistical training which explains why statistical methodologies have not been fully considered in such applications.

There are three primary objectives behind this book article:

(A) review and categorize the literature within the field of industrial statistics examining how CPS technologies can be used in modeling human performance;
(B) present an overview of the types of statistical modeling approaches that can be considered in the context CPS analysis; and
(C) highlight how industrial statistics methodologies can be used/developed to advance the reviewed literature.

We focus on the general area of human performance modeling since it has been largely ignored by the industrial statistics community and can be considered as an important pillar in many application domains (e.g., advanced manufacturing, motor vehicle safety, and cybersecurity where humans continue to be the most important and vulnerable link).

The remainder of this book chapter is organized as follows. In Sect. 2, we provide a data-driven review of industrial statistics (quality control/engineering, reliability, and experimental design) papers and highlight that our literature has yet to focus on CPS technology applications. Then, we examine how specific statistical methodologies can give insights to three CPS applications in Sects. 3–5. Our goals in these sections are to: (a) summarize main research streams within these applications; (b) provide an example to explain the data structure in detail; and (c) discuss future opportunities for statistical methodologies. Finally, we present our concluding remarks in Sect. 6.

2 Background

Prior to examining CPS technologies and models, it is important to understand how the field of industrial statistics has evolved over the past few years. The goal here is to determine whether statistical approaches that focus on monitoring, experimental design, and reliability have been applied to the data structures and application domains associated with CPS. To achieve this goal, we extracted titles, abstracts, citations, keywords, and author information from all articles published in *Technometrics*, *Journal of Quality Technology*, *Quality and Reliability Engineering International*, and *Quality Engineering* from January 2014 to August 1, 2019. The time span and journals were selected to capture the most recent developments in popular journals associated with the areas of statistical surveillance/monitoring, experimental design, and reliability. Additionally, we have limited the results to the following document types: (a) Article, (b) Early Access, or (c) Review since we did not want to include results from letters to the editor, book reviews, etc. The search was conducted on August 1, 2019 and resulted in 1576 journal papers. Capitalizing on techniques from bibliometrics and natural language processing (NLP), we present an overview of those papers in the paragraphs below.

From a bibliometric standpoint, the process of analyzing a large corpus of papers typically follows the following sequential steps: (Börner et al. 2003) (1) data extraction, which we performed using the Web of Science's portal; (2) definition of unit of analysis (in our analysis, we focused on the 1576 papers); (3) selection of metrics for evaluation and computing similarity among units of analysis (we used keywords and title words to evaluate the subject of the papers, and similarity metrics to evaluate the relationships among the papers; the reader can refer to our code in the *Supplementary Materials* section for more details on how we computed similarity); and (4) data visualization and analytics. Based on this framework, we present our results for three different metrics in Fig. 1, which was generated using the *bibliometrix* **R** package (Aria and Cuccurullo 2017).

Figure 1a depicts the 50 most frequently used keywords among the 1576 papers. One can easily see that the majority of these keywords correspond to fundamental concepts/techniques in industrial statistics. Examples include: statistical process control, estimated parameters, computer experiments, regression, and control charts. Note that the size of the node corresponds to the frequency by which it is used. Thus, the terms: *model*, *models*, *design*, *performance*, and *statistical process control* are among the most used to describe or summarize the research papers. These should not be surprising, given the methodological nature of our journals. It is, however, surprising to find that there are no application domains in the list of the top 50 keywords. Although we do see the term *construction*, this was primarily used to describe statistical concepts (e.g., construction of control limits) and not the actual field where buildings are built. Initially, we thought we would find terms such as *Industry 4.0*, *advanced manufacturing*, and/or *public health* among the top fifty keywords; however, none of these terms appear among the most used keywords. Another noteworthy feature is that the co-occurrence of keywords indicates the presence of two large clusters: (a) top/blue cluster, which captures statistical process control (SPC) methodologies, and (b) bottom/red cluster, which captures other sub-domains within industrial statistics (e.g., experimental design, reliability, dimension reduction, and model selection).

To capture an alternative representation of the major concepts represented in this corpus of recent papers, we have extracted keywords from the titles of the 1576 articles. Here, we have utilized Porter's stemming algorithm (Porter 2006) to combine keywords/concepts that have a similar root (e.g., measur is used in lieu of measurement and measuring) and *k-means clustering* to identify how the concepts should be grouped. Additionally, we have limited our analysis to stemmed terms that appeared

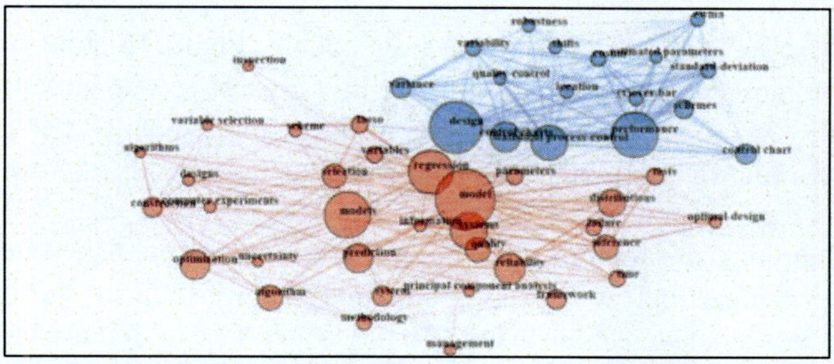

(a) The co-occurrence of the 50 most frequently used keywords

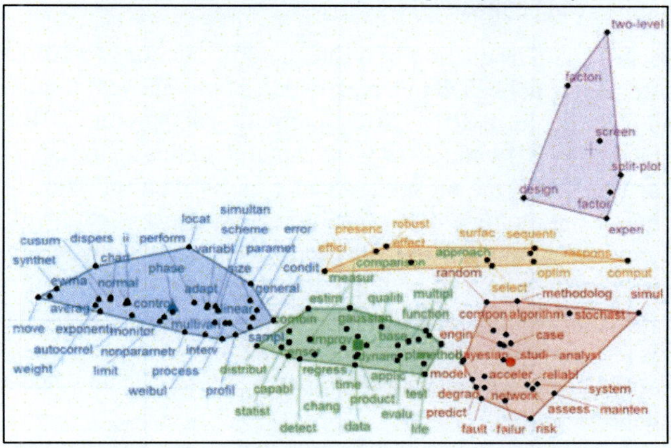

(b) A concept map of the literature, organizing stemmed title words into five clusters

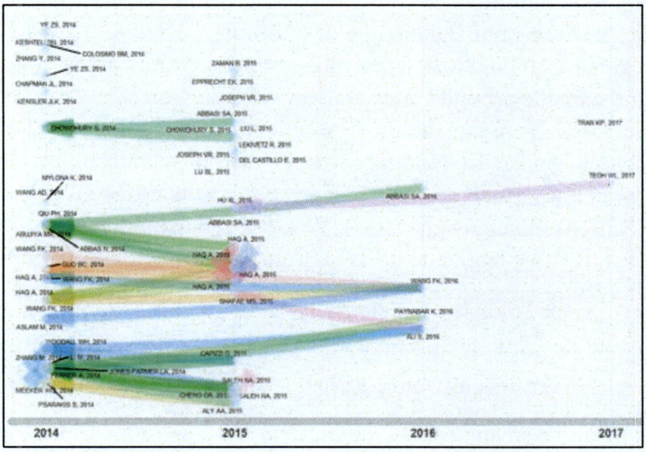

(c) Direct citation network of papers with at least 10 citations (among our 1576 papers)

Fig. 1 A bibliometric analysis of 1576 journal papers, published in four popular industrial statistics journals between 2014–2019 at least 20 times such that the graph can be readable. Figure 1b shows that there are five main topics that are commonly being investigated in our literature. When compared to Fig. 1a, we can see that Fig. 1b presents a more detailed perspective on the literature. Similar to Fig. 1a, the stemmed words do not directly capture the types of applications being examined.

We also examined the relationships among the recently published articles. To facilitate the graphical representation of this analysis, we reduced the number of papers to only those that were cited at least 10 times (per Web of Science calculations on August 1, 2019). The results of this analysis is depicted in Fig. 1c. From the figure, several observations can be made:

(a) As is expected, the introduction of a threshold of citations (>10) provides an advantage to papers published in the early part of our 5+ year time span. Thus, we see more papers from 2014–2015 timeframe compared to 2016–2017. No papers published in 2018 and 2019 were captured in this analysis.
(b) Ye and Chen (2014) is the most cited paper in our data set, with 140 Web of Science citations at the time of our analysis. This paper is captured as Ye ZS, 2014 in our visualization (see top left paper). From the figure, one can see that there are no arrows associated with this paper. We did not expect this result. In our view, there are two possible explanations for the lack of arrows: (i) most of its citations are captured in journals that are not included in our analysis and/or (ii) none of the citing papers within our data set have accumulated 10 Web of Science citations.
(c) Self-cites dominate the arrows in the visualization (see e.g., the works of Haq, Chowdhury, and Wang). It is important to note that, in the context of this analysis, this statement is not intended to be a negative comment since papers included in this list had to be published in one of our four selected journals in/after 2014 and have at least 10 citations. Thus, this merely captures potentially important work (limited by the pros and cons of using citations as a proxy to a work's quality/importance), where the author(s) continued pursuing this research stream.
(d) The majority of papers in the figure were related to statistical process monitoring/control charting methodologies. Examples include: Psarakis et al. (2014); Haq et al. (2014); Zhang et al. (2014); Woodall and Montgomery (2014); Jones-Farmer et al. (2014); Haq et al. (2015); Capizzi (2015); Paynabar et al. (2016); Teoh et al. (2017).
(e) Among these papers, CPS-related contributions were limited to: Colosimo et al. (2014) and Del Castillo et al. (2015), who investigated how profile surfaces can be monitored in advanced manufacturing scenarios.

Based on this data-driven analysis of the literature and the importance of CPS technology to the future of industry and society, we believe that our research community needs to be more active in developing methodologies for CPS technologies. As a part of the methodology development, it is important to showcase when our approaches can be used and their benefits/disadvantages when compared to machine learning methodologies.

3 Wearables for Occupational Fatigue Management

Given that manufacturing applications have traditionally been a building block for theory development and evaluation in industrial statistics and statistical process control, we start by examining physical fatigue in manufacturing workplaces. It is important to note that the new paradigm(s) of advanced manufacturing/Industry 4.0 are conceptually different from the computer-integrated manufacturing (CIM) paradigm of the 1990s. The goals of Industry 4.0 are to maximize the impact of a worker's skills by integrating him/her as an integral component of the cyber-physical infrastructure; however, the end goal of CIM was to achieve a worker-less manufacturing environment (Gorecky et al. 2014). Additionally, recent publications from the *ergonomics* and *manufacturing systems* literature are showing that the transition to advanced manufacturing is increasing the workload on skilled labor (Brocal and Sebastián 2015; Romero et al. 2016; Ferjani et al. 2017) and consequently, increasing fatigue levels.

3.1 Importance of the Domain

In a recent survey of the U.S. advanced manufacturing workers, 57.9% of respondents indicated that they were somewhat fatigued during the past work week (Lu et al. 2017). The high prevalence of occupational fatigue is problematic since fatigue results in detrimental health outcomes (both short- and long-term), increases work-errors, and reduces workers' productivity (see Cavuoto and Megahed 2017; Lu et al. 2017; Maman et al. 2017; Baghdadi et al. 2018; and references within). Moreover, Ricci et al. (2007) estimated that the annual cost of lost production time due to occupational fatigue for U.S. workers exceeds $136 billion.

Wearable devices (hereafter wearables) provide the opportunity to "unobtrusively capture physical exposure information in the workplace, a problem that has challenged the field for several decades" (Schall Jr et al. 2018, p. 351). More specifically, information extracted from wearable devices can be used to: (a) measure body angles; (b) quantify the intensity of workload/physical activity; and (c) capture a time-series of heart rate values (Maman et al. 2017; Baghdadi et al. 2018; Schall Jr et al. 2018). These are important predictors in attempting to quantify physical fatigue (Cavuoto and Megahed 2017; Maman et al. 2017).

3.2 An Illustrative Example

The example presented in this chapter is part of the broader study published by Maman et al. (2017) and further reported on in Baghdadi et al. (2018, 2019). In this example, we utilize the freely available data set Baghdadi et al. (2019), which can be accessed at: https://github.com/fmegahed/fatigue-changepoint/tree/master/Data/Raw/. The data correspond to a 3 h manual materials handling (MMH) lab study, which involved a significant period of continuous walking that would allow for analysis of changes in gait over the duration of the session. The study was completed by fifteen subjects, who were instrumented with four small inertial measurement units (IMU of size 51 mm × 34 mm × 14 mm) located at the ankle, hip, torso, and wrist and coupled with a heart rate sensor. The IMU is a wearable device that has three sensors: (a) accelerometer, measuring a body's specific force, (b) gyroscope to measure the angular rate, and (c) magnetometer for measuring the magnetic field, which is useful for determining directions based on a global reference field. The IMU captured data at 51.2 Hz, and the heart rate sensor recorded data 1 Hz throughout the task. Moreover, each of the IMU's sensors, captured data in the x, y, and z directions of the Cartesian coordinate system. We refer the reader to https://fmegahed.github.io/fatigue_case_jqt.html for information pertaining how this data can be preprocessed to extract features that can be used for fatigue modeling.

3.2.1 On the Role of Experimental Design in Data Collection

In order to understand physical fatigue development, Maman et al. (2017) designed a cross-sectional lab study using one-factor within subjects design. The one-factor corresponded to the physical workload of the task, divided into three levels: (a) low, which involved an assembly task in a standing position at a workstation, (b) medium, which simulated a supply pickup and delivery task, and (c) high, which simulated a MMH task where participants picked cartons of varying weights. The tasks were selected based on the range of tasks reported in the survey of Lu et al. (2017). Each task level was performed in a separate 3 h of continuous work session. Per Sect. 3.2, we only focus on the MMH task in our discussion.

3.2.2 Data Description

While each of the three IMU sensors can capture data at a high-frequency rate in the x, y, and z local channels (i.e., relative to body part positioning), the data needs to be preprocessed prior to use. The goals of preprocessing are to: (a) transform the data to the global X, Y, and Z frameworks, i.e., to make them independent of body positioning, (b) overcome the sensor drift problem associated with accelerometer data, and (c) generate/engineer features that can be used for either monitoring or prediction. The reader is referred to the thorough discussion of Baghdadi et al. (2019) for more details on this step.

In general, the prediction/monitoring can be applied to two different aspects of the data: (a) adjusted and filtered acceleration data or (b) features extracted from the acceleration data (e.g., statistical summaries of acceleration, jerk, stride length, stride height, and stride duration). To assist the viewer to visualize the difference between both "levels" of data analysis, we provide an animated example showing ten

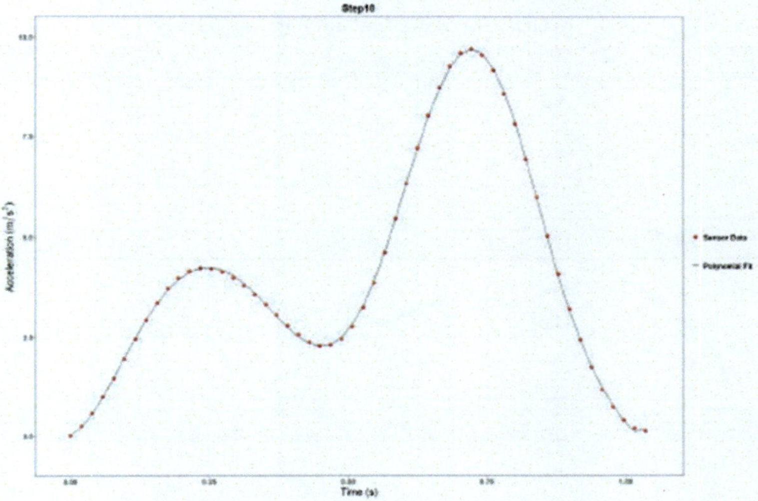

Fig. 2 Ten consecutive gait cycle segments, which can be viewed through interacting with this figure in *Adobe Acrobat Reader* (not through a web browser). The x-axis represents time, and the y-axis captures the magnitude of acceleration

consecutive strides in Fig. 2. Then, in Fig. 3, we show how three features (cumulative sum of stride length, height, and duration) vary across participants over the course of the three-hour experimental session.

3.2.3 Strategies to Analyzing the Data

The scoping/framing of this data set can lead to a number of different data analysis strategies. For example, Maman et al. (2017) performed a preliminary analysis (using only eight subjects) using penalized linear and logistic regression to determine whether statistical features extracted from the five sensors' data (e.g., mean, standard deviation, max, and min in a non-overlapping 10-minute time window) can be used to explain the variability in users' ratings of perceived exertion (RPE). Their test results showed that the RPE can be predicted with a geometric mean value, $\sqrt{\text{sensitivity} \times \text{specificity}} = 0.88$. It is also interesting to note that their selected features for their model, based on LASSO, included features from all five sensors.

In a follow-up study with a larger sample size, Maman et al. (2019) attempted to answer two main research questions: (a) whether five sensors are truly needed for modeling fatigue occurrence, and (b) whether the introduction of kinematic-driven features (e.g., stride length, height, and duration as well as back angle rotation) can improve the prediction. The results showed that: (a) with only one IMU, they can achieve a geometric mean of 0.85 (compared to 0.87 when all sensors were used), and (b) kinematic features can improve the prediction performance of the model.

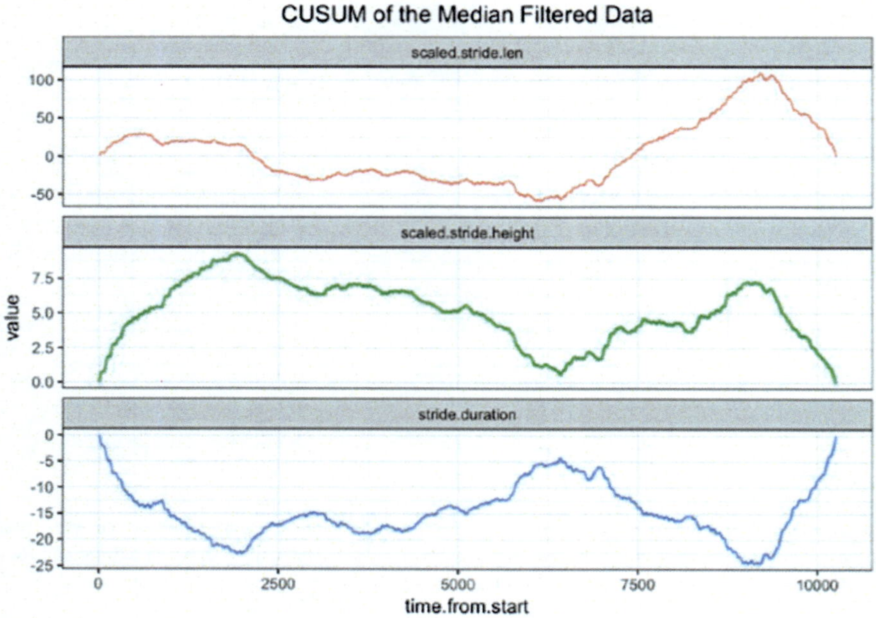

Fig. 3 The CUSUMs of stride length, height, and duration for each participant, which can be viewed through interacting with this figure in *Adobe Acrobat Reader*

Deviating from the statistical/machine learning paradigm, Baghdadi et al. (2019) examined whether the combination of multivariate change-point models and clustering can help in understanding how different subjects fatigue over the course of the experimental task. In their analysis, they chose to investigate changes in stride length, height, and duration. Based on their analysis, the multivariate change-point method revealed systematic changes in walking patterns. From the clustering analysis, the participants were divided into four groups, which reflected changes in both the magnitude and pattern of fatigue development. The participants adjusted their stride to mitigate the effects of fatigue. While maintaining the required pace, some participants elected to have shortened and faster slides, and others had longer but slower strides. This observation has not been observed in the ergonomics literature in the past since: (a) wearables have not been thoroughly examined for modeling fatigue development, and (b) the limited number of papers that examined the use of wearables (or vision-based systems) utilized regression or classification approaches, which would not allow for the obtained insights.

3.3 Opportunities for Statistical (Process Control) Research

There are many opportunities for methodological research in human performance monitoring. Much research is currently being conducted on the use of wearable sensors for human performance monitoring in occupational, athletic, health-related, and even leisure settings; however, there remains little consensus on how to address issues of validity and reliability of the data that originate from these sensors. Sensor devices, often with limited computational power and battery life, often lead to erroneous values and missing data. Industrial statisticians have a long history of conducting a measurement system analysis prior to establishing a monitoring scheme. These methodologies need to be extended for sensor data, especially multiple sensors that are mobile. For example, the recent work of Tsung et al. (2018) applies transfer learning methods to the multiple sensor problem in manufacturing, seismic, and rail transit domains. Modifying these methods for use with mobile sensors could be of benefit in the event of sensor failures or missing observations. In addition to poor data reliability, sensor data is generally unusable in raw form and requires substantial preprocessing in order to be usable.

Once the data from sensors can be considered reliable and valid for measurement, it is important to establish a baseline sample of typical performance. This seemingly retrospective analysis may be impossible as monitoring is required in "real time", and the process may change frequently depending on the task that is being conducted. Thus, it is important to consider adaptive methods that will take into account the changing nature of the process. For example, one of the ultimate goals when monitoring gait is to determine when a worker begins to experience fatigue. An obstacle to achieving this goal is that different people exhibit fatigue in different ways. Models would have to be developed that use sensor data to determine whether one among several profiles is being exhibited. The adaptive monitoring methods should include consistent methodologies to transform the data into usable forms, to continuously evaluate the quality and reliability of the sensor data, and evaluate the process for changes.

4 The Use of On-Board Vehicular Sensors to Capture Changes in Driver's Safety Performance

With the emergence of on-board vehicular sensors and technology, an increasing number of naturalistic driving studies (NDS) have been initialized by research teams around the world. NDS continuously records details of the driver, the vehicle, and surrounding environment via unobstructive on-board vehicular sensors and not having experimental control (Regan et al. 2012; Eenink et al. 2014). The first large-scale NDS was the 100-Car NDS, pioneered by the Virginia Tech Transportation Institute (Dingus et al. 2006). Following this 100-Car study, NDS has also been explored in other countries, such as the Second Highway Research Program (SHRP 2) in the

United States and the European Naturalistic Driving Study (UDRIVE), as well as within specific target populations, such as teenagers, older drivers, and commercial truck drivers (Guo 2019). The NDS supplies researchers with high-volume, high-resolution, and high-dimensional driving data, which create opportunities and pose challenges to data transformation techniques and existing statistical models.

4.1 Importance of the Domain

Traditional truck crash prediction studies heavily rely on retrospective data that ultimately trace back to post-crash police reports, interviews with witnesses and survivors, and vehicle inspection (Hickman et al. 2018; Stern et al. 2019). Although these post-crash data can be thorough and detailed, they inherently suffer from several limitations.

(a) In real-life data, crashes are *extremely rare* compared with non-crashes. The fatality rate of traffic accidents is 1.13 per 100 million miles driven in the United States, and the property-damage-only crash rate is 313 per 100 million miles driven (National Highway Traffic Safety Administration 2017). Considering this rareness nature of crashes, hundreds of years of data are needed to achieve the sufficient statistical power to conclude on the difference in fatality rates between autonomous vehicles and human drivers (Kalra and Paddock 2016; Guo 2019).
(b) Post hoc reports, interviews, and inspections are often problematic. Retrospective data sources collected hours, days, or even weeks after the occurrence of the crash are subject to *recall bias*, so the accuracy and validity of these data are heterogeneous across different sources. In addition, as the data were collected by police officers, some critical factors in a meaningful time period leading up to the crash, such as distraction, were not regularly collected or reported due to various reasons (Dingus et al. 2011); this is called the *low-resolution* issue.
(c) Crashes are generally *under-reported*, particularly for minor accidents and those without injuries. Savolainen et al. (2011) estimated that 25% of minor-injury and 50% of non-injury crashes were not reported in the data collected after accident, compared with 100% of fatality-involved crashes were reported. High under-reporting rates of non- and minor-injury crashes may cause bias to statistical inference.

In view of these limitations, NDS has been developed by proactively and continuously collecting high-resolution driving data without obtrusive interference. Therefore, NDS has several strengths compared to traditional retrospective data sources. First, NDS collects safety critical events (SCE) such as hard braking events, which have significantly higher incidence rates than crashes. These unsafe incidents have been suggested to be indicative of near-crashes, collision, and crashes (Dingus et al. 2006; Guo et al. 2010). Second, NDS records high-frequency and detailed traveling data, including but not limited to speed, GPS, and multidimensional accelerometers. Unsafe incidents are often recorded once a kinematic threshold is triggered.

Therefore, researchers can accurately trace back to several seconds prior to an incident and identify risk factors associated with that event. Since all events and data were collected by automated sensors and devices, recall bias, information bias, and under-reporting are minimized in NDS.

4.2 An Illustrative Example

Here, we demonstrate an example of transforming and fusing NDS on-board sensor data with externally obtained weather and road geometry data from online APIs, and fitting statistical models based on fused data. For the purpose of illustration, we used a sample of a 10-driver NDS data set collected by a commercial trucking and transportation company in the United States using on-board vehicular sensors.

4.2.1 Data

The data for this demonstrating example include five sources. Three of them were collected by the NDS study (real-time ping, SCEs, and driver characteristics), while two of them were from online API sources (weather and road geometry).

(a) *Real-time pings*: A small device was installed in each sample truck, and it will ping irregularly (typically every 1–10 min). Each ping will collect real-time data such as the vehicle number, date and time, latitude, longitude, driver identification number (ID), and speed at that time. A sample of the ping data is shown in Table 1.

(b) *SCEs*: Real-time, time-stamped SCEs and associated GPS locations were collected by the truck company and provided as outcome variables. Specifically, four types of safety critical events were recorded: (1) hard brake (HB), (2) head-

Table 1 Ping data

Ping_time	Speed	Latitude	Longitude	Driver
2015-10-23 08:00:00	0	33.9	−118.1	canj1
2015-10-23 08:08:10	0	33.9	−118.1	canj1
2015-10-23 08:09:26	5	33.9	−118.1	canj1
2015-10-23 08:22:58	4	33.9	−118.1	canj1
2015-10-23 08:23:12	8	33.9	−118.1	canj1
…	…	…	…	…

Table 2 Safety critical events

Driver	Event_time	Event_type
canj1	2015-10-23 14:46:08	HB
canj1	2015-10-26 15:06:03	HB
canj1	2015-10-28 11:58:24	HB
canj1	2015-10-28 17:42:36	HB
canj1	2015-11-02 07:13:56	HB
...	...	...

way (HW), (3) collision mitigation (CM), and (4) rolling stability (RS). Once some kinematic thresholds with regard to the driving behavior were met, the sensor will be automatically triggered and the information of these SCEs (latitude, longitude, speed, and driver ID) will be recorded. A sample of SCE data is shown in Table 2.

(c) *Driver demographics*: A table that includes the birth date of each driver was provided by the commercial truck company, and the age of the driver can be calculated. The driver's age table is shown in Table 3.

(d) *Road geometry data from the OpenStreetMap API*: Two road geometry variables for the sample drivers will be queried from the OpenStreetMap (OSM) project: *speed limits* and *the number of lanes*. The OSM data are collaboratively collected by over two million registered users via manual survey, GPS devices, aerial photography, and other open-access sources (Wikipedia contributors 2019). The OpenStreetMap Foundation supports a website to make the data freely available to the public under the Open Database License, and could be queried using the osmar package in statistical computing environment **R**. A sample of road geometry data retrieved from the OpenStreetMap is demonstrated in Table 4.

Table 3 Drivers

Driver	Age
canj1	46
farj7	54
gres0	55
hunt	48
kell0	51
lewr10	27
rice30	34
smiv	49
sunc	37
woow59	24
...	...

Table 4 Road geometry from the OpenStreetMap API

Latitude	Longitude	Speed_limit	Num_lanes
30.3	−89.8	65	2
30.3	−91.7	65	2
30.3	−91.7	60	2
30.3	−91.6	60	2
30.3	−91.6	60	2
…	…	…	…

Table 5 Weather from the DarkSky API

Ping_time	Longitude	Latitude	Precip_intensity	Precip_probability	Wind_speed	Visibility
2015-10-23 08:09:26	−118.1	33.9	0	0	0.21	9.82
2015-10-23 08:22:58	−118.1	33.9	0	0	0.22	9.81
2015-10-23 08:23:12	−118.1	33.9	0	0	0.22	9.81
2015-10-23 08:23:30	−118.1	33.9	0	0	0.22	9.81
2015-10-23 08:38:00	−118.1	34.0	0	0	0.24	9.81
…	…	…	…	…	…	…

(e) *Weather data from the Dark Sky API*: Wseather variables, including precipitation intensity, precipitation probability, wind speed, and visibility, were retrieved from the Dark Sky API. The Dark Sky API allows the users to query historic minute-by-minute weather data anywhere on the globe (The Dark Sky Company, LLC 2019). According to the official document, the Dark Sky API is supported by a wide range of weather data sources, which are aggregated together to provide the most precise weather data possible for a given location (The Dark Sky API 2019). Among several different weather data APIs we tested, the Dark Sky API provides the most accurate and complete weather variables. A sample of the weather data retrieved from the DarkSky API is shown in Table 5.

4.2.2 Data Transformation and Merging

The research question in this example is *whether the truck driver's cumulative driving time is associated with unsafe driving behaviors*. To answer the question, we transformed the original ping data in the following ways to fit in various statistical models; the SCEs, age of the drivers, road geometry, and weather were joined back to the transformed data using different combinations of keys.

(a) *Trips*: for each of the truck drivers, if the real-time ping data showed that the truck was not moving for at least 20 min, the ping data will be separated into two different trips. A trip is continuous driving intervals with a mean length of 1.8 h.
(b) *Half-hour intervals*: since the length of the trips is heterogeneous (it varies from 5 min to more than 8 h), making it difficult to conduct statistical analysis, we further divide trips into half-an-hour fixed intervals.
(c) *Shifts*: the trips will be further divided into different shifts if the driver was not moving for at least eight hours. A shift is, therefore, a long driving time with intermittent short rests (20 min to 8 h) within shifts.
(d) *A proxy of driver fatigue*: we took the cumulative summation of interval time within a shift for each driver as the cumulative driving time, and used it as a proxy of driver fatigue. The rest time between trips was not counted in the cumulative driving time calculation.

After the transformed data sets were created, different sources of data sets were merged for statistical analyses. Our statistical analyses were based on the following two merged data sets: *merged half-hour intervals (MHHI)* and *merged shifts*.

- *MHHI*: SCEs were left joined to half-hour intervals if the two data sets had a common driver ID and the event time of SCE was between the start and end time of the half-hour interval. A binary variable of whether SCEs occurred and a count variable of the number of SCEs in each half-hour interval were created using the merged SCEs table. Driver's table was merged to MHHI using a common driver ID. Road geometry data were merged to the ping data by the latitude and longitude coordinates, and weather data were merged to the ping data by the latitude and longitude coordinates, date, and time. The road geometry and weather at ping level were then aggregated to MHHI by taking the mean of each variable.
- *Merged shifts*: SCEs, driver's age, weather, and road geometry were merged to transformed shifts data in a similar way as described in MHHI. The only difference is that the four tables were merged to transformed shifts, instead of transformed half-an-hour intervals.

4.2.3 Statistical Models

To answer the question of *whether the truck driver's cumulative driving time is associated with unsafe driving behaviors*, we first consider a Bayesian hierarchical logistic regression, where the response is $Y_i = 1$ if a crash occurred in a given segment/time period, and $Y_i = 0$ if no crash occurred. Logistic regression is the most popular statistical model used in traffic safety studies. This hierarchy will be performed based on the transformed MHHI data. The outcome is:

$$Y_i \sim \text{Bernoulli}(p_i)$$

where the probability p_i of the outcome being $Y_i = 1$, that is, a crash occurred, is

$$\log \frac{p_i}{1 - p_i} = \beta_{0,d(i)} + \beta_{1,d(i)} \cdot \mathrm{CT}_i + \sum_{j=1}^{J} x_{ij}\eta_j. \tag{1}$$

In other words, the log-odds of p_i is dependent on the predictors through the model in (1). We further assume that the intercepts $\beta_{0,d(i)}$ and the slopes $\beta_{1,d(i)}$ are random effects because they vary from driver to driver. They are treated as if they are random samples from two normal distributions with unknown means and variances; that is,

$$\beta_{0,d} \sim \text{i.i.d. } N(\mu_0, \sigma_0^2), \quad d = 1, 2, \cdots, D \tag{2}$$

and

$$\beta_{1,d} \sim \text{i.i.d. } N(\mu_1, \sigma_1^2), \quad d = 1, 2, \cdots, D \tag{3}$$

Here i is an index of the ith observation and $d(i)$ is the driver's index of the ith observation. We assume that each driver has a different baseline probability of having SCEs, which is the random intercept $\beta_{0,d}$. We also assume that the probability change of SCEs as a consequence of cumulative driving time (CT_i) is different among drivers, which is the random slope $\beta_{1,d(i)}$. We assume exchangeable priors for the random intercepts and slopes, respectively. The parameters μ_0, σ_0, μ_1, and σ_1 are hyper-parameters with priors. Since we usually have little prior knowledge on the hyper-parameters, we assigned diffuse priors for these hyper-parameters.

Figure 4 presents the primary results of the logistic regression for the sample 10 drivers. The x-axis shows the cumulative driving time in hours and y-axis shows the estimated probability of SCEs. The gray lines indicate the estimated curve for the sample 10 drivers, while the blue line is the estimated curve for hyper-parameters. The figure indicates that despite the heterogeneity among the 10 drivers, there was a common trend of negative association between cumulative driving time and the probability of SCEs. It is to be noted that this is an illustrating example of 10 drivers and may not capture the true relationship.

Logistic regression considers the probability of an event during a given interval, but it ignores the frequency of events. Therefore, we considered a Poisson regression, another widely used statistical model in traffic safety studies, to account for the frequency of events. In a Poisson regression, the response is the number of events Y_i in a given time interval T_i. We assume that the number of events has a Poisson distribution with the mean of λ_i.

$$Y_i \sim \text{Poisson}(T_i \lambda_i) \tag{4}$$

where

$$\log \lambda_i = \beta_{0,d(i)} + \beta_{1,d(i)} \cdot \mathrm{CT}_i + \sum_{j=1}^{J} x_{ij}\eta_j. \tag{5}$$

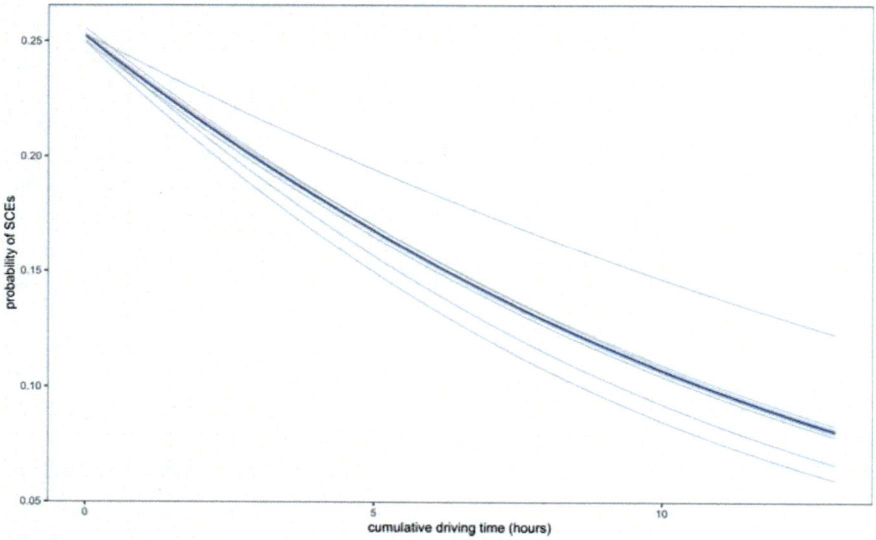

Fig. 4 Cumulative driving time and estimated probability of SCEs from the hierarchical logistic model. The gray lines are estimated curves for individual drivers and the blue line is the estimated curve for hyper-parameters

We assume random effects for the intercept and slope for cumulative driving time similar to the logistic regression.

Figure 5 presents the primary results of Poisson regression model for the sample 10 drivers. In contrast to the logistic regression, the y-axis is the rate of SCEs (the number of SCEs in a certain amount of time). The interpretation is similar to the logistic regression: despite the heterogeneity among the 10 drivers, there was a negative association between the rate of SCEs and cumulative driving time. This figure demonstrates less variability among drivers compared to the logistic regression.

The Poisson regression model assumes that the intensity of events is a constant, which may not be true in real-life transportation practice. Here we present a reliability model, a time-truncated non-homogeneous Poisson process (NHPP) with an intensity function of the form $\lambda(t) = (\beta/\theta)(t/\theta)^{\beta-1}$, called the power law process (PLP). Using the merged shift data set described previously, we aim to determine whether SCEs occurred more frequently at early stages of shifts, toward the end of shifts, or neither. Figure 6 presents the idea of the NHPP model. The x-axis is the cumulative driving time in minutes and y-axis is the shift number. Each arrow represents a shift, while a red cross is a SCE. The figure shows a pattern of more events in the early stages of shifts, indicating reliability deteriorating.

We can use mathematical notations to formulate this NHPP. Let $T_{d,s,i}$ denote the time to the dth driver's sth shift's ith critical event. The total number critical events of dth driver's sth shift is $n_{d,s}$. The ranges of the indices are:

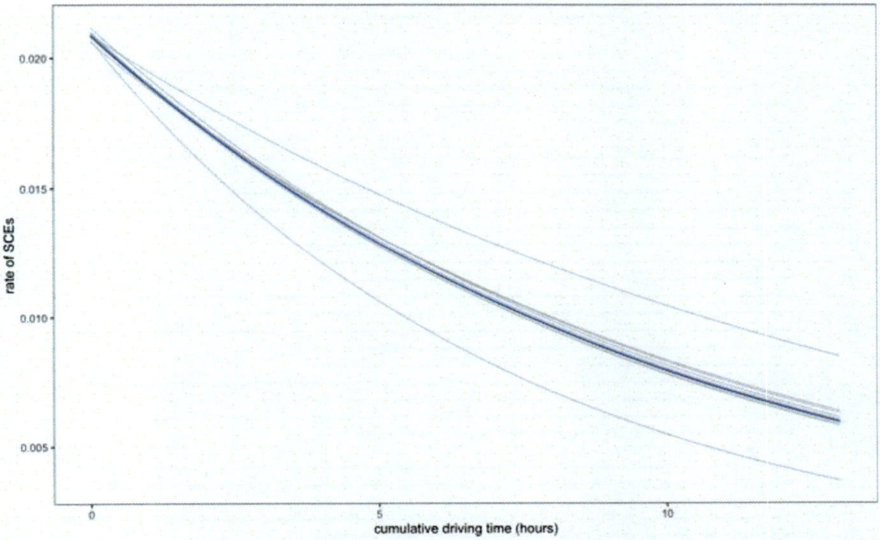

Fig. 5 Cumulative driving time and estimated rate of SCEs from the hierarchical Poisson model. The gray lines are estimated curves for individual drivers and the blue line is the estimated curve for hyper-parameters

$$
\begin{aligned}
i &= 1, 2, \ldots, n_{d,S_d}, \\
s &= 1, 2, \ldots, S_d, \\
d &= 1, 2, \ldots, D.
\end{aligned}
$$

We assume that the times of critical events within the dth driver's sth shift were generated from a PLP, with a fixed shape parameter β and varying scale parameters $\theta_{d,s}$ across drivers d and shifts s. The model can be described as Eq. 6.

$$
\begin{aligned}
T_{d,s,1}, T_{d,s,2}, \cdots, T_{d,s,n_{d,s}} &\sim \text{PLP}(\beta, \theta_{d,s}), \qquad s = 1, 2, \ldots, S_d; \;\; d = 1, 2, \ldots, D \\
\log \theta_{d,s} &= \gamma_{0d} + \gamma_1 x_{d,s,1} + \gamma_2 x_{d,s,2} + \cdots + \gamma_k x_{d,s,k} \\
\gamma_{01}, \gamma_{02}, \cdots, \gamma_{0D} &\sim \text{i.i.d. } N(\mu_0, \sigma_0^2)
\end{aligned}
\tag{6}
$$

As before, we assume random intercept and slope effects for the scale parameter θ that vary from driver to driver. The shape parameter β shows the reliability changes of drivers. When $\beta > 1$, the intensity function $\lambda(t)$ is increasing, the reliability of drivers is decreasing, and SCEs are becoming more frequent; when $\beta < 1$, the intensity function $\lambda(t)$ is decreasing, the reliability of drivers is increasing, and SCEs are becoming less frequent; when $\beta = 1$, the NHPP is simplified as a homogeneous Poisson process with the intensity of $1/\theta$. We assume diffuse priors for the fixed parameters.

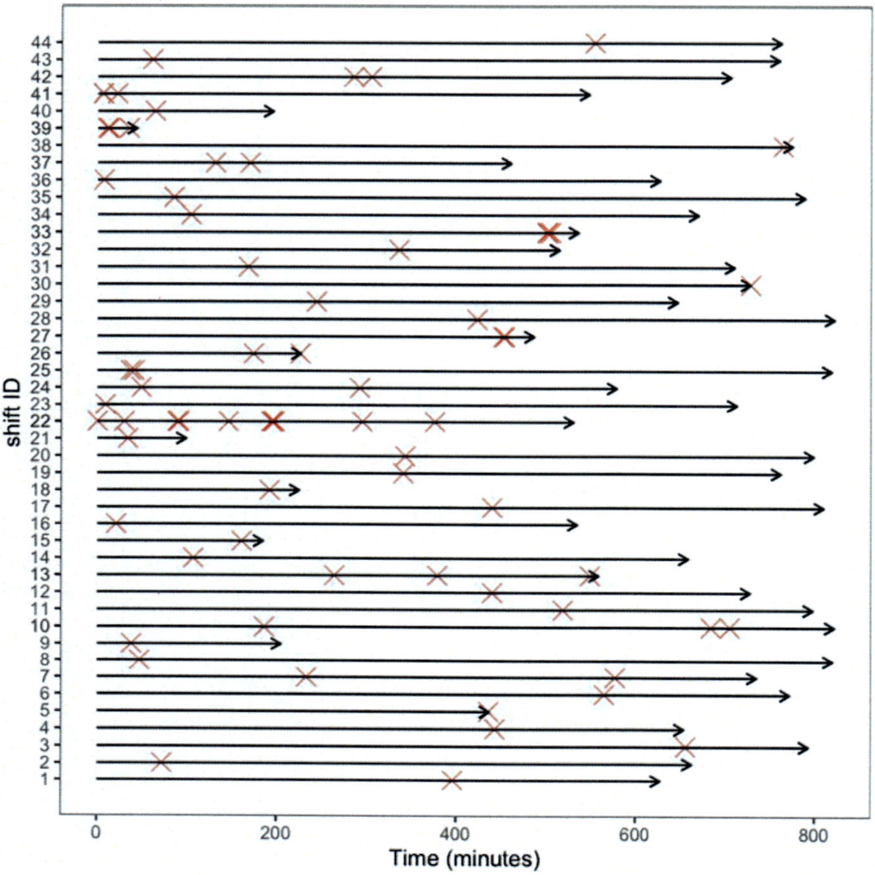

Fig. 6 An arrow plot of time to SCEs and shift length at different shifts

Figure 7 presents the primary results of the NHPP model. Each line is a shift and the colors indicate different drivers. The y-axis is the intensity of the PLP and x-axis is the cumulative driving time. The figure suggests that the intensity of SCEs is negatively associated with cumulative driving time.

4.3 Opportunities for Statistical (Process Control) Research

NDS data provide a unique opportunity to understand the risk factors of driving, but also present challenges in statistical analyses. The challenges are in two aspects: (1) high-resolution, high-dimension, and sparse data nature push for faster and less computationally intensive estimation methods; (2) the fact that multiple SCEs can

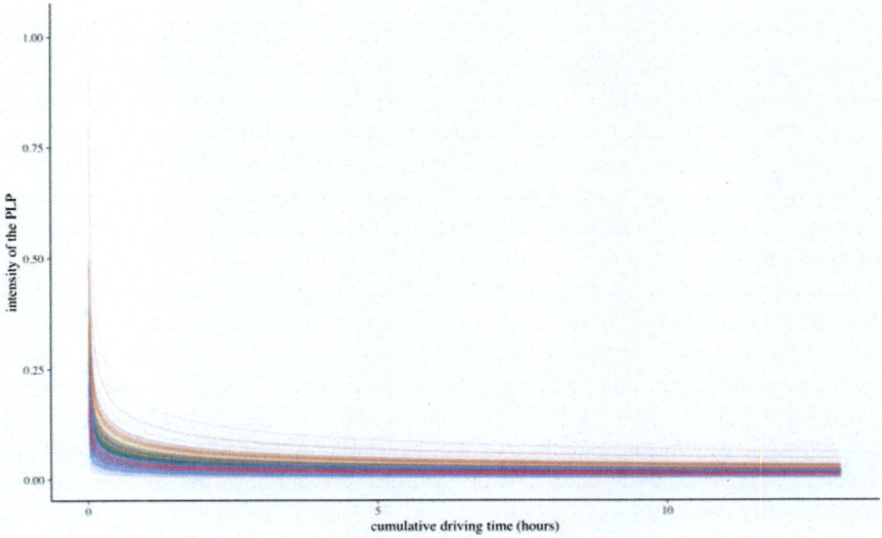

Fig. 7 Cumulative driving time and estimated intensity of SCEs from the hierarchical NHPP. Each line represents a shift and each color represents a driver. In total, there are 196 shifts for the 10 sample drivers

occur in one trip or shift requires more application of recurrent event models or reliability models.

Although SCEs are more frequent than crashes, they are still rare events compared to the total miles driven. Bayesian estimation, such as hierarchical Bayes, is especially useful and powerful in the context of sparse data by placing informative or weakly informative priors on parameters and hyper-parameters. However, modern Bayesian estimation is empowered by Markov Chain Monte Carlo (MCMC), which is not scalable in the context of high-volume and high-dimensional NDS data. There are interesting and more efficient MCMC estimation strategies for high-volume or high-dimensional data, such as the Firefly Monte Carlo (Maclaurin and Adams 2015), Pseudo-Marginal MCMC (Quiroz et al. 2019), and energy conserving subsampling Hamiltonian Monte Carlo (Dang et al. 2019), but these algorithms requires coding and hyper-parameter tuning, and are applied among a limited number of statisticians.

Recurrent event models or reliability models fit naturally with the event generating process of NDS (Guo 2019). With high-resolution NDS data, recurrent event or reliability models could uncover the patterns of non-homogeneous process, which are important for improving traffic safety. There are several studies that used recurrent event models, such as risk change-point for novice teenage drivers (Li et al. 2017, 2018) and random-effects frailty models (Chen and Guo 2016). We presented an application of NHPP with PLP among truck drivers in this paper, which could be further improved by adding one more recovery parameter that accounts for the reliability recovery at each rest within a shift. The goal of such models is to determine when a driver begins to exhibit symptoms that can be a precursor to risky driving behavior. If such a task can be accomplished, it can be used to help schedule rest times thereby leading to safer driving conditions. Just as was the case for the wearable sensors, we have found that different drivers behave differently with regard to cumulative driving time and other variables not associated directly with the driver, such as weather. It may be that a personalized rest strategy would have to be developed for every driver in order to minimize that driver's chance of a crash or other critical event.

In addition to the use of reliability models, there exists several opportunities in the realm of naturalistic driving performance monitoring (both for commercial driving and commuters alike). Commonly available data sources include: (a) vehicular speed and location data from mobile phones (Hosseinioun et al. 2015); (b) mirror check data from driver assistance systems (Li and Busso 2015); and (c) driver drowsiness detection systems through the use of either on-board cameras or lane departure warning systems. The key questions in each of these applications are: (1) how to define/identify "normal" driving behavior and (2) what is an "optimal" approach to quickly detect changes from this normal baseline. Addressing these questions typically would entail designing an experiment/simulation to collect some data and possibly, capitalizing/developing statistical monitoring procedures for addressing both questions. The challenge here is to develop methodologies that can possibly account for the inherent high-frequency (i.e., highly autocorrelated data structures), non-stationary nature of the collected data without loss of important driving-related information. Furthermore, the problem is complicated since the performance is affected by external, often non-observed, conditions that relate to traffic conditions and other commuters on the road. From a monitoring perspective, these systems should be considered in the context of processes that exhibit transient shifts or transient degradation state(s) since, in the absence of a crash, breaks at the end of a trip would result in a recovered and reinvigorated driver.

5 Biometric-Driven Computer Security

5.1 Importance of the Domain

User authentication is a process used to verify that someone seeking access to a computing device (remote or otherwise) is who they claim to be. The primary goal of user authentication is to ascertain that only a legitimate user is granted access. The increasing popularity of personal devices and internet services and the sensitivity of information they often store (i.e, banking) prompt the need for secure authentication mechanisms.

Although various user authentication mechanisms are widely deployed by almost all devices and web-services, finding a secure mechanism remains a challenging problem. As almost all of the deployed mechanisms suffer from well-known security issues. For example, passwords which are the most widely deployed authentication approach nowadays tend to be insecure as the users normally pick weak passwords (Florencio and Herley 2007), or share their password (Singh et al. 2004). Moreover, traditional biometrics, such as fingerprints, often have high error rates, and susceptible to impersonation or spoofing attacks.

Behavioral biometrics is the study of using the human unique behavioral patterns to authenticate a person. For a long time, handwriting and signature have been used as a mechanism to identify the users. Over the last several decades, several studies have investigated the use of behavioral biometrics to authenticate the user to web-services or devices in order to improve the security of the authentication process. Examples of behavioral biometrics include gait, GUI interaction, key stroke dynamic, mouse dynamics, and tapping (Yampolskiy and Govindaraju 2008).

With the increase of the data that can be collected from various sensors on mobile devices, smart-watches or keyboard and mouse on desktops/laptops, a lot of work has been done on investigating the ability of using these data to identify the unique behavior of the user and use it to authenticate the user; either as a stand-alone mechanism or combined with another mechanism. In contrast with passwords and traditional biometrics, behavioral biometrics cannot be forgotten, stolen, or shared and are noninvasive.

5.2 *An Illustrative Example*

The example in this chapter is part of a study published by Mohamed and Saxena (2016). Gametrics (Mohamed and Saxena 2016) is a game-based behavioral biometric system based on simple drag and drop games used to capture the unique user's cognitive ability as well as her unique mouse interactions with the game.

5.2.1 Task Design

Gametrics is based on a simple drag and drop game challenge. A sample of the games used in the Gametrics study is shown in Fig. 8. The game has three target objects and six moving objects and the user's task is to drag a subset of the moving objects (answer objects) to their corresponding target objects. In order to solve the challenge, the user needs to understand the content of the images, find the semantic relationship between the answer objects and the target objects, and drag the answer objects to their corresponding targets to solve the challenge.

The game starts by placing the moving objects in random locations on the image. Then, each moving object picks a random direction in which it will move. The object continues moving in its current direction until it collides with another object or with the challenge border. A collision results in an object moving in a new random direction. The user needs to press a "Start" button to start the game and drag and drop all the answer objects to their corresponding target objects in order to complete the

Fig. 8 Gametrics Challenge Instance. Targets, on the Left, are Static; Moving Objects, on the Right, are Mobile. The User Task is to Drag–Drop a Subset of the Moving Objects (Answer Objects) to their Corresponding Targets. *Republished with permission of ACM (Association for Computing Machinery), from* Mohamed and Saxena (2016); *permission conveyed through Copyright Clearance Center, Inc.*

game. In order to utilize the user's gameplay as an authentication mechanism, all the user's interaction with the game is recorded. Specifically, at each time interval, the gameplay's log stores the objects locations, the mouse location, and status (up/down).

5.2.2 Data Collection

In order to evaluate the applicability of identifying the user based on the way she interacts with the game challenges, Mohamed and Saxena (2016) collected data from human users. They recruited the participants using Amazon Mechanical Turk (AMT) service. For the purpose of their study, they created three human intelligence tasks (HITs) distributed over three days. The first HIT was created with 100 assignments to have 100 unique workers. Ninety eight valid submissions were gathered until the HIT expired. On the next two days, emails were sent to the participants asking them to participate in the follow-up study. On the first day, the participates were asked to solve 60 challenges and on the second and third day, they were asked to solve 36 challenges. Sixty two participants performed the study on the second day and twenty nine performed the study on the third day. In total, the participants successfully completed a total of 9,076 challenges. The average time the participants took to complete a game challenge was around 7.5 s.

5.2.3 Feature Extraction

From each gameplay log, a total of 64 features were extracted. The features can be categorized into three categories: (1) features that capture the cognitive characteristics of individuals, (2) features that capture the mouse interaction characteristics of the participants, and (3) features that are related to both the cognitive abilities as well as mouse interaction. The different mechanisms of solving the game challenges are related to the cognitive characteristics of individuals. These characteristics were captured based on the following features:

(a) The time between the user pressing the start button and the first mouse event and the time of the first click/drag. These timing measures capture the time the participants take to understand the challenge and start solving it.
(b) The average, standard deviation, minimum, and maximum of the times between each of the drops and the start of the next drag (these capture the time the user takes to find the next answer object).
(c) The total time taken by the user to complete the challenge.

The mouse movement characteristics of the users were captured through the following features:

(a) The average, standard deviation, minimum, and maximum of the speed and acceleration while the user is searching for an answer object and while the user is dragging the object.
(b) The average, standard deviation, minimum, and maximum of the duration between each two consecutively generated timestamps and the "silence" during move and during drag.
(c) The average, standard deviation, minimum and maximum of time duration between reaching an object and clicking on it, and the time duration between approaching a target object and dropping an answer object on it.
(d) The average, standard deviation, minimum, and maximum of the angles between the lines that connect each three consecutive points in the mouse movement trajectory.

Other mixed features were also extracted that relate to both cognitive and mouse movement characteristics of the participants such as:

(a) The total distance the mouse moved within a game challenge.
(b) The average, standard deviation, minimum, and maximum of the difference between the straight line connecting the start and the end of a move or a drag and the real distance traveled.
(c) The average, standard deviation, minimum, and maximum of the distance between a click and the object center, and a drop and the target center.

5.3 *Data Analysis*

Mohamed and Saxena (2016) utilized the random forest classifier in their analysis as it is efficient, can estimate the importance of the features, and is robust against noise (Maxion and Killourhy 2010). A random forest is an ensemble approach based on the generation of many classification trees, where each tree is constructed using a separate bootstrap sample of the data. In order to classify a new input, the new input is run down all the trees and the result is determined based on majority voting.

In the classification task, the positive class corresponds to the legitimate user's gameplay and the negative class corresponds to the impersonator (other user/zero-effort attacker). Therefore, true positive (TP) represents the number of times the legitimate user is granted access, true negative (TN) represents the number of times the impersonator is rejected, false positive (FP) represents the number of times the impersonator is granted access, and false negative (FN) represents the number of times the correct user is rejected.

As performance measures for the classifier, false positive rate (FPR) and false negative rate (FNR) were used. The FPR measures the security of the proposed system, i.e., the accuracy of the system in rejecting impersonators. The FNR measures the usability of the proposed system as high FNR which means that the system has a high rejection rate to the legitimate users. To make the system both usable and secure, ideally, both FPR and FNR need to be as close to 0 as possible.

To improve the accuracy of the classification, they ran a program to find the subset of features that produces the best classification results, because using many features can cause over fitting of the classifier and therefore reduce the accuracy of the future prediction; thus removing some features may improve the accuracy. Therefore, they report the results obtained from selecting the best subset of features per user. Moreover, they studied the identification of the user based on a single game challenge and on combining two challenges.

Inter-session Analysis

As mentioned above, Mohamed and Saxena (2016) collected data from 98 AMT workers during the first day of our data collection experiment. Each of them completed 60 challenges. The data was divided into 98 sets based on the users' identities (ids). In order to build a classifier to authenticate a user based on her gameplay biometrics, Mohamed and Saxena (2016) defined two classes. The first class contains the gameplay data from a given user (to be identified), and the other class contains randomly selected gameplay data from other users. Then, the data was divided into two sets, one for training and the other for testing. The first 40 gameplay instances of each participant and 40 gameplay instances of the randomly selected set were used to train the classifier, while the other 20 are used for testing. The results are shown in the first row ("Day 1") of each block in Table 6. We see that utilizing two gameplay instances is better than using a single instance. The best results are acquired by merging two challenge instances in which both the FPR and FNR are 2%.

Table 6 Study results: performance for the classifier. We show the results of using a single challenge and merging of two challenges

		FPR	FNR
Single	Day 1	0.06 (0.06)	0.02 (0.04)
	Day 2	0.09 (0.09)	0.07 (0.10)
	Day 3	0.07 (0.06)	0.07 (0.10)
Merge	Day 1	0.02 (0.05)	0.02 (0.05)
	Day 2	0.05 (0.09)	0.04 (0.09)
	Day 3	0.04 (0.06)	0.03 (0.05)

Intra-session Analysis

The other main goal of the study was to check the accuracy of the classifier over multiple sessions. As mentioned above, 62 AMT workers participated in the study on the second day and 36 participated in the study on the third day. For each of these users, the data of the gameplay of the previous day(s) was used to train the classifier, and then the classifier was tested with the data collected in the next day(s). The results are shown in the second and third rows ("Day 2" and "Day 3") in each block in Table 6. We find that the performance of the classifier degrades slightly compared to the first day, inter-session analysis. Also, we still found that merging two instances provides better results than using a single instance. The best results are again acquired by merging two instances. For the second day, False Positive Rate = 0.05 and False Negative Rate = 0.04 and for the third day, False Positive Rate = 0.04 and False Negative Rate = 0.03.

Summary of Results

The results obtained from the classification models show that Gametrics is a viable form of behavioral biometrics. The results show that the classifier can identify the users and reject a zero-effort attacker with a high overall accuracy, especially when two game instances are merged together.

5.4 Opportunities for Statistical (Process Control) Research

In this section, we provided an example that was first shown by Mohamed and Saxena (2016) to illustrate how cybersecurity researchers and professionals approach existing open questions in computer security. The provided example presented a scenario, where one would like to authenticate a single user. Behavioral biometrics can also be applied to other scenarios. For example, in a consulting project by the first author, a client who utilizes a *User and Entity Behavior Analytics (UEBA)* platform was interested in developing thresholds for detecting compromised insiders who may be leaking and/or engaging in unauthorized computer systems based activities. The problem is complex given that a user's risk score is generated accounting for: (a) their behavior over the past several sessions, which means that users engaging in unauthorized behavior can "normalize" their score if not detected quickly; (b) the user's computer activity can be affected by cyclical/periodic tasks that are not captured in their expected score calculation; and (c) both internal (mergers and acquisitions) and external (changes in regulatory requirements) events by the client company also influence their observed risk scores. Thus, the risk score should be adjusted to account for such events, which tend to overlap. From an SPC perspective, this problem can be framed as a multivariate (ideally self-starting and distribution-free) control charting problem. The use of SPC should be superior to the classification approach presented herein since its performance cannot be calibrated using out-of-control signals. Unlike the approach presented herein, which implicitly assumed that there are only a few out-of-control behaviors that need to be accounted for, in our estimation, the success of deploying and standardizing the use of SPC in such scenarios would heavily depend on its translation to packages (in multiple programming languages) that would reduce the statistical burden on practitioners.

6 Concluding Remarks

In this paper, we presented an overview of human performance modeling and discussed three application domains where our methodologies and research efforts can result in significant contributions and impact. To encourage the readers to build on these example applications, we have made the data and code available through links in the *Supplementary Materials* section. In our estimation, the future of many "industrial" and "work" systems would depend heavily on how to optimize the synergies across machines, humans, and computing technologies. This clearly fits within the traditional driving forces of quality and productivity for statistical quality control research and practice. The readers should note that our perspectives shared in this article are also reflected by other research communities, where the role of "humans" and how to capitalize on their data/performance represent the next research frontier. For example, "human-in-the-loop" models are widely considered to be the next frontier in artificial intelligence research (Rea 2018; Oneto et al. 2018; Bavaresco et al. 2019; Zanzotto 2019). Thus, as a community, we need to play a major role on devising statistical procedures for monitoring and modeling such data.

Acknowledgements This work was supported in part by: the National Science Foundation (CMMI-1635927 and CMMI-1634992); the Ohio Supercomputer Center (PMIU0138 and PMIU0162); the American Society of Safety Professionals (ASSP) Foundation; the University of Cincinnati Education and Research Center Pilot Research Project Training Program; the Transportation Informatics Tier I University Transportation Center (TransInfo); a Google Cloud Platform research grant for data management; and a Dark Sky grant for extended API access (i.e., they increased the number of possible queries per day). Dr. Megahed's research was also partially supported by the Neil R. Anderson Endowed Assistant Professorship at Miami University.

Supplementary Materials

To promote future work in this area, we have created the following GitHub repository, https://github.com/caimiao0714/ISQC2019, which hosts the data and/or code that are associated with the three application examples presented in this paper. In addition, a **R** Markdown summarizing our analyses and main results is also hosted at: https://caimiao0714.github.io/ISQC2019/.

References

Aria, M., & Cuccurullo, C. (2017). bibliometrix: An r-tool for comprehensive science mapping analysis. *Journal of Informetrics*, *11*(4), 959–975.

Baghdadi, A., Megahed, F. M., Esfahani, E. T., & Cavuoto, L. A. (2018). A machine learning approach to detect changes in gait parameters following a fatiguing occupational task. *Ergonomics*, *61*(8), 1116–1129.

Baghdadi, A., Cavuoto, L.A., Jones-Farmer, L.A., Rigdon, S.E., Esfahani, E.T., & Megahed, F.M. (2019). Monitoring worker fatigue using wearable devices: A case study to detect changes in gait parameters. *Journal of Quality Technology* (to appear).

Bavaresco, M.V., D'Oca, S., Ghisi, E., & Lamberts, R. (2019). Technological innovations to assess and include the human dimension in the building-performance loop: A review. *Energy and Buildings*, 109365.

Börner, K., Chen, C., & Boyack, K. W. (2003). Visualizing knowledge domains. *Annual Review of Information Science and Technology*, *37*(1), 179–255.

Brocal, F., & Sebastián, M. A. (2015). Identification and analysis of advanced manufacturing processes susceptible of generating new and emerging occupational risks. *Procedia Engineering*, *132*, 887–894.

Capizzi, G. (2015). Recent advances in process monitoring: Nonparametric and variable-selection methods for phase I and phase II. *Quality Engineering*, *27*(1), 44–67.

Cavuoto, L., & Megahed, F. (2017). Understanding fatigue: Implications for worker safety. *Professional Safety*, *62*(12), 16–19.

Chen, C., & Guo, F. (2016). Evaluating the influence of crashes on driving risk using recurrent event models and naturalistic driving study data. *Journal of Applied Statistics*, *43*(12), 2225–2238.

Chen, M., Qu, J., Xu, Y., & Chen, J. (2018). Smart and connected health: What can we learn from funded projects? *Data and Information Management*, *2*(3), 141–152.

Colosimo, B. M., Cicorella, P., Pacella, M., & Blaco, M. (2014). From profile to surface monitoring: SPC for cylindrical surfaces via gaussian processes. *Journal of Quality Technology*, *46*(2), 95–113.

Dang, K. D., Quiroz, M., Kohn, R., Tran, M. N., & Villani, M. (2019). Hamiltonian Monte Carlo with energy conserving subsampling. *Journal of Machine Learning Research*, *20*(100), 1–31.

Del Castillo, E., Colosimo, B. M., & Tajbakhsh, S. D. (2015). Geodesic gaussian processes for the parametric reconstruction of a free-form surface. *Technometrics*, *57*(1), 87–99.

Dingus, T.A., Klauer, S.G., Neale, V.L., Petersen, A., Lee, S.E., Sudweeks, J., Perez, M.A., Hankey, J., Ramsey, D., Gupta, S., et al. (2006). The 100-car naturalistic driving study, Phase 2: Results of the 100-car field experiment. Technical report, United States, Department of Transportation, National Highway Traffic Safety.

Dingus, T. A., Hanowski, R. J., & Klauer, S. G. (2011). Estimating crash risk. *Ergonomics in Design*, *19*(4), 8–12.

Eenink, R., Barnard, Y., Baumann, M., Augros, X., & Utesch, F. (2014). Udrive: The European naturalistic driving study. In *Proceedings of Transport Research Arena*. IFSTTAR.

Ferjani, A., Ammar, A., Pierreval, H., & Elkosantini, S. (2017). A simulation-optimization based heuristic for the online assignment of multi-skilled workers subjected to fatigue in manufacturing systems. *Computers & Industrial Engineering, 112*, 663–674.

Florencio, D., & Herley, C. (2007). A large-scale study of web password habits. In *Proceedings of the 16th International Conference on World Wide Web* (pp. 657–666). ACM.

Gorecky, D., Schmitt, M., Loskyll, M., & Zühlke, D. (2014). Human-machine-interaction in the industry 4.0 era. In *2014 12th IEEE International Conference on Industrial Informatics (INDIN)* (pp. 289–294). https://doi.org/10.1109/INDIN.2014.6945523.

Gubbi, J., Buyya, R., Marusic, S., & Palaniswami, M. (2013). Internet of things (IoT): A vision, architectural elements, and future directions. *Future Generation Computer Systems, 29*(7), 1645–1660.

Guo, F. (2019). Statistical methods for naturalistic driving studies. *Annual Review of Statistics and Its Application, 6*, 309–328.

Guo, F., Klauer, S. G., Hankey, J. M., & Dingus, T. A. (2010). Near crashes as crash surrogate for naturalistic driving studies. *Transportation Research Record, 2147*(1), 66–74.

Haq, A., Brown, J., & Moltchanova, E. (2014). Improved fast initial response features for exponentially weighted moving average and cumulative sum control charts. *Quality and Reliability Engineering International, 30*(5), 697–710.

Haq, A., Brown, J., Moltchanova, E., & Al-Omari, A. I. (2015). Improved exponentially weighted moving average control charts for monitoring process mean and dispersion. *Quality and Reliability Engineering International, 31*(2), 217–237.

Hickman, J. S., Hanowski, R. J., & Bocanegra, J. (2018). A synthetic approach to compare the large truck crash causation study and naturalistic driving data. *Accident Analysis & Prevention, 112*, 11–14.

Hosseinioun, S.V., Al-Osman, H., & El Saddik, A. (2015). Employing sensors and services fusion to detect and assess driving events. In *2015 IEEE International Symposium on Multimedia (ISM)* (pp. 395–398). IEEE.

IDC. (2019). The growth in connected IoT devices is expected to generate 79.4zb of data in 2025, according to a new IDC forecast. International Data Corporation. https://www.idc.com/getdoc.jsp?containerId=prUS45213219. Accessed 1 Aug 2019.

Jeschke, S., Brecher, C., Meisen, T., Özdemir, D., & Eschert, T. (2017). Industrial internet of things and cyber manufacturing systems. In *Industrial internet of things* (pp. 3–19). Berlin: Springer.

Jones-Farmer, L. A., Woodall, W. H., Steiner, S. H., & Champ, C. W. (2014). An overview of phase I analysis for process improvement and monitoring. *Journal of Quality Technology, 46*(3), 265–280.

Kalra, N., & Paddock, S. M. (2016). Driving to safety: How many miles of driving would it take to demonstrate autonomous vehicle reliability? *Transportation Research Part A: Policy and Practice, 94*, 182–193.

Lasi, H., Fettke, P., Kemper, H. G., Feld, T., & Hoffmann, M. (2014). Industry 4.0. *Business & Information Systems Engineering, 6*(4), 239–242.

Lee, E. A., & Seshia, S. A. (2017). *Introduction to embedded systems: A cyber-physical systems approach*. Cambridge: MIT Press.

Leroy, G., Chen, H., & Rindflesch, T. C. (2014). Smart and connected health [guest editors' introduction]. *IEEE Intelligent Systems, 29*(3), 2–5.

Li, N., & Busso, C. (2015). Detecting drivers' mirror-checking actions and its application to maneuver and secondary task recognition. *IEEE Transactions on Intelligent Transportation Systems, 17*(4), 980–992.

Li, Q., Guo, F., Klauer, S. G., & Simons-Morton, B. G. (2017). Evaluation of risk change-point for novice teenage drivers. *Accident Analysis & Prevention, 108*, 139–146.

Li, Q., Guo, F., Kim, I., Klauer, S. G., & Simons-Morton, B. G. (2018). A Bayesian finite mixture change-point model for assessing the risk of novice teenage drivers. *Journal of Applied Statistics, 45*(4), 604–625.

Lu, L., Megahed, F. M., Sesek, R. F., & Cavuoto, L. A. (2017). A survey of the prevalence of fatigue, its precursors and individual coping mechanisms among us manufacturing workers. *Applied Ergonomics*, *65*, 139–151.

Maclaurin, D., & Adams, R.P. (2015). Firefly Monte Carlo: Exact MCMC with subsets of data. In *24th International Joint Conference on Artificial Intelligence*.

Maman, Z. S., Yazdi, M. A. A., Cavuoto, L. A., & Megahed, F. M. (2017). A data-driven approach to modeling physical fatigue in the workplace using wearable sensors. *Applied Ergonomics*, *65*, 515–529.

Maman, Z.S., Chen, Y.J., Baghdadi, A., Lombardo, S., Cavuoto, L.A., & Megahed, F.M. (2019). A data analytic framework for physical fatigue management using wearable sensors. *Expert Systems with Applications* (under review).

Maxion, R. A., & Killourhy, K. S. (2010). Keystroke biometrics with number-pad input. In *2010 IEEE/IFIP International Conference on Dependable Systems & Networks (DSN)* (pp. 201–210). IEEE.

Mohamed, M., & Saxena, N. (2016). Gametrics: Towards attack-resilient behavioral authentication with simple cognitive games. In *Proceedings of the 32nd Annual Conference on Computer Security Applications* (pp. 277–288). ACM.

National Highway Traffic Safety Administration. (2017). Traffic safety facts 2015: A compilation of motor vehicle crash data from the fatality analysis reporting system and the general estimates system.

National Science Foundation. (2019). Cyber physical systems (CPS) — NSF 19-553. https://www.nsf.gov/pubs/2019/nsf19553/nsf19553.htm. Accessed 4 Aug 2019.

Oneto, L., Navarin, N., Donini, M., & Anguita, D. (2018). Emerging trends in machine learning: Beyond conventional methods and data. In *ESANN*.

Paynabar, K., Zou, C., & Qiu, P. (2016). A change-point approach for phase-I analysis in multivariate profile monitoring and diagnosis. *Technometrics*, *58*(2), 191–204.

Porter, M. F. (2006). An algorithm for suffix stripping. *Program*, *40*.

Psarakis, S., Vyniou, A. K., & Castagliola, P. (2014). Some recent developments on the effects of parameter estimation on control charts. *Quality and Reliability Engineering International*, *30*(8), 1113–1129.

Quiroz, M., Kohn, R., Villani, M., & Tran, M. N. (2019). Speeding up MCMC by efficient data subsampling. *Journal of the American Statistical Association*, *114*(526), 831–843.

Rea, B. (2018). Ai, robotics, and automation: Putting humans in the loop. Deloitte. Dbriefs Webcast. https://www2.deloitte.com/us/en/pages/dbriefs-webcasts/events/october/2018/dbriefs-ai-robotics-automation-putting-humans-in-loop.html. Accessed 28 Nov 2019.

Regan, M., Williamson, A., Grzebieta, R., & Tao, L. (2012) Naturalistic driving studies: Literature review and planning for the australian naturalistic driving study. In *Australasian College of Road Safety Conference (2012), Sydney, New South Wales, Australia*.

Ricci, J. A., Chee, E., Lorandeau, A. L., & Berger, J. (2007). Fatigue in the us workforce: prevalence and implications for lost productive work time. *Journal of Occupational and Environmental Medicine*, *49*(1), 1–10.

Romero, D., Bernus, P., Noran, O., Stahre, J., & Fast-Berglund, Å. (2016). The operator 4.0: Human cyber-physical systems & adaptive automation towards human-automation symbiosis work systems. In *APMS (Advances in Production Management Systems)*.

Savolainen, P. T., Mannering, F. L., Lord, D., & Quddus, M. A. (2011). The statistical analysis of highway crash-injury severities: A review and assessment of methodological alternatives. *Accident Analysis & Prevention*, *43*(5), 1666–1676.

Schall, M. C, Jr., Sesek, R. F., & Cavuoto, L. A. (2018). Barriers to the adoption of wearable sensors in the workplace: A survey of occupational safety and health professionals. *Human Factors*, *60*(3), 351–362.

Singh, S., Cabraal, A., Demosthenous, C., Astbrink, G., & Furlong, M. (2007). Password sharing: Implications for security design based on social practice. In *Proceedings of the SIGCHI Conference on Human Factors in Computing Systems* (pp. 895–904). ACM.

Stern, H. S., Blower, D., Cohen, M. L., Czeisler, C. A., Dinges, D. F., Greenhouse, J. B., et al. (2019). Data and methods for studying commercial motor vehicle driver fatigue, highway safety and long-term driver health. *Accident Analysis & Prevention*, *126*, 37–42.

Teoh, W. L., Chong, J. K., Khoo, M. B., Castagliola, P., & Yeong, W. C. (2017). Optimal designs of the variable sample size chart based on median run length and expected median run length. *Quality and Reliability Engineering International*, *33*(1), 121–134.

The Dark Sky API. (2019). Data sources. https://darksky.net/dev/docs/sources. Accessed 20 Jun 2019.

The Dark Sky Company, LLC. (2019). Dark Sky API – Overview. https://darksky.net/dev/docs. Accessed 20 Feb 2019.

Tsung, F., Zhang, K., Cheng, L., & Song, Z. (2018). Statistical transfer learning: A review and some extensions to statistical process control. *Quality Engineering*, *30*(1), 115–128.

Wikipedia Contributors. (2019). Openstreetmap — Wikipedia, the free encyclopedia. https://en.wikipedia.org/w/index.php?title=OpenStreetMap&oldid=900226891. Accessed 5 Jun 2019.

Wolfert, S., Ge, L., Verdouw, C., & Bogaardt, M. J. (2017). Big data in smart farming-a review. *Agricultural Systems*, *153*, 69–80.

Woodall, W. H., & Montgomery, D. C. (2014). Some current directions in the theory and application of statistical process monitoring. *Journal of Quality Technology*, *46*(1), 78–94.

Yampolskiy, R. V., & Govindaraju, V. (2008). Behavioural biometrics: A survey and classification. *International Journal of Biometrics*, *1*(1), 81–113.

Ye, Z. S., & Chen, N. (2014). The inverse gaussian process as a degradation model. *Technometrics*, *56*(3), 302–311.

Zanzotto, F. M. (2019). Human-in-the-loop artificial intelligence. *Journal of Artificial Intelligence Research*, *64*, 243–252.

Zhang, M., Megahed, F. M., & Woodall, W. H. (2014). Exponential CUSUM charts with estimated control limits. *Quality and Reliability Engineering International*, *30*(2), 275–286.

Monitoring Performance of Surgeons Using a New Risk-Adjusted Exponentially Weighted Moving Average Control Chart

Fah F. Gan, Wei L. Koh, and Janice J. Ang

Abstract Risk-adjusted charting procedures have been developed in the literature. One important class of risk-adjusted procedures is based on the likelihood ratio statistic obtained by testing the odds ratio of mortality. The likelihood ratio statistic essentially converts the binary surgical outcomes of death and survival into penalty and reward scores, respectively, that are dependent on the predicted risk of death of a patient. For cardiac operations, the risk distribution is highly right skewed resulting in penalty and reward scores in a narrow range for a majority of the patients. This means effectively there is little risk adjustment for the majority of the patients. We propose a risk-adjusted statistic which is the ratio of surgical outcome to the estimated probability of death as the monitoring statistic. The main characteristic of this statistic is that the resulting penalty score is substantially higher if a patient with low risk dies, and the penalty score decreases sharply as the risk increases. We compare our chart with the original risk-adjusted cumulative sum chart in terms of average run length. Finally, we will perform a retrospective study using data from two surgeons.

Keywords Cumulative sum chart · Logistic regression model · Odds ratio · Parsonnet scores · Standardized mortality ratio · Surgical outcomes

1 Introduction

Risk-adjusted control charting procedures have been developed for monitoring surgical processes in the health care industry. Monitoring surgical outcomes is important as early detection of deterioration of a surgeon's performance could lead to a reduction in surgical failures. The success of a surgical outcome depends on both the surgeon and the health condition of a patient. A surgeon who operates on mostly low-risk patients is more likely to have a greater proportion of successful operations. It is thus naive to use non risk-adjusted charting procedures that were developed for

F. F. Gan (✉) · W. L. Koh · J. J. Ang
National University of Singapore, 6 Science Drive 2, Singapore 117546, Singapore
e-mail: staganff@nus.edu.sg

S. Knoth and W. Schmid (eds.), *Frontiers in Statistical Quality Control 13*,
Frontiers in Statistical Quality Control,
https://doi.org/10.1007/978-3-030-67856-2_13

monitoring manufacturing processes to monitor the performance of a surgeon. In order to monitor the performance of a surgeon effectively, the health conditions of patients must be taken into consideration.

In order to use a risk-adjusted charting procedure, the probability of death of a patient from an operation will first have to be estimated. The Parsonnet score developed by Parsonnet et al. (1989) provides an estimate of the risk of death of a patient from a cardiac operation. Another one is the EuroScoreII developed by Nashef et al. (2012). A surgical outcome is usually represented by one if a patient dies within 30 days of an operation and zero otherwise. The binary surgical outcomes are non risk-adjusted and any charting procedure based directly on them could lead to incorrect inferences about the performance of a surgeon. The earliest risk-adjusted charting procedure developed in the literature is the variable life-adjusted display (VLAD) developed by Lovegrove et al. (1997). It accumulates the difference between estimated probability of death and surgical outcome of a patient. A plot that shows a change in slope provides evidence that the performance has changed. Treasure et al. (2004) provided an example of VLADs of six surgeons. One of them shows a horizontal plot for approximately the first 200 patients and then followed by a steady drop for the next 80 patients, resulting from the deaths of many of these 80 patients. The surgeon involved was later found to have a cortical visual handicap. Early detection of this assignable cause could have saved many lives. In the past, the VLAD lacked a simple signaling rule but Wittenberg et al. (2018) provided a simple signaling rule based on the V-mask.

An important class of risk-adjusted charting procedures is based on the likelihood ratio statistic derived by testing the odds ratio of death of a patient. Risk-adjusted cumulative sum (RA-CUSUM) chart based on this statistic was developed by Steiner et al. (2000). The likelihood ratio statistic essentially converts the binary surgical outcomes into penalty and reward scores that are dependent on the health condition of a patient. The resulting penalty score is the highest when a patient with zero Parsonnet score dies from an operation. The penalty score decreases with increasing Parsonnet score. Similarly, the reward score is the lowest if a patient with zero Parsonnet score survives an operation. The reward score increases with increasing Parsonnet score. The penalty-reward score derived from the likelihood ratio approach is logical. However, for monitoring cardiac operations, the risk distribution is highly skewed to the right and a hospital data set shows that only about 18% of the patients are in the very high risk group. This results in penalty scores that are close to the maximum penalty score and reward scores that are close to the minimum reward score about 82% of the time. This means effectively there is little risk adjustment about 82% of the time. Consequently, any risk-adjusted charting procedure based on the likelihood ratio statistic would be making inferences using a risk-adjusted statistic that is little risk-adjusted about 82% of the time.

In addition to the RA-CUSUM chart, risk-adjusted exponentially weighted moving average (RA-EWMA) charts have also been proposed. One such chart is the RA-EWMA chart developed by Grigg and Spiegelhalter (2007) for the exponential family data. Another risk-adjusted procedure is an updating RA-EWMA chart developed by Steiner and Jones (2010). It uses penalty-reward scores that are determined by the survival times of patients. It is updated using the latest penalty-reward scores at regular time intervals or when a death occurs. Therefore, this charting procedure uses the latest information of all patients continuously. Cook et al. (2011) have also proposed an RA-EWMA charting scheme that is based on the EWMAs of observed and predicted values. This scheme displays the EWMA of observed values and the EWMA of expected values separately. Tang and Gan (2018) have developed a RA-EWMA charting procedure based on multi-responses. An extensive literature review of risk-adjustment methods for monitoring surgical outcomes is given in Woodall et al. (2015). A more recent review is given in Sachlas et al. (2019).

In Sect. 2, we introduce a data set consisting of 6994 cardiac patients that we will use for analysis. We then review the risk-adjustment mechanism based on the likelihood ratio statistic. Our main objective is to develop a two-sided RA-EWMA chart for detecting both deterioration and improvement. We will use the ratio of surgical outcome to estimated probability of death of a patient as the monitoring statistic. This risk-adjusted statistic is selected because it gives a much heavier penalty when a patient with low risk dies as compared to that based on the likelihood ratio approach. Our proposed RA-EWMA chart is developed in Sect. 3. The CUSUM chart is known to have a certain optimality property (Moustakides 1986). We therefore compare our chart with the RA-CUSUM chart (Steiner et al. 2000) in terms of average run length (ARL) in Sect. 4. These two charts are then used to analyze two surgeons' data in Sect. 5. Finally, a conclusion is given in the last section.

2 Real Dataset and Review of Risk-Adjustment Mechanism

A surgical outcome depends on two main factors: health condition of a patient and surgical skills of the surgeon who performs the operation. The health condition of a patient varies from patient to patient. If a surgeon operates on a higher proportion of high risk patients, a higher proportion of deaths will likely occur. If the monitoring statistic is the raw surgical outcome, any charting procedure that uses this statistic without taking the patient's health condition into consideration would likely show a deterioration. But this inference is incorrect. For any charting procedure to infer correctly the performance of a surgeon, the monitoring statistic must, therefore, take into account the health condition.

One popular measure of the health condition of a cardiac patient is the Parsonnet score developed by Parsonnet et al. (1989). It is based on an additive scoring system in which every risk factor such as gender, age, morbid obesity, hypertension, etc. is given a risk score. The sum of all these scores is the Parsonnet score. The Parsonnet score ranges from 0 to 148 but in practice, it rarely exceeds 70 as observed from our data set of 6994 cardiac patients. The higher the Parsonnet score, the higher the risk of death. The Parsonnet score is categorized into five groups (Parsonnet et al. 1989; Keogh and Kinsman 2004). A Parsonnet score of 0–4 means low risk (1% mortality), a score of 5–9 means elevated risk (5% mortality), a score of 10–14 means significantly elevated risk (9% mortality), a score of 15–19 means high risk

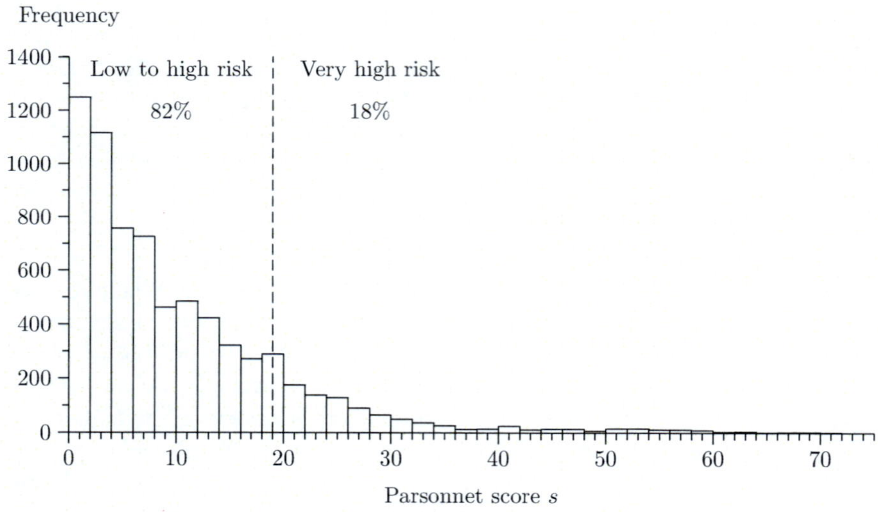

Fig. 1 Frequency distribution of the Parsonnet scores of 6994 patients who underwent cardiac surgeries from 1992 to 1998

(17% mortality), and a score that exceeds 19 means very high risk (31% mortality). Figure 1 shows the frequency distribution of 6994 patients who underwent cardiac surgeries from 1992 to 1998. The data set is highly right skewed with 82% of the patients in the low to high-risk categories and only 18% in the very high-risk category.

We next review how risk adjustment is done using the likelihood ratio approach. Let the Parsonnet score be denoted by S and its probability mass function $f(s)$. The surgical outcome is usually determined after 30 days of an operation and it can be represented by a discrete random variable Y which takes the value one if a patient dies within 30 days of the operation and zero otherwise. Conditional on a patient's Parsonnet score $S = s$, the probability of death is denoted as $P(Y = 1|S = s) = x(s)$. The joint probability mass function (pmf) of (S, Y) is then given as $f(s, y) = x(s)^y(1 - x(s))^{1-y} f(s), y = 0, 1$. We consider testing the null hypothesis $H_0 : f(s, y) = f_0(s, y)$ against the alternative hypothesis $H_A : f(s, y) = f_A(s, y)$ where $x(s) = x_0(s)$ under the null hypothesis, and $x(s) = x_A(s)$ under the alternative hypothesis. The nth log-likelihood ratio statistic is given by

$$W_n = W(S_n) = \log(f_A(S_n, Y_n)/f_0(S_n, Y_n)). \quad (1)$$

Hence, the statistic W_n is obtained by risk-adjusting the surgical outcome Y_n using the Parsonnet score S_n. The joint pmfs under the null and alternative hypotheses are given by $f_0(s_n, y_n) = x_0(s_n)^{y_n}[1 - x_0(s_n)]^{1-y_n} f(s_n)$ and $f_A(s_n, y_n) = x_A(s_n)^{y_n}[1 - x_A(s_n)]^{1-y_n} f(s_n)$, respectively, thus

$$w_n = w(s_n) = \log\left(\left[\frac{x_A(s_n)}{x_0(s_n)}\right]^{y_n}\left[\frac{[1-x_A(s_n)]}{[1-x_0(s_n)]}\right]^{1-y_n}\right). \tag{2}$$

The statistic w_n does not contain $f(s_n)$ because the risk distribution is assumed to be the same for both hypotheses. This is a special result of that given in Tang et al. (2015a).

The probability of death $x(s)$ of a patient from an operation can be estimated using a binary logistic regression model fitted with a historical data set of Parsonnet scores and surgical outcomes. A model fitted with the data set of 6994 patients who underwent cardiac operations is given by

$$\log\left(\frac{x(s)}{1-x(s)}\right) = -3.63321 + 0.7367s. \tag{3}$$

The probability $x(s)$ is the estimated probability of death of a patient with a Parsonnet score s from an operation assuming the average performance of surgeons in the data set. This is referred to as the reference surgeon. A particular surgeon's performance can be defined in terms of this average performance using the odds ratio of death Q_* associated with the surgeon as

$$\frac{x_*(s)}{1-x_*(s)} = Q_*\frac{x(s)}{1-x(s)}. \tag{4}$$

The quantity $x_*(s)$ is the estimated probability of death of a patient with Parsonnet score s if the patient were to be operated on by this surgeon. The earlier hypotheses can also be stated in terms of Q as $H_0 : Q = Q_0$ versus $H_A : Q = Q_A$. The odds ratio of death associated with a surgeon better than one with Q_0 will have $Q < Q_0$. On the other hand, $Q > Q_0$ for a surgeon worse than one with Q_0. By setting $Q_0 = 1$, $x_0(s)$ will be the estimated probability of death of a patient assuming the average performance of surgeons in the data set.

Suppose we want to derive the likelihood ratio statistic for detecting a deterioration in performance, we would test $H_0 : Q = Q_0$ versus $H_A : Q = Q_A$ with $Q_A > Q_0$. Using Eq. (4), we obtain

$$x_0(s_n) = Q_0\, x(s_n)/[1 - x(s_n) + Q_0\, x(s_n)], \tag{5}$$

and

$$x_A(s_n) = Q_A\, x(s_n)/[1 - x(s_n) + Q_A\, x(s_n)]. \tag{6}$$

Using Eqs. (2), (5) and (6), the statistic w_n can be written as

$$w_n = \log\left(\frac{1 - x(s_n) + Q_0\, x(s_n)}{1 - x(s_n) + Q_A\, x(s_n)}\left[\frac{Q_A}{Q_0}\right]^{y_n}\right). \tag{7}$$

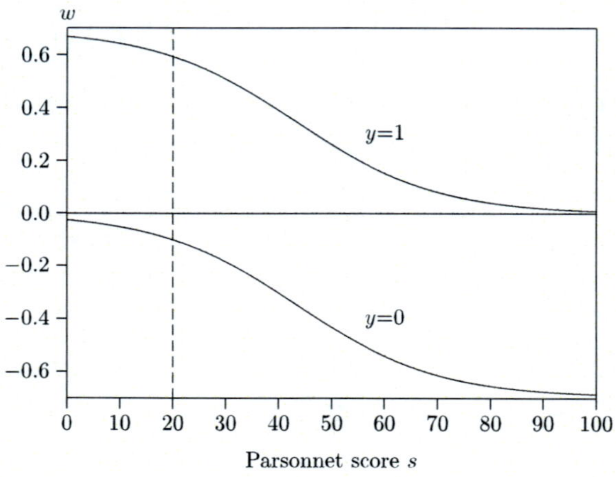

Fig. 2 Plots of log-likelihood ratio statistic w against the Parsonnet score s for surgical outcome $y = 0$ (survive), 1 (death)

With the logistic regression model in Eq. (3), the probability of death $x(s_n)$ can be estimated, a plot of w against s for $Q_0 = 1$ and $Q_A = 2$ can be constructed and it is displayed in Fig. 2. The statistic W_n is a risk-adjusted statistic of the surgical outcome Y_n and can be interpreted as the penalty-reward score given to a surgeon for an operation done. Figure 2 shows that a patient with a low Parsonnet score who died will be given a high penalty score and the penalty score decreases as the Parsonnet score increases. Similarly, a patient with a low Parsonnet score who survived an operation will be given a small reward score (near zero) and the reward score increases (a more negative reward is better) as the Parsonnet score increases. The risk-adjusted cumulative sum (RA-CUSUM) chart developed by Steiner et al. (2000) uses this statistic as the monitoring statistic.

The log-likelihood ratio approach is appealing because it converts non risk-adjusted binary surgical outcomes into penalty-reward scores that are risk-adjusted by the health conditions of patients. This risk-adjustment mechanism results in little risk adjustment for most of the patients because the underlying risk distribution is highly right skewed. The reason becomes clearer when we examine the penalty-reward score with respect to the distribution of the Parsonnet scores. Figure 1 shows that the distribution of Parsonnet scores is highly right skewed such that only about 18% of the patients are in the very high risk category. This means that for about 82% of the patients, a patient who died will only result in a risk-adjusted score in the narrow range of (0.593, 0.669). For a patient who survived, the range of risk-adjusted score is (−0.093, −0.024). In other words, after risk-adjustment using the log-likelihood ratio procedure, about 82% of the surgical outcomes still result in approximately binary outcomes. Thus, any charting procedure that uses the log-likelihood ratio statistic as the monitoring statistic will not be able to take full advantage of the risk adjustment mechanism. We propose a new risk-adjustment mechanism in the next section, one that would transform binary surgical outcomes into a penalty-reward scheme that gives more differentiation even for patients with low risk. We will develop a RA-EWMA charting procedure based on this new risk-adjusted statistic.

3 Risk-Adjusted EWMA Chart

The traditional standardized mortality ratio (SMR) is a risk-adjusted statistic that compares the observed number of deaths in a study population to the expected number of deaths in the study population assuming the death rate of a reference population. Tang et al. (2015b) considered a new SMR in which the observed number of deaths in a sample instead of a population, is compared with the estimated number of deaths in the sample assuming mortality rates in different subgroups of the sample. For a surgical process, the new SMR is given as

$$\text{SMR} = \sum_{i=1}^{n} Y_i / \sum_{i=1}^{n} X_i, \tag{8}$$

where Y_i is the surgical outcome and X_i is the estimated probability of death of the ith patient. More details of this SMR can be found in Tang et al. (2015b).

In this paper, we develop an EWMA chart using the risk-adjusted statistic Y/X as the monitoring statistic. One attractive feature of an EWMA chart is that the resulting RA-EWMA will continuously provide updated estimates of μ_Y/μ_X where $\mu_Y = E(Y)$ and $\mu_X = E(X)$, as more and more patients are observed. Just like the log-likelihood ratio statistic, Y/X can also be viewed as a penalty-reward score. Let x be estimated using the model in Eq. (3). A plot of Y/X against the Parsonnet score is shown in Fig. 3. The figure shows that the penalty score for death is severe, especially when the Parsonnet score is small. The penalty score decreases sharply as the Parsonnet score increases. Unlike the log-likelihood ratio statistic which makes little risk adjustment for Parsonnet score less than 20, risk adjustment using the statistic y/x is much more differentiated across the range [0, 20). For a patient who survives, a reward score of zero is given.

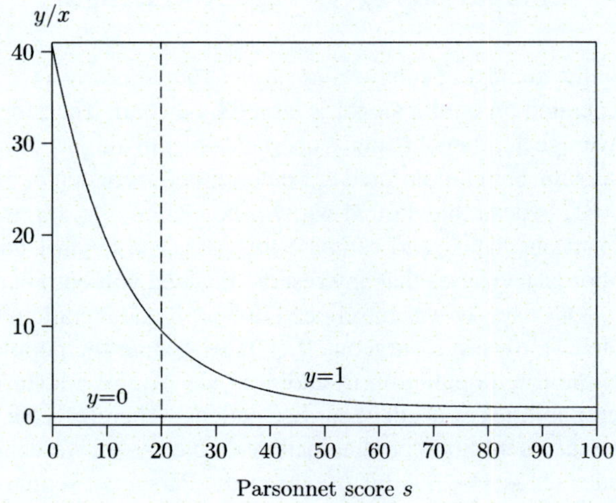

Fig. 3 Plots of y/x against the Parsonnet score s for $y = 0$ (survive), 1 (death)

The EWMA chart was first developed by Roberts (1959) for detecting small process shifts. Our RA-EWMA chart is obtained by plotting

$$I_n = (1-\lambda)I_{n-1} + \lambda Y_n/X_n, \tag{9}$$

against the patient number n where λ is a smoothing constant such that $0 < \lambda \leq 1$. The starting value I_0 can be set as $E(I_n)$ when the process is in control, which is 1. It must be emphasized that "in control" here means the performance of a surgeon being monitored has a performance that is similar to the average performance of all the surgeons ("the reference surgeon") whose data were used to fit the logistic regression model (3). A signal is issued when $I_n < h$ for detecting improvement or $I_n > H$ for detecting deterioration. Using Taylor's approximation, we can show that

$$E(I_n) = E(Y_n/X_n) \approx \mu_Y/\mu_X. \tag{10}$$

It can also be shown using Taylor's approximation that

$$\begin{aligned} Var(I_n) =& \frac{\lambda}{2-\lambda}[1-(1-\lambda)^{2n}]Var(Y_n/X_n) \\ \approx& \frac{\lambda}{2-\lambda}[1-(1-\lambda)^{2n}]\left(\sigma_X^2\frac{\mu_Y^2}{\mu_X^4} + \frac{\sigma_Y^2}{\mu_X^2} - 2\sigma_{YX}\frac{\mu_Y}{\mu_X^3}\right), \end{aligned} \tag{11}$$

where $\sigma_{YX} = Cov(X_n, Y_n)$. The asymptotic variance of I_n is given by

$$\lim_{n\to\infty} Var(I_n) \approx \frac{\lambda}{2-\lambda}\left(\sigma_X^2\frac{\mu_Y^2}{\mu_X^4} + \frac{\sigma_Y^2}{\mu_X^2} - 2\sigma_{YX}\frac{\mu_Y}{\mu_X^3}\right). \tag{12}$$

4 Comparison of Average Run Lengths

In the context of monitoring surgical outcomes, we let R be the number of patients operated on until a signal is issued by a chart. Therefore, the ARL is defined as the average number of patients operated on until a signal is issued. If the performance of a surgeon being monitored has deteriorated from that of the reference surgeon, a small ARL is desirable so that we will be alerted as soon as possible. We will resample from our data set to estimate the ARL and control limits of a control chart. This approach ensures that any results obtained will be reflective of practical scenarios.

Our data set was introduced in Sect. 2 and it contains 6994 cardiac patients operated on by seven surgeons. Let P_{6994} denote the population and assume that P_{6994} is the true population. In addition, we require a historical data set to estimate the probabilities of death of cardiac patients. Therefore, we first sampled randomly with replacement 5000 cardiac patients from P_{6994}. We denote this simulated historical data set as P_{5000}. A histogram of the Parsonnet scores contained in this data set is

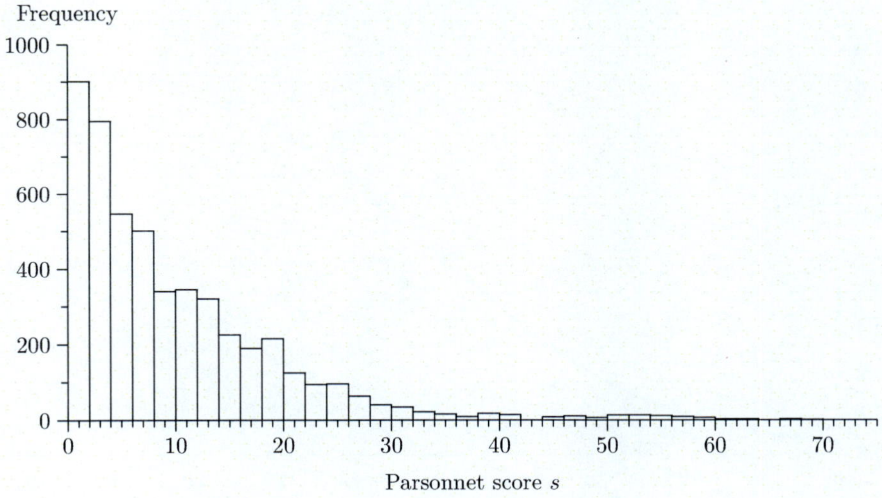

Fig. 4 Frequency distribution of the Parsonnet scores of 5000 patients, P_{5000} sampled randomly with replacement from the real data set P_{6994}

shown in Fig. 4. This figure shows that the distribution of the Parsonnet scores is very similar to that of P_{6994} as shown in Fig. 2. The logistic regression model fitted using the data set P_{5000} is given as

$$\log\left(\frac{x_1}{1-x_1}\right) = -3.49650 + 0.6986s. \tag{13}$$

Figure 5 shows the plots of x_0 and x_1 against s. The figure shows that the x_1 provides reasonably good estimate of x_0, especially for Parsonnet score less than 40 which covers almost the entire Parsonnet score population.

When we use the resampling approach to determine the control limits of a chart, we are assuming P_{6994} is the true population. The probability x_0 calculated using the logistic regression model in Eq. (3) is thus the true probability of death of a patient operated on by a surgeon of average performance as characterized by P_{6994}. We consider a two-sided RA-EWMA chart with lower control limit h and upper control limit H. In order to use the resampling approach to estimate the ARL of a RA-EWMA with parameter λ, and control limits h and H for monitoring a surgeon with true odds ratio of death $Q = Q_*$, we use the following procedure:

Procedure A—Resampling procedure to obtain a run length of the RA-EWMA chart

Step 0. Set λ, Q_*, h, H, $I_0 = 1$ and run length $R = 0$.
Step 1. Sample randomly a Parsonnet score s with replacement from P_{6994}.
Step 2. Given the s sampled in Step 1, calculate the true probability of death x_0 using Eq. (3).

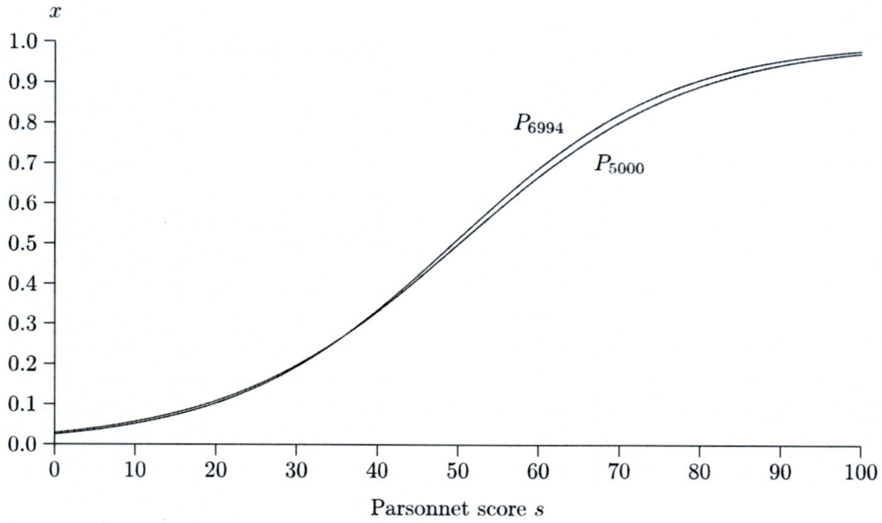

Fig. 5 Plots of probability of death x against the Parsonnet score s, calculated using the logistic regression models based on the data sets P_{6994} and P_{5000}

Step 3. Determine the true probability of death x_* of the patient using Eq. (4) assuming the performance of a surgeon with $Q = Q_*$.

Step 4. Generate a number u from the standard uniform distribution. If $u \leq x_*$, set the surgical outcome $y = 1$, otherwise set $y = 0$.

Step 5. Given the s sampled in Step 1, estimate the probability of death x_1 using Eq. (13).

Step 6. With y and x_1, determine the RA-EWMA, I using Eq. (9).

Step 7. If $h < I < H$, add one to the run length R and repeat Steps 1–7. Otherwise, stop the procedure to obtain a run length.

Procedure A can be repeated M times to obtain M run lengths from which the ARL and standard error of the ARL can be found. Procedure A was used to simulate the ARL profiles of RA-EWMA charts with parameters $\lambda = 0.05, 0.10, 0.15$ and 0.20 for $Q = 1/4, 1/3, 2/5, 1/2, 2/3, 5/6, 1.0, 1.2, 1.5, 2.0, 2.5, 3.0$, and 4.0. They are displayed in Table 1. The control limits of these charts are chosen such that the charts have the same in-control ARL of 100 when $Q = 1.0$. The control limits h and H are chosen such that both one-sided charts have the same ARL.

We would like to compare the performance our RA-EWMA chart with the RA-CUSUM chart developed by Steiner et al. (2000) based on Page's CUSUM chart (Page 1954). A one-sided RA-CUSUM chart is obtained by plotting

$$C_n = \max(0, C_{n-1} + W_n), \tag{14}$$

against n where $C_0 = u, 0 \le u < H$. A signal is issued when $C_n > H$. We use the following procedure to simulate a run length of the RA-CUSUM chart:
Procedure B - Resampling procedure to obtain a run length of the RA-CUSUM chart

Step 0. Set Q_0, Q_A, Q_*, H, $C_0 = 0$ and run length $R = 0$.
Step 1. Sample randomly a Parsonnet score s with replacement from P_{6994}.
Step 2. Given the s sampled in Step 1, calculate the true probability of death x_0 using Eq. (3).

Table 1 ARL profiles of two-sided RA-EWMA and RA-CUSUM charts. The number within a bracket is the standard error of the ARL

Q	RA-EWMA chart				RA-CUSUM chart		
	$\lambda = 0.05$	0.10	0.15	0.20	$Q_0 = 1$; $Q_A = 5/6$	1/2	1/4
					$Q_0 = 1$; $Q_A = 1.2$	2.0	4.0
	$h = 0.144$	0.023	0.00318	0.000372	$h = 0.526$	1.439	2.043
	$H = 2.515$	3.934	5.342	6.844	$H = 0.463$	1.442	2.337
1/4	54.1	53.0	53.5	53.6	60.8	56.7	54.6
	(0.08)	(0.09)	(0.09)	(0.09)	(0.06)	(0.07)	(0.08)
1/3	60.4	59.7	60.4	60.0	67.3	63.2	61.6
	(0.10)	(0.11)	(0.11)	(0.11)	(0.08)	(0.09)	(0.10)
2/5	66.0	65.4	65.7	65.4	72.7	68.7	67.3
	(0.13)	(0.13)	(0.13)	(0.13)	(0.09)	(0.11)	(0.12)
1/2	74.0	73.8	74.6	73.9	80.6	77.6	76.2
	(0.15)	(0.16)	(0.17)	(0.17)	(0.12)	(0.14)	(0.15)
2/3	87.8	87.4	87.4	86.4	93.4	91.4	89.9
	(0.21)	(0.21)	(0.22)	(0.21)	(0.16)	(0.19)	(0.20)
5/6	97.5	97.0	97.2	95.7	100.6	100.5	98.8
	(0.25)	(0.25)	(0.26)	(0.25)	(0.19)	(0.22)	(0.24)
1.0	100.8	100.9	100.5	100.2	100.4	100.6	100.1
	(0.27)	(0.28)	(0.28)	(0.28)	(0.20)	(0.24)	(0.26)
1.2	96.7	97.1	97.8	98.6	92.3	92.7	93.0
	(0.27)	(0.28)	(0.28)	(0.28)	(0.20)	(0.23)	(0.25)
1.5	80.5	83.7	86.0	87.9	73.9	74.2	76.5
	(0.23)	(0.25)	(0.25)	(0.26)	(0.17)	(0.19)	(0.21)
2.0	56.5	60.0	63.0	66.6	51.0	50.1	52.0
	(0.16)	(0.18)	(0.19)	(0.20)	(0.11)	(0.13)	(0.14)
2.5	40.8	44.3	47.1	50.4	37.8	36.4	37.3
	(0.12)	(0.13)	(0.14)	(0.15)	(0.08)	(0.09)	(0.10)
3.0	31.5	34.2	36.7	40.0	30.2	28.5	28.8
	(0.09)	(0.10)	(0.11)	(0.12)	(0.06)	(0.07)	(0.08)
4.0	21.6	23.2	25.0	27.5	21.9	20.4	20.0
	(0.06)	(0.07)	(0.07)	(0.08)	(0.04)	(0.04)	(0.05)

Step 3. Determine the true probability of death x_* of the patient using Eq. (4) assuming the performance of a surgeon with $Q = Q_*$.

Step 4. Generate a number u from the standard uniform distribution. If $u \leq x_*$, set the surgical outcome $y = 1$, otherwise set $y = 0$.

Step 5. Given the s sampled in Step 1, estimate the probability of death x_1 using Eq. (13).

Step 6. With y and x_1, determine W using Eq. (7) and RA-CUSUM, C using Eq. (14).

Step 7. If $C < H$, add one to the run length R and repeat Steps 1–7. Otherwise, stop the procedure to obtain a run length.

Procedure B can be repeated M times to obtain M run lengths from which the ARL and standard error of the ARL can found. A similar procedure was developed for a two-sided RA-CUSUM scheme with control limit h for detecting improvement and control limit H for detecting deterioration. The two one-sided charts of the two-sided scheme are chosen such that they have the same in-control ARL. This procedure was used to generate the ARL profiles of two-sided RA-CUSUM schemes with parameters determined by $H_0 : Q = Q_0 = 1$ versus $H_1 : Q = Q_A = 1/4, 1/2$, for detecting improvement, and $H_0 : Q = Q_0 = 1$ versus $H_1 : Q = Q_A = 1.2, 2.0$, and 4.0 for detecting deterioration. They are also displayed in Table 1. The control limits of these charts are also chosen such that the in-control ARL of a two-sided scheme is approximately 100 when $Q = 1$. A total of $M = 100,000$ run lengths were simulated for each case to estimate the ARL and standard error. The ARL profiles of the various RA-CUSUM schemes are also displayed in Table 1.

Table 1 reveals that the RA-CUSUM scheme is more sensitive than the RA-EWMA scheme in detecting deterioration. For detecting improvement, the RA-EWMA scheme seems to be slightly better.

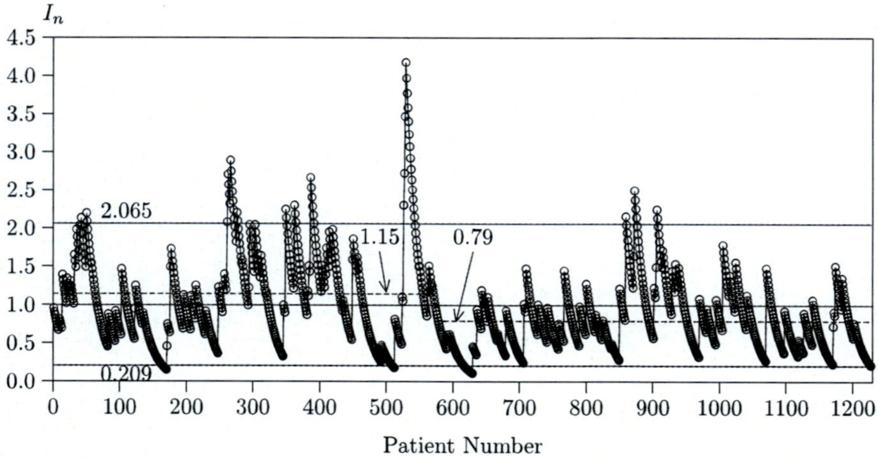

Fig. 6 RA-EWMA control chart of surgeon A's data

5 Analyses of Surgeons' Data

In this section, we analyze the data of two surgeons retrospectively using the RA-EWMA chart that we have developed. The RA-CUSUM chart developed by Steiner et al. (2000) is also included for comparison. We use the first two years of data of a surgeon as the phase I data for estimating the probability of death of a patient given the Parsonnet score. We analyze the same surgeon's data from the last five years. The control charts constructed for surgeon A are displayed in Figs. 6 and 7 and charts constructed for surgeon B are displayed in Figs. 8 and 9. The RA-EWMA charts are constructed using $\lambda = 0.05$ and $Q_0 = 1$ while the RA-CUSUM charts are based on $Q_0 = 1$ and $Q_A = 1/2$ and 2. The control limits displayed are for two-sided schemes with an in-control ARL of 100.

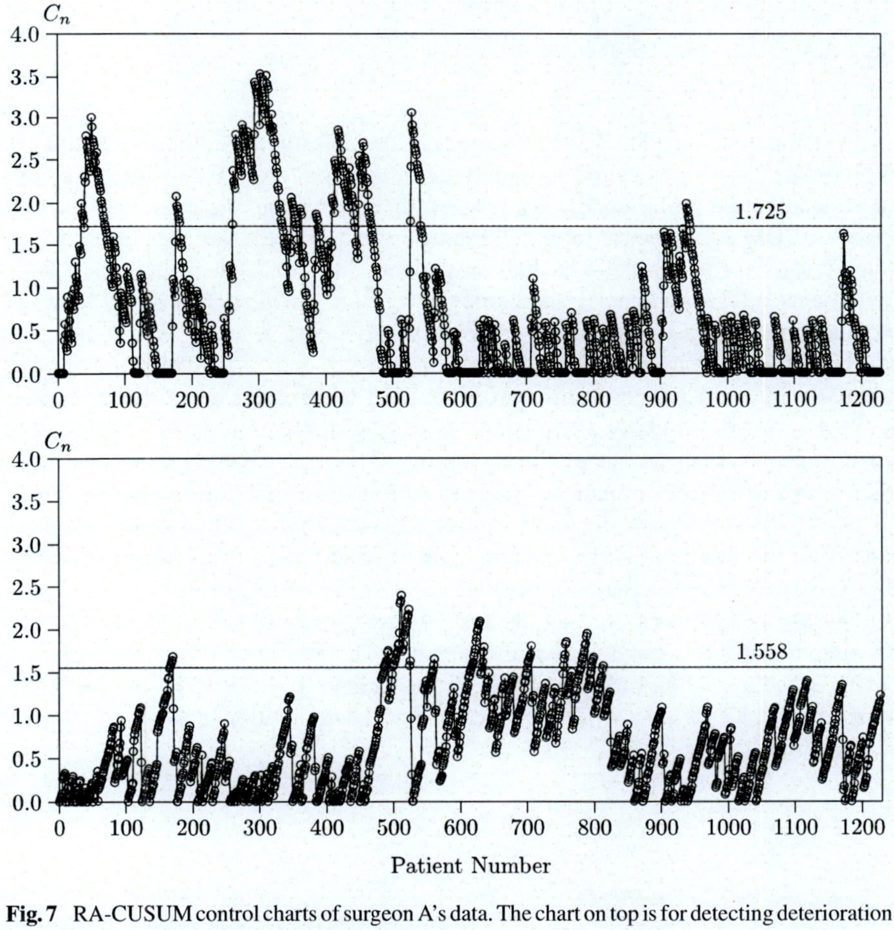

Fig. 7 RA-CUSUM control charts of surgeon A's data. The chart on top is for detecting deterioration and the chart below is for detecting improvement

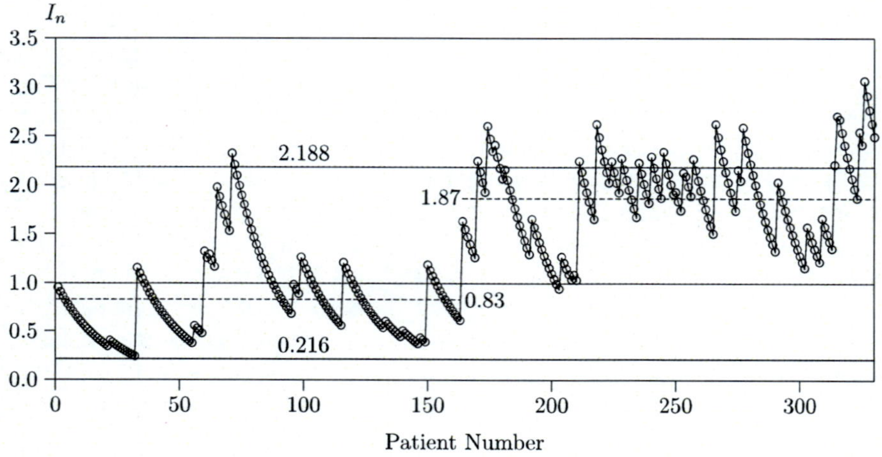

Fig. 8 RA-EWMA control chart of surgeon B's data

A point plotted on a RA-EWMA chart provides an estimate of the mean ratio of observed deaths to predicted deaths, and hence the chart provides a simple way of understanding the performance of a surgeon. In comparison, the points plotted on a RA-CUSUM are harder to interpret. Figure 6 shows that the variance of points is much greater for patients 1–573 than for patients 574–1227. The average EWMA for patients 1–573 is about 1.15 and drops to 0.79 for patients 574–1227. All these provide evidence that surgeon A's performance is below average for patients 1–573 but better than average for patients 574–1227. A more detailed study shows that there are short periods within patients 1–573 when surgeon's A performance is above average. Similarly, there is also a short period within patients 574–1227 when surgeon A's performance is below average. The RA-CUSUM provides similar inferences about the performance of surgeon A. The most notable difference between the two charts is the peak at patient number 529 of the RA-EWMA chart which is caused by the deaths of a few patients with very low risks. This shows that the RA-EWMA chart reacts more to the deaths of very low-risk patients.

Figures 8 and 9 provide strong evidence that the performance of surgeon B has deteriorated for the second half of the patients. The average EWMA for patients 1–163 is about 0.83 and increases to 1.87 for patients 164–330. The big increase from 0.83 to 1.87 provides evidence of a significant deterioration in the performance of surgeon B.

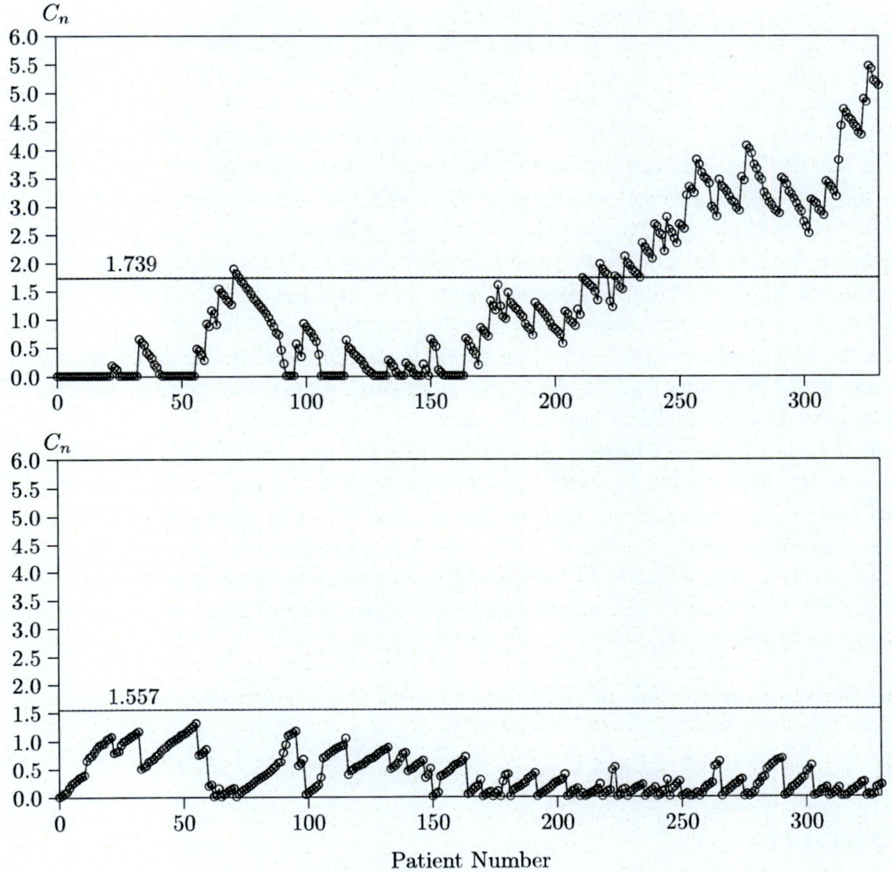

Fig. 9 RA-CUSUM control charts of surgeon B's data. The chart on top is for detecting deterioration and the chart below is for detecting improvement

6 Conclusions

In order to account for the heterogeneity of patients who undergo cardiac operations, any effective charting procedure for monitoring the performance of a surgeon must be based on a risk-adjusted statistic. The RA-CUSUM charting procedure developed by Steiner et al. (2000) is one such procedure. It is based on the likelihood ratio statistic obtained by testing the odds ratio of mortality. The likelihood ratio statistic essentially converts the binary surgical outcomes into penalty and reward scores based on the health condition of a patient. For cardiac operations, the risk distribution is highly skewed to the right and this results in a penalty score that is close to the maximum penalty and a reward score that is close to the minimum reward most of the time. Effectively, there is little risk adjustment done most of the time. In this paper, we have developed a RA-EWMA charting procedure based on y/x which is a risk-adjusted statistic that is much more differentiated across the range of Parsonnet scores. This allows the RA-EWMA chart to react much more strongly when a patient with a low Parsonnet score dies. A comparison between the RA-CUSUM and RA-EWMA schemes in terms of ARL shows that in general, the RA-CUSUM scheme has better performance than the RA-EWMA scheme in detecting deterioration. For detecting improvement, the RA-EWMA scheme seems to be slightly better. It seems the statistic y/x has not improved the ability to detect a deterioration in the performance. However, an analysis of two surgeons reveals an important difference between the two procedures in their signaling mechanisms. The RA-EWMA chart reacts much more strongly to deaths of very low-risk patients. Such patients are more likely to provide evidence of a deterioration in performance than patients with high Parsonnet scores who died. One could also use a RA-CUSUM chart based on our risk-adjusted statistic but we have chosen the RA-EWMA chart because the RA-EWMA is easier to understand for practitioners and the RA-EWMA provides an estimate of the current performance of a surgeon. Cook et al. (2011) commented that the evidence accumulated by the RA-CUSUM chart is discarded when the RA-CUSUM chart resets to zero while the influence of previous observations is gradually reduced in a RA-EWMA chart. This seems more logical and hence easier to be accepted by health care practitioners.

Acknowledgements The first author is supported by the Singapore Ministry of Education Academic Research Fund Tier 1 (R-155-000-200-114).

References

Cook, D. A., Coory, M., & Webster, R. A. (2011). Exponentially weighted moving average charts to compare observed and expected values for monitoring risk-adjusted hospital indicators. *BMJ Quality and Safety, 20,* 469–474.

Grigg, O., & Spiegelhalter, D. (2007). A simple risk-adjusted exponentially weighted moving average. *Journal of the American Statistical Association, 102,* 140–152.

Keogh, B. E., & Kinsman, R. (2004). *Fifth national adult cardiac surgical database report, 215.* Oxfordshire, UK: Dendrite Clinical Systems Ltd.

Lovegrove, J., Valencia, O., Treasure, T., Sherlaw-Johnson, C., & Gallivan, S. (1997). Monitoring the results of cardiac surgery by variable-life-adjusted display. *The Lancet, 350,* 1128–1130.

Moustakides, G. V. (1986). Optimal stopping times for detecting changes in distribution. *The Annals of Statistics, 14,* 1379–1387.

Nashef, S. A. M., Roques, F., Sharples, L. D., Nilsson, J., Smith, C., Goldstone, A. R., & Lockowandt, U. (2012). EuroSCOREII. *European Journal of Cardio-Thoracic Surgery, 41*(4), 734–745.

Page, E. S. (1954). Continuous inspection schemes. *Biometrika, 41,* 100–115.

Parsonnet, V., Dean, D., & Bernstein, A. D. (1989). A method of uniform stratification of risk for evaluating the results of surgery in acquired adult heart disease. *Circulation, 79,* I3–I12.

Roberts, S. W. (1959). Control chart tests based on geometric moving averages. *Technometrics, 1,* 239–250.

Sachlas, A., Bersimis, S., & Psarakis, S. (2019). Risk-adjusted control charts: Theory, methods, and applications in health. *Statistics in Biosciences, 11,* 630–658.

Steiner, S. H., Cook, R. J., Farewell, V. T., & Treasure, T. (2000). Monitoring surgical performance using risk-adjusted cumulative sum charts. *Biostatistics, 1,* 441–452.

Steiner, S. H., & Jones, M. (2010). Risk-adjusted survival time monitoring with an updating Exponentially Weighted Moving Average (EWMA) control chart. *Statistics in Medicine, 29,* 444–454.

Tang, X., & Gan, F. F. (2018). Risk-adjusted exponentially weighted moving average charting procedure based on multi-responses. In Sven Knoth & Wolfgang Schmid (Eds.), *Frontiers in statistical quality control 12* (pp. 113–131). Berlin: Springer.

Tang, X., Gan, F. F., & Zhang, L. (2015a). Risk-adjusted cumulative sum charting procedure based on multi-responses. *Journal of the American Statistical Association, 110,* 15–26.

Tang, X., Gan, F. F., & Zhang, L. (2015b). Standardized mortality ratio for an estimated number of deaths. *Journal of Applied Statistics, 42,* 1348–1366.

Treasure, T., Galliven, S., & Sherlaw-Johnson, C. (2004). Monitoring cardiac surgical performance: A commentary. *The American Association for Thoracic Surgery, 128,* 823–825.

Wittenberg, P., Gan, F. F., & Knoth, S. (2018). A simple signaling rule for variable life-adjusted display derived from an equivalent risk-adjusted CUSUM chart. *Statistics in Medicine, 37*(16), 2455–2473.

Woodall, W. H., Fogel, S. L., & Steiner, S. H. (2015). The monitoring and improvement of surgical outcome quality. *Journal of Quality Technology, 47*(4), 383–399.

Exploring the Usefulness of Functional Data Analysis for Health Surveillance

Zezhong Wang and Inez Maria Zwetsloot

Abstract Health surveillance is the process of ongoing systematic collection, analysis, interpretation, and dissemination of health data for the purpose of preventing and controlling disease, injury, and other health problems. Health surveillance data is often recorded continuously over a selected time interval or intermittently at several discrete time points. These can often be treated as functional data, and hence functional data analysis (FDA) can be applied to model and analyze these types of health data. One objective in health surveillance is early event detection. Statistical process monitoring tools are often used for online event detecting. In this paper, we explore the usefulness of FDA for prospective health surveillance and propose two strategies for monitoring using control charts. We apply these strategies to monthly ovitrap index data. These vector data are used in Hong Kong as part of its dengue control plan.

Keywords Health surveillance · Functional Data Analysis (FDA) · Functional Principal Component Analysis (FPCA) · Statistical Process Monitoring (SPM) · Control chart · Dengue surveillance

1 Introduction

Health surveillance is the systematic, ongoing assessment of the health of a community, based on the collection, interpretation, and analysis of health data, and provides information necessary for public health decision-making (Teutsch and Churchill 2000). A public health system can both provide an overall understanding of the current condition and serve as an early warning system for upcoming health emergencies.

Z. Wang · I. M. Zwetsloot (✉)
Department of Systems Engineering and Engineering Management, City University of Hong Kong, Kowloon, Hong Kong
e-mail: i.m.zwetsloot@cityu.edu.hk

Z. Wang
e-mail: zezhowang3-c@my.cityu.edu.hk

S. Knoth and W. Schmid (eds.), *Frontiers in Statistical Quality Control 13*,
Frontiers in Statistical Quality Control,
https://doi.org/10.1007/978-3-030-67856-2_14

The surveillance effort often involves both retrospective and prospective studies. In the retrospective studies, historical data is analyzed to find any correlations between environments and diseases and to identify clustering of diseases (Tsui et al. 2008). In the prospective studies, statistical techniques are used to identify anomalies as they arise (Unkel et al. 2012). One of these techniques for early event detection is the control chart. Woodall (2006) discussed the application of control charts in healthcare and public health surveillance. In addition, Sonesson and Bock (2003) and Yuan et al. (2019) provided excellent reviews of prospective health surveillance.

Shmueli and Burkom (2010) and Sparks (2013) discussed challenges of applying classical statistical techniques to health surveillance. Three identified challenges are temporal correlation, seasonal influences, and regional clustering, since health related data are generally collected in different locations over time. In addition, it is challenging to model the stable normal state of the variable under surveillance, an essential first step in any statistical surveillance scheme. Characterizing health related data using univariate distributions or a more general multivariate version is often not effective. In this paper, we propose to use functional data analysis (FDA) to deal with these challenges, as health data can often be represented by a curve (or profile). Woodall (2006) also recommended the application of profile monitoring in health surveillance.

Frequently, health related data are collected intermittently at several discrete time points. Due to the seasonality, a single function cannot fully capture these patterns. An alternative is to model health data using FDA, which treats each profile as a linear combination of multiple functions, the set of basis functions (Ramsay et al. 2009). Figure 1 shows two examples of data that can be modeled using FDA techniques. In Fig. 1a, the total cholesterol level, an important risk factor of stroke, was recorded seven times for each of 27 stroke patients (Qiu and Xiang 2014). Figure 1b shows the heights of 10 girls measured at a set of 31 unequally spaced ages in the Berkeley Growth Study (Ramsay et al. 2009). Ullah and Finch (2013) provided a systematic review of functional data analysis applications. About 40 out of the 84 included articles are in the health domain. The authors classify the articles by the FDA features as well as by the objectives of the study. The included papers focus on classification and forecasting based on historical data, but none considered online prospective monitoring. Hence, prospective methods are not as developed as retrospective methods in FDA applications. Online monitoring is necessary to detect abnormal behavior in disease data as early as possible so that actions can be taken to reduce the influence of emergencies. Therefore, in this paper, we propose an online monitoring method based on FDA, as far as we are aware this is the first work using FDA for prospective health surveillance.

Statistical process monitoring tools, including the control chart, are useful for prospective studies. Originally used in industry, these techniques have been applied in health surveillance as well (Woodall 2006; Qiu and Xiang 2014). An underlying assumption of control charts is that the process under surveillance is stable except for when anomalies occur. As illustrated in Fig. 1a, health data are rarely stable. In this paper, we propose a two-step procedure; first we model the health data using

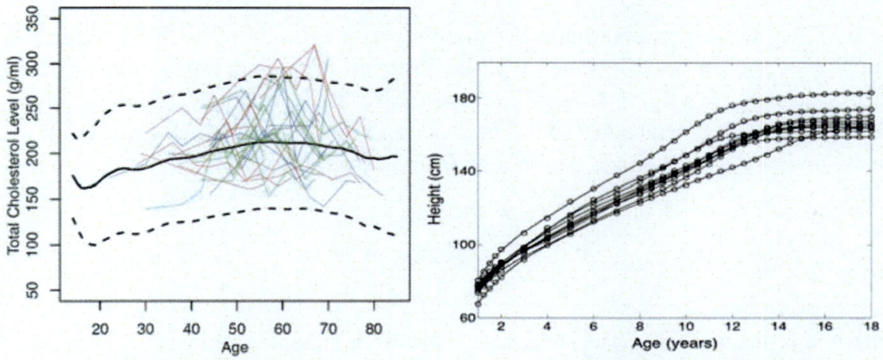

(a) Total Cholesterol Level (b) Growth Data

Fig. 1 **a** Total cholesterol level of 27 stroke patients, *reproduced from* Qiu and Xiang (2014), *published with permission of* © *Taylor & Francis Ltd* (www.tandfonline.com); **b** Growth data of 54 girls, *reproduced from* Ramsay and Silverman (2005), *p2, reprinted by permission of* © *Springer Science+Business Media, Inc. 2005*

FDA, and next we set up a control chart based on either the functional principal components or the residuals to detect changes online.

We illustrate our proposed FDA control chart using a case study from Hong Kong. We are interested in prospective surveillance of the monthly ovitrap index (MOI). This index measures the presence of adult Aedine mosquitoes in Hong Kong. These mosquitoes transmit dengue. Hence, surveillance of MOI is a part of the Hong Kong dengue prevention strategy. We are able to detect an increase in MOI in July 2018. In August 2018, Hong Kong suffered a dengue outbreak.

This paper is organized as follows: In Sect. 2, we introduce FDA. In Sect. 3, we develop the FDA based monitoring methods. In Sect. 4, we showcase our methods using the MOI case study. In Sect. 5, we conclude and give some potential future research directions.

2 Functional Data Analysis

Annual incidence and monthly number of cases are typical health data with seasonal variation. FDA methods are able to model the trend and periodicity of data using a set of basis functions. FDA has been widely used in retrospective studies. Erbas et al. (2007) used FDA and functional principal components analysis (FPCA) to model the age-specific breast cancer mortality time trends as curves. Based on that, they estimated the mortality by forecasting the coefficients of the fitted function. In this section, we provide a brief introduction to FDA and FPCA methods according to Ramsay et al. (2009), all the computations can be achieved using the package "fda" in R.

In FDA, the response variable Y can be represented by a set of time depended functions $x(t)$. For one single curve $\boldsymbol{y}$, the observation value at time point t_j can be expressed as: $y_j = x(t_j) + \varepsilon_j$, $j = 1, 2, ..., n$. Error ε_j is assumed to follow a normal distribution with a mean equal to 0. The function $x(t)$ is the weighted summation of basis functions $\phi_k(t)$ defined on the time interval from t_1 to t_n

$$x(t) = \sum_{k=1}^{K} c_k \phi_k(t) = \boldsymbol{c}'\boldsymbol{\phi}(t) \tag{1}$$

where K is the number of basis function, $\boldsymbol{c}'$ is the coefficient vector with K elements, and $\boldsymbol{\phi}(t)$ is the column vector containing K basis functions. Various basis functions are used in FDA; such as Fourier basis, B-spline, and Wavelets. Fourier basis and Wavelets are effective to model data containing seasonality. The Fourier basis is a set of sinusoidal functions and cosine functions, $\phi_1(t) = 1,\ \phi_2(t) = \sin \omega t,\ \phi_3(t) = \cos \omega t,\ \phi_4(t) = \sin 2\omega t,\ \phi_5(t) = \cos 2\omega t\ ...$, where $\omega = 2\pi/n$. Ratcliffe et al. (2002) applied Fourier basis functions to model the periodically stimulated fetal heart rate data since the data was roughly periodic.

Coefficient vector $\boldsymbol{c}'$ is estimated by minimizing the sum of square error $\sum_{j=1}^{n} [y_j - x(t_j)]^2$. But when the number of basis functions K is large, overfitting becomes a risk. To avoid this, Ramsay et al. (2009) recommended adding a roughness penalty to the least squares criterion

$$F(c) = \sum_j [y_j - x(t_j)]^2 + \lambda \int [Lx(t)]^2 dt \tag{2}$$

where λ is the smoothing parameter and $\int [Lx(t)]^2 dt$ is the roughness penalty. L indicates an operator of $x(t)$. One popular roughness penalty is the square of the second derivative (D^2), that is $\int [Lx(t)]^2 dt = \int [D^2 x(t)]^2 dt$. For periodic functions expressed by Fourier basis with known ω, harmonic acceleration operator $Lx(t) = \omega^2 Dx(t) + D^3 x(t)$ is recommended (Ramsay et al. 2009). The value of λ is usually selected using generalized cross-validation (GCV), some details are given in Appendix A. Now, $\boldsymbol{c}$ can be estimated by minimizing Eq. (2) subject to Eq. (1). This gives us the following fitted function:

$$x(t) = \hat{\boldsymbol{c}}'\boldsymbol{\phi}(t) \tag{3}$$

Generally, multiple curves are observed and modeled as $y_{ij} = x_i(t_j) + \varepsilon_{ij}$ by FDA. Where $i = 1, 2, ..., N$, $j = 1, 2, ..., n$, y_{ij} is the j-th observation on the i-th curve $\boldsymbol{y}_i$. It can be represented in matrix notation as

$$\boldsymbol{Y} = \boldsymbol{\Phi C} + \boldsymbol{\varepsilon} \tag{4}$$

where $\boldsymbol{Y}$ is the $n \times N$ observation matrix, $\boldsymbol{\Phi}$ is the $n \times K$ matrix of basis functions, $\boldsymbol{C}$ is the $K \times N$ coefficient matrix, and $\boldsymbol{\varepsilon}$ is the error matrix. The mean curve and variance among N samples can be computed as follows:

$$\bar{\boldsymbol{x}}(t) = N^{-1} \sum_i x_i(t) \text{ and } S^2_{x(t)} = (N-1)^{-1} \sum_i [x_i(t) - \bar{\boldsymbol{x}}(t)]^2 \tag{5}$$

And the covariance between two fitted values, $x_i(s)$ at time s and $x_i(t)$ at time t, is estimated by

$$\upsilon(s,t) = (N-1)^{-1} \sum_i [x_i(s) - \bar{x}(s)][x_i(t) - \bar{x}(t)] \tag{6}$$

Note that N here is the number of curves in the baseline sample, also referred to as phase I sample. These are used to obtain the model which can be used for prospective surveillance of those curves index $N+1, N+2, \ldots$

Functional principal component analysis (FPCA) is a commonly used dimension reduction technique to capture the primary modes of variation in functional data. It shares similar functions and calculation steps with principal component analysis (PCA). But FPCA is different from PCA in some critical aspects. First, FPCA can deal with high-dimensional data and even with infinite dimension. Second, in FPCA, a principal component is represented by an eigenfunction instead of an eigenvector. Third, FPCA is used to deal with time series data, so the order of data is important and immutable. After applying FPCA, infinite dimensional functional data is transformed to a finite dimensional vector of random scores (Wang et al. 2015). Leng and Müller (2005) modeled individual temporal gene expression profile by using FPCA.

After modeling all samples using Eq. (4), FPCA can be applied to explain the most significant variation among the curves. FPCA is defined as searching for eigenvalues μ_p and corresponding eigenfunctions $\xi_p(t),\ p = 1, 2, 3\ldots.$ for covariance function $\upsilon(s,t)$ by calculating

$$\int \upsilon(s,t)\xi_p(t)dt = \mu_p \xi_p(s) \tag{7}$$

subject to the orthogonality constraints $\int \xi_p(t)\xi_\ell(t)dt = 0$, where ℓ and p are two unequal positive integers, and size restriction $\int \xi_p^2(t)dt = 1$. The value of μ_p indicates the importance of corresponding principal component $\xi_p(t)$.

The principal component score for curve i based on $\xi_p(t)$ is calculated as

$$\rho_{ip}(x_i - \bar{\boldsymbol{x}}) = \int \xi_p(t)[x_i(t) - \bar{\boldsymbol{x}}(t)]dt. \tag{8}$$

We simplify $\rho_{ip}(x_i - \bar{\boldsymbol{x}})$ into ρ_{ip} in the following chapter.

3 FDA Based Statistical Process Monitoring

After defining the normal pattern of the baseline data using FDA, we need to construct an online monitoring system to compare the incoming observations with this normal behavior. Control charts, one popular technique in prospective studies, have been widely used to detect abnormalities in health surveillance. Joner et al. (2008) proposed a one-sided MEWMA control chart for spatial surveillance under the assumption of no seasonal effects. Jackson et al. (2007) compared the performances of three control chart based statistics, two EWMA methods, and a generalized linear model (GLM) in health surveillance. Their results showed a relatively poor performance of the control chart methods since it excluded the day-of-week trends. Whenever the seasonality and trend can be described by FDA, we proposed to use control charts to detect any deviations within or between the functional data samples. In this section, we propose two different monitoring strategies, one for a whole curve, another for the individual observations, to detect both global and local changes.

3.1 *Control Chart for a Whole Curve*

The first strategy is for monitoring a whole curve. For example, in our case study, we might wish to know if the MOI levels in a whole year are different than the baseline. Since the samples are curves, they may vary in amplitude and may have varying deviations compared to the normal state. By using FPCA, the diversities among curves are summarized and expressed using eigenfunctions $\xi_p(t)$ and eigenvalues μ_p. From the N baseline curves, we compute the eigenfunctions by Eq. (7). Next for the prospective monitoring, we use the same basic functions to model new curves out of the N baselines as $x_i(t), i \geq N+1$. And the principal component scores of each new curve are computed by Eq. (8).

Since the number of principal components for functional data is infinite, after calculating enough eigenpairs, we recommend sorting the principal components by their eigenvalues in descending order. Then selecting the first ℓ principal components, that is, the cumulative sum of ℓ eigenvalues that account for more than 90% of the summation of all eigenvalues should be kept. This step is necessary to reduce the dimension and simplify the sequential analysis.

Now, each profile can be estimated as a linear combination of the first ℓ eigenfunctions, where the coefficients are corresponding FPCA scores ρ_{ip}, $p = 1, 2, ...\ell$:

$$\hat{y}_{i(\ell)} = \bar{y} + \rho_{i1}\xi_1 + \rho_{i2}\xi_2 + \cdots + \rho_{ip}\xi_p + \cdots + \rho i\ell\xi_\ell \tag{9}$$

where $\bar{y} = \sum_{i=1}^{N} y_i$ is the mean curve of N baselines. Colosimo and Pacella (2010) applied FPCA to monitor roundness profile. They used a T^2 control chart for monitoring the eigenvalues μ_p and scores ρ_{ip} of FPCA. This T^2 chart was originally proposed by Jackson (2005), and can detect shifts related to the first ℓ principal components. The charting statistic is equal to

$$T_i^2 = \frac{\rho_{i1}^2}{\mu_1} + \frac{\rho_{i2}^2}{\mu_2} + \cdots + \frac{\rho_{ip}^2}{\mu_p} + \cdots + \frac{\rho_{i\ell}^2}{\mu_\ell}. \tag{10}$$

The upper control limit of the T^2 statistic in Eq. (10) can be calculated as

$$UCL = \chi^2_{\alpha,\ell} \tag{11}$$

Once the T^2 chart signals an outbreak, that means the shift occurs in one or more principal components of the data. Since we only keep the first ℓ principal components, the T^2 is not able to detect changes that occur in an orthogonal direction to those components. In order to handle this issue, a control chart based on the Q statistic (Squared Prediction Error) is used complementary to the T^2 chart (Jackson 2005). The Q statistic, can be calculated by summing the squared errors, and is defined as

$$Q_i = \int (y_i - \hat{y}_{i(\ell)})^2 dt \tag{12}$$

The sample mean $\bar{Q}$ and sample variance $\hat{\sigma}_Q^2$ of the Q statistics are computed from the N baseline curves. The upper control limit of the Q chart is

$$UCL = g\chi^2_{\alpha,h} \tag{13}$$

where g and h are computed as $\hat{g} = \hat{\sigma}_Q^2/(2\bar{Q})$, $\hat{h} = 2\bar{Q}^2/\hat{\sigma}_Q^2$. For more details, see Colosimo and Pacella (2010).

3.2 *Control Chart for Individual Observations*

Two downsides of monitoring based on a whole curve, as proposed above, are that (1) you need to collect the whole curve of data causing potential detection delay, and (2) when shifts occur locally they may go undetected as they get "absorbed" in the whole curve. In these situations, monitoring incoming individual observations can possibly speed up change detection. We propose a Shewhart control chart for monitoring observations as they come in. We apply the control chart in three steps. Firstly, we need to remove the systematic effects from the data by calculating the residuals $r_{ij} = y_{ij} - \bar{x}_i(t_j)$. Next, the standard deviations σ for the residual needs to be estimated, one can either use the sample standard deviation or a robust estimator. The choice depends on the application. Finally, a control chart is designed for monitoring the residuals. The control limits are decided to achieve a specific false alarm rate α, for more details see Sect. 3.3. Here we assume the residuals follow approximately a normal distribution with mean 0. So the control limits are $\pm Z_{\alpha/2}\hat{\sigma}$, where $Z_{\alpha/2}$ is the $100(1-\alpha/2)$ percentile of the standard normal distribution.

3.3 Performance Measurement

In order to set up the chart introduced above, the control limits need to be set. To decide on the control limits for the surveillance system, Fricker (2013) recommended a metric called the average time between false signals (ATFS). ATFS is the mean number of time periods it takes for the early event detection method to resignal after a signal, given that there are no outbreaks. Which is similar to the average run length ARL in SPM. The relationship between the ATFS and the false alarm rate α is $ATFS = 1/\alpha$. We illustrate the usefulness of this method in our case study.

There are two other metrics for measuring the performance of temporal surveillance system, they are the conditional expected delay (CED) and the probability of successful detection (PSD). CED represents the mean number of time periods it takes for the method to first signal, given that an outbreak is occurring and that the method signals during the outbreak. PSD is the probability the method signals during an outbreak, where the probability of detection is both a function of the early event detection method and the type of outbreak. For more discussion about these two concepts see Fricker (2013).

4 Case Study

In this section, we illustrate the proposed methods using dengue surveillance in Hong Kong. Dengue is an arboviral disease, which is transmitted by female mosquitoes. According to the World Health Organization (2019), the number of dengue cases reported increased from 2.2 million in 2010 to 3.2 million in 2015, and the number of the affected countries increased from 9 in 1970 to 100 in 2018. 'There were 2.35 million dengue cases reported in America in 2015. Among them, 10,200 cases were diagnosed as severe dengue and 1181 of them died. In Hong Kong, the Department of Health (2019) reports the monthly statistics on dengue fever. In August 2018, they reported 29 local cases, which is the largest outbreak since the government began keeping records. There is no effective dengue vaccine, so the real-time surveillance and reliable prediction of an outbreak becomes crucial.

Current research in dengue surveillance focuses on retrospective studies. Generally, meteorological data and incidence data are used to analyze influence factors of dengue and forecast the future outbreaks. Buczak et al. (2012, 2014) used fuzzy association rules to analyze the relationship between dengue incidences and meteorological data in Philippines and Peru, and to predict outbreaks. Althouse et al. (2011) used various methods to predict incidence and forecast outbreaks in Singapore and Bangkok. Ramadona et al. (2016) used dengue cases and climate data in Indonesia to project outbreaks by generalized linear regression model. Goto et al. (2013) analyzed the effects of meteorological factors on dengue incidence in Sri Lanka and showed that temperatures and rainfalls did not significantly affect dengue incidences. Overall conclusions on the relationship between incidence and meteorological data are inconsistent. Lin (2018) established a dengue surveillance system based on spatio-temporal scan statistics to track dengue outbreaks in Taiwan with risk factors. Also in Taiwan, Yuan et al. (2019) developed a Poisson regression model with extreme weather parameters for the prediction of annual dengue incidence. Chen et al. (2019) used an EWMA control chart to monitor the dengue incidence in Singapore. In our case, we use a new type of data, MOI, as an underlying factor of dengue outbreaks in Hong Kong. We apply the proposed prospective method to detect and signal any abnormal behaviors.

4.1 Data Description

Since 2000, the Food and Environmental Hygiene Department in Hong Kong (2019) started using Oviposition Trap (Ovitrap) to detect the presence of adult Aedine mosquitoes in selected areas. Aedes albopictus is the main vector in the transmission of dengue virus. These devices can be used to estimate the population of adult Aedes mosquitoes in the selected areas and act as an early warning signal of impending dengue outbreaks (Ai-Leen and Song 2000). Area ovitrap index for Aedes albopictus (AOI), which indicates the extensiveness of the distribution of Aedine mosquitoes in a predefined area, can be calculated as

$$Ovitrap\ Index\ for\ Aedes\ albopictus\ (AOI) = \frac{Number\ of\ the\ Aedes - positive\ ovitraps}{Total\ number\ of\ ovitraps\ retrieved\ from\ a\ particular\ area} \times 100\%$$

In this case, we are more interested in the overall situation of Aedes albopictus in Hong Kong, so we average the AOI across areas and name the new index as monthly ovitrap index for Aedes albopictus (MOI).

$$Monthly\ Ovitrap\ Index\ for\ Aedes\ albopictus\ (MOI) = \sum AOI \div Number\ of\ Areas$$

The MOI data is available from 2005 to 2018. Figure 2a shows the annual curves of MOI. They show significant seasonality, the MOI maintains a low level in spring and winter and approach their maximum value in summer. This trend is consistent with the habits of Aedes albopictus and the Hong Kong climate. The monthly local cases of dengue are recorded and published by the Department of Health of Hong

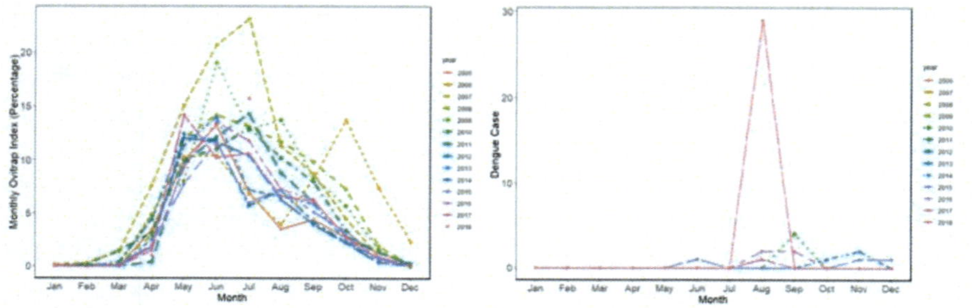

(a) MOI Data (b) Local Dengue Case

Fig. 2 **a** Monthly Ovitrap Index from 2005 to 2018; **b** Local dengue cases from 2006 to 2018

Kong. From Fig. 2b, we find that the local dengue cases kept at a low level from 2006 to 2017. But there are 29 local cases in 2018, the largest outbreak in Hong Kong.

4.2 *Fitted MOI by Functional Data Analysis*

The inherent seasonality in MOI is obvious, so we apply a Fourier basis to the MOI data. We collect 14 years of MOI data and use the "fda" package in R to generate 7 Fourier basis to smooth these data. As for the roughness penalty in Eq. (2), harmonic acceleration $Lx(t) = \omega^2 Dx(t) + D^3 x(t)$ recommended in Ramsay and Silverman (2005) is used for our periodic MOI data. The smoothing parameter is set at $\lambda = 10^{-1.4}$ through minimizing the cross-validation, more details on the selection of λ can be found in Appendix A. The fitted MOI curves displayed in Fig. 3a are the monitoring objects.

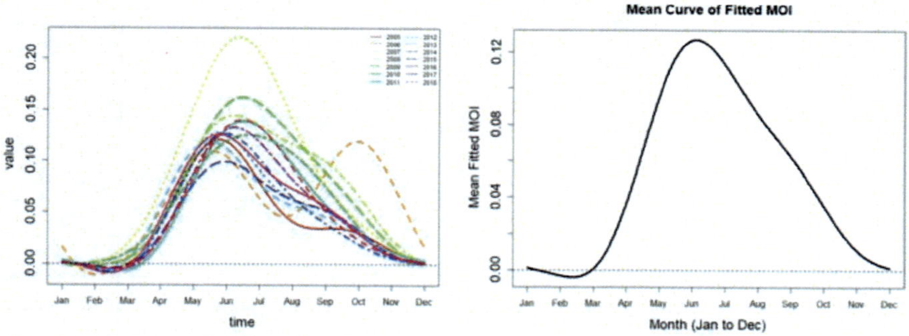

(a) Fitted MOI per Year (b) Fitted Mean MOI

Fig. 3 **a** Fitted MOI data in Hong Kong from 2005 to 2018; **b** Mean Curve of Fitted MOI data in Hong Kong from 2008 to 2016

From Figs. 2a and 3a, the peak value of MOI in 2007 obviously exceeds other years, and the 2006 curve shows a small peak in October. These two abnormal curves could affect the estimation of normal state of MOI, so we use the mean curve from 2008 to 2016 to model the normal MOI pattern, the resulting mean curve is shown in Fig. 3b. We treat it as a baseline in the following monitoring process.

4.3 *Implementation of Control Chart*

Before constructing control charts for whole curves, we apply variance-covariance analysis (Eq. (6)), the 12 × 12 variance-covariance matrix is plotted as a surface in Fig. 4a, and as a contour plot in Fig. 4b. These two figures show that the most significant variance-covariance appears in June to September. Other than these months, the covariance between two neighboring months is negligible, which means the temporal correlation has almost no influence in winter and spring.

Based on the variance-covariance matrix, the eigenfunctions $\xi_p(t)$ and eigenvalues μ_p can be calculated. The first three eigenfunctions and the type of variance they explain are shown in Fig. 5. These three figures display the mean curve along with +'s and −'s indicating the consequences of adding and subtracting a small amount of corresponding principal components. The first principal component shows that the variance in the second half of the year contributes 77.2% to the overall variance. The second and third principal component are not as significant as the first one, but they totally explain about 20% variance. These results are consistent with the variance-covariance plots in Fig. 4.

These three principal components totally represent 96.5% variance, so we build a T^2 and Q control chart based on them. We retrospectively monitor the MOI profiles of 2008–2016 and also include prospectively 2017 and 2018. To set the control limits we set ATFS at 10 years, as we believe this is a reasonable average time between two

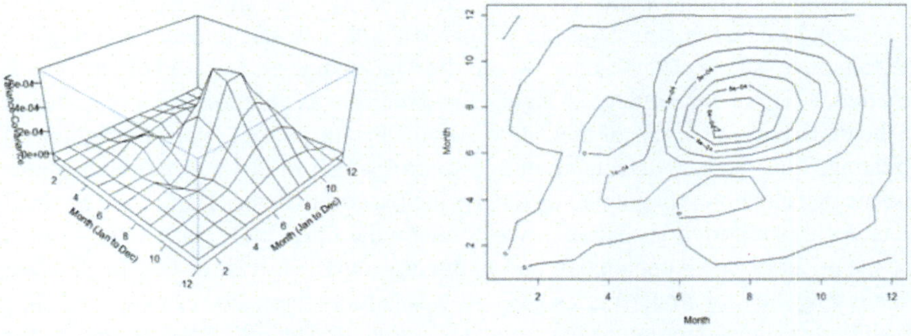

(a) Variance-Covariance Surface (b) Contour Plot

Fig. 4 **a** Monthly variance-covariance across years; **b** Contour plot, as a complement of variance-covariance surface

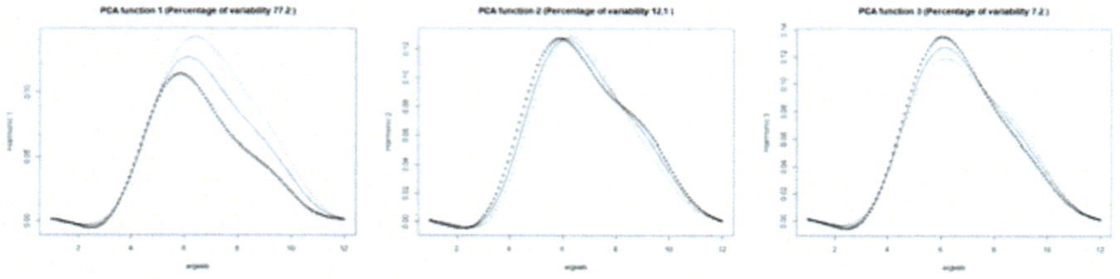

(a) PCA 1 (b) PCA2 (c) PCA3

Fig. 5 First three principal components for the MOI data 2008–2016

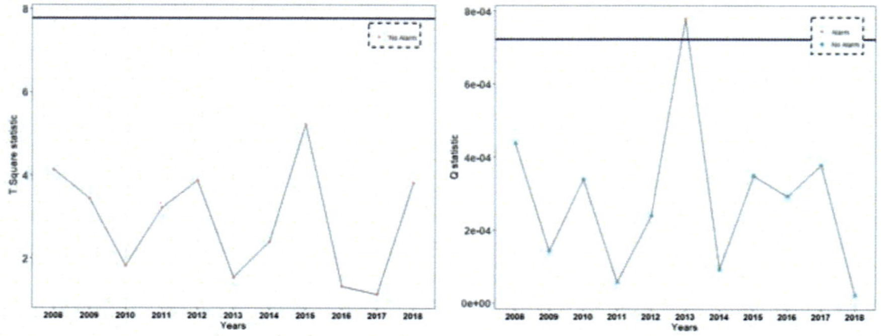

(a) T Square Chart (b) Q Chart

Fig. 6 Control chart for monitoring the full MOI curve per year

false signals, the false alarm rate, therefore, is set at $\alpha = 0.1$. The false alarm rate for each control chart is $\alpha = 1 - \sqrt{1 - 0.1} \approx 0.051$ (Colosimo and Pacella 2010). The T^2 control chart with control limit $UCL = \chi^2_{0.051,3} = 7.77$ is shown in Fig. 6a. The control limit of the Q chart is $UCL = 0.000722$ calculated by Eq. (13).

As shown in Fig. 6, neither the T^2 chart nor the Q chart show a signal in 2018. If we would detect a signal on the T^2 or Q chart, it would indicate a change in the functions' shape. The T^2 chart would detect a change associated with the $\ell = 3$ selected principal components. A signal could indicate an abnormal year or a change in the underlying structure of the data which means the phase I analysis might need updating. The Q chart detects an out-of-control signal in 2013, which indicates a change that can not be explained by the 3 principal components. There were no local dengue cases reported in that year. We compare the fitted MOI curve in 2013 with the mean curve, see Appendix B, and find a downward shift from July to October. That is why the Q chart detects a signal. As we are more interested in an upward shift in MOI, this signal does not have practical importance. The FPCA based strategy for a whole curve is ineffective in this case in detecting the 2018 outbreak.

We also apply the Shewhart chart to monitor individual observations. The histogram and Q-Q plot for the residuals from 2008 to 2016 are shown in Fig. 7.

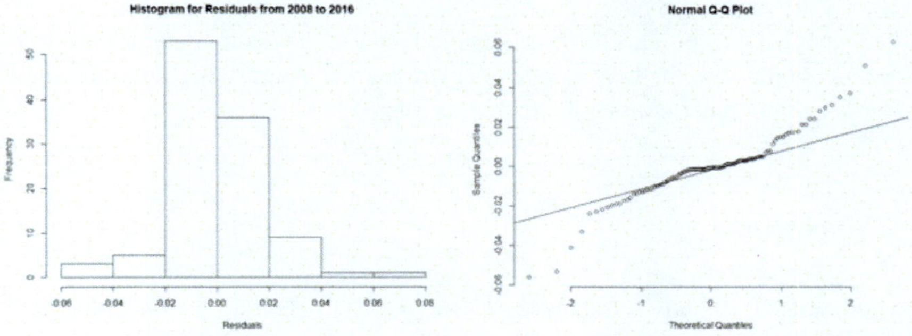

(a) Histogram of Residual (b) Q-Q Plot

Fig. 7 Histogram and Q-Q plot for the residuals from 2008 to 2016

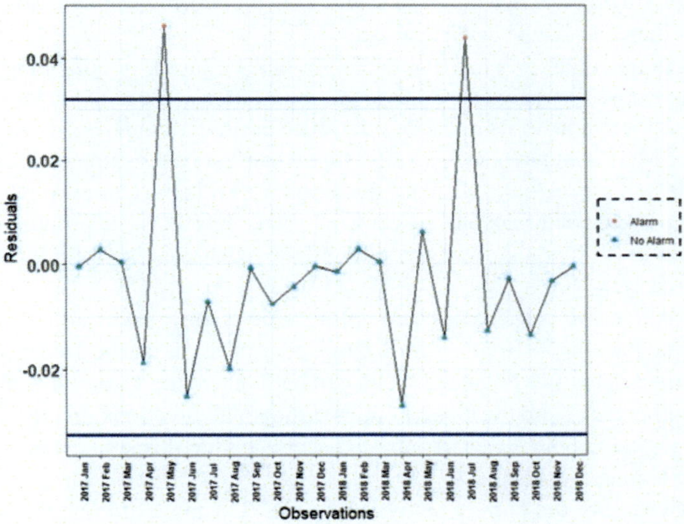

Fig. 8 The Shewhart control chart based on robust standard deviation from 2017–2018

In Fig. 7b, the residuals show heavy tails. So we estimate the standard deviation by a robust estimator proposed by Tatum (1997). We implement this estimator by setting the tuning constant at 10, and the normalizing constant at 1.035. The estimated standard deviation $\hat{\sigma}$ is 0.012. In this case, the expected ATFS is set at 120 months (10 years), so $\alpha = \frac{1}{120}$, and the control limits are $\pm Z_{\alpha/2}\hat{\sigma} = \pm 0.0321$.

The Shewhart control chart in Fig. 8 shows two signals, one in May 2017, another in July 2018. However, there is only one local dengue case reported in August 2017. We also check the original AOI data and find that some locations have significantly higher AOI values than expected. So we can classify the alarm as legitimate, because of the higher MOI level. This method is an indirect way to monitor dengue outbreaks. As we mentioned before, there were 29 dengue cases in August 2018, and the signal appears one month earlier than the real outbreak.

5 Conclusions

The purpose of this study is to develop a prospective method for health surveillance specifically for dengue monitoring. We have developed and described two functional data analysis-based strategies for temporal health surveillance. Since FDA can treat missing observations easily, and the balance between smoothness and accuracy can be controlled manually, it is a powerful method to model unstable health related data. In this study, we propose two different FDA based monitoring strategies to cope with continuous data and sparse data separately.

For the sparse data, to avoid the detection delay caused by collecting a whole curve, we propose a strategy for individual observations. We construct a Shewhart control chart for residuals, using FDA to estimate the underlying baseline model. The residuals of MOI data have heavy trails, so we use a robust estimator proposed by Tatum (1997) to estimate the standard deviation of residuals. Our proposed method can detect a dengue outbreak one month in advance (July 18). The proposed surveillance system for individual observations is effective in detecting a dengue outbreak one month early.

If the whole curve can be collected within a short time, such as the example in Colosimo and Pacella (2010), the FPCA based control charts can be used to detect the shifts of curves. This surveillance system can both detect the variation explained by principal components and the changes in its orthogonal directions. In the case study, there is no signal in both T^2 chart and Q chart indicating this strategy is insensitive to local shifts in the MOI data.

One drawback of the current study is that the FDA model for MOI has a small negative value in February (see Fig. 3b). Obviously, negative MOI values are impossible. The model could be adjusted to fit only for positive values through, for example, a logarithm transformation of the data. However, this strategy would influence the set-up and evaluation of the proposed control charts, because of the corresponding changes in PCA's and the MOI model. Therefore, we leave this issue for future research.

Another issue in this method is that there is not a one-to-one correspondence between changes in AOI level and dengue outbreaks. Outbreaks can occur with normal levels of AOI, also high levels of AOI do not mean outbreaks are going to happen. Nevertheless, vector monitoring can give a strong indication of possible outbreaks and can give us an earlier warning of possible dengue outbreaks than monitoring dengue incidence directly. In addition to monitoring the deviations from the normal AOI pattern, multiple constant thresholds can be used to monitor different AOI levels. This implies that in winter months, an increase in AOI level would not be of interest.

One benefit of using FDA is that it can treat missing observations easily. For the follow-up study, we will focus on the usefulness of FPCA based control chart for sparse health data. And a limitation of this study is that it only considers average AOI data. Incorporation of spatial data into the FDA model may make the method more sensitive as well.

Acknowledgements We would like to express our sincere thanks to Dr. Rob Goedhart from the University of Amsterdam for his expert advice. We would also like to thank all the participants in the ISQC workshop for their constructive comments on our work. And the reviewer of this paper for his/her constructive comments.
The work of Inez M. Zwetsloot was partially supported by a grant from the City University of Hong Kong (project 7005090).

Appendix A: Choosing Smooth Parameters

The generalized cross-validation method, developed by Craven and Wahba (2013), is used to determine the smoothing parameter in (2). The GCV for one curve is defined as

$$GCV(\lambda) = \left(\frac{n}{n - df(\lambda)}\right)\left(\frac{SSE}{n - df(\lambda)}\right), \tag{14}$$

where SSE is the sum of squared error calculated by

$$SSE(x) = \sum_{j}^{n}[y_j - x(t_j)]^2$$

and where $df(\lambda)$ is the degree of freedom of the fit defined by λ and computed as $df(\lambda) = trace[\boldsymbol{H}(\lambda)]$. Where $\boldsymbol{H}(\lambda) = \boldsymbol{\Phi}(\boldsymbol{\Phi}^T\boldsymbol{\Phi} + \lambda\boldsymbol{R})^{-1}\boldsymbol{\Phi}^T$, and $\boldsymbol{R} = \int L\phi(t)L\phi'(t)dt$ is the symmetric roughness penalty matrix with order K.

In the case study, we predefined $K = 7$, and choose λ subject to it minimized the GCV as given in (14). Figure 9 illustrates how to choose smoothing parameter λ given the GCV values. Based on this, we select $\lambda = 10^{-1.4}$ (as $log(\lambda) = -1.4$).

Appendix B: Verify the Signal in Q Chart

We plot the mean cure shown in Fig. 3b and the fitted MOI curve of 2013 on Fig. 10. We can see the significant differences between these two lines showed in July and August, but reviewing the principal components in Fig. 5, it seems none of three principal components can capture the shifts in 2013 perfectly.

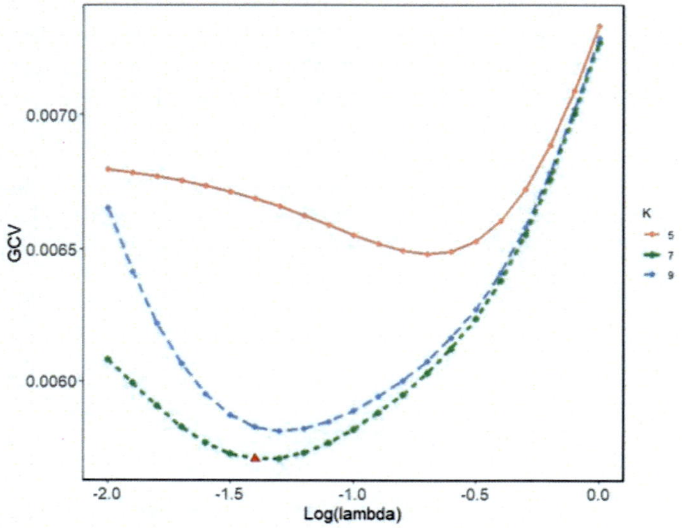

Fig. 9 The value of the GCV as a function of λ the smoothing parameter for fitting the MOI curve

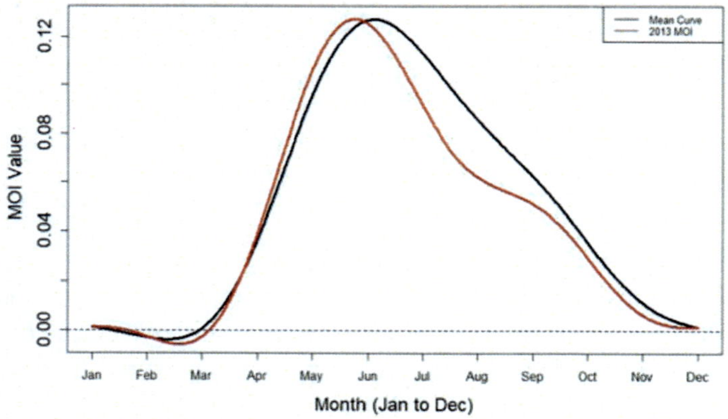

Fig. 10 The comparison between mean curve and fitted MOI in 2013

References

Ai-Leen, G. T., & Jin Song, R. (2000). The use of GIS in ovitrap monitoring for dengue control in Singapore. WHO Regional Office for South-East Asia

Althouse, B. M., Ng, Y. Y., & Cummings, D. A. (2011). Prediction of dengue incidence using search query surveillance. *PLoS Neglected Tropical Diseases*, *5*(8), e1258.

Buczak, A. L., Koshute, P. T., Babin, S. M., Feighner, B. H., & Lewis, S. H. (2012). A data-driven epidemiological prediction method for dengue outbreaks using local and remote sensing data. *BMC Medical Informatics and Decision Making*, *12*(1), 124.

Buczak, A. L., Baugher, B., Babin, S. M., Ramac-Thomas, L. C., Guven, E., Elbert, Y., et al. (2014). Prediction of high incidence of dengue in the Philippines. *PLoS Neglected Tropical Diseases*, *8*(4), e2771.

Chen, P., Fu, X., & Goh, R. S. M. (2019). Determining epidemic threshold for dengue incidences in Singapore based on extreme value theory. Presentation at the 11th Mathematical Methods in Reliability S34

Colosimo, B. M., & Pacella, M. (2010). A comparison study of control charts for statistical monitoring of functional data. *International Journal of Production Research*, *48*(6), 1575–1601.

Department of Health in Hong Kong(2019). Dengue Fever. https://www.fehd.gov.hk/english/pestcontrol/dengue_fever/index.html Accessed 30 June 2019.

Erbas, B., Hyndman, R. J., & Gertig, D. M. (2007). Forecasting age-specific breast cancer mortality using functional data models. *Statistics in Medicine*, *26*(2), 458–470.

Department of Health in Hong Kong(2019). Statistics on Dengue fever. https://data.gov.hk/en-data/dataset/hk-dh-chpsebcdde-dengue-fever-cases. Accessed 30 June 2019.

Fricker, R. D. (2013). *Introduction to statistical methods for biosurveillance: With an emphasis on syndromic surveillance*. Cambridge: Cambridge University Press.

Golub, G. H., Heath, M., & Wahba, G. (1979). Generalized cross-validation as a method for choosing a good ridge parameter. *Technometrics, 21*(2), 215–223.

Goto, K., Kumarendran, B., Mettananda, S., Gunasekara, D., Fujii, Y., & Kaneko, S. (2013). Analysis of effects of meteorological factors on dengue incidence in Sri Lanka using time series data. *PLoS One, 8*(5), e63,717.

Jackson, J. E. (2005). *A user's guide to principal components* (Vol. 587). New York: Wiley.

Jackson, M. L., Baer, A., Painter, I., & Duchin, J. (2007). A simulation study comparing aberration detection algorithms for syndromic surveillance. *BMC Medical Informatics and Decision Making*, *7*(1), 6.

Joner, M. D, Jr., Woodall, W. H., Reynolds, M. R, Jr., & Fricker, R. D, Jr. (2008). A one-sided mewma chart for health surveillance. *Quality and Reliability Engineering International*, *24*(5), 503–518.

Leng, X., & Müller, H. G. (2005). Classification using functional data analysis for temporal gene expression data. *Bioinformatics*, *22*(1), 68–76.

Lin, P. S. (2018). Cluster-temporal models for disease surveillance with an application to dengue fever infection in Taiwan. Presentation at the Joint Statistical Meeting 21

Qiu, P., & Xiang, D. (2014). Univariate dynamic screening system: An approach for identifying individuals with irregular longitudinal behavior. *Technometrics*, *56*(2), 248–260.

Ramadona, A. L., Lazuardi, L., Hii, Y. L., Holmner, Å., Kusnanto, H., & Rocklöv, J. (2016). Prediction of dengue outbreaks based on disease surveillance and meteorological data. *PloS One, 11*(3), e0152,688.

Ramsay, J. O., Hooker, G., & Graves, S. (2009). *Functional data analysis with R and MATLAB*. New York: Springer.

Ramsay, J. O., & Silverman, B. W. (2005). *Functional data analysis* (2nd ed.). New York: Springer.

Ratcliffe, S. J., Leader, L. R., & Heller, G. Z. (2002). Functional data analysis with application to periodically stimulated foetal heart rate data. I: functional regression. *Statistics in Medicine*, *21*(8), 1103–1114.

Shmueli, G., & Burkom, H. (2010). Statistical challenges facing early outbreak detection in biosurveillance. *Technometrics*, *52*(1), 39–51.

Sonesson, C., & Bock, D. (2003). A review and discussion of prospective statistical surveillance in public health. *Journal of the Royal Statistical Society: Series A (Statistics in Society)*, *166*(1), 5–21.

Sparks, R. (2013). Challenges in designing a disease surveillance plan: What we have and what we need? *IIE Transactions on Healthcare Systems Engineering*, *3*(3), 181–192.

Tatum, L. G. (1997). Robust estimation of the process standard deviation for control charts. *Technometrics*, *39*(2), 127–141.

Teutsch, S., & Churchill, R. (2000). *Principles and practice of public health surveillance* (2nd ed.). Oxford: Oxford University Press.

Tsui, K. L., Chiu, W., Gierlich, P., Goldsman, D., Liu, X., & Maschek, T. (2008). A review of healthcare, public health, and syndromic surveillance. *Quality Engineering*, *20*(4), 435–450.

Ullah, S., & Finch, C. F. (2013). Applications of functional data analysis: A systematic review. *BMC Medical Research Methodology*, *13*(1).

Unkel, S., Farrington, C., Garthwaite, P. H., Robertson, C., & Andrews, N. (2012). Statistical methods for the prospective detection of infectious disease outbreaks: a review. *Journal of the Royal Statistical Society: Series A (Statistics in Society)*, *175*(1), 49–82.

Wang, J. L., Chiou, J. M., & Müller, H. G. (2015). Review of functional data analysis. arXiv:150705135.

Woodall, W. H. (2006). The use of control charts in health-care and public-health surveillance. *Journal of Quality Technology*, *38*(2), 89–104.

World Health Organization. (2019). Dengue and severe dengue. https://www.who.int/news-room/fact-sheets/detail/dengue-and-severe-dengue Accessed 30 June 2019.

Yuan, M., Boston-Fisher, N., Luo, Y., Verma, A., & Buckeridge, D. L. (2019). A systematic review of aberration detection algorithms used in public health surveillance. *Journal of Biomedical Informatics* 103181

Yuan, H. Y., Wen, T. H., Kung, Y. H., Tsou, H. H., Chen, C. H., Chen, L. W., et al. (2019). Prediction of annual dengue incidence by hydro-climatic extremes for southern Taiwan. *International Journal of Biometeorology*, *63*(2), 259–268.

Rapid Detection of Hot-Spot by Tensor Decomposition with Application to Weekly Gonorrhea Data

Yujie Zhao, Hao Yan, Sarah E. Holte, Roxanne P. Kerani, and Yajun Mei

Abstract In many bio-surveillance and healthcare applications, data sources are measured from many spatial locations repeatedly over time, say, daily/weekly/monthly. In these applications, we are typically interested in detecting hot-spots, which are defined as some structured outliers that are sparse over the spatial domain but persistent over time. In this paper, we propose a tensor decomposition method to detect when and where the hot-spots occur. Our proposed methods represent the observed raw data as a three-dimensional tensor including a circular time dimension for daily/weekly/monthly patterns, and then decompose the tensor into three components: smooth global trend, local hot-spots, and residuals. A combination of LASSO and fused LASSO is used to estimate the model parameters, and a CUSUM procedure is applied to detect when and where the hot-spots might occur. The usefulness of our proposed methodology is validated through numerical simulation and a real-world dataset in the weekly number of gonorrhea cases from 2006 to 2018 for 50 states in the United States.

Y. Zhao · Y. Mei (✉)
School of Industrial and Systems Engineering, Georgia Institute of Technology, Atlanta, GA, USA
e-mail: yajun.mei@isye.gatech.edu

Y. Zhao
e-mail: yzhao471@gatech.edu

H. Yan
School of Computing, Informatics, and Decision Systems Engineering, Arizona State University, Tempe, AZ, USA
e-mail: haoyan@asu.edu

S. E. Holte
Division of Public Health Sciences, Fred Hutchinson Cancer Research Center, Seattle, WA, USA
e-mail: sholte@fredhutch.org

R. P. Kerani
Department of Medicine, University of Washington, and Public Health-Seattle & King County, Seattle, WA, USA
e-mail: rkerani@uw.edu

S. Knoth and W. Schmid (eds.), *Frontiers in Statistical Quality Control 13*,
Frontiers in Statistical Quality Control,
https://doi.org/10.1007/978-3-030-67856-2_15

Keywords Circular time · CUSUM · Hot-spot · Spatio-temporal model · Tensor decomposition

1 Introduction

In many bio-surveillance and healthcare applications, data sources are measured from many spatial locations repeatedly over time, say, daily, weekly, or monthly. In these applications, we are typically interested in detecting *hot-spots*, which are defined as some structured outliers that are sparse over the spatial domain but persistent over time. A concrete real-world motivating application is the weekly number of gonorrhea cases from 2006 to 2018 for 50 states in the United States, also see the detailed data description in the next section. From the monitoring viewpoint, there are two kinds of changes: one is the global-level trend, and the other is the local-level outliers. Here we are more interested in detecting the so-called hot-spots, which are local-level outliers with the following two properties: (1) spatial sparsity, i.e., the local changes are sparse over the spatial domain; and (2) temporal persistence, i.e., the local changes last for a reasonably long time period unless one takes some actions.

Generally speaking, the hot-spot detection can be thought as detecting sparse anomaly in spatio-temporal data, and there are three different categories of methodologies and approaches in the literature. The first one is LASSO-based control chart that integrates LASSO estimators for change point detection and declares non-zero components of the LASSO estimators as the hot-spot, see Zou and Qiu (2009), Zou et al. (2012), Saltyte Benth and Saltyte (2011). Unfortunately, the LASSO-based control chart lacks the ability to separate the local hot-spots from the global trend of the spatio-temporal data. The second category of methods is the dimension reduction based control chart where one monitors the features from PCA or other dimension reduction methods, see De Ketelaere et al. (2015), Louwerse and Smilde (2000), Hu and Yuan (2009). The drawback of PCA or other dimension reduction methods is that it fails to detect sparse anomalies and cannot take full advantage of the spatial location of hot-spot. The third category of anomaly detection methods is the decomposition-based method that uses the regularized regression methods to separate the hot-spots from the background event, see Tran et al. (2012), Yan et al. (2017, 2018). However, these existing approaches investigate structured images or curves data under the assumption that the hot-spots are independent over the time domain.

In this paper, we propose a decomposition-based anomaly detection method for spatial-temporal data when the hot-spots are autoregressive, which is typical for time series data. Our main idea is to represent the raw data as a 3-dimensional tensor: states, weeks, years. To be more specific, at each year, we observe a 50×52 data matrix that corresponds to 50 states and 52 weeks (we ignore the leap years). Next, we propose to decompose the 3-dimension tensor into three components: Smooth global trend, Sparse local hot-spot, and Residuals, and term our proposed decomposition model as SSR-Tensor. When fitting the observed raw data to our proposed SSR-Tensor model, we develop a penalized likelihood approach by adding two penalty functions: one is the LASSO type penalty to guarantee the sparsity of hot-spots, and the other is the fused-LASSO type penalty for the autoregressive properties of hot-spots or time-series data. By doing so, we are able to (1) detect when the hot-spots occur (i.e., the change point detection problem); and (2) localize where and which type of the hot-spots occur (i.e., the spatial localization problem).

We would like to acknowledge that much research has been done on modeling and prediction of the spatio-temporal data. Some popular time series models are AR, MA, ARMA model, etc., and the parameter can be estimated by Yule–Walker method (Hannan and Quinn 1979), maximum likelihood estimation or least square method (Hamilton 1994). In addition, spatial statistics have also been extensively investigated on its own right, see Reynolds and Madden (1988), Lichstein et al. (2002), Lan et al. (2014), Elhorst (2014), Call and Voss (2019) for examples. When one combines time series with spatial statistics, the corresponding spatio-temporal models generally become more complicated, see (Zhu et al. 2005; Lai and Lim 2015; Diggle 2013) for more discussions.

In principle, it is possible to represent the spatio-temporal process as a sequence of random vector $\mathbf{Y}_t$ with weekly observation t, where $\mathbf{Y}_t$ is p-dimensional vector that characterize the spatial domain (i.e., spatial dimension $p = 50$ in our case study). However, such an approach might not be computationally feasible in the context of hot-spot detection, in which one needs to specify the covariance structure of $\mathbf{Y}_t$, not only over the spatial domain, but also over the time domain. If we write all data into a vector, then the dimension of such vector is $50 \times 52 \times 13 = 33,800$, and thus the covariance matrix is of dimension $33,800 \times 33,800$, which is not computationally feasible, see Reinsel (2003), Tsay (2013) for more details. Meanwhile, under our proposed SSR-Tensor model, we essentially conduct a dimensional reduction by assuming that such a covariance matrix has a nice sparsity structure, as we reduce the dimensions 50, 52 and 13 to much smaller numbers, e.g., AR(1) model over the week or year dimension, and local correlation over the spatial domain.

We acknowledge that here we follow the standard time series literature to conduct Phase I analysis when the full data set is available to detect the hot-spots. We feel that our proposed tensor model has a potential to be extended from Phase I to Phase II analysis by extending our algorithms from off-line to online/recursive forms. However, such extension is non-trivial, and is beyond the scope of this paper. In addition, while our paper focuses only on 3-dimensional tensor due to our motivating application in gonorrhea, our proposed SSR-Tensor model can easily be extended to any d-dimensional tensor or data with $d \geq 3$, e.g., when we have further information, such as the unemployment rate, economic performance, and so on. As the dimension d increases, we can simply add more corresponding bases, as our proposed model uses *basis* to describe correlation within each dimension, and utilizes *tensor product* for interaction between different dimensions. The capability of extending to high-dimensional data is one of the main advantages of our proposed SSR-Tensor model. Furthermore, our proposed SSR-Tensor model essentially involves block-wise diagonal covariation matrix, which allows ut to develop computationally efficient methodologies by using tensor decomposition algebra, see Sect. 5.2 for more technical details.

The remainder of this paper is as follows. Section 2 discusses and visualizes the gonorrhea dataset, which is used as our motivating example and in our case study. Section 3 presents our proposed SSR-Tensor model, and discusses how to estimate model parameters from observed data. Section 4 describes how to use our proposed SSR-Tensor model to find hot-spots, both for temporal detection and for spatial localization. Efficient numerical optimization algorithms are discussed in Sect. 5. Our proposed methods are then validated through extensive simulations in Sect. 6 and a case study in gonorrhea dataset in Sect. 7.

2 Data Description

To protect Americans from serious disease, the National Notifiable Disease Surveillance System (NNDSS) at the Centers for Disease Control and Prevention (CDC) helps public health monitor, control, and prevent about 120 diseases, see its website https://wwwn.cdc.gov/nndss/infectious-tables.html. One disease that receives intensive attention in recent years is gonorrhea, due to the possibility of multi-drug resistances. Historically the instances of antibiotic resistance (in gonorrhea) have first been in the west and then move across the country. Since 1965, the CDC has collected the number of cumulative new infected patients every week in a calendar year. There are several changes on report policies or guidelines, and the latest one is year 2006. As a result, we focus on the weekly numbers of new gonorrhea patients during January 1, 2006 and December 31, 2018. The new weekly gonorrhea cases are computed as the difference of the cumulative cases in two consecutive weeks. The last week is dropped during this calculation.

Let us first discuss the spatial patterns of the gonorrhea data among 50 states. For this purpose, we consider the cumulative number of gonorrhea cases from week 1 to week 52 by sum up all data during years 2006–2018. Figure 1 plots some selected weeks (#1, #11, #21, #31, #41, #51). In Fig. 1, if the state has a deeper and bluer color, then it experiences a higher number of gonorrhea cases.

One obvious pattern is that, California and Texas have generally higher number of gonorrhea cases as compared to other states. In addition, the number of gonorrhea cases in the northern US is smaller than that in the southern US.

Next, we consider the temporal pattern of the gonorrhea data set. Figure 2 plots the annual number of gonorrhea cases over the years 2006–2018 in the US. It can be seen that there is a decrease during 2007–2009, and then the number of gonorrhea cases become to increase. The increasing trend from 2010 to 2014 is very gentle, but the increasing trend after 2015 become severe. One possible explanation for the different increase speed is the Affordable Care Act, which signed into law by President Barack Obama on March 23, 2010. This policy may help to stabilize the increase of gonorrhea disease. As we mentioned before, we are not interested in detecting this type of global changes, and we focus on the detection of the changes on the local patterns, which are referred to as hot-spots in our paper.

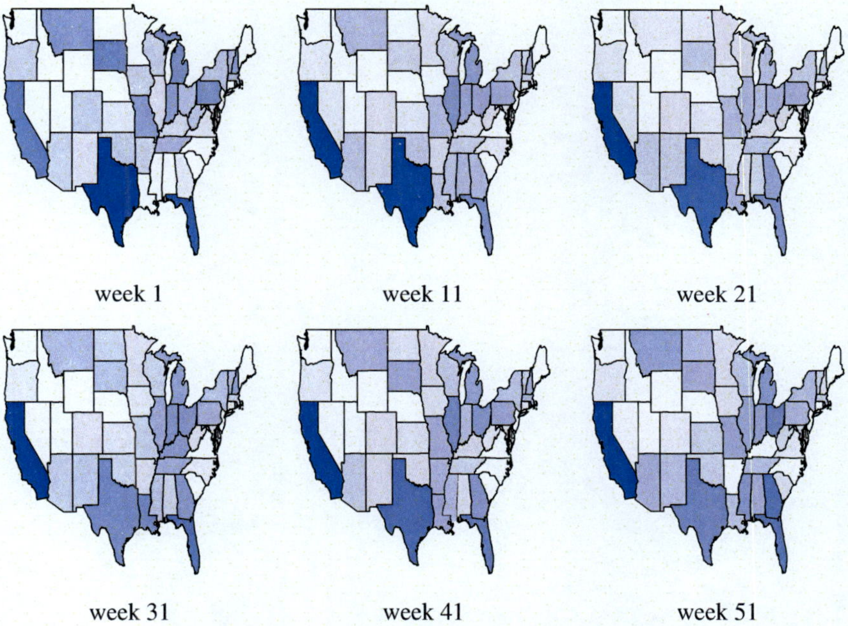

Fig. 1 The cumulative number of gonorrhea cases at some selected weeks during years 2006–2018. The deeper the color, the higher number of gonorrhea cases

Fig. 2 Annual number of gonorrhea cases (in thousands) over the years 2006–2018 in the US

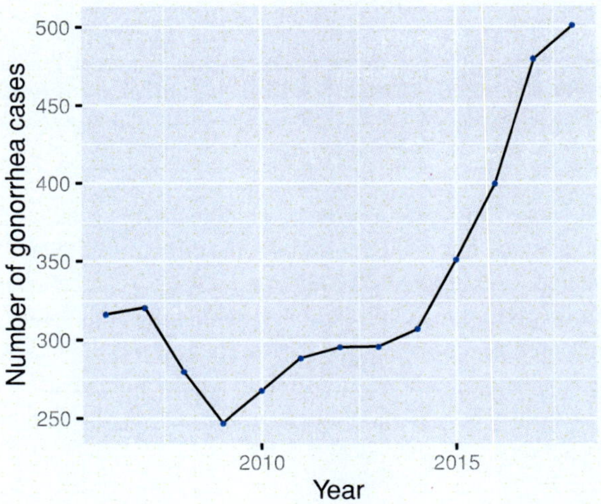

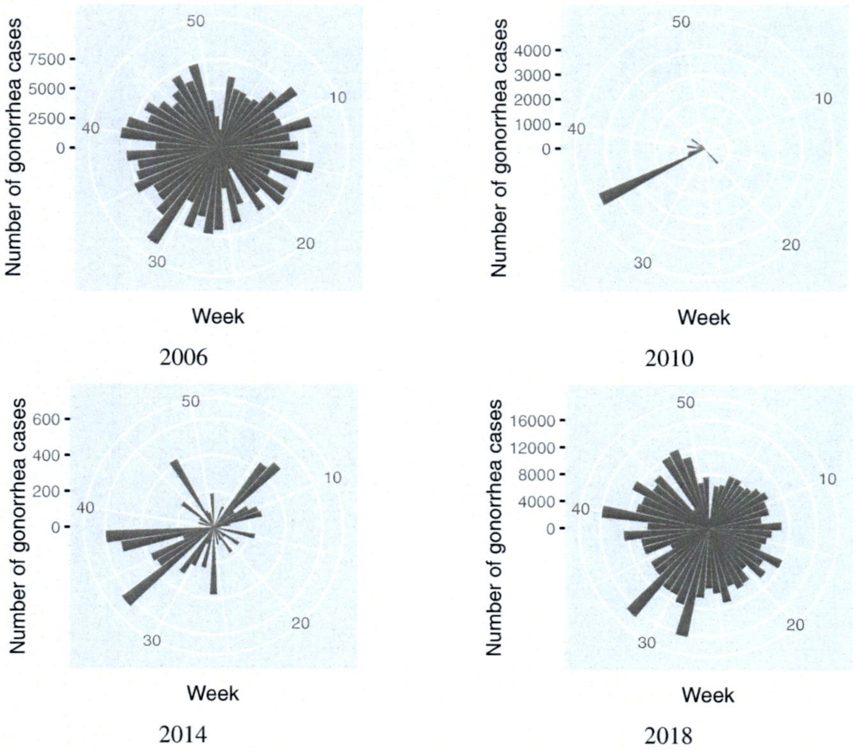

Fig. 3 Histograms of the number of gonorrhea cases of Year 2006, 2010, 2014, 2018. The y-axis is the number of gonorrhea cases, and the circular x-axis is the 51 weeks. Each bar represents a given week, and the length represents the number of gonorrhea cases for a given week in US

Moreover, the gonorrhea data consists of weekly data, and thus it is necessary to address the circular patterns over the direction of "week". Figure 3 shows the country-scaled weekly gonorrhea case in the form of "rose" diagram for some selected years. In this figure, each direction represents a given week, and the length represents the number of gonorrhea cases for a given week. It reveals differences in the number of gonorrhea cases across a different week of the year. For instance, in July and August (in the direction of 8 o'clock on the circle), the number of gonorrhea case tends to be larger than other weeks.

3 Proposed Model

In this section, we present our proposed SSR-Tensor model, and postpone the discussion of hot-spot detection methodology to the next section. Owing to the fact that the gonorrhea data is of three dimensions, namely, {state, week, year}, it will likely have complex "within-dimension" and "between-dimension" interaction/correlation relationship. Within-dimension relationship includes within-state correlation, within-week correlation, and within-year correlation. Between-dimension relationship includes between-state-and-week interaction, between-state-and-year interaction, as well as between-week-and-year interaction. In order to handle these complex "within" and "between" interaction structures, we propose to use the tensor decomposition method, where bases are used to address "within-dimension" correlation, and the tensor product is used for "between-dimension" interaction. Here, the basis is a very important concept where different basis can be chosen for different dimensions. Detailed discussions of the choice of bases are presented in Sect. 6.2.

For the convenience of notation and easy understanding, we first introduce some basic tensor algebra and notation in Sect. 3.1. Then Sect. 3.2 presents our proposed model that is able to characterize the complex correlation structures.

3.1 Tensor Algebra and Notation

In this section, we introduce basic notations, definitions, and operators in tensor (multi-linear) algebra that are useful in this paper. Throughout the paper, scalars are denoted by lowercase letters (e.g., θ), vectors are denoted by lowercase boldface letters ($\boldsymbol{\theta}$), matrices are denoted by uppercase boldface letter ($\boldsymbol{\Theta}$), and tensors by curlicue letter (ϑ). For example, an order-N tensor is represented by $\vartheta \in \mathbb{R}^{I_1 \times \cdots \times I_N}$, where I_n represent the mode-n dimension of ϑ for $n = 1, \ldots, N$.

The mode-n product of a tensor $\vartheta \in \mathbb{R}^{I_1 \times \ldots \times I_N}$ by a matrix $\mathbf{B} \in \mathbb{R}^{J_n \times I_n}$ is a tensor $\mathscr{A} \in \mathbb{R}^{I_1 \times \ldots I_{n-1} \times J_n \times I_{n+1} \times \ldots I_N}$, denoted as $\mathscr{A} = \vartheta \times_n \mathbf{B}$, where each entry of $\mathscr{A}$ is defined as the sum of products of corresponding entries in $\mathscr{A}$ and $\mathbf{B}$: $\mathscr{A}_{i_1,\ldots,i_{n-1},j_n,i_{n+1},\ldots,i_N} = \sum_{i_n} \vartheta_{i_1,\ldots,i_N} \mathbf{B}_{j_n,i_n}$. Here we use the notation $\mathbf{B}_{j_n,i_n}$ to refer the (j_n, i_n)-th entry in matrix $\mathbf{B}$. The notation $\vartheta_{i_1,\ldots,i_N}$ is used to refer to the entry in tensor ϑ with index $(i_1, \ldots, i_N)$. The notation $\mathscr{A}_{i_1,\ldots,i_{n-1},j_n,i_{n+1},\ldots,i_N}$ is used to refer the entry in tensor $\mathscr{A}$ with index $(i_1, \ldots, i_{n-1}, j_n, i_{n+1}, \ldots, i_N)$.

The mode-n unfold of tensor $\vartheta \in \mathbb{R}^{I_1 \times \ldots \times I_N}$ is noted by $\vartheta_{(n)} \in \mathbb{R}^{I_n \times (I_1 \times \ldots I_{n-1} \times I_{n+1} \times I_N)}$, where the column vector of $\vartheta_{(n)}$ are the mode-n vector of ϑ. The mode-n vector of ϑ are defined as the I_n dimensional vector obtained from ϑ by varying the index i_n while keeping all the other indices fixed. For example, $\vartheta_{:,2,3}$ is a model-1 vector.

A very useful technique in the tensor algebra is the Tucker decomposition, which decomposes a tensor into a core tensor multiplied by matrices along each mode: $\mathscr{Y} = \vartheta \times_1 \mathbf{B}^{(1)} \times_2 \mathbf{B}^{(2)} \cdots \times_N \mathbf{B}^{(N)}$, where $\mathbf{B}^{(n)}$ is an orthogonal $I_n \times I_n$ matrix and is a principal component mode-n for $n = 1, \ldots, N$. Tensor product can be represented equivalently by a Kronecker product, i.e., $\text{vec}(\mathscr{Y}) = (\mathbf{B}^{(N)} \otimes \cdots \otimes \mathbf{B}^{(1)})\text{vec}(\vartheta)$, where $\text{vec}(\cdot)$ is the vectorized operator. Finally, the definition of Kronecker product is as follow: Suppose $\mathbf{B}_1 \in \mathbb{R}^{m \times n}$ and $\mathbf{B}_2 \in \mathbb{R}^{p \times q}$ are matrices, the Kronecker product of these matrices, denoted by $\mathbf{B}_1 \otimes \mathbf{B}_2$, is an $mq \times nq$ block matrix defined by

$$\mathbf{B}_1 \otimes \mathbf{B}_2 = \begin{bmatrix} b_{11}\mathbf{B}_2 & \cdots & b_{1n}\mathbf{B}_2 \\ \vdots & \ddots & \vdots \\ b_{m1}\mathbf{B}_2 & \cdots & b_{mn}\mathbf{B}_2 \end{bmatrix}.$$

3.2 Our Proposed SSR-Tensor Model

Our proposed SSR-Tensor model is built on tensors of order three, as it is inspired by the gonorrhea data, which can be represented as a three-dimension tensor $\mathscr{Y}_{n_1 \times n_2 \times T}$ with $n_1 = 50$ states, $n_2 = 51$ weeks, and $T = 13$ years. Note that the i-th, j-th, and k-th slice of the 3-D tensor along the dimension of state, week, and year can be achieved as $\mathscr{Y}_{i::}, \mathscr{Y}_{:j:}, \mathscr{Y}_{::k}$ correspondingly, where $i = 1 \cdots n_1$, $j = 1 \cdots n_2$ and $k = 1 \cdots T$. For simplicity, we denote $\mathbf{Y}_k = \mathscr{Y}_{::k}$. We further denote $\mathbf{y}_k$ as the vectorized form of $\mathbf{Y}_k$, and $\mathbf{y}$ as the vectorized form of $\mathscr{Y}$.

The key idea of our proposed model is to separate the global trend from the local pattern by decomposing the tensor $\mathbf{y}$ into three parts, namely the smooth global trend $\boldsymbol{\mu}$, local hot-spot $\mathbf{h}$, and residual $\mathbf{e}$, i.e. $\mathbf{y} = \boldsymbol{\mu} + \mathbf{h} + \mathbf{e}$. For the first two of the components (e.g. the global trend mean and local hot-spots), we introduce basis decomposition framework to represent the structure of the within correlation in the global background and local hot-spot, also see Yan et al. (2018).

To be more concrete, we assume that global trend mean and local hot-spot can be represented as $\boldsymbol{\mu} = \mathbf{B}_m\boldsymbol{\theta}_m$ and $\boldsymbol{h} = \mathbf{B}_h\boldsymbol{\theta}_h$, where $\mathbf{B}_m$ and $\mathbf{B}_h$ are two bases that will discussed below, and $\boldsymbol{\theta}_m$ and $\boldsymbol{\theta}_h$ are the model coefficients vector of length $n_1 n_2 T$ and needed to be estimated (see Sect. 5). Here the subscript of m and h are abbreviations for mean and hot-spot. Next, it is useful to discuss how to choose the bases $\mathbf{B}_m$ and $\mathbf{B}_h$, so as to characterize the complex "within" and "between" correlation or interaction structures. For the "within" correlation structures, we propose to use pre-specified bases, $\mathbf{B}_{m,s}$ and $\mathbf{B}_{h,s}$, for within-state correlation in global trend and hot-spot, where the subscript of s is an abbreviation for states. Similarly, $\mathbf{B}_{m,w}$ and $\mathbf{B}_{h,w}$ are the pre-specified bases for within-correlation of the same week, whereas $\mathbf{B}_{m,y}$ and $\mathbf{B}_{h,y}$ are the bases for within-time correlation over time. As for the "between" interaction, we use tensor product to describe it, i.e, $\mathbf{B}_m = \mathbf{B}_{m,s} \otimes \mathbf{B}_{m,w} \otimes \mathbf{B}_{m,y}$ and $\mathbf{B}_h = \mathbf{B}_{h,s} \otimes \mathbf{B}_{h,w} \otimes \mathbf{B}_{h,y}$. This Kronecker product has been proved to have better computational efficiency in the tensor response data (Kolda and Bader 2009). Mathematically speaking, all these bases are matrices, which is pre-assigned in our paper. And the choice of bases in shown in Sect. 6.2. With the well-structured "within" and "between" interaction, our proposed model can be written as:

$$\mathbf{y} = (\mathbf{B}_{m,s} \otimes \mathbf{B}_{m,w} \otimes \mathbf{B}_{m,y})\boldsymbol{\theta}_m + (\mathbf{B}_{h,s} \otimes \mathbf{B}_{h,w} \otimes \mathbf{B}_{h,y})\boldsymbol{\theta}_h + \mathbf{e}, \tag{1}$$

where $\mathbf{e} \sim N(0, \sigma^2\mathbf{I})$ is the random noise. Mathematically speaking, both $\mathbf{B}_{m,s}$ and $\mathbf{B}_{h,s}$ are $n_1 \times n_1$ matrix, $\mathbf{B}_{m,w}$ and $\mathbf{B}_{h,w}$ are $n_2 \times n_2$ matrix and $\mathbf{B}_{m,y}$ and $\mathbf{B}_{h,y}$ are $T \times T$ matrix, respectively.

Mathematically, our proposed model in (1) can be rewritten into a tensor format:

$$\mathscr{Y} = \vartheta_m \times_3 \mathbf{B}_{m,y} \times_2 \mathbf{B}_{m,w} \times_1 \mathbf{B}_{m,s} + \vartheta_h \times_3 \mathbf{B}_{h,y} \times_2 \mathbf{B}_{h,w} \times_1 \mathbf{B}_{h,s} + \mathbf{e}, \quad (2)$$

where ϑ_m and ϑ_h is the tensor format of $\boldsymbol{\theta}_m$ and $\boldsymbol{\theta}_h$ with dimensional $n_1 \times n_2 \times T$. Accordingly, the $((k-1)n_1n_2 + (i-1)n_1 + j)$-th entry of $\boldsymbol{\theta}_h, \boldsymbol{\theta}_m$ can estimate the global mean and hot-spot in i-th state and j-th week in k-th year respectively. The tensor representation in Eq. (2) allows us to develop computationally efficient methods for estimation and prediction.

3.3 *Estimation of Hot-Spots*

With the proposed SSR-Tensor model above, we can now discuss the estimation of hot-spot parameters $\boldsymbol{\theta}$'s (including $\boldsymbol{\theta}_m, \boldsymbol{\theta}_h$) in our model in (1) or (2) from the data via the penalized likelihood function. We propose to add two penalties in our estimation. First, because hot-spots rarely occur, we assume that $\boldsymbol{\theta}_h$ is sparse and the majority of entries in the hot-spot coefficient $\boldsymbol{\theta}_h$ are zeros. Thus we propose to add the penalty $R_1(\boldsymbol{\theta}_h) = \lambda\|\boldsymbol{\theta}_h\|_1$ to encourage the sparsity property of $\boldsymbol{\theta}_h$. Second, we assume there is temporal continuity of the hot-spots, as the usual phenomenon of last year is likely to affect the performance of hot-spot in this year. Thus, we add the second penalty $R_2(\boldsymbol{\theta}_h) = \lambda_2\|\mathbf{D}\boldsymbol{\theta}_h\|_1$ to ensure the yearly continuity of the hot-spot, where $\mathbf{D} = \mathbf{D}_s \otimes \mathbf{D}_w \otimes \mathbf{D}_y$ with $\mathbf{D}_s$ as identical matrix of dimension $n_1 \times n_1$, and $T \times T$ matrix $\mathbf{D}_y = \begin{bmatrix} 1 & -1 & & \\ & \ddots & \ddots & \\ & & 1 & -1 \\ & & & 1 \end{bmatrix}$, $n_2 \times n_2$ matrix $\mathbf{D}_w = \begin{bmatrix} 1 & -1 & & \\ & \ddots & \ddots & \\ & & 1 & -1 \\ -1 & & & 1 \end{bmatrix}$. With the formula of $\mathbf{D}_y$, the hot-spot has the property of yearly continuity. By the formula of $\mathbf{D}_w$, the hot-spot has a weekly circular pattern.

By combining both penalties, we propose to estimate the parameters via the following optimization problem:

$$\begin{aligned} &\arg\min_{\boldsymbol{\theta}_m, \boldsymbol{\theta}_h} \|\boldsymbol{e}\|^2 + \lambda_1\|\boldsymbol{\theta}_h\|_1 + \lambda_2\|\mathbf{D}\boldsymbol{\theta}_h\|_1 \\ &\text{subject to } \boldsymbol{y} = (\mathbf{B}_{m,s} \otimes \mathbf{B}_{m,w} \otimes \mathbf{B}_{m,y})\boldsymbol{\theta}_m + (\mathbf{B}_{h,s} \otimes \mathbf{B}_{h,w} \otimes \mathbf{B}_{h,y})\boldsymbol{\theta}_h + \mathbf{e}, \end{aligned} \quad (3)$$

where $\boldsymbol{\theta}_m = \text{vec}(\boldsymbol{\theta}_{m,1}, \ldots, \boldsymbol{\theta}_{m,t}, \ldots, \boldsymbol{\theta}_{m,T})$ and $\boldsymbol{\theta}_h = \text{vec}(\boldsymbol{\theta}_{h,1}, , \ldots, \boldsymbol{\theta}_{h,t}, \ldots, \boldsymbol{\theta}_{h,T})$. The choice of the turning parameters λ_1, λ_2 will be discussed in Sect. 4.

Note that there are two penalties in Eq. (3): $\lambda_1\|\boldsymbol{\theta}_h\|_1$ is the LASSO penalty to control both the sparsity of the hot-spots and $\lambda_2\|\mathbf{D}\boldsymbol{\theta}_h\|_1$ is the fused LASSO penalty (Tibshirani et al. 2005) to control the temporal consistency of the hot-spots. Traditional algorithms often involve the storage and computation of the matrix $\mathbf{B}_m$ and $\mathbf{B}_h$, which is of the dimension $n_1n_2n_3 \times n_1n_2n_3$. Thus they might work to solve the optimization problem in Eq. (3) when the dimensions are small, but they will be computationally infeasible as the dimensions grow. To address this computational challenge, we propose to simplify the computational complexity by modifying the matrix algebra in traditional algorithm into tensor algebra, and will discuss how to optimize the problem in Eq. (3) computationally efficiently in Sect. 5.

4 Hot-Spot Detection

This section focuses on the detection of the hot-spot, which includes the detection and identification of the year (when), the state (where) and the week (which) of the hot-spots. In our case study, we focus on the upward shift of the number of gonorrhea cases, since the increasing gonorrhea is generally more harmful to the societies and communities. Of course, one can also detect the downward shift with a slight modification of our proposed algorithms by multiplying -1 to the raw data.

For the purpose of easy presentation, we first discuss the detection of the hot-spot, i.e., detect when hot-spot occurs in Sect. 4.1. Then, in Sect. 4.2, we consider the localization of the hot-spot, i.e., determine which states and which weeks are involved for the detected hot-spots.

4.1 Detect When the Hot Spot Occurs

To determine when the hot-spot occurs, we consider the following hypothesis test and set up the control chart for the hot-spot detection (4).

$$H_0 : \widetilde{\mathbf{r}}_t = 0 \quad v.s. \quad H_1 : \widetilde{\mathbf{r}}_t = \delta \widehat{\mathbf{h}}_t \quad (\delta > 0), \tag{4}$$

where $\widetilde{\mathbf{r}}_t$ is the expected residuals after removing the mean. The essence of this test is that, we want to detect whether $\widetilde{\mathbf{r}}_t$ has a mean shift in the direction of $\widehat{\mathbf{h}}_t$, estimated in Sect. 5. To test this hypotheses, the likelihood ratio test is applied to the residual $\mathbf{r}_t$ at each time t, i.e. $\mathbf{r}_t = \mathbf{y}_t - \boldsymbol{\mu}_t$, where it assumes that the residuals $\mathbf{r}_t$ is independent after removing the mean and its distribution before and after the hot-spot remains the same. Accordingly, the test statistics monitoring upward shift is designed as $P_t^+ = \widehat{\mathbf{h}}_t^{\prime +} \mathbf{r}_t / \sqrt{\widehat{\mathbf{h}}_t^{\prime +} \widehat{\mathbf{h}}_t^+}$ (Hawkins 1993), where $\widehat{\mathbf{h}}_t^+$ only takes the positive part of $\widehat{\mathbf{h}}_t$ with other entries as zero. Here we put a superscript "+" to emphasize that it aims for upward shift.

The choices of the penalty parameters λ_1, λ_2 are described as follows. In order to select the one with the most power, we propose to calculate a series of P_t^+ under different combination of (λ_1, λ_2) from the set $\Gamma = \{(\lambda_1^{(1)}, \lambda_2^{(1)}) \cdots (\lambda_1^{(n_\lambda)}, \lambda_2^{(n_\lambda)})\}$. For better illustration, we denote the test statistics under penalty parameter (λ_1, λ_2) as $P_t^+(\lambda_1, \lambda_2)$. The test statistics (Zou and Qiu 2009) with the most power to detect the change, noted as $\widetilde{P}_t^+$, can be computed by

$$\widetilde{P}_t^+ = \max_{(\lambda_1,\lambda_2)\in\Gamma} \frac{P_t^+(\lambda_1,\lambda_2) - E(P_t^+(\lambda_1,\lambda_2))}{\sqrt{Var(P_t^+(\lambda_1,\lambda_2))}}, \quad (5)$$

where $E(P_t^+(\lambda_1,\lambda_2))$, $Var(P_t^+(\lambda_1,\lambda_2))$ respectively are the mean and variance of $P_t(\lambda_1,\lambda_2)$ under H_0 (e.g. for phase-I in-control samples).

Note that the penalty parameter (λ_1,λ_2) to realize the maximization in Eq. (5) is generally different under different time t. To emphasize such dependence of time t, denote by $(\lambda_{1,t}^*, \lambda_{2,t}^*)$ the parameter pair that attains the maximization in Eq. (5) at time t, i.e,

$$(\lambda_{1,t}^*, \lambda_{2,t}^*) = \arg\max_{(\lambda_1,\lambda_2)\in\Gamma} \frac{P_t^+(\lambda_1,\lambda_2) - E(P_t^+(\lambda_1,\lambda_2))}{\sqrt{Var(P_t^+(\lambda_1,\lambda_2))}}. \quad (6)$$

Thus, the series of the test statistics for the hot-spot at time t is $\widetilde{P}_t^+(\lambda_{1,t}^*, \lambda_{2,t}^*)$ where $t = 1\cdots T$.

With the test statistic available, we design a control chart based on the CUSUM procedure due to the following reasons: (1) we are interested in detecting the change with the temporal continuity, therefore, aligns with the objective of CUSUM. (2) In the view of social stability, we want to keep gonorrhea at a target value without sudden changes, which makes the CUSUM chart is a natural better fit.

To be more specific, in the CUSUM procedure, we compute the CUSUM statistics recursively by

$$W_t^+ = \max\{0, W_{t-1}^+ + \widetilde{P}_t^+(\lambda_{1,t}^*, \lambda_{2,t}^*) - d\},$$

and $W_{t=0}^+ = 0$, where d is a constant and can be chosen according to the degree of the shift that we want to detect. Next, we set the control limit L to achieve a desirable ARL for in-control samples. Finally, whenever $W_t^+ > L$ at some time $t = t^*$, we declare that a hot-spot occurs at time t^*.

4.2 *Localize Where and Which the Hot Spot Occur?*

After the hot-spot t^* has been detected by the CUSUM control chart in the previous section, the next step is to localize where and which week may account for this hot-spot. To do so, we propose to utilize the vector

$$\widehat{\mathbf{h}}_{\lambda_{1,t^*}^*,\lambda_{2,t^*}^*} = \mathbf{B}_h \widehat{\boldsymbol{\theta}}_{h,\lambda_{1,t^*}^*,\lambda_{2,t^*}^*}$$

at the declared hot-spot time t^* and the corresponding parameter λ_{1,t^*}^*, λ_{2,t^*}^* in Eq. (6). For the numerical computation purpose, it is often easier to directly work with the tensor format of the hot-spot $\widehat{\mathbf{h}}_{\lambda_{1,t^*}^*,\lambda_{2,t^*}^*}$, denoted as $\widehat{\mathscr{H}}_{\lambda_{1,t^*}^*,\lambda_{2,t^*}^*}$, which is a tenor of dimension $n_1 \times n_2 \times T$. If the (i, j, t^*)-th entry in $\widehat{\mathscr{H}}_{\lambda^*_{1,t^*}, \lambda^*_{2,t^*}}$ is non-zero, then we declare that there is a hot-spot for the j-th week in the i-th state in t^*-th year.

5 Optimization Algorithm

In this section, we will develop an efficient optimization algorithm for solving the optimization problem in Eq. (3). For notion convenience, we adjust the notation above a little bit. Because $\boldsymbol{\theta}_m, \boldsymbol{\theta}_h$ in Eq. (3) is solved under penalty $\lambda_1 R_1(\boldsymbol{\theta}_h) + \lambda_2 R_2(\boldsymbol{\theta}_h)$, we change $\boldsymbol{\theta}_m, \boldsymbol{\theta}_h$ into $\boldsymbol{\theta}_{m,\lambda_1,\lambda_2}, \boldsymbol{\theta}_{h,\lambda_1,\lambda_2}$ to emphasize the penalty parameter λ_1 and λ_2. Accordingly, $\boldsymbol{\theta}_{h,0,\lambda_2}$ refers to the estimator only under the second penalty $\lambda_2 R_2(\boldsymbol{\theta}_h)$, i.e,

$$\boldsymbol{\theta}_{h,0,\lambda_2} = \arg\min_{\boldsymbol{\theta}_m, \boldsymbol{\theta}_h} \{\|\mathbf{e}\|_2^2 + \lambda R_2(\boldsymbol{\theta}_h)\}. \tag{7}$$

The structure of this section is that, we first develop the procedure of our proposed method in Sect. 5.1 and then give the computational complexity in Sect. 5.2.

5.1 *Procedure of Our Algorithm*

In the optimization problem shown in Eq. (3), there are two unknown vectors, namely $\boldsymbol{\theta}_{m,\lambda_1,\lambda_2}, \boldsymbol{\theta}_{h,,\lambda_1,\lambda_2}$. To simplify the optimization above, we first figure out the close-form correlation between $\boldsymbol{\theta}_{m,\lambda_1,\lambda_2}$ and $\boldsymbol{\theta}_{h,\lambda_1,\lambda_2}$. Then, we solve the optimization by modifying the matrix algebra in FISTA (Beck and Teboulle 2009) into tensor algebra. The key to realize it is the proximal mapping of $\lambda_1 R_1(\boldsymbol{\theta}_{h,\lambda_1,\lambda_2}) + \lambda_2 R_2(\boldsymbol{\theta}_{h,\lambda_1,\lambda_2})$. To address it, we first aim at the proximal mapping of $\lambda_2 R_2(\boldsymbol{\theta}_{h,0,\lambda_1})$, where SFA via gradient descent (Liu et al. 2010) is used. And then the proximal mapping of $\lambda_1 R_1(\boldsymbol{\theta}_{h,\lambda_1,\lambda_2}) + \lambda_2 R_2(\boldsymbol{\theta}_{h,\lambda_1,\lambda_2})$ can be solved with a close-form correlation between it and the proximal mapping of $\lambda_2 R_2(\boldsymbol{\theta}_{h,0,\lambda_2})$.

There are three subsections in this section, where each subsection represents one step in our proposed algorithm.

5.1.1 Estimate the Mean Parameter

To begin with, we first simplify the optimization problem in Eq. (3), i.e., figure out the close-form correlation between $\boldsymbol{\theta}_{m,\lambda_1,\lambda_2}$ and $\boldsymbol{\theta}_{h,\lambda_1,\lambda_2}$.

Although there are two sets of parameters $\boldsymbol{\theta}_{m,\lambda_1,\lambda_2}$ and $\boldsymbol{\theta}_{h,\lambda_1,\lambda_2}$ in the model, we note that given $\boldsymbol{\theta}_{h,\lambda_1,\lambda_2}$, the parameter $\boldsymbol{\theta}_{m,\lambda_1,\lambda_2}$ is involved in the standard least squared estimation and thus can be solved in the closed-form solution, see Eq. (8) in the proposition below.

Proposition 1 *Given* $\boldsymbol{\theta}_{h,\lambda_1,\lambda_2}$*, the closed-form solution of* $\boldsymbol{\theta}_{m,\lambda_1,\lambda_2}$ *is given by:*

$$\boldsymbol{\theta}_{m,\lambda_1,\lambda_2} = (\mathbf{B}'_m\mathbf{B}_m)^{-1}(\mathbf{B}'_m y - \mathbf{B}'_m\mathbf{B}_h\boldsymbol{\theta}_{h,\lambda_1,\lambda_2}). \tag{8}$$

It remains to investigate how to estimate the parameter $\boldsymbol{\theta}_{h,\lambda_1,\lambda_2}$. After plugging in (8) into (3), the optimization problem for estimating $\boldsymbol{\theta}_{h,\lambda_1,\lambda_2}$ becomes

$$\arg\min_{\boldsymbol{\theta}_{h,\lambda_1,\lambda_2}} \|\mathbf{y}^* - \mathbf{X}\boldsymbol{\theta}_{h,\lambda_1,\lambda_2}\|_2^2 + \lambda_1\|\boldsymbol{\theta}_{h,\lambda_1,\lambda_2}\|_1 + \lambda_2\|\mathbf{D}\boldsymbol{\theta}_{h,\lambda_1,\lambda_2}\|_1, \tag{9}$$

where $\mathbf{y}^* = [\mathbf{I} - \mathbf{H}_m]\,\mathbf{y}, \mathbf{X} = [\mathbf{I} - \mathbf{H}_m]\,\mathbf{B}_h$ and $\mathbf{H}_m = \mathbf{B}_m(\mathbf{B}'_m\mathbf{B}_m)^{-1}\mathbf{B}'_m$ is the projection matrix.

Due to the high dimension, we need to develop an efficient and precise optimization algorithm to optimize (3). Obviously, (9) is a typical sparse optimization problem. However, most of the sparse optimization frameworks focus on optimizing:

$$\arg\min_{\boldsymbol{\theta}_{h,0,\lambda_2}} \|\mathbf{y}^* - \mathbf{X}\boldsymbol{\theta}_{h,\lambda_1,0}\|_2^2 + \lambda_1\|\boldsymbol{\theta}_{h,\lambda_1,0}\|_1, \tag{10}$$

such as Daubechies et al. (2004), Beck and Teboulle (2009), Friedman et al. (2010) and so on, where iterative updating rule are used base either on the gradient information or the proximal mapping. In most cases, the algorithms above works, however, two challenges occur in our paper:

1. When the dimension of $\mathbf{X}$ (of size $n_1n_2T \times n_1n_2T$) become increasingly large, it is difficult for the computer to store and memorize it.
2. When the penalty term is $\lambda_1\|\boldsymbol{\theta}_{h,\lambda_1,\lambda_2}\|_1 + \lambda_2\|\boldsymbol{D\theta}_{h,\lambda_1,\lambda_2}\|_1$, instead of only $\lambda_1\|\boldsymbol{\theta}_{h,\lambda_1,\lambda_2}\|_1$, direct application of the proximal mapping of $\lambda_1\|\boldsymbol{\theta}_{h,\lambda_1,\lambda_2}\|_1$ is not workable.

Therefore, directly applying these above algorithms (Beck and Teboulle 2009; Daubechies et al. 2004; Friedman et al. 2010) to our case is not feasible. To extend the existing research, we proposed an iterative algorithm in Algorithm 1 and we explain the approach to solve the proximal mapping of $\lambda_1\|\boldsymbol{\theta}_{h,\lambda_1,\lambda_2}\|_1 + \lambda_2\|\boldsymbol{D\theta}_{h,\lambda_1,\lambda_2}\|_1$ in Sect. 5.1.2.

5.1.2 Proximal Mapping

The main tool we use to solve the optimization problem in Eq. (9) is a variation of proximal mapping. Denote that $F(\boldsymbol{\theta}_{h,\lambda_1,\lambda_2}) = \frac{1}{2}\|\mathbf{y}^* - \mathbf{X}\boldsymbol{\theta}_{h,\lambda_1,\lambda_2}\|_2^2$. And in the i-th iteration, the according recursive estimator of $\boldsymbol{\theta}_{h,\lambda_1,\lambda_2}$ is noted as $\boldsymbol{\theta}^{(i)}_{h,\lambda_1,\lambda_2}$. Besides,an auxiliary variable $\boldsymbol{\eta}^{(i)}$ is introduced to update from $\boldsymbol{\theta}^{(i)}_{h,\lambda_1,\lambda_2}$ to $\boldsymbol{\theta}^{(i+1)}_{h,\lambda_1,\lambda_2}$ through

$$\begin{aligned}\boldsymbol{\theta}_{h,\lambda_1,\lambda_2}^{(i+1)} &= \arg\min_{\boldsymbol{\theta}} F(\boldsymbol{\eta}^{(i)}) + \frac{\partial}{\partial \boldsymbol{\theta}_{h,\lambda_1,\lambda_2}} F(\boldsymbol{\eta}^{(i)}) \left(\boldsymbol{\theta} - \boldsymbol{\eta}^{(i)}\right) + \\ &\quad \lambda_1 \|\boldsymbol{\theta}\|_1 + \lambda_2 \|\mathbf{D}\boldsymbol{\theta}\|_1 + \frac{L}{2} \|\boldsymbol{\theta} - \boldsymbol{\eta}^{(i)}\|_2^2 \\ &= \arg\min_{\boldsymbol{\theta}} \left[\frac{1}{2} \left[\boldsymbol{\theta} - \left(\boldsymbol{\eta}^{(i)} - \frac{\partial}{L\partial \boldsymbol{\theta}} F(\boldsymbol{\eta}^{(i)}) \right) \right]^2 + \lambda_1 \|\boldsymbol{\theta}\|_1 + \lambda_2 \|\mathbf{D}\boldsymbol{\theta}\|_1 \right] \\ &\triangleq \pi_{\lambda_2}^{\lambda_1}(\mathbf{v})\end{aligned}$$

where $\mathbf{v} = \boldsymbol{\eta}^{(i)} - \frac{\partial}{L\partial \boldsymbol{\theta}} F(\boldsymbol{\eta}^{(i)})$, $\boldsymbol{\eta}^{(i)} = \boldsymbol{\theta}_{h,\lambda_1,\lambda_2}^{(i)} + \frac{t_{i-2}-1}{t_{i-1}} (\boldsymbol{\theta}_{h,\lambda_1,\lambda_2}^{(i)} - \boldsymbol{\theta}_{h,\lambda_1,\lambda_2}^{(i-1)})$ and $t_{-1} = t_0 = 1$, $t_{i+1} = \frac{1+\sqrt{1+4t_i^2}}{2}$.

Because it is difficult to solve $\pi_{\lambda_2}^{\lambda_1}(\mathbf{v})$ directly, we aim to solve $\pi_{\lambda_2}^{0}(\mathbf{v})$ first. And proved by Liu et al. (2010), there is a close-form correlation between $\pi_{\lambda_2}^{\lambda_1}(\mathbf{v})$ and $\pi_{\lambda_2}^{0}(\mathbf{v})$, which is shown in Proposition 2.

Proposition 2 *The close form relationship between $\pi_{\lambda_2}^{\lambda_1}(\mathbf{v})$ and $\pi_{\lambda_2}^{0}(\mathbf{v})$ is*

$$\pi_{\lambda_2}^{\lambda_1}(\mathbf{v}) = sign(\pi_{\lambda_2}^{0}(\mathbf{v})) \odot \max\{|\pi_{\lambda_2}^{0}(\mathbf{v})| - \lambda_1, 0\}. \tag{11}$$

where $\odot$ is an element-wise product operator.

With the proximal mapping function in Proposition 2, we can now develop the algorithm shown in Algorithm 1.

Algorithm 1: Iterative updating based on tensor decomposition

---

Input: $\mathbf{y}^*, \mathbf{B}_s, \mathbf{B}_w, \mathbf{B}_y, \mathbf{D}_s, \mathbf{D}_w, \mathbf{D}_y, K, L, \lambda_1, \lambda_2, L_0, M_1, M_2$

Output: $\boldsymbol{\theta}_{h,\lambda_1,\lambda_2}$

initialization;

$\boldsymbol{\Theta}^{(1)} = \boldsymbol{\Theta}^{(0)}, t_{-1} = 1, t_0 = 1, L = L_0$

for $i = 1 \cdots M_1$ **do**

$\quad \mathcal{N}^{(i)} = \mathcal{N}^{(i)} + \frac{t_{i-2}-1}{t_{i-1}}(\boldsymbol{\Theta}^{(i)} - \boldsymbol{\Theta}^{(i-1)})$

$$\mathcal{V} = \mathcal{N}^{(i)} - \frac{1}{L}\mathcal{N}^{(i)} \times_1 (\mathbf{P}_s'\mathbf{P}_s) \times_2 (\mathbf{P}_w'\mathbf{P}_w) \times_3 (\mathbf{P}_y'\mathbf{P}_y) - \frac{1}{L}\mathcal{Y}^* \times_1 \mathbf{P}_s' \times_2 \mathbf{P}_w' \times_3 \mathbf{P}_y'$$

$\quad$ **for** $j = 0 \cdots M_2$ **do**

$$\mathcal{G}^{(i)} = \left(\mathcal{Z}^{(j)} \times_1 (\mathbf{D}_s'\mathbf{D}_s) \times_2 (\mathbf{D}_w'\mathbf{D}_w) \times_3 (\mathbf{D}_y'\mathbf{D}_y))\right) - \left(\mathcal{V} \times_1 \mathbf{D}_s \times_2 \mathbf{D}_w \times_3 \mathbf{D}_y\right)$$

$\quad\quad \mathcal{Z}^{(j+1)} = P\left(\mathcal{Z}^{(j)} - \mathcal{G}^{(j)}/L\right)$

$\quad \pi_{\lambda_2}^{0}(\mathcal{V}) = \mathcal{V} - (\mathcal{Z}^{(M_2)}) \times_1 \mathbf{D}_s \times_2 \mathbf{D}_w \times_3 \mathbf{D}_y$

$\quad \pi_{\lambda_2}^{\lambda_1}(\mathcal{V}) = \mathbf{sign}(\pi_{\lambda_2}^{0}(\mathcal{V})) \odot \mathbf{max}\{\left|\pi_{\lambda_2}^{0}(\mathcal{V})\right| - \lambda_1, 0\}$

$\quad t_{i+1} = \frac{1+\sqrt{1+4t_i^2}}{2}$

$\widehat{\boldsymbol{\Theta}}_{h,\lambda_1,\lambda_2} = \pi_{\lambda_2}^{\lambda_1}(\mathcal{V})$

$\widehat{\boldsymbol{\theta}}_{h,\lambda_1,\lambda_2} = \mathbf{vector}(\widehat{\boldsymbol{\Theta}}_{h,\lambda_1,\lambda_2})\ \boldsymbol{v} = \mathbf{vector}(\mathcal{V})$

---

vector$(\cdot)$ is a function that unfolding a order-3 tensor of dimension $n_1 \times n_2 \times n_3$ into a vector $n_1 n_2 n_3$.

5.2 *Computational Complexity*

This section discusses the computational complexity of our proposed algorithm. Suppose the raw data is structured into a tensor of order three with dimensional $n_1 \times n_2 \times n_3$, then the computation complexity of our propose method is of order $O\left(n_1 n_2 n_3 \max\{n_1, n_2, n_3\}\right)$ (see Proposition 3).

Proposition 3 *The computational complexity of our proposed algorithm (see Algorithm 1) is of order* $O\left(n_1 n_2 n_3 \max\{n_1, n_2, n_3\}\right)$.

Proof The main computational load in Algorithm 1 is on the calculation of $\mathbf{v}$ (line 4), $\mathbf{g}^{(i)}$(line 5) and $\pi^0_{\lambda_2}(\mathbf{v})$ (line 7). We will take the calculation of $\mathbf{v}$ in line 4 in the algorithm as an example. To begin with, we focus on the computational complexity of

$$\mathcal{N}^{(i)} \times_1 (\mathbf{P}_s'\mathbf{P}_s) \times_2 (\mathbf{P}_w'\mathbf{P}_w) \times_3 (\mathbf{P}_y'\mathbf{P}_y)). \tag{12}$$

For better illustration, we denote tensor($\boldsymbol{\eta}^{(i)}$) as $\mathcal{N}^{(i)}$ and $\mathcal{N}^{(i)} \times_1 (\mathbf{P}_s'\mathbf{P}_s)$ as tensor $\mathcal{L}_1$. According to the tensor algebra (Kolda and Bader 2009, Sect. 2.5),

$$\mathcal{L}_1 = \mathcal{N}^{(i)} \times_1 (\mathbf{P}_s'\mathbf{P}_s) \Longleftrightarrow \mathcal{L}_{1(1)} = \mathbf{P}_s'\mathbf{P}_s\mathcal{N}^{(i)}_{(1)}.$$

Therefore, the computational complexity of Eq. (12) is the same as two-matrix multiplication with order $n_1 \times n_1$ and $n_1 \times n_1n_2$, which is of order $O\ (n_1n_2n_3(2n_1 - 1))$.

After the calculation of $\mathcal{L}_1$, Eq. (12) is reduced to

$$\mathcal{L}_1 \times_2 (\mathbf{P}_w'\mathbf{P}_w) \times_3 (\mathbf{P}_y'\mathbf{P}_y)). \tag{13}$$

Similarly, denotes $\mathcal{L}_2 = \mathcal{L}_1 \times_2 (\mathbf{P}_w'\mathbf{P}_w)$, then

$$\mathcal{L}_2 = \mathcal{L}_1 \times_2 (\mathbf{P}_w'\mathbf{P}_w) \Longleftrightarrow \mathcal{L}_{2(2)} = \mathbf{P}_w'\mathbf{P}_w\mathcal{N}_{(2)}.$$

Therefore, the computational complexity of Eq. (13) is the same as two-matrix multiplication with order $n_2 \times n_2$ and $n_2 \times n_1n_3$, which is of order $O\ (n_1n_2n_3(2n_2 - 1))$.

After the calculation of $\mathcal{L}_2$, Eq. (13) is reduced to

$$\mathcal{L}_2 \times_3 (\mathbf{P}_y'\mathbf{P}_y)). \tag{14}$$

Similarly, denotes $\mathcal{L}_3 = \mathcal{L}_2 \times_2 (\mathbf{P}_y'\mathbf{P}_y)$, then

$$\mathcal{L}_3 = \mathcal{L}_2 \times_3 (\mathbf{P}_y'\mathbf{P}_y) \Longleftrightarrow \mathcal{L}_{3(3)} = \mathbf{P}_w'\mathbf{P}_w\mathcal{N}_{(3)}.$$

Therefore, the computational complexity of Eq. (13) is the same as two-matrix multiplication with order $n_3 \times n_3$ and $n_3 \times n_1n_2$, which is of order $O\ (n_1n_2n_3(2n_3 - 1))$.

By combining all these blocks built above, we conclude that the computational complexity of Eq. (12) is of order $O(n_1n_2n_3\ (\max\{n_1, n_2, n_3\}))$.

In the same way, the computational complexity in line 5 and 7 of Algorithm 1 is also of order $O(n_1n_2n_3\ (\max\{n_1, n_2, n_3\}))$. Thus, the computational complexity of Algorithm is of order $O(n_1n_2n_3\ (\max\{n_1, n_2, n_3\}))$.

6 Simulation

In this section, we conduct simulation studies to evaluate our proposed methodologies by comparing with several benchmark methods in the literature. The structure of this section is as follows. We first present the data generation mechanism for our simulations in Sect. 6.1, then discuss the performance of hot-spot detection and localization in Sect. 6.2.

6.1 *Generative Model in Simulation*

In our simulation, at each time index $t\,(t = 1 \cdots T)$, we generate a vector $\mathbf{y}_t$ of length $n_1 n_2$ by

$$\mathbf{y}_{i,t} = (\mathbf{B}\boldsymbol{\theta}_t)_i + \delta \mathbb{1}\{t \geq \tau\}\mathbb{1}_i\{i \in S_h\} + \mathbf{w}_{i,t}, \tag{15}$$

where $\mathbf{y}_{i,t}$ denotes the i-th entry in vector $\mathbf{y}_t$, $(\mathbf{B}\boldsymbol{\theta}_t)_i$ denotes the i-th entry in vector $\mathbf{B}\boldsymbol{\theta}_t$, and δ denotes the change magnitude. Here $\mathbb{1}(A)$ is the indicator function, which has the value 1 for all elements of A and the value 0 for all elements not in A, and $\mathbf{w}_{i,t}$ is the i-th entry in the white noise vector whose entries are independent and follow $N(0, 0.1^2)$ distribution.

Next, after the temporal detection of hot-spots, we need to further localize the hot-spots in the sense that we need to find out which state and which week may lead to the occurrence of temporal hot-spot. Because the baseline methods, PCA and T2, can only realize the detection of temporal changes, we only show the localization of spatial hot-spot by SSR-Tensor, SSD (Yan et al. 2018), ZQ lasso (Zou and Qiu 2009). For the anomaly setup, $\mathbb{1}\{t \geq \tau\}$ indicates that the spatial hot-spots only occur after the temporal hot-spot τ. This ensures that the simulated hot-spot is temporal consistent. The second indicator function $\mathbb{1}_i\{i \in S_h\}$ shows that only those entries whose location index belongs set S_h are assigned as local hot-spots. This ensures that the simulated hot-spot is sparse. Here we assume the change happens at $\tau = 50$ among total $T = 100$ years. And the spatial hot-spots index set is formed by the combination of states Conn, Ohio, West Va, Tex, Hawaii and week from 1–10 and 41–51.

To match the dimension in the case study, we choose $n_1 = 50, n_2 = 51$. As for the three terms on the right side of Eq. (15), they serve for the global trend mean, local sparse anomaly and white noise respectively. In our simulation, the matrix $\mathbf{B}$ is $\mathbf{B}_{m,s} \otimes \mathbf{B}_{m,w} \otimes \mathbf{B}_{m,y}$ with the same choice as that in Sect. 3.2.

Besides, in each of these two scenarios, we further consider two sub-cases, depending on the value of change magnitude δ in Eq. (15): one is $\delta = 0.1$ (small shift) and the other is $\delta = 0.5$ (large shift).

6.2 Hot-Spot Detection Performance

In this section, we compare the performance of our proposed method (denoted as 'SSR-tensor') for detection of hot-spot with some benchmark methods. Specifically, we compare our proposed method with Hotelling T^2 control chart (Qiu 2013) (denoted as 'T2'), LASSO-based control chart proposed by Zou and Qiu (2009) (denoted as 'ZQ LASSO'), PCA-based control chart proposed by De Ketelaere et al. (2015) (denoted as 'PCA') and SSD proposed by Yan et al. (2018) (denoted as 'SSD'). Note that there are two main differences between our SSR-tensor method and the SSD method in Yan et al. (2018). First, SSR-Tensor has the autoregressive or fussed LASSO penalty in Eq. (3) so as to ensure the temporal continuity of the hot-spot. Second, SSD uses the Shewhart control chart to monitor temporal changes, while SSR-Tensor utilizes CUSUM instead, which is more sensitive for a small shift.

For the choices of basis matrices in our proposed model and method, we choose $\mathbf{B}_{m,1}$ as the Gaussian kernel matrix to describe the spatial structure of the global trend, i.e., the (i, j) entry is $\exp\{-d^2/(2c^2)\}$ where d is the distance between the i-th state and j-th state and c is the bandwidth chosen by cross-validation. In addition, we choose identical matrices for the yearly basis and weekly basis, since we do not have any prior information. Moreover, we use the identity matrix for the spatial and temporal basis of the hot-spots, since the L_1-penalty in (3) has already addressed the temporal continuity properties of our hot-spot estimation. For SSD in Yan et al. (2018), we will use the same spatial and temporal basis in order to have a fair comparison.

For evaluation, we will compute the following four criteria: (i) precision, defined as the proportion of detected anomalies that are true hot-spots; (ii) recall, defined as the proportion of the anomalies that are correctly identified; (iii) F measure, a single criterion that combines the precision and recall by calculating their harmonic mean; and (iv) the corresponding average run length (ARL_1), a measure on the average detection delay in the special scenario when the change occurs at time $t = 1$. All simulation results below are based on 1000 Monte Carlo simulation replications.

Table 1 shows the merits of our methodology mainly lies on the higher precision and shorter ARL_1. For example, when the shift is very small, i.e., $\delta = 0.1$, the ARL_1 of our SSR-Tensor method is only 1.6420 compared with 7.4970 of SSD and 9.5890 of ZQ-LASSO. The reason for SSR-Tensor has shorter ARL_1 than that of SSD is that, SSD uses Shewhart control chart to detect temporal changes, which make it insensitive for a small shift. While for SSR-Tensor, it applies the CUSUM control chart, which is capable to detect the shift of small size. The reason for both SSR-Tensor and SSD have shorter ARL_1 than that of ZQ-LASSO, PCA and T2 is that ZQ-LASSO fails to capture the global trend mean. Yet, the data generated in our simulation has both decreasing and circular global trend, which makes it hard for ZQ-LASSO to model well.

Table 1 Scenario 1 (decreasing global trend): Comparison of hot-spot detection under small shift and large shift

methods	Small shift $\delta = 0.1$				Large shift $\delta = 0.5$			
	Precision	Recall	F measure	ARL	Precision	Recall	F measure	ARL
SSR-tensor	**0.0824**	**0.9609**	**0.5217**	**1.6420**	**0.0822**	**0.9633**	**0.5228**	**1.0002**
	(0.0025)	(0.0536)	(0.0270)	(0.7214)	(0.0022)	(0.0549)	(0.0277)	(0.0144)
SSD	0.0404	0.9820	0.5112	7.4970	0.0412	1.0000	0.5206	1.0000
	(0.0055)	(0.1330)	(0.0692)	(9.4839)	(0.0000)	(0.0000)	(0.0000)	(0.0000)
ZQ LASSO	0.0412	1.000	0.5206	9.5890	0.0412	1.0000	0.5206	8.8562
	(0.0000)	(0.0000)	(0.0000)	(7.5414)	(0.0000)	(0.0000)	(0.0000)	(7.1169)
PCA	–	–	–	28.7060	–	–	–	32.0469
	–	–	–	(16.9222)	–	–	–	(17.4660)
T2	–	–	–	50.0000	–	–	–	50.0000
	–	–	–	(0.0000)	–	–	–	(0.0000)

7 Case Study

In this section, we apply our proposed SSR-tensor model and hot-spot detection/localization method to the weekly gonorrhea dataset in Sect. 2. For the purpose of comparison, we also consider other benchmark methods mentioned in Sect. 6, and consider two performance criteria: one is the temporal detection of hot-spots (i.e., which year it occurs) and the other is the localization of the hot-spots (i.e., which state and which week might involve the alarm).

7.1 When the Temporal Changes Happen?

Here we consider the performance on the temporal detection of hot-spots of our proposed method and other benchmark methods. For our proposed SSR-Tensor method, we build a CUSUM control chat utilizing the test statistic in Sect. 4.1, which is shown in Fig. 4. From this plot, we can see that the hot-spots are detected at 10th year, i.e., 2016.

For the purpose of comparison, we also apply the benchmark methods, SSD (Yan et al. 2018), ZQ LASSO (Zou and Qiu 2009), PCA (De Ketelaere et al. 2015) and T2 (Qiu 2013), into the gonorrhea dataset. Unfortunately, all benchmark methods are unable to raise any alarms, but our proposed SSR-tensor method raises the first hot-spot alarm in year 2016.

Fig. 4 CUSUM Control chart of gonorrhea dataset during years 2006–2018

7.2 *In Which State and Week Do the Spatial Hot-spots Occur?*

Next, after the temporal detection of hot-spots, we need to further localize the hot-spots in the sense that we need to find out which state and which week may lead to the occurrence of temporal hot-spot. Because the baseline methods, SSD, ZQ-LASSO, PCA, and T2, can only realize the detection of temporal changes, we only show the localization of spatial hot-spot by SSR-Tensor, which is visualized in Fig. 5.

There are some circular patterns in specific areas. For example, CENTRAL(Ark, La, Okla, Tex) tends to have a circular pattern every 11 weeks, which is shown in Fig. 5. Besides, there are also some circular pattern for a certain state, for instance, Kansas has the bi-weekly pattern as shown in Fig. 6. To validate the bi-weekly circular pattern of Kansas, we plot the time series plot of Kansas in 2016 as well as the auto-correlation function plot in Fig. 5. Besides, the auto-correlation function plot in the left panel of Fig. 6 serves as a baseline. It can be seen from the middle and right plot of Fig. 6 that, Kansas has some bi-weekly or tri-weekly circular pattern.

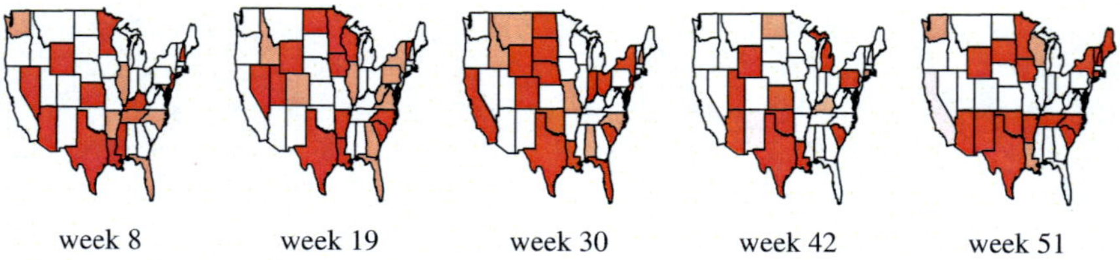

Fig. 5 Hot-spot detection result of circular pattern of W.S. CENTRAL(Arkansas, Louisiana, Oklahoma, Texas)

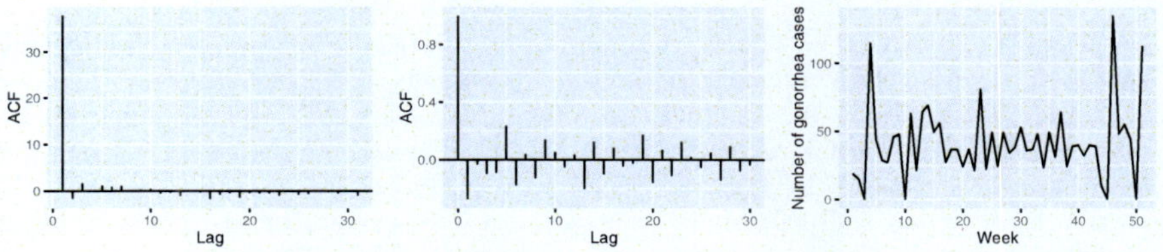

Fig. 6 Auto-correlation of all US (left) & Kans.(middle) in 2016 and time series plot of Kansas in 2016 (right)

Acknowledgements The authors thank an anomalous reviewer for her or his thoughtful and constructive comments that greatly improved the quality and presentation of this article, and appreciate Professors Sven Knoth, Kwok Tsui, and Bill Woodall for help and encouragements during the process. Y. Zhao and Y. Mei were supported in part by NSF grant DMS-1830344, H. Yao was supported in part by NSF DMS-1830363, and S. E. Holte was supported in part by NSF DMS-1830372 and NIH #AI027757 (Baeten).

References

Beck, A., & Teboulle, M. (2009). A fast iterative shrinkage-thresholding algorithm for linear inverse problems. *SIAM Journal on Imaging Sciences*, *2*(1), 183–202.

Call, M. A., & Voss, P. R. (2016). Spatio-temporal dimensions of child poverty in America, 1990–2010. *Environment and Planning A*, *48*(1), 172–191.

Daubechies, I., Defrise, M., & De Mol, C. (2004). An iterative thresholding algorithm for linear inverse problems with a sparsity constraint. *Communications on Pure and Applied Mathematics: A Journal Issued by the Courant Institute of Mathematical Sciences*, *57*(11), 1413–1457.

De Ketelaere, B., Hubert, M., & Schmitt, E. (2015). Overview of pca-based statistical process-monitoring methods for time-dependent, high-dimensional data. *Journal of Quality Technology*, *47*(4), 318–335.

Diggle, P. J. (2013). *Statistical analysis of spatial and spatio-temporal point patterns*. New York: CRC Press.

Elhorst, J. P. (2014). Spatial panel data models. In: *Spatial Econometrics*, Springer (pp. 37–93)

Friedman, J., Hastie, T., & Tibshirani, R. (2010). Regularization paths for generalized linear models via coordinate descent. *Journal of Statistical Software*, *33*(1), 1.

Hamilton, J. D. (1994). *Time series analysis* (Vol. 2). Princeton: Princeton University Press.

Hannan, E. J., & Quinn, B. G. (1979). The determination of the order of an autoregression. *Journal of the Royal Statistical Society Series B (Methodological)*, *41*(2), 190–195.

Hawkins, D. M. (1993). Regression adjustment for variables in multivariate quality control. *Journal of Quality Technology*, *25*(3), 170–182.

Hu, K., & Yuan, J. (2009). Batch process monitoring with tensor factorization. *Journal of Process Control*, *19*(2), 288–296.

Kolda, T. G., & Bader, B. W. (2009). Tensor decompositions and applications. *SIAM Review*, *51*(3), 455–500.

Lai, T. L., Lim, J. (2015). Asymptotically efficient parameter estimation in hidden markov spatio-temporal random fields. Statistica Sinica pp. 403–421

Lan, H., Zhou, C., Wang, L., Zhang, H., & Li, R. (2004). Landslide hazard spatial analysis and prediction using gis in the xiaojiang watershed, yunnan, china. *Engineering Geology*, *76*(1–2), 109–128.

Lichstein, J. W., Simons, T. R., Shriner, S. A., & Franzreb, K. E. (2002). Spatial autocorrelation and autoregressive models in ecology. *Ecological Monographs*, *72*(3), 445–463.

Liu, J., Yuan, L., Ye, J. (2010). An efficient algorithm for a class of fused lasso problems. In: *Proceedings of the 16th ACM SIGKDD International Conference on Knowledge Discovery and Data Mining*, ACM (pp. 323–332).

Louwerse, D., & Smilde, A. (2000). Multivariate statistical process control of batch processes based on three-way models. *Chemical Engineering Science*, *55*(7), 1225–1235.

Qiu, P. (2013). *Introduction to statistical process control*. New York: Chapman and Hall/CRC.

Reinsel, G. C. (2003). *Elements of multivariate time series analysis*. Berlin: Springer Science & Business Media.

Reynolds, K., & Madden, L. (1988). Analysis of epidemics using spatio-temporal autocorrelation. *Phytopathology*, *78*(2), 240–246.

Šaltytė Benth, J., & Šaltytė, L. (2011). Spatial-temporal model for wind speed in lithuania. *Journal of Applied Statistics*, *38*(6), 1151–1168.

Tibshirani, R., Saunders, M., Rosset, S., Zhu, J., & Knight, K. (2005). Sparsity and smoothness via the fused lasso. *Journal of the Royal Statistical Society Series B (Methodological)*, *67*(1), 91–108.

Tran, L., Navasca, C., Luo, J. (2012). Video detection anomaly via low-rank and sparse decompositions. In: *2012 Western New York Image Processing Workshop*, IEEE (pp. 17–20).

Tsay, R. S. (2013). *Multivariate time series analysis: With R and financial applications*. New York: Wiley.

Yan, H., Paynabar, K., & Shi, J. (2017). Anomaly detection in images with smooth background via smooth-sparse decomposition. *Technometrics*, *59*(1), 102–114.

Yan, H., Paynabar, K., & Shi, J. (2018). Real-time monitoring of high-dimensional functional data streams via spatio-temporal smooth sparse decomposition. *Technometrics*, *60*(2), 181–197.

Zhu, J., Huang, H. C., & Wu, J. (2005). Modeling spatial-temporal binary data using markov random fields. *Journal of Agricultural, Biological, and Environmental Statistics*, *10*(2), 212.

Zou, C., & Qiu, P. (2009). Multivariate statistical process control using lasso. *Journal of the American Statistical Association*, *104*(488), 1586–1596.

Zou, C., Ning, X., & Tsung, F. (2012). Lasso-based multivariate linear profile monitoring. *Annals of Operations Research*, *192*(1), 3–19.

An Approach to Monitoring Time Between Events When Events Are Frequent

Ross Sparks, Aditya Joshi, Cecile Paris, and Sarvnaz Karimi

Abstract This paper focuses on monitor plans aimed at the early detection of the increase in the frequency of events. The literature recommends either monitoring the Time Between Events (TBE), if events are rare, or counting the number of events per unit non-overlapping time intervals, if events are not rare. Recent monitoring work has suggested that monitoring counts in preference to TBE is not recommended even when counts are low (less than 10). Monitoring TBE is the real-time option for outbreak detection, because outbreak information is accumulated when an event occurs. This is preferred to waiting for the end of a period to count events if outbreaks are large and occur in a short time frame. If the TBE reduces significantly, then the incidence of these events increases significantly. This paper explores monitoring TBE when the daily counts are quite high. We consider the case when TBEs are Weibull distributed.

Keywords Statistical process control · False discovery rate · Early outbreak detection · Weibull distribution

1 Introduction

When monitoring events over time there are two competing approaches. The first and more common approach is to monitor counts over a defined period (usually a day). The second is to monitor the time between events (TBE) over time. Recent papers by Sparks et al. (2019, 2020) demonstrate that monitoring the time between events is more efficient for the early detection of very large outbreaks that occur in a very short space of time. This paper aims to extend these methods to larger counts (e.g., up to 100,000 events per day) with particular emphasis on outbreaks occurring in a very short time period.

R. Sparks (✉) · A. Joshi · C. Paris · S. Karimi
CSIRO Data61, Corner of Vimiera and Pembroke Roads, Marsfield, NSW 2122, Australia
e-mail: ross.sparks@csiro.au

S. Knoth and W. Schmid (eds.), *Frontiers in Statistical Quality Control 13*,
Frontiers in Statistical Quality Control,
https://doi.org/10.1007/978-3-030-67856-2_16

In fact, we aim to show that, in such circumstances, we should almost always monitor the time between events in preference to monitoring counting processes, if our aim is early outbreak detection. Real-time decision support for outbreak detection of events involves a decision whenever an event occurs, and therefore the time between events is the natural statistic to consider to achieve this aim. The literature adequately deals with the simple case where the mean time between events is homogeneous. For example, Borror et al. (2003) examine the robustness of the Cumulative Sum (CUSUM) statistic for the TBE. They deal with the situation where the events are quite rare. They consider the log-normal and Weibull distributions. Sürücü and Sazak (2009) look at monitoring reliability using a three-parameter Weibull distribution where the failure time is a random variable greater than some positive value. Shafae et al. (2015) examine the CUSUM statistic for Weibull distributed data. They compare three CUSUM statistics for detecting decreases in the TBE values and demonstrate that the in-control behaviour of the exponential CUSUM is not robust. Aslam (2016) and Aslam et al. (2018) introduce a chart that first applied the Exponentially Weighted Moving Average (EWMA) and then the CUSUM for Weibull distributed data. This seems a sensible robust alternative in cases where the distribution is highly skewed. Panza and Vargas (2016) monitor the shape parameter of a Weibull regression model by taking the natural logarithm of the Weibull distributed response and then using an extreme value linear regression model. Panza and Vargas (2017) monitor profiles in time to event situations by comparing two CUSUM approaches to a multivariate EWMA approach. Erto et al. (2018) use a semi-empirical Bayesian control charts for monitoring Weibull data. Wang et al. (2017) compare two CUSUM charts to two EWMA charts for Weibull distributed time between events and found that the performance of these charts depends on the shape parameter. None of these approaches dominate in the performance assessment. All of these and other papers deal with the case where the in-control TBE data are homogeneous across time. In fact, besides Sparks et al. (2019, 2020), we could not find papers on monitoring for outbreaks when the in-control TBE values are non-homogeneous.

The Average Number of Events (ANE) before an out-of-control false alarm (i.e., a flagged outbreak when no outbreak has occurred) is used to control the out-of-control false discovery rate. This is usually made acceptably large without adversely influencing the early detection of outbreaks too much. When out-of-control, the ANE is used to assess the performance of the approach, and the approach with the smallest ANE is the better approach given that all approaches have the same false discovery rate. Designing plans with very large ANE values can be computationally tedious, particularly for large in-control ANE values. Therefore, this paper aims to come up with a computationally feasible approach that is used when the in-control ANE values are greater than the 2000 derived in Sparks et al. (2020). To this end, we apply run rules on the number of consecutive signals. This strategy is used to derive plans for processes with a larger in-control ANE value than 2000 in Sparks et al. (2020), particularly when the frequency of events is very much higher than 20 per unit time interval (say daily).

The plans are recommended for both homogeneous and non-homogeneous processes. Section 2 deals with homogeneous processes, while Sect. 3 suggests ways this can be extended to cover slow changing non-homogeneous processes. Section 4 covers examples of application. Section 5 looks at diagnosing the nature of the outbreaks in sickness. Section 6 summarises the results in the paper.

2 Monitoring TBE for Homogeneous Processes

For convenience, throughout this section, we will use days as the measure of the time period over which the rates of events are measured, but this can be any unit of time. If we use the results of Sparks et al. (2020) for an out-of-control false discovery rate of an average of one in 2000 events, then for daily counts averaging 20 per day when in-control would result in roughly 3 to 4 out-of-control false discoveries in a year on average. This may be an acceptable false discovery rate in many cases, but if the in-control average per day is 100 or more, this would result in on average at least 15 to 20 out-of-control false discoveries in a year. This level of false signals may be too high and build complacency into the decision making process with flagged outbreaks. Therefore we need a simple process for extending the TBE monitoring plans to having a more acceptable out-of-control false discovery rate for processes with larger daily counts than 20.

In this paper, we consider the Weibull distribution with two parameters: scale and shape. We use the same adaptive EWMA statistic applied in Sparks et al. (2019). We use the Weibull distribution threshold estimates for an out- of-control false discovery rate of an average of one in every 2000 events (see Sparks et al. 2020), but now we plan for an out-of-control false discovery rate for higher counts than 20 per day, by insisting that there are m consecutive signals in a row before an outbreak is flagged. We use the threshold estimates derived assuming a Weibull distribution for TBE and with an out-of-control false discovery rate of an average of every 2000 events. This is convenient as we want to avoid training the thresholds for each different ANE higher than 2000, because such training of plans using the approach in Sparks et al. (2020) takes a considerable amount of time and computational effort. If we have on average 10,000 events per day, then to have an average false alarm every 100 days we need to have an ANE value of 1,000,000. Training the thresholds for this when the process is non-homogeneous involves an onerous computational effort, because we need to do this for a range of acceptable scale and shape parameters of the Weibull distribution. The higher the daily counts, the higher the value of m that needs to be selected to deliver an acceptable fixed number of out-of-control false alarms per annum (taken here as roughly 2 to 4 per annum in Table 1). Table 1 provides plans using 2 to 4 false discoveries per annum for $m = 2 - 10$, 13, 16, 20, 25, 30, 35, 40, 45, 50, 55 and 60, when the daily counts range from 50 to 100,000 per day. The plans in Table 1 leads to roughly one in 100 days or more out-of-control false alarm rate. The way to interpret the results in Tables 2 and 3 are as follows: if the scale is q then there are roughly on average $1/q$ events per day, and the out-of-control false discovery rate is one in $ANE \times q$ days. For example, the out-of-control false discovery rate for $scale = 0.4$ and $shape = 0.6$ for 2 consecutive signals, the false out-of-control

Table 1 The plans that approximately provide an out-of-control false discovery rate of 1 in 100 (or more)

Scale	Shape	Average out-of-control false discover rate per day	Plan (consecutive signals)	Average number of events per day
0.02000	0.6	1 in 113.9 days	6	33
0.01000	0.6	1 in 108.9 days	10	66
0.00500	0.6	1 in 126.6 days	16	133
0.00200	0.6	1 in 104.2 days	21	333
0.00100	0.6	1 in 106.3 days	28	666
0.00050	0.6	1 in 109.2 days	33	1330
0.00020	0.6	1 in 104.4 days	40	3333
0.00010	0.6	1 in 105.8 days	46	6666
0.00005	0.6	1 in 106.9 days	52	13300
0.00002	0.6	1 in 101.1 days	59	33300
0.00001	0.6	1 in 102.4 days	66	66600
0.02000	1.0	1 in 117.9 days	5	50
0.01000	1.0	1 in 100.8 days	8	100
0.00500	1.0	1 in 99.4 days	13	200
0.00200	1.0	1 in 97.4 days	18	500
0.00100	1.0	1 in 103.9 days	23	1000
0.00050	1.0	1 in 101.7 days	28	2000
0.00020	1.0	1 in 114.8 days	35	5000
0.00010	1.0	1 in 106.3 days	40	10000
0.00005	1.0	1 in 99.8 days	45	20000
0.00002	1.0	1 in 98.6 days	52	50000
0.00001	1.0	1 in 108.8 days	57	100000
0.02000	1.4	1 in 117.9 days	5	55
0.01000	1.4	1 in 96.0 days	7	110
0.00500	1.4	1 in 112.3 days	12	220
0.00200	1.4	1 in 102.5 days	17	550
0.00100	1.4	1 in 108.6 days	22	1100
0.00050	1.4	1 in 98.2 days	26	2200
0.00020	1.4	1 in 108.1 days	33	5500
0.00010	1.4	1 in 101.9 days	37	11000
0.00005	1.4	1 in 104.1 days	42	22000
0.00002	1.4	1 in 102.8 days	48	55000
0.00001	1.4	1 in 127.5 days	55	110000

Table 2 The average number of events to an out-of-control false discovery. NCS: Number of Consecutive Signals ($m = 2, 3, \ldots, 16$)

Shape, scale											
NCS(m)	2	3	4	5	6	7	8	9	10	13	16
0.6, 0.4	2707	2879	3877								
0.6, 0.1	2702	2955	3014	4721	5631	6588					
0.6, 0.05	2768	2940	3261	4851	5636	6699					
0.6, 0.02	2808	3095	3196	4605	5696	6891	8162				
0.6, 0.01					5797	6635	7899	9714	10865		
0.6, 0.005							7977	10299	10963	17514	25331
0.6, 0.002							8120	9913	11522	17141	24106
0.6, 0.001									10979	17376	24906
1.0, 0.4	3219	4426	6657	9014							
1.0, 0.1	3364	3632	4537	5831	7097	8696					
1.0, 0.05	3401	3608	4590	6167	6895	8441					
1.0, 0.02	3241	3591	4676	5894	6997	8503	10003				
1.0, 0.01					6744	8387	10078				
1.0, 0.005						8594	10380	12569	14492	23896	38045
1.0, 0.002							10175	12117	14304	24497	37881
1.0, 0.001								12366	14241	23885	37292
1.0, 0.0005								12266	14901	24762	38189
1.4, 0.4	3894	4933	8187								
1.4, 0.1	3435	3894	5088	6167							
1.4, 0.05	3492	3852	4933	6402							
1.4, 0.02	3311	3972	4973	6451	7623						
1.4, 0.01			5257	6344	7786	9610					
1.4, 0.005					7821	9602	11325	14660	17054	28049	45657
1.4, 0.002					7952	9709	11412	13993	16946	28952	45854
1.4, 0.001						9581	11207	14103	17028	28597	46679
1.4, 0.0005							11754	14095	17206	27920	47013

Table 3 The average number of events to an out-of-control false discovery. NCS: Number of Consecutive Signals ($m = 20, 25, \ldots, 65$)

Shape, scale										
NCS(m)	20	25	30	35	40	45	50	55	60	65
0.6, 0.005	45069									
0.6, 0.002	46967	72111								
0.6, 0.001	47542	68726	144039	298011	532268					
0.6, 0.0005		67937	151541	294382	498749					
0.6, 0.0002			160705	301347	503226	939184				
0.6, 0.0001				317317	498749	945749	1696473			
0.6, 0.00005					488333	931677	1653067	3129708	5499948	
0.6, 0.00002								3470506	5522649	
0.6, 0.00001								2938066	5549975	9676299
1.0, 0.002	67179	139842								
1.0, 0.001	65902	139842	301347	535388	1018742					
1.0, 0.0005	69779	145802	282070	589398	1075667					
1.0, 0.0002			293539	574275	1024533					
1.0, 0.0001			279529	541849	1062582	1992889				
1.0, 0.00005					1354164	2000055	3961106			
1.0, 0.00002							4283289	8272879	14706930	
1.0, 0.00001							4418121	8192138	15039951	
1.4, 0.002	77391	168188								
1.4, 0.001	78899	162874	311347	689053	1672705					
1.4, 0.0005	81339	160684	360820	695138	1780021					
1.4, 0.0002			366243	668311	1743456					
1.4, 0.0001			345039	712356	1762528					
1.4, 0.00005					1601708	3054244	6023702	13015910		
1.4, 0.00002						2942189	5768183	13332261	25956883	
1.4, 0.00001						3092074	5620318	12755414	25315429	

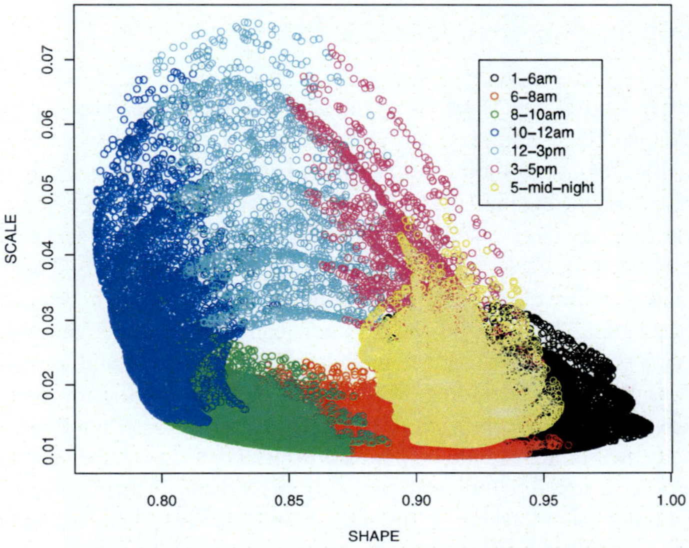

Fig. 1 The estimated parameters for the fitted Weibull regression model

discover rate is one in roughly $2707 \times 0.4 = 1082.8$ days. There is the potential to increase this to an average of larger than 100,000 per day. The results in Tables 2 and 3 report the average number of events to an out-of-control false discovery for processes with m consecutive signals. For example, for shape and scale parameters of 0.6 and 0.4, respectively, with $m = 2$ signals in a row, the monitoring plan has an average false discovery rate of one in 2707 events. Figure 1 provides the estimated parameters for the fitted Weibull regression model, thus indicating the daily trends in these parameters. Note from Fig. 1 that, as the scale decreases, the frequency of rate of events increases, and therefore more consecutive signals of events are required to deliver a one in hundred day out-of-control false alarm rate; for example, for $shape = 0.6$ and $scale = 0.0001$, we need 50 consecutive signal of events, but we have roughly on average 10,000 events per day. This provides us with an approach of using the threshold results in Sparks et al. (2020) to design the monitoring plan using the TBE for up to an average of several thousand events daily. This delivers a plan with 9950 opportunities to flag an outbreak on the first day at the start, and ten thousand events per day thereafter. The start-up requirement of the needing 50 consecutive events to signal before an outbreak can be flagged is a disadvantage but not as large a disadvantage as waiting for the end of the day when monitoring daily counts.

3 Monitoring Non-homogeneous Processes

As we can see from Table 1, the plan changes very little if the mean rate changes a small amount. For example, if we were monitoring the warrantee claims for a product, by looking at its quality from batch to batch, the quality is likely to change only slightly from one batch to the next. In these cases, little will need to be changed in the monitoring plan from one batch to the next. With batch processes, we will need a Phase I approach for each batch to train the monitoring thresholds. This threshold may be found by using the usual stress testing results for the batch, or early Phase I data from the batch. However, this Phase I data can be used to derive adjustment to the monitoring plan relative to the batch before. If the process is continuous and non-homogeneous, then the model forecasts will be needed to define the non-homogeneity of the process during any time with the day and across days. When there are many events per day, this model should be quite accurate. Then the number of consecutive events that need to signal before an outbreak is flagged can change depending on the local expect mean rate of events. A simulation study can be carried out to determine when the false discovery rate is acceptably low over time. This strategy is appropriate as long as the mean rate never goes below 20 per day (the situation where we would flag an outbreak if one event occurred).

For example, if we are dealing with sales of a product, this may vary from one time slot to the next within the shops open hours. In this case, we can expect the time between sales to vary across the day based on the purchasing behaviour of the public and across days of the week. In these cases, the plans can be adjusted for this behaviour by adjusting the number of consecutive signals needed on any day or time of the day using simulation techniques. We now carry this out for people expressing that they or someone else is currently feeling sick, using Twitter data in Queensland, Australia.

4 Example of Application

We consider Twitter data for Queensland, Australia, for any one mentioning that they are sick with one of the following symptoms: diarrhoea, vomiting, upset stomach, headache, head cold, cough, fever or generally feeling unwell. The data runs from early 2015 to early 2018. The TBE values between the times these were first mentioned by a person with a personal unwell mention (Joshi et al. 2019) are calculated. This resulted in roughly an average of 50 events per day, and the Weibull regression model fitted the data adequately. The fitted Weibull regression model is given in the Appendix. The model delivered fitted values for the scale parameter of the Weibull distribution ranged in values:

Min.	1st Qu.	Median	Mean	3rd Qu.	Max.
0.009078	0.013330	0.016769	0.019356	0.021836	0.075807

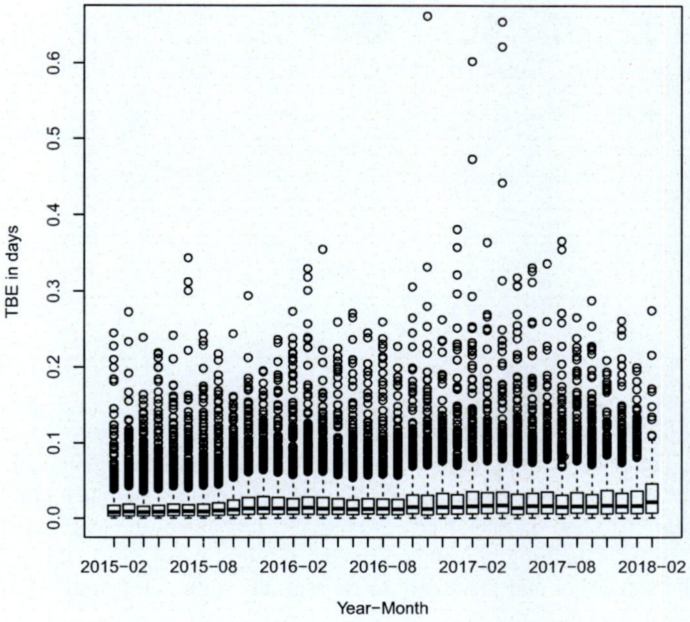

Fig. 2 Monthly boxplots of the time series of time between events

and the fitted shape parameter of the Weibull distribution were

Min.	1st Qu.	Median	Mean	3rd Qu.	Max.
0.7746	0.8675	0.9094	0.8959	0.9305	0.9919

Figure 1 illustrates the trends in the scale and shape parameter estimates within the time of the day. The daily trends for scale and shape is a trend that cycles round the same trajectory within a day of people reporting that they are sick. However, the different magnitude of these trends are basically driven by differences in the days of the week. This indicates the non-homogeneous nature of the TBE values within and across days.

The monthly distribution of TBE values are reported in Fig. 2. Note that it is not easy to identify any outbreaks from this plot, because, generally, the width of the boxes increases with each increasing month, indicating that there are more events over time. Monitoring these TBE values is the only way to identify a significant increase in incidences.

The plans in Table 4 have approximately a one in 100 days out-of-control false outbreak discovery rate (approximately 3 to 4 outbreak false discoveries per annum) with the number of consecutive signals as in the Table 4. The in- control ANE is the average number of events before an outbreak false signal is flagged.

The last plan in Table 4 as applied to the data gives roughly a false discovery rate of one in 122.6 days. This translates to approximately three false outbreak discoveries

Table 4 Plans with approximately a one in 100 out-of-control false outbreak discovery rate

Scale closest to	0.073	0.058	0.044	0.04	0.036	0.02	0.01	In-control ANE
No. consecutive signals	1	2	3	5	7	9	10	4892.5
No. consecutive signals	1	2	3	5	8	9	10	5088.7
No. consecutive signals	1	2	3	5	8	9	11	5120.6
No. consecutive signals	1	2	3	5	8	10	11	5944.6

per annum. The results of applying this approach to our application is given in Fig. 3. The blue circles in this figure indicate the flagged outbreaks of people feeling unwell. Note that year 2015 had six dispersed days flagging unusual outbreaks, year 2016 had five dispersed days flagging unusual outbreaks. Each of these events flagged several times on the same day and are likely to be real day outbreak events. However, in

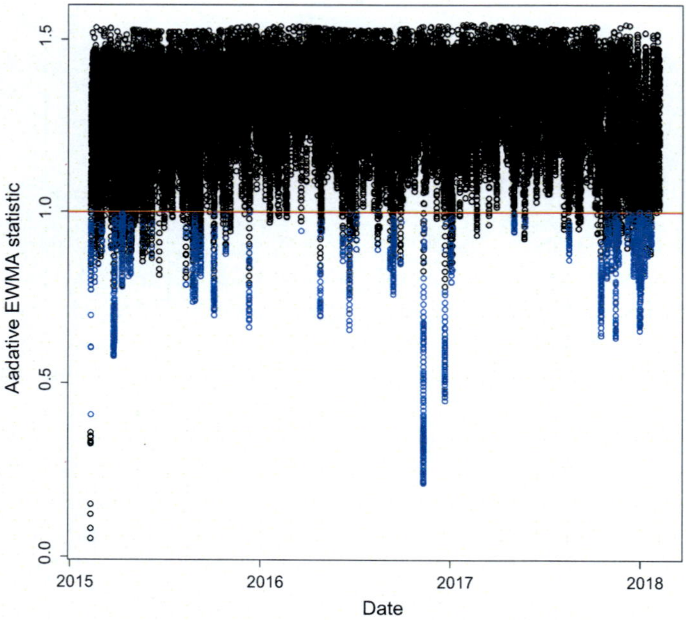

Fig. 3 Flagging outbreaks of people feeling sick with one of the following symptoms diarrhoea, vomiting, upset stomach, headache, head cold, cough, fever or generally feeling unwell using run rules

2017, there was a sustained outbreak period flagged, and this is likely to correspond to a very unusual influenza outbreak for that year. In 2017 there were two other dispersed days flagging unusual outbreaks.

5 Diagnosing the Nature of the Outbreak

In this section, we look at which counts are unusually high for particular days to diagnose the nature of the flagged outbreaks. We explore the unusual symptoms across days to decide the nature of the people who said were sick in the twitter messages on particular days. These are:

Diarrhoea
2015-02-10 2015-10-27 2016-09-07 2016-09-13 2016-09-14 2016-11-10 2016-12-22 2017-01-03 2017-12-20

Vomiting
2015-02-10 2015-02-11 2015-04-10 2015-04-11 2015-04-13 2015-04-23 2015-08-12 2015-08-28 2015-10-04 2016-04-28 2016-11-10 2017-05-03 2017-10-18 2017-11-17 2018-01-08

Upset stomach
2015-02-10 2015-02-17 2015-03-21 2015-04-13 2015-04-28 2015-05-20 2015-06-07 2015-08-12

Cough
2015-04-27 2015-06-08 2015-08-29 2015-09-09 2015-12-11 2016-06-08 2016-06-21 2016-09-07 2016-09-13 2016-09-14 2016-09-27 2017-10-31 2017-11-07 2017-12-20 2018-01-02

Headache
2015-04-27 2015-05-20 2015-05-21 2015-09-09 2016-08-14 2016-09-13 2016-09-27 2016-11-10 2017-01-01 2017-10-29 2017-10-31 2017-12-07 2017-12-19 2017-12-31

Feeling unwell
2015-02-17 2015-02-23 2015-02-24 2015-03-03 2015-03-26 2015-03-27 2015-03-28 2015-03-31 2015-04-11 2015-04-23 2015-04-28 2015-08-25 2015-10-05 2016-04-26 2016-11-10 2016-12-22

Headcold
2015-06-08 2015-08-25 2016-06-21 2017-01-02 2017-01-05 2017-11-13 2017-11-14 2017-11-17 2017-11-20 2017-12-07 2017-12-16 2017-12-18 2017-12-20 2017-12-29 2017-12-30 2017-12-31 2018-01-01 2018-01-02 2018-01-03 2018-01-06 2018-01-08 2018-01-09 2018-01-13 2018-01-14 2018-01-15 2018-01-18 2018-01-22

Note the following interpretations:

1. On the 10, 11 and 17 February 2015 there seemed to be a stomach related events involving diarrhoea, vomiting, upset stomach and feeling unwell. No other symptom was unusual on those days.
2. From 2015-03-26 to 2015-03-31 there is evidence of feeling unwell.
3. From 2015-04-10 to 2015-04-23 there is evidence of a vomiting, upset stomach event and feeling unwell.
4. From 2015-08-12 to 2015-08-28 there is evidence of a vomiting, upset stomach event, headcold and feeling unwell.
5. From 2016-09-07 to 2016-09-27 there is a cough and headache outbreak.
6. On 2016-11-10 there are outbreaks of vomiting, headache and feeling unwell.
7. From 2017-01-01 to 2017-01-05 there is Headache, diarrhoea and headcold outbreaks.

8. Headcold most occur in the Southern Hemisphere winter or Spring period but towards the end of 2017 and early 2018 this runs into the summer period because a large influenza outbreak in 2017. This outbreak period is also intermingled with periods of cough, headache and vomiting outbreaks.

6 Conclusions

The paper has demonstrated that TBE monitoring plans can be derived for events ranging from 20 events per day to 100,000 events per day. When outbreaks are large over a short time period, these plans are likely to flag outbreaks earlier than monitoring the daily counts of these events, when the plans are trained to have the same out-of-control false discovery rate. This suggests that we should never monitor counting processes for detecting outbreaks of events when events are large and occur in a short time-frame.

7 Appendix A

```
*******************************************************************
Family:  c("WEI", "Weibull")

Call:  gamlss(formula = TBE ~ cos(2 * pi * time/365.25) + sin(2 * pi *
     time/365.25) + (wd == "Sunday") + (wd == "Saturday") + time +
     hr + cos(hr * 2 * pi/24) + sin(hr * 2 * pi/24) + cos(hr *
     2 * pi/12) + sin(hr * 2 * pi/12), sigma.formula = ~cos(2 *
     pi * time/365.25) + sin(2 * pi * time/365.25) + (wd == "Sunday") +
     time + hr + cos(hr * 2 * pi/24) + sin(hr * 2 * pi/24) + cos(hr *
     2 * pi/12) + sin(hr * 2 * pi/12), family = WEI(), data = TBE)

Fitting method: RS()

-------------------------------------------------------------------
Mu link function:  log
Mu Coefficients:
                            Estimate Std. Error t value Pr(>|t|)
(Intercept)               -7.748e+00  3.652e-01 -21.219   <2e-16 ***
cos(2 * pi * time/365.25)  9.761e-03  9.614e-03   1.015   0.3100
sin(2 * pi * time/365.25)  4.030e-03  9.498e-03   0.424   0.6714
wd == "Sunday"TRUE         3.961e-02  1.952e-02   2.029   0.0425 *
wd == "Saturday"TRUE       2.981e-02  1.960e-02   1.521   0.1283
time                       2.261e-04  2.149e-05  10.522   <2e-16 ***
hr                         2.864e-02  1.941e-03  14.752   <2e-16 ***
cos(hr * 2 * pi/24)       -1.586e-01  1.022e-02 -15.518   <2e-16 ***
sin(hr * 2 * pi/24)        1.087e-02  1.689e-02   0.644   0.5198
cos(hr * 2 * pi/12)        2.231e-02  9.523e-03   2.343   0.0191 *
sin(hr * 2 * pi/12)        1.278e-01  1.242e-02  10.294   <2e-16 ***
---
Signif. codes:  0 '***' 0.001 '**' 0.01 '*' 0.05 '.' 0.1 ' ' 1

-------------------------------------------------------------------

Sigma link function:  log
Sigma Coefficients:
                            Estimate Std. Error t value Pr(>|t|)
```

```
(Intercept)                  -1.617e+00  1.037e-01 -15.601  < 2e-16 ***
cos(2 * pi * time/365.25)     1.343e-03  2.721e-03   0.494    0.622
sin(2 * pi * time/365.25)     9.511e-05  2.682e-03   0.035    0.972
wd == "Sunday"TRUE            8.864e-03  5.498e-03   1.612    0.107
time                          2.652e-05  6.104e-06   4.344 1.40e-05 ***
hr                            1.771e-02  5.210e-04  33.993  < 2e-16 ***
cos(hr * 2 * pi/24)          -5.234e-02  2.921e-03 -17.918  < 2e-16 ***
sin(hr * 2 * pi/24)           3.793e-02  4.706e-03   8.060 7.66e-16 ***
cos(hr * 2 * pi/12)          -1.425e-02  2.693e-03  -5.291 1.22e-07 ***
sin(hr * 2 * pi/12)           3.701e-02  3.545e-03  10.441  < 2e-16 ***
---
Signif. codes:  0 '***' 0.001 '**' 0.01 '*' 0.05 '.' 0.1 ' ' 1

------------------------------------------------------------------
No. of observations in the fit:  160035
Degrees of Freedom for the fit:  21
      Residual Deg. of Freedom:  160014
                      at cycle:  5

Global Deviance:     -843570.6
            AIC:     -843528.6
            SBC:     -843319
******************************************************************
```

References

Aslam, M. (2016). A mixed EWMA-CUSUM control chart for Weibull-distributed quality characteristics. *Quality and Reliability Engineering International*, *32*(8), 2987–2994.

Aslam, M., Azam, M., & Jun, C. H. (2018). A HEWMA-CUSUM control chart for the Weibull distribution. *Communications in Statistics-Theory and Methods*, *47*(24), 5973–5985.

Borror, C. M., Keats, J. B., & Montgomery, D. C. (2003). Robustness of the time between events CUSUM. *International Journal of Production Research*, *41*(15), 3435–3444.

Erto, P., Pallotta, G., Palumbo, B., & Mastrangelo, C. M. (2018). The performance of semi-empirical Bayesian control charts for monitoring Weibull data. *Quality Technology and Quantitative Management*, *15*(1), 69–86.

Joshi, A., Karimi, S., Sparks, R., Paris, C., & MacIntyre, C. R. (2019). Survey of text-based epidemic intelligence: A computational linguistics perspective. *ACM Computing Surveys (CSUR)*, *52*(6), 1–19.

Panza, C. A., & Vargas, J. A. (2016). Monitoring the shape parameter of a Weibull regression model in phase II processes. *Quality and Reliability Engineering International*, *32*(1), 195–207.

Panza, C. A., & Vargas, J. A. (2017). Monitoring the parameter vector of regression models with time-to-event response in phase ii processes. *Journal of Statistical Computation and Simulation*, *87*(14), 2779–2798.

Shafae, M. S., Dickinson, R. M., Woodall, W. H., & Camelio, J. A. (2015). Cumulative sum control charts for monitoring Weibull-distributed time between events. *Quality and Reliability Engineering International*, *31*(5), 839–849.

Sparks, R., Jin, B., Karimi, S., Paris, C., & MacIntyre, C. (2019). Real-time monitoring of events applied to syndromic surveillance. *Quality Engineering*, *31*(1), 73–90.

Sparks, R., Joshi, A., Paris, C., Karimi, S., & MacIntyre, C. (2020). Monitoring time between events using social media data. *Engineering Reports, 2*(5), e12152, 23 pages.

Sürücü, B., & Sazak, H. S. (2009). Monitoring reliability for a three-parameter Weibull distribution. *Reliability Engineering and System Safety*, *94*(2), 503–508.

Wang, F. K., Bizuneh, B., & Abebe, T. H. (2017). A comparison study of control charts for Weibull distributed time between events. *Quality and Reliability Engineering International*, *33*(8), 2747–2759.

Selected Topics from Statistical Quality Control

Analysis of Measurement Precision Experiment with Ordinal Categorical Variables

Tomomichi Suzuki, Jun-ichi Takeshita, Mayu Ogawa, Xiao-Nan Lu, and Yoshikazu Ojima

Abstract Many collaborative studies are run to evaluate precision of measurement methods. The main focus is on estimating repeatability and reproducibility, which are the variation within a laboratory and the overall variation of the measurement method, respectively. ISO 5725 provides how to design and analyze such precision experiments for quantitative cases where the measurement results follow a continuous distribution, namely a normal distribution. However, there are cases where the measurement results are qualitative such as binary or categorical. In this paper, the cases with ordinal categorical variables are considered. Using methods that can be applied to qualitative data, an analysis of a measurement precision experiment with measurements involving ordinal categorical variables is investigated. The data analysed are from an actual precision experiment of intratracheal administration testing whose objectives were to study the precision of a standardized test method for evaluating nanomaterial pulmonary toxicity.

Keywords Repeatability · Reproducibility · Item response theory · Dose-response relationship

T. Suzuki (✉) · M. Ogawa · Y. Ojima
Department of Industrial Administration, Tokyo University of Science, 2641 Yamazaki, Noda, Chiba 278-8510, Japan
e-mail: szk@rs.tus.ac.jp

M. Ogawa
e-mail: 7417605@ed.tus.ac.jp

Y. Ojima
e-mail: ojima@rs.tus.ac.jp

J. Takeshita
Research Institute of Science for Safety and Sustainability, National Institute of Advanced Industrial Science and Technology (AIST), 16-1 Onogawa, Tsukuba, Ibaraki 305-8569, Japan
e-mail: jun-takeshita@aist.go.jp

X.-N. Lu
Department of Computer Science and Engineering, Faculty of Engineering, University of Yamanashi, 4-3-11 Takeda, Kofu, Yamanashi 400-8511, Japan
e-mail: xnlu@yamanashi.ac.jp

S. Knoth and W. Schmid (eds.), *Frontiers in Statistical Quality Control 13*,
Frontiers in Statistical Quality Control,
https://doi.org/10.1007/978-3-030-67856-2_17

1 Introduction

Evaluating performance of a measurement method is essential in metrology. Concepts of repeatability and reproducibility are introduced in the International Organization for Standardization (1994) series, including how to run and analyse experiments (usually collaborative studies) to obtain these precision measures. International Organization for Standardization (1994) describes precision evaluation in quantitative measurements but not in qualitative measurements. Some methods such as Wilrich (2010), de Mast and van Wieringen (2010), and Bashkansky et al. (2012) have been proposed for qualitative measurement cases. Item response theory (Muraki 1992) is another methodology that can be used to analyse qualitative data. Using these methods, an analysis of a measurement precision experiment with measurements involving ordinal categorical variables is investigated.

The data analysed are from a precision experiment of intratracheal administration testing (AIST 2018) whose objectives were to study the precision of a standardized test method for evaluating nanomaterial pulmonary toxicity. In such experiments, the dose-response relationship must also be considered, making the situation more complicated. Thus, this paper's objective is to discuss how these data should be analysed using actual data.

The remainder of this paper is organized as follows. Section 2 explains the data used in this paper. Section 3 introduces the methods for analysing qualitative data, Sect. 4 describes the results of the analyses, and Sect. 5 provides a summary of the paper.

2 Data

The data used in this paper are from a precision experiment of intratracheal administration testing (AIST 2018). The main objective of this experiment was to study the precision of a standardized test method for evaluating nanomaterial pulmonary toxicity. An overview of the experiment is as follows:

(a) Three nanomaterials were used in the study.
(b) Four dose levels (none, low, middle, and high) were designed for each nanomaterial.
(c) The experiment was conducted by five laboratories.
(d) The number of replicates was five.
(e) The replications were conducted using rats.
(f) The same test method was used for all laboratories.
(g) The equipment used in each laboratory may differ.
(h) Each rat went through a pathological examination.
(i) There were 19 characteristics to be examined by experts.
(j) The result of the examination revealed the inflammation grade of the response.
(k) The response was given one of five grades (−, +−, +, ++, +++).

Therefore, we obtain ordinal categorical data (five categories) of 19 characteristics for each of five rats, four level of doses, and three nanomaterials in each laboratory.

3 Methods

3.1 *ISO 5725*

In the ISO 5725 series, accuracy of a measurement result, measurement method, or measurement system is a general term involving trueness and precision. Trueness is the closeness in agreement between the average value obtained from a large series of measurement results and an accepted reference value. Trueness is usually expressed in terms of bias, which is the difference between the expectation of the measurement results, and the accepted reference value. Precision is the closeness in agreement between independent measurements results obtained under stipulated conditions and is usually expressed in terms of the standard deviations of the measurement results.

Generally, two measures of accuracy, repeatability and reproducibility, are required. Repeatability is measurement results under repeatability conditions, where the independent measurement results are obtained using the same method on the identical test items in the same laboratory by the same operator using the same equipment within short intervals of time. Reproducibility is measurement results under reproducibility conditions, where the measurement results are obtained using the same method on identical test items in different laboratories with different operators using different equipment..

The basic model to estimate accuracy of measurement method a measurement result y in the ISO 5725 series is given by

$$y = m + B + e \tag{1}$$

where m is the general mean (expectation), B is the laboratory component of variation (under repeatability conditions), and e is random error (under repeatability conditions). The expectation of B is assumed to be 0, and the variance of B, which is the between-laboratory variance, is denoted by σ_L^2. The expectation of e is also assumed to be 0, and the variance of e, which is the within-laboratory variance, is assumed to be equal in all laboratories and is denoted as the repeatability variance σ_r^2. The reproducibility variance σ_R^2 can be expressed as the sum of the between-laboratory variance and the repeatability variance

$$\sigma_R^2 = \sigma_L^2 + \sigma_r^2. \tag{2}$$

3.2 Ordinal Analysis of Variance (ORDANOVA)

ORDANOVA (Bashkansky et al. 2012) is a method for investigating measurement result differences among laboratories when the results consist of ordered categorical data. The null hypothesis in ORDANOVA is defined as "no measurement result differences exist among all laboratories" and the alternative hypothesis is defined as "differences in measurement results exist among some of the laboratories".

The within-laboratory variation $\widehat{h}^2_{m(W)}$, is given as

$$\widehat{h}^2_{m(W)} = \frac{1}{(K-1)/4} \sum_{k=1}^{K-1} \widehat{F}_{km}(1 - \widehat{F}_{km}), \tag{3}$$

where $\widehat{F}_{km}$ denotes the cumulative frequency of data for laboratory $m = 1, 2, \ldots, M$ and category $k = 1, 2, \ldots, K$. Measure of the between samples variation per kth category $\widehat{S}^2_{k(B)}$, is given as

$$\widehat{S}^2_{k(B)} = \frac{1}{M} \sum_{m=1}^{M} \left(\widehat{F}_{km} - \widehat{F}_{k\cdot}\right)^2. \tag{4}$$

The total variation $\widehat{h}^2_{(T)}$ is given as

$$\begin{aligned} \widehat{h}^2_{(T)} &= \widehat{h}^2_{(W)} + \widehat{S}^2_{(B)} \\ &= \frac{1}{M} \sum_{m=1}^{M} \widehat{h}^2_{m(W)} + \frac{1}{(K-1)/4} \sum_{k=1}^{K-1} \widehat{S}^2_{k(B)}. \end{aligned} \tag{5}$$

Here, $\widehat{h}_{(W)}$, $\widehat{S}_{(B)}$ and $\widehat{h}_{(T)}$ are analogous to repeatability variance, reproducibility variance and total variance. The test statistic can be expressed as

$$I = \frac{\widehat{S}^2_{(B)}/df_B}{\widehat{h}^2_{(T)}/df_T} \tag{6}$$

with degrees of freedom $df_B = M - 1, df_T = N - 1$. The null hypothesis is rejected when

$$I > I_{cr} = \frac{\chi^2_{1-\alpha}}{M-1}, \tag{7}$$

where I_{cr} is the critical value and α is the significance level.

3.3 Attribute Agreement Analysis (AAA)

AAA (International Organization for Standardization (ISO) 2010) is a method for analysing agreement among nominal data. Fleiss' κ statistic is applied to investigate the between-laboratory and within-laboratory agreement of measurements, and the estimate of κ is given by

$$\widehat{\kappa} = \frac{\widehat{P_o} - \widehat{P_e}}{1 - \widehat{P_e}}, \tag{8}$$

where $\widehat{P_o}$ is the probability the measurement results matched and $\widehat{P_e}$ is the probability that the measurement results match by chance. κ takes the value between -1 and $+1$, and indicates greater agreement when κ is nearer to $+1$.

To obtain within-laboratory agreement, $\widehat{P_o}$ and $\widehat{P_e}$ are expressed as

$$\widehat{P_o} = \frac{1}{Nl(l-1)} \left(\sum_{i=1}^{N} \sum_{k=1}^{K} x_{ik}^2 - Nl \right),$$

$$\widehat{P_e} = \sum_{k=1}^{K} p_k^2, \qquad p_k = \frac{1}{Nl} \sum_{i=1}^{N} x_{ik}, \tag{9}$$

where x_{ik} denotes the frequency of item $i = 1, 2, \ldots, N$ classified as category $k = 1, 2, \ldots, K$, and l express the number of replications. To obtain between-laboratory agreement, $\widehat{P_o}$ and $\widehat{P_e}$ are expressed as

$$\widehat{P_o} = \frac{1}{NMl(Ml-1)} \left(\sum_{i=1}^{N} \sum_{k=1}^{K} \sum_{j=1}^{M} x_{ijk}^2 - NMl \right),$$

$$\widehat{P_e} = \sum_{k=1}^{K} p_k^2, \qquad p_k = \frac{1}{NMl} \sum_{i=1}^{N} \sum_{j=1}^{M} x_{ijk}, \tag{10}$$

where x_{ijk} denotes the frequency of item $i = 1, 2, \ldots, N$ classified as category $k = 1, 2, \ldots, K$ for laboratory $j = 1, 2, \ldots, M$.

3.4 Item Response Theory (IRT)

IRT was developed in the education and psychology fields, and is widely used in tests and examinations. IRT enables users to estimate both examinee abilities and question difficulty.

Many models have been developed in IRT to accommodate various situations. The generalized partial credit model (GPCM) is a model applied to ordinal polyto-

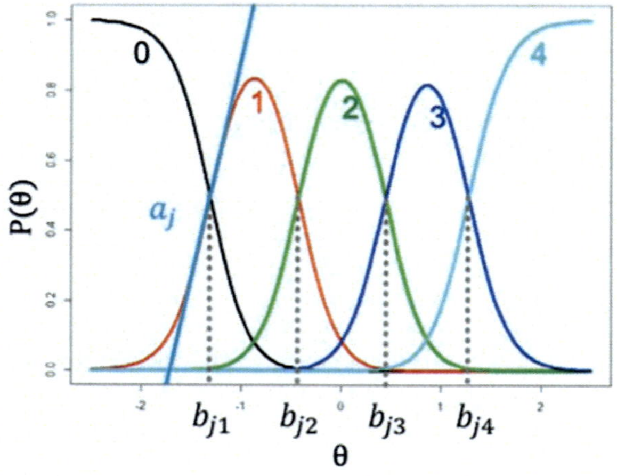

Fig. 1 Item characteristic curve for item j

mous data, in which a partial point can be possible for each test question. GPCM is expressed as

$$P_{jh}(\theta) = \frac{\exp\left\{a_j \sum_{m=0}^{h}(\theta - b_{jm})\right\}}{\sum_{h=0}^{H}\exp\left\{a_j \sum_{m=0}^{h}(\theta - b_{jm})\right\}}, \tag{11}$$

where $P_{jh}(\theta)$ denotes the probability of an examinee with ability θ obtaining partial points h for a test question (known as an item in IRT). The plots of the probabilities regarding ability are known as the item characteristic curve (ICC), and an example is shown in Fig. 1. In Fig. 1, b_{jm} represents ability values of the intersections (thresholds) of adjacent points, and a larger b_{jm} value means that a question is more difficult. In Fig. 1, a_j expresses the slope of the tangents at the intersections, and a larger a_j value means that a question discriminates examinees better.

When IRT is applied to precision experiments, the model can be expressed (de Mast and van Wieringen 2010) as

$$q_j(h|x) = \frac{\exp\left\{\alpha_j \sum_{m=0}^{h}(x - \delta_{jm})\right\}}{\sum_{h=0}^{H}\exp\left\{\alpha_j \sum_{m=0}^{h}(x - \delta_{jm})\right\}}. \tag{12}$$

Here, $q_j(h|x)$ denotes the probability that laboratory j assigns category h when the toxicity of the nanomaterial is x. α_j represents discrimination parameters and δ_{jm} represents threshold parameters. For any laboratory j, the relation $\delta_{j,h} < \delta_{j,h+1}$ is assumed for any value of h. In precision experiments, laboratories are regarded as test questions in IRT, and the toxicity of the nanomaterial in precision experiments are regarded as examinee ability in IRT.

Precision measures for within-laboratory variation and between-laboratory variation can be derived using Eqs. (13) and (14), respectively (de Mast and van Wieringen 2010).

$$\pi_j^w(h) = P\left(Y_{ij} = h \mid \delta_{j,h} < X_i < \delta_{j,h+1}\right) \tag{13}$$

$$\pi_{j_1,j_2}^b = \sum_{h=0}^{H} P\left(\delta_{j_1,h-1} < X < \delta_{j_1,h} \wedge \delta_{j_2,h-1} < X < \delta_{j_2,h}\right) \tag{14}$$

Here, $\pi_j^w(h)$ is the probability that a laboratory assigns category h based on its threshold, and can be interpreted as consistency within a laboratory. π_{j_1,j_2}^b is the probability that the measurement results of an identical item by two arbitrary laboratories match, and can be interpreted as consistency among laboratories.

Repeatability can be calculated using the average of $\pi_j^w(h)$ and reproducibility can be calculated using the average of π_{j_1,j_2}^b.

4 Results and Discussion

4.1 *Graphical Presentation*

The toxicity experiment had a complicated data structure, thus we started by drawing bubble charts, which were applied to clarify the dose-response relationship. Figure 2 shows an example of the bubble charts. Bubble size corresponds to frequency, which is the number of rats. These graphs were drawn for all possible cases, that is, 285 (3 nanomaterials × 19 characteristics × 5 laboratories).

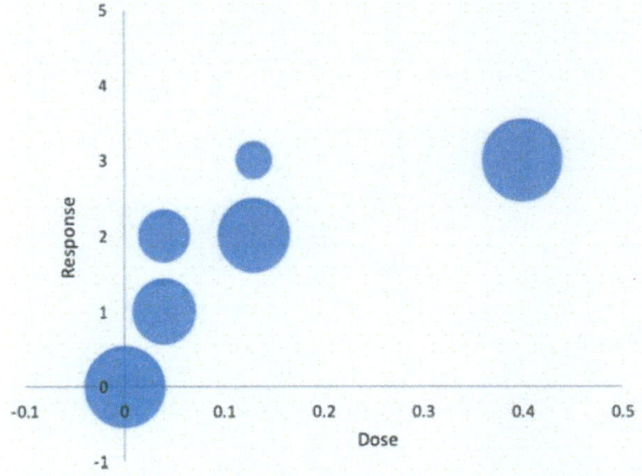

Fig. 2 Bubble chart (Nanomaterial C, characteristic No. 1, lab. No. 1)

4.2 Estimation of Precision Measures

In this subsection, precision measures are estimated using (1) ISO 5725 method, (2) ORDANOVA, (3) AAA and (4) IRT. For each of these method, repeatability and reproducibility measures are estimated. The measures themselves differ among the methods as in described in Sect. 3, so the estimated values cannot directly be compared. The comparison is made through focusing on the changes of the estimated values.

4.2.1 Estimation Using ISO 5725 Method

We calculated repeatability and reproducibility measures for each combination of nanomaterials, doses, and characteristics using the ISO 5725 method, which applies one-way ANOVA layout. The value of the category is treated as quantitative values, so the repeatability variance and the reproducibility variance are directly estimated. Significance tests were conducted to check for the existence of between-laboratory variance. The statistics calculated are the F-values that is the mean squares between laboratories divided by the mean squares within labratories as in one-way ANOVA. Table 1 shows an example of the summarized results (for nanomaterial A). More significant results are obtained for higher doses, and if we focus on F-values, they are larger for characteristics Nos. 8, 9, 11 and 18.

Table 1 Results of ISO 5725 (F-values of between-lab variance testing; Nanomaterial A)

Char	None	Low	Medium	High
1	0.0	8.6[a]	4.1	4.8
2	0.0	3.4	5.7	10.9[a]
6	0.0	2.4	9.6[a]	6.0
7	0.0	6.0[a]	5.7	4.3
8	2.0	5.1	22.0[a]	17.6[a]
9	1.0	2.1	19.6[a]	12.9[a]
10	4.0	3.4	3.8	4.8
11	0.0	18.6[a]	5.7	14.7[a]
16	0.0	1.1	0.0	2.7
18	0.0	1.0	7.5[a]	23.5[a]
19	0.0	0.0	2.7	1.0

[a]: Statistically significant

Table 2 Results of ORDANOVA (Nanomaterial A)

	None		Low		Medium		High	
Char.	r	R	r	R	r	R	r	R
1	0.00	0.00	0.08	0.25	0.08	0.27	0.00	0.40
2	0.00	0.00	0.08	0.25	0.11	0.52	0.00	0.72
6	0.13	0.28	0.06	0.28	0.13	0.29	0.10	0.39
7	0.00	0.00	0.10	0.30	0.22	0.49	0.16	0.51
8	0.05	0.07	0.18	0.32	0.18	0.60	0.18	0.48
9	0.10	0.20	0.08	0.26	0.19	0.42	0.16	0.49
10	0.13	0.36	0.21	0.23	0.26	0.31	0.26	0.36
11	0.05	0.07	0.16	0.21	0.24	0.45	0.26	0.50
16	0.06	0.08	0.06	0.08	0.03	0.04	0.16	0.22
18	0.05	0.11	0.03	0.20	0.08	0.38	0.10	0.48
19	0.00	0.00	0.06	0.08	0.21	0.38	0.19	0.42

r: repeatability, R: reproducibility

4.2.2 Estimation Using ORDANOVA

Repeatability and reproducibility measures are calculated for each combination of nanomaterials, doses, and characteristics using ORDANOVA. Here, $\widehat{h^2_{(W)}}$ and $\widehat{S^2_{(B)}}$ descried in Sect. 3.2 are the measures for repeatability and reproducibility, respectively. Table 2 shows summarized results for nanomaterial A. The repeatability measures are larger for characteristics Nos. 10 and 11, and the reproducibility measures are larger with higher doses, especially for characteristic No. 2.

4.2.3 Estimation Using AAA

Repeatability and reproducibility measures are calculated for each combination of nanomaterials, doses, and characteristics using nominal AAA. The statistics estimated here are Kappa statistics described in Sect. 3.3. The summarized results for nanomaterial A are shown in Fig. 3, with variation becoming larger (Kappa statistic being smaller) with higher doses.

4.2.4 Comparison Among ISO 5725, ORDANOVA and AAA

The above results are aggregated in Table 3 and ISO 5725, ORDANOVA, and AAA results were compared. In Table 3, the estimated measures are (1) repeatability variance and reproducibility variance for ISO 5725 method, (2) $\widehat{h^2_{(W)}}$ and $\widehat{S^2_{(B)}}$ for ORDANOVA, and (3) Kappa statistics for AAA. Table 3 clearly shows that the precision measure peaks are the same among all methods and variations were smallest with no dose in all cases.

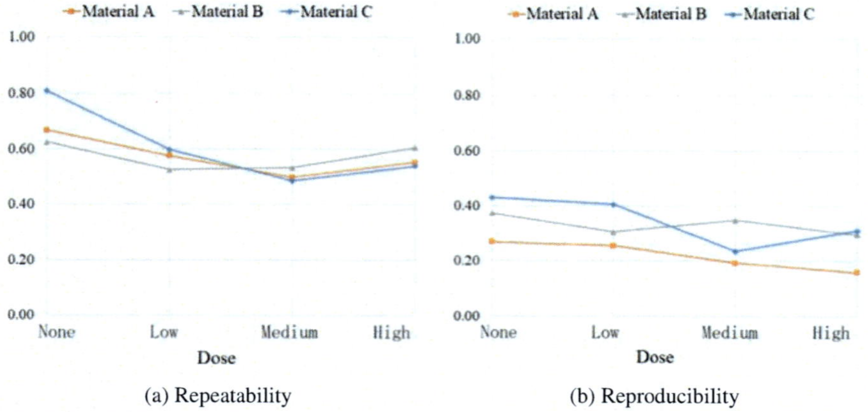

Fig. 3 Results of AAA

Table 3 Estimated repeatability and reproducibility measures

		Repeatability			Reproducibility		
		5725	ORD.	AAA	5725	ORD.	AAA
Material A	None	0.08	0.05	0.67	0.14	0.11	0.27
	Low	0.15	0.10	0.58	0.32	0.22	0.26
	Medium	0.28	0.16	0.50	0.64	0.38	0.19
	High	0.21	0.14	0.55	0.67	0.45	0.16
Material B	None	0.09	0.05	0.63	0.08	0.08	0.37
	Low	0.33	0.16	0.53	0.62	0.32	0.31
	Medium	0.25	0.15	0.53	0.61	0.32	0.35
	High	0.30	0.15	0.61	0.73	0.36	0.30
Material C	None	0.03	0.02	0.81	0.04	0.05	0.43
	Low	0.16	0.10	0.60	0.26	0.18	0.41
	Medium	0.27	0.16	0.49	0.69	0.35	0.24
	High	0.23	0.15	0.54	0.67	0.34	0.31

4.2.5 Estimation Using IRT

Repeatability and reproducibility measures were calculated for each nanomaterial using IRT. Also, ICCs were drawn for all laboratories and each nanomaterial. Figure 4 shows the ICC of laboratory No. 1 for nanomaterial A. The vertical axis expresses the probability of the measurement result being a specific category h, which is denoted by $q_j(h|x)$. The horizontal axis expresses the true toxicity, which is denoted by x. From Fig. 4, the probability of the measurement result being category 1 is more than 95% when $x = 0$. Additionally, when $x = 2$, the probability of being in category 3 is approximately 40%, and the category 4 probability is about 60%.

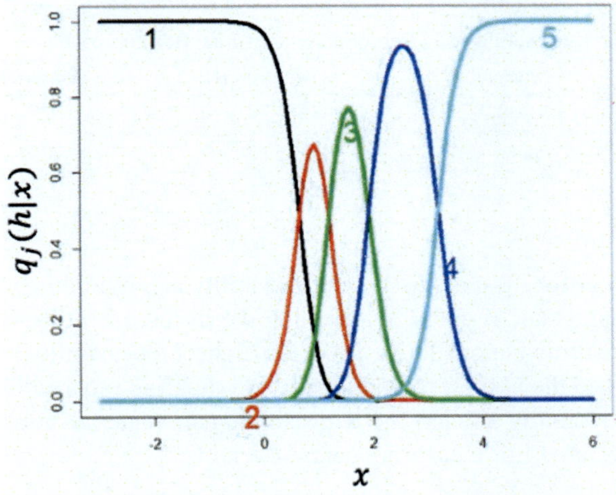

Fig. 4 Item category curve for nanomaterial A and laboratory No.1

Table 4 shows the results of estimated parameters for nanomaterial A. An ideal situation for the discrimination parameters would be if they are equivalent among all the laboratories and the values are large enough although it need to be judged subjectively. From Table 4, we can understand this is not the case, and we can conclude that there exist between-laboratory variation. The order of the estimates of discrimination parameter α_j were laboratories No. 1, No. 2, No. 3, No. 4 and No. 5, in descending order. This means that appropriate classification was also performed in this order from better to worse. Laboratory No.1 had the best discrimination ability.

An ideal situation for the thresholds values would be if they are equivalent among laboratories for each of the categories. The threshold parameter estimates were generally similar among the laboratories except for laboratory No. 3 which suggests the existence of some kind of bias. Looking closely, the threshold values for laboratory No. 3 were larger, which means that laboratory No. 3 tends to give lower category numbers compared to other laboratories. Finally, repeatability and reproducibility are calculated using Eqs. (13) and (14) with the threshold parameters shown in Table 4. Since these values are based on probabilities of results matching, the higher value

Table 4 Parameter estimates of α_j (discrimination) and δ_{jh} (threshold)

	Lab 1	Lab 2	Lab 3	Lab 4	Lab 5
δ_{j1}	0.65	0.74	1.01	0.68	0.75
δ_{j2}	1.18	1.29	1.68	0.86	1.00
δ_{j3}	1.91	1.80	3.04	1.70	1.33
δ_{j4}	3.19	3.24	3.00	3.38	2.19
α_j	5.25	4.22	3.10	3.06	1.61

Table 5 Precisions (Repeatability and Reproducibility) of the IRT approach

	Material A	Material B	Material C
Repeatability $\overline{\pi_j^w(h)}$	0.808	0.817	0.829
Reproducibility $\overline{\pi_{j_1,j_2}^b}$	0.847	0.854	0.777

means higher precision. Table 5 lists the repeatability and reproducibility measures for all nanomaterials (A, B, and C). Although the difference in these values were not so large, measuring nanomaterial C had the highest repeatability and measuring nanomaterial B had the highest reproducibility. Comparing repeatability and reproducibility, reproducibility was larger for measuring nanomaterials A and B, whereas repeatability was larger for measuring nanomaterial C.

4.3 *Estimation of Toxicity*

The toxicity of the nanomaterials was estimated using IRT. Figure 5 shows box and whisker plots of estimated toxicity for material A. From Fig. 5, we can see the effect of doses. Although the distribution of toxicity among the dosage levels overlaps considerably, dosage effects are apparent as a whole, that is, the average of the toxicity increases as the dose become larger.

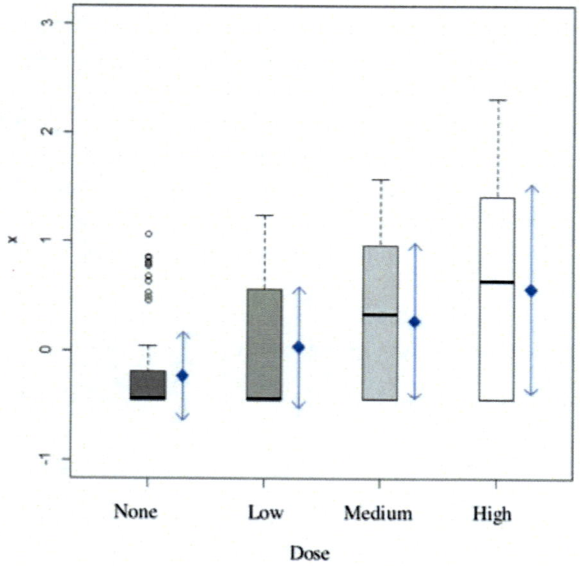

Fig. 5 Box and whisker plot of estimated toxicity (nanomaterial A)

5 Conclusions

Precision measures of ordinal categorical data were estimated for actual intratracheal administration testing experiment data, and dose-response relationships were also investigated. It was not possible to perform an analysis that considers all factors as there were many factors in the experiment. Therefore, analyses were performed considering appropriate factors for each analysis. Precision measures for characteristics were well clarified by ISO 5725 and ORDANOVA and characteristics with large variations were identified. Precision measures for each dose were estimated using ISO 5725, ORDANOVA, and AAA. It became clear that the variation increases as the dose becomes larger. The condition that gave the maximum and minimum values was the same. Precision measures for each nanomaterial were estimated using IRT. Measuring material B had greater repeatability and measuring material C had greater reproducibility. The features of each laboratory could be observed using the ICC of IRT. Dose-response relationships were also examined using estimated toxicity using IRT. The relationships were then investigated for each characteristic for each nanomaterial. Because of the inherent variation in the data, it was not possible to obtain a precise dose-response relationship. Nevertheless, the existence of a dose-response relationship could be verified using a rank correlation coefficient.

References

International Organization for Standardization. (1994). *ISO 5725–1 Accuracy (trueness and precision) of measurement methods and results - Part 1: General principles and definitions*. Geneva: ISO.

International Organization for Standardization. (1994). *ISO 5725–2 Accuracy (trueness and precision) of measurement methods and results - Part 2: Basic method for the determination of repeatability and reproducibility of a standard measurement method*. Geneva: ISO.

Wilrich, P-Th. (2010). The determination of precision of qualitative measurement methods by interlaboratory experiments. *Accreditation and Quality Assurance*, *15*(8), 439–444.

de Mast, J., & van Wieringen, W. (2010). Modeling and evaluating repeatability and reproducibility of ordinal classifications. *Technometrics*, *52*(1), 94–106.

Bashkansky, E., Gadrich, T., & Kuselman, I. (2012). Interlaboratory comparison of test results of an ordinal or nominal binary: Analysis of variation. *Accreditation and Quality Assurance*, *17*(3), 239–444.

Muraki, E. (1992). A generalized partial credit model: Application of an EM algorithm. *Applied Psychological Measurement*, *16*(2), 159–176.

The National Institute of Advanced Industrial Science and Technology (AIST), (2018). Annual Report on a Project 'Survey on standardization of intratracheal administration study for nanomaterials and related issues'. In Japanese, Accessed 2019-05-31: http://www.meti.go.jp/meti_lib/report/H29FY/000102.pdf.

International Organization for Standardization (ISO). (2010). *ISO/TR 14468 Selected illustrations of attribute agreement analysis*. Geneva: ISO.

Assessing a Binary Measurement System with Operator and Random Part Effects

Stefan H. Steiner, R. Jock MacKay, and Kevin Fan

Abstract Consider the assessment of a binary measurement system with multiple operators when a gold standard measurement system is also available (for the assessment study). Data are collected as in a gauge repeatability and reproducibility plan for a continuous measurement system and each operator in the study measures a number of parts multiple times. We characterize the performance of the measurement system by estimating the probabilities of accepting a non-conforming part and of rejecting a conforming part. To model the data, we assume that some parts are more difficult to correctly classify than others and so choose to use random part effects. We consider two cases, modeling the operator effects as fixed or random. For each, we study a conditional and marginal model and their corresponding estimates of the parameters of interest. We also provide guidance on the planning of the assessment study in terms of the number of parts, number of operators and number of repeated measurements.

Keywords Binary measurement system assessment · Marginal and conditional models · Planning of assessment studies · Random and fixed effects · Sensitivity and specificity

1 Introduction

For a continuous measurement system in industry, the standard assessment plan is a gauge repeatability and reproducibility study (GR&R) in which a number of operators measures a number of parts repeatedly. See the AIAG Reference Manual AIAG (2010). The goal of the assessment is to estimate

S. H. Steiner (✉) · R. Jock MacKay · K. Fan
Department of Statistics and Actuarial Science, University of Waterloo, Waterloo, ON, Canada
e-mail: shsteiner@uwaterloo.ca

R. Jock MacKay
e-mail: jock.mackay@uwaterloo.ca

S. Knoth and W. Schmid (eds.), *Frontiers in Statistical Quality Control 13*,
Frontiers in Statistical Quality Control,
https://doi.org/10.1007/978-3-030-67856-2_18

- the overall precision of the measurement system relative to the part to part variation
- the reproducibility (variability among operators) and repeatability (variability within operators).

The statistical model used in the analysis of the data from a GR&R study treats the part effects and the repeatability as random. Operator effects are modelled as fixed or, more typically, random. See Burdick et al. (2005).

Binary measurement systems (BMS) are widely used in industry to check that parts conform to specification. We study the BMS using the GR&R plan (or a simple extension, see Steiner et al. 2011), i.e. a number of operators repeatedly classify a number of parts. We assume that the BMS is non-destructive so that parts can be re-measured/classified without changing their properties. We also assume the availability of a gold standard measurement system to determine the true conforming status of any part used in the assessment study.

We assess the performance of the BMS in terms of the two misclassification errors. To model the probability of such an error, we suppose that parts in the study are selected at random from the population of parts and so treat part effects as random. When only a small number of operators use the BMS, we treat the operator effects as fixed. In other instances, such as a large inter-laboratory trial, we use random operator effects.

To introduce some notation, suppose the operator effects are random. Then, for a randomly selected operator and part, let $Y = 1$ be the event that the part is passed by the BMS and $Y = 0$ the event that the part fails inspection. Also let $X = 1$ if the part is conforming and $X = 0$ otherwise. To quantify the performance of the measurement system, we estimate the misclassification probabilities

$$\alpha = P(\text{pass} \mid \text{non-conforming}) = P(Y = 1 \mid X = 0) \quad \text{and}$$
$$\beta = P(\text{fail} \mid \text{conforming}) = P(Y = 0 \mid X = 1)\,.$$

In medical contexts, $1 - \alpha$ and $1 - \beta$ correspond to the specificity and sensitivity of the BMS (or vice versa depending on the definition of X). When operator effects are fixed, we define and estimate α_j and β_j for each operator $j = 1, \ldots, n$ separately.

In our proposed plan and analysis we assume the availability of two separate populations, one consisting of conforming parts and another of non-conforming parts as determined by the gold standard. As a result, we conduct two similar but separate studies using samples of conforming and non-conforming parts to estimate the misclassification probabilities. In practice, the two studies are likely combined. However, from this point forward for simplicity, we consider only the estimation of α or α_j, the probability of passing a part that is non-conforming. We consider the availability and selection of conforming and non-conforming parts in the Discussion section.

To mirror a GR&R study, suppose we select m non-conforming parts at random and n operators from the populations of interest. Each operator measures each part r times and, for each operator/part combination, we record the number of times the measurement system classifies the part as conforming. We denote this plan by

$L(m, n, r)$. For fixed operator effects, n is small, typically 2 or 3. When operator effects are random, both m and n are large and the $L(m, n, r)$ plan is logistically cumbersome. Alternately, we consider multiple copies of smaller plans. In terms of the notation, if we have independent replications of a plan, say 25 independent copies of $L(2, 2, 5)$, we write $L(2, 2, 5)^{25}$ or, more generally, $L(m, n, r)^T$ to indicate we have T copies of $L(m, n, r)$. For example, when there are three operators with fixed effects, we might consider a $L(m, 1, r)^3$ plan where, for logistic reasons, each operator measures different parts. Steiner et al. (2011) examine similar plans for the assessment of a continuous measurement system with random part and operator effects.

The goal of the paper is to look at the plan, the modelling and the estimation of the misclassification error rate(s) from a $L(m, n, r)^T$ plan for both fixed and random operator effects. This extends the existing literature in a number of important ways as described in more detail in the literature review to follow. In addition, by considering the case with both random parts and operators, it provides the natural extension of the standard gauge R&R analysis for a continuous measurement system. We believe this natural extension has not been considered previously probably due to the complexity of fitting the corresponding conditional model (as described in Sect. 3).

If there are no operator effects (e.g., we have a single operator or an automated BMS), there is substantial literature, referenced below, dealing with a BMS assessment. When there is no gold standard, we can use a latent class model (a mixture over the unknown distribution of X) to estimate the parameters of interest. See Akkerhuis et al. (2017), Erdmann et al. (2016), Danila et al. (2012), Beavers et al. (2011), van Wieringen and de Mast (2008) for some recent references to this approach in an industrial context. Akkerhuis et al. (2017), Albert and Dodd (2004, 2008), Severn et al. (2016) show that the estimates of α and β from latent class models are highly sensitive to untestable assumptions so that this approach cannot be recommended. Akkerhuis et al. (2017) suggest alternative parameters to assess the BMS when there is no gold standard system available.

There are numerous references to estimating the probability of a misclassification error when a gold standard system is available to verify the status of each part in the study. See Danila et al. (2013) for a list of recent references for the no operator case. Alternately, Severn et al. (2016) show that using the gold standard to verify the true status of a carefully selected small subset of the parts can resolve the lack of robustness with latent class methods while providing estimates with precision close to those available from the complete verification plan. See also Albert and Dodd (2008) for a discussion of partial verification.

Other related work includes the large literature on assessing a diagnostic (binary) test in a medical context. See, for example, Pepe's (2003) book for a thorough coverage of estimating sensitivity and specificity in a variety of situations. She also discusses at length the receiver operating characteristic (ROC) curve where the BMS passes or fails a part depending whether or not a continuous variate is less than or greater than a specified constant c. In this context, unlike in our proposed work, due to ethical and other constraints, subjects (parts) are only measured once by each operator. In the psychological and educational context there are also many papers devoted to Item Response Theory (IRT) that, in its simplest form, connects an underlying unobservable continuous trait to the observable response to one or more binary questions. See for example, de Boeck (2008). Here the focus is not on estimating the misclassification rates and thus the goals are very different than what we consider. Finally, in a manufacturing context, the measurement system manual AIAG (2010) proposes assessing a BMS using a combination of measures of agreement between operators (see Gwet 2014) and measures of effectiveness based on the proportion of correct decisions. Here also the focus is not on estimating the misclassification rates and also the latter method does not recognize the difference between repeated measurements on the same part and individual measurement on different parts. See de Mast and Erdmann (2011) for further criticism of the AIAG approach.

We organized the paper in the following manner. Section 2 deals with fixed operator effects. We start with two models. The first treats the random part effect as a latent variable and the second, a moment-based model, captures the dependency among measurements on the same part. We show that the estimates of the parameters of interest from the two models are essentially equal so we adopt the simpler moment-based model for estimation (though retaining the other model for planning). Next, we look at an example and to end the section, consider the choice of plan. In Sect. 3, we repeat these sub-sections for the case of random operator effects. We end in Sect. 4 with a summary and a number of discussion points.

2 Fixed Operator Effects

Suppose the BMS has a small number of operators and we use an $L(m, n, r)$ plan to collect the data. Assuming fixed operator effects, in most cases we do not use multiple copies of the plan (i.e. we choose $T = 1$) since it is advantageous for each part to be measured by every operator, especially if there are strong part effects. We treat the operator effects as fixed in the following models and analysis. We assume some parts are more difficult to correctly classify than others. Since parts are sampled from the population of non-conforming parts, we use random part effects. We describe the models for $n = 2$ operators and indicate how they can be extended to larger values of n.

2.1 Models

When part effects are random and operator effects are fixed, the basic assumption is that measurements on different parts are independent but measurements on the same part are not. We build a model for each part separately. So here we temporarily delete the indices that distinguish among parts.

Suppose each operator measures a part randomly selected from the population of non-conforming parts. We denote the unobservable random part effect by P and assume $P \sim G(0, \sigma_P)$. We model the r measurements assuming conditional independence by

$$Y_{j1}, \ldots, Y_{jr} \mid (P = p) \sim \mathrm{B}\big(1, g(\mu_j, p)\big), \quad j = 1, 2 \text{ (operators)}$$

where B indicates a Bernoulli distribution, $0 < g(.) < 1$ is a suitable link function and μ_j is the fixed effect for operator j. Equivalently, because of the conditional independence, we have

$$S_j = Y_{j1} + \ldots + Y_{jr} \mid (P = p) \sim \mathrm{Bin}\big(r, g(\mu_j, p)\big)$$

where S_j is the total number of passes in the r measurements due to operator j and Bin() denotes a binomial distribution. We can, for example, specify $g(.)$ as a logistic or normal cumulative distribution function and, for simplicity, suppose that the effects act additively, i.e. $g(\mu_j, p) = g(\mu_j + p)$. The joint probability function of S_1 and S_2 is then

$$P(S_1 = s_1, S_2 = s_2) = k \int g(\mu_1 + \sigma_p z)^{s_1} \big(1 - g(\mu_1 + \sigma_p z)\big)^{r-s_1} \ldots \\ \ldots g(\mu_2 + \sigma_p z)^{s_2} \big(1 - g(\mu_2 + \sigma_p z)\big)^{r-s_2} \phi(z)\, dz \qquad (1)$$

where $0 \leq s_1, s_2 \leq r$, k is the product of the binomial coefficients and $\phi(z)$ is the standard Gaussian density. In (1) the latent random variable Z is used to represent the random part effect. Note that S_1 and S_2 are conditionally independent given $P = p$ but marginally dependent. McCulloch et al. (2008) call this a conditional model.

The connection between the misclassification error probability α_j and the model parameters is

$$\alpha_j = P(Y_j = 1) = E\big[g(\mu_j + \sigma_p Z)\big] \qquad (2)$$

We cannot interpret the parameters in the conditional model directly. As an alternative, we consider a simpler model partially specified in terms of α_1 and α_2, the parameters of interest, and other moments of the marginal distributions. Suppose the two operators repeatedly measure a randomly selected part. Then given a randomly selected part $\hat{\alpha}_j = S_j/r$ is an unbiased estimate of α_j with variance denoted by λ_j. Since S_1 and S_2 are dependent, let δ be the covariance of $\hat{\alpha}_1$ and $\hat{\alpha}_2$. Note that

$$Cov[\hat{\alpha}_1, \hat{\alpha}_2] = \frac{1}{r^2} Cov[Y_{11} + \ldots + Y_{1r}, Y_{21} + \ldots + Y_{2r}] = Cov[Y_{11}, Y_{21}] = \delta \qquad (3)$$

so δ is also the covariance between single measurements on the same part by different operators.

This so-called marginal model McCulloch et al. (2008) can stand on its own or may arise from the integration of a latent conditional model. Note that the conditional model described here has three parameters whereas the marginal model has five so the two models are not equivalent.

2.2 Estimation

Suppose we have data collected with a $L(m, 2, r)$ plan. Note that now we have m parts and we add a subscript i to represent the different parts. Using the conditional model, we can write the log-likelihood as $l(\mu_1, \mu_2, \sigma_P) = \sum_{i=1}^{m} \ln \left[P(S_{i1} = s_{i1}, S_{i2} = s_{i2}) \right]$ where $i = 1, \ldots, m$ indexes the parts and each term in the sum is given by (1), a one-dimensional integral. We use the MATLAB (2012) functions *integral* and *fmincon* respectively to calculate the integrals for given parameter values and to maximize the log-likelihood. We find the maximum likelihood estimates (MLEs) for α_1 and α_2 by numerically evaluating the expectations (2) after substituting the MLEs $\hat{\mu}_1, \hat{\mu}_2$ and $\hat{\sigma}_P$.

Alternatively, for the marginal model, we estimate α_1 and α_2 by the average over all parts.

$$\hat{\alpha}_1 = \frac{1}{m} \sum_{i=1}^{m} \hat{\alpha}_{i1} , \quad \hat{\alpha}_2 = \frac{1}{m} \sum_{i=1}^{m} \hat{\alpha}_{i2}$$

The variances of these estimates are

$$Var[\hat{\alpha}_j] = \lambda_j / m , \quad j = 1, 2, \quad \text{and} \quad Var[\hat{\alpha}_1 - \hat{\alpha}_2] = \frac{\lambda_1 + \lambda_2 - 2\delta}{m} \tag{4}$$

We estimate the variances λ_1, λ_2 and the covariance δ by the sample variances and covariance of the m pairs $(\hat{\alpha}_{i1}, \hat{\alpha}_{i2})$.

We found the covariance δ difficult to interpret and so we reparametrized the marginal model as follows:

$$\begin{aligned}
\delta &= P(Y_{11} = 1, Y_{21} = 1) - P(Y_{11} = 1)P(Y_{21} = 1) \\
&= P(Y_{11} = 1 \mid Y_{21} = 1)P(Y_{21} = 1) - P(Y_{11} = 1)P(Y_{21} = 1) \\
&= \alpha_1 \alpha_2 \left(\frac{P(Y_{11} = 1 \mid Y_{21} = 1)}{P(Y_{11} = 1)} - 1 \right) \\
&= \alpha_1 \alpha_2 (\theta - 1)
\end{aligned}$$

Note that if $\theta = 1$ (or correspondingly $\delta = 0$), then Y_{11} and Y_{21} are independent. On the other hand, if θ is large, say $\theta = 3$, then given that Operator 2 misclassifies a part, Operator 1 is three times as likely (compared to the overall misclassification rate α_1) to also misclassify the part.

Based on earlier work Danila et al. (2013), we suspected that the simple estimates from the marginal model will be close to the MLEs from the conditional model. Accordingly we conducted a $3^4 \times 4$ factorial experiment with factors defined by the sampling plan ($m = 100, 500, 1000$, $r = 1, 3, 5$) and the underlying model parameters ($\alpha_1, \alpha_2 = 0.05, 0.10, 0.15$, $\theta = 1, 2, 3, 4$).

To generate the data for each simulation run, we used the conditional model. That is, given α_1, α_2 and θ, we found the corresponding values for μ_1, μ_2 and σ_P. Then

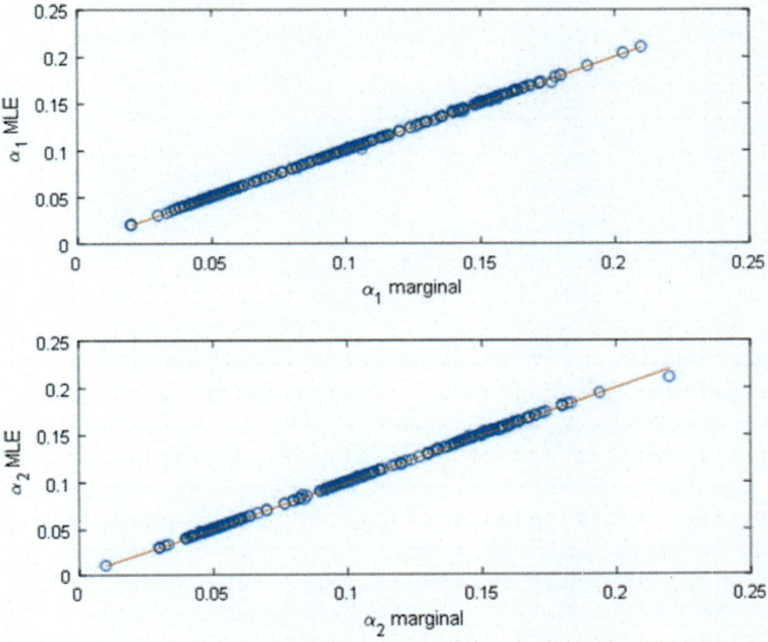

Fig. 1 Scatterplots for the conditional model MLEs and marginal model estimates for α_1 and α_2

we generated a single set of data from $L(m, 2, r)$ with these parameter values. Next, we found the MLEs from the conditional model and the moment estimates from the marginal model.

We show the results in Fig. 1. There is remarkable agreement, so much so that we omitted the MLEs in conditional model from further consideration (though we use the conditional model in the planning). Because the two estimates are almost equal for every run, so are their standard errors and we lose little in using the simple moment estimates. Importantly, the moment estimates retain their properties whether or not the conditional model is appropriate. Note that outside of the design space in the experiment, we suspect that the agreement may not be so strong, especially if α_1 and α_2 are much larger. However, in an industrial context this is unlikely. Given the complexity of the log-likelihood for the conditional model, we are unable to show that the two estimation procedures are mathematically identical.

2.3 Example

The example is based on a real context; the data are artificial. Credit card blanks are 100% inspected by an automated BMS for a large number of different flaws and each card is either rejected or passed. Human inspectors act as the gold standard.

Table 1 Response Pattern (Number of Misclassifications) for Credit Card Example

s_1 (New)	s_2 (Old)			
	0	1	2	3
0	222	41	7	3
1	9	10	4	2
2	1	0	1	0
3	0	0	0	0

There have been customer complaints that suggest a proportion of the passed cards have one or more flaws. In an attempt to improve the inspection system, a new fixturing procedure was developed. Three hundred flawed cards and 300 good cards are used in a study to compare the new fixture to the old. Here we report only the results with the flawed cards. Note that there are no operator effects. Instead, we want to compare the old and new fixturing with fixed effects (i.e. the fixtures take the role of operators). Each card is measured three times by each procedure. The data are summarized in Table 1 as a response pattern. For example, 222 cards were correctly classified (i.e. failed) all three times by each fixturing procedure. Using the marginal model, we have $\hat{\alpha}_{new} = 0.032$, $\hat{\alpha}_{old} = 0.100$ with corresponding standard errors 0.019 and 0.026 respectively. The estimated covariance is $\hat{\delta} = 0.007$ and thus $\hat{\theta} = 3.2$. To compare the two procedures, we have $\hat{\alpha}_{old} - \hat{\alpha}_{new} = 0.068$ with standard error 0.032 by substituting the estimate into (4). As a result, there is evidence that the new fixturing is better than the old in detecting defective cards. As expected, the MLEs from the conditional model agree.

2.4 Choice of Plan

Since the goal of the assessment is to look at differences between operators, we consider the selection of the number of parts m and the number of repeated measurements r so that the estimate $\hat{\alpha}_1 - \hat{\alpha}_2$ has pre-specified standard deviation. Since we expect $\theta \geq 1$ (or equivalently $\delta \geq 0$), there is little value in considering $L(m, 1, r)^2$ where operators do not measure the same parts. As with most planning problems, we need to elicit reasonable parameter values, here α_1, α_2 and θ (or equivalently δ) to select the design. We do not attempt to guess at the variances λ_1 and λ_2.

From (4), we see that $Var[\hat{\alpha}_1 - \hat{\alpha}_2]$ does not explicitly depend on r. However, if we revert to the conditional model and calculate $\lambda_j = Var[S_j/r]$ by conditioning on the random effect we get

$$\lambda_j = \frac{1}{r^2}\left(Var[E[S_j \mid P]] + E[Var[S_j \mid P]]\right)$$

Using the assumption that $S_j \mid P \sim B(r, g(P))$ and rearranging terms, we get

$$\lambda_j = \frac{\alpha_j(1-\alpha_j)}{r} + \frac{r-1}{r} Var[g(\mu_j + \sigma_p Z)]$$

We showed earlier (see (3)) that δ does not depend on r and so $Var[\hat{\alpha}_1 - \hat{\alpha}_2]$, as given in (4) depends on r only through λ_1 and λ_2. As r gets large, $Var[\hat{\alpha}_1 - \hat{\alpha}_2]$ does not go to zero. In other words, increasing r (the number of repeated measurements) with m (number of parts) fixed has diminishing returns.

To investigate possible plans, we start with the elicited values for α_1 and α_2. Given these values, for any value of σ_p, we determine the corresponding values for μ_1 and μ_2 in the conditional model, now using a logistic link. Then we vary σ_p until the conditional model covariance matches the specified δ (θ). Next, as given in Table 2, we calculate $v_1 = Var[g(\mu_1 + \sigma_p Z)]$ and $v_2 = Var[g(\mu_2 + \sigma_p Z)]$.

Combining the results we have

$$Var[\hat{\alpha}_1 - \hat{\alpha}_2] = \frac{[\alpha_1(1-\alpha_1) + \alpha_2(1-\alpha_2)]/r + (r-1)[v_1 + v_2]/r - 2\delta}{m} \quad (5)$$

We can investigate choices for m and r using (5) to achieve a desired precision for $\hat{\alpha}_1 - \hat{\alpha}_2$.

For example, suppose we guess parameter values $\alpha_1 = 0.05$, $\alpha_2 = 0.10$ and $\theta = 3$ (covariance $\delta = 0.010$). We want to distinguish between operators so we aim for $StDev[\hat{\alpha}_1 - \hat{\alpha}_2] = 0.02$ or $Var[\hat{\alpha}_1 - \hat{\alpha}_2] = 0.0004$. From Table 2, we have $Var[g(\mu_1 + \sigma_p Z)] = 0.0062$ and $Var[g(\mu_1 + \sigma_p Z)] = 0.0166$. Substituting we have

Table 2 $v_1 = Var[g(\mu_1 + \sigma Z)]$ and $v_2 = Var[g(\mu_2 + \sigma Z)]$ as functions of $\alpha_1, \alpha_2, \theta$

θ	α_1	$\alpha_2 = 0.01$	$\alpha_2 = 0.05$	$\alpha_2 = 0.10$
1.5	0.01	0.0000, 0.0000	0.0001, 0.0011	0.0001, 0.0043
	0.05	0.0011, 0.0001	0.0013, 0.0013	0.0014, 0.0045
	0.10	0.0042, 0.0001	0.0045, 0.0014	0.0050, 0.0050
2.0	0.01	0.0000, 0.0000	0.0001, 0.0020	0.0001, 0.0078
	0.05	0.0021, 0.0001	0.0025, 0.0025	0.0029, 0.0087
	0.10	0.0078, 0.0001	0.0087, 0.0029	0.0100, 0.0100
3.0	0.01	0.0000, 0.0000	0.0003, 0.0041	0.0003, 0.0139
	0.05	0.0041, 0.0003	0.0050, 0.0050	0.0062, 0.0166
	0.10	0.0139, 0.0003	0.0166, 0.0063	0.0200, 0.0200
4.0	0.01	0.0000, 0.0000	0.0004, 0.0058	0.0006, 0.0194
	0.05	0.0058, 0.0004	0.0075, 0.0075	0.0099, 0.0241
	0.10	0.0194, 0.0006	0.0241, 0.0099	0.0300, 0.0300

$$
\begin{aligned}
Var(\hat{\alpha}_1 - \hat{\alpha}_2) &= \frac{(0.0475 + 0.0900)/r + (r - 1)(0.0062 + 0.0166)/r - 0.02}{m} \\
&= \frac{0.1375/r + (r - 1) \times 0.0228/r - 0.02}{m}
\end{aligned}
$$

For $r = 1$, we need $m = 294$ parts to attain the required precision. If we increase r to 5, then we can reduce the number of parts substantially to $m = 64$. As r increases, we can decrease m to 7 but the total number of measurements $2mr$ increases. There is considerable value in the repeated measurements, especially if non-conforming parts are difficult to find. We can investigate the sensitivity of the choices of r and m for different values for α_1, α_2 and θ.

Note that for planning, we use both the conditional and marginal models. For the analysis, we use only the marginal model.

2.5 *Three or More Operators*

For each additional operator, the number of parameters in the conditional model increases by one. The evaluation of the likelihood for each part still involves a single one dimensional integral. The number of parameters in the marginal model increases rapidly as the number of operators increases. For $n = 3$ operators, there are nine parameters in the marginal model (three αs, three λs and three covariances) and four parameters in the conditional model. There is no equivalence. However, the marginal model can be easily applied to get estimates of all of the parameters and corresponding standard errors.

With a larger number of operators, there is no obvious metric to summarize the operator effect. We suggest comparing the error rates pairwise. For planning a measurement assessment study with three or more operators, we suggest using the above approach and choosing a plan where all the pairwise comparisons meet the desired maximum standard error.

3 Random Operator Effects

We now consider a BMS with both part and operator effects random. In this context the primary parameter of interest is α the misclassification error probability of a randomly selected part measured by a randomly selected operator.

3.1 Models

To model the measurement process, we propose similar conditional and marginal models as in the previous section. Suppose we select a non-conforming part and an operator at random from the populations of interest and the operator measures the part r times. Denote the two random effects by p (parts) and o (operators) and conditionally model the number of passes S by a binomial distribution.

$$S = Y_1 + \ldots + Y_r \mid (P = p, O = o) \sim \text{Bin}\big(r, g(o + p)\big)$$

We assume that the random effects are independent Gaussians with mean zero and standard deviations σ_O and σ_P respectively. In what follows, we also assume a logistic link function, that is

$$\ln\left(\frac{g(.)}{1 - g(.)}\right) = \mu + p + o \tag{6}$$

The random effects cannot be observed. We derive the marginal probability distribution of S depending on the three unknown parameters μ, σ_O and σ_P by integrating the binomial probability $P(S = s|O = o, P = p)$ in (4) over the distribution of the two independent random effects. The observed measurements $Y_1, Y_2, \ldots, Y_r$ for a single part/operator combination are conditionally independent given $(P = p, O = o)$ but dependent marginally. The connection between the misclassification error rate α and the random effects is

$$\alpha = P(Y = 1) = E[g(\mu + O + P)] \tag{7}$$

By assuming that the random effects are independent Gaussian, we have $O + P \sim G(\mu, \sigma_t)$ where the total variation is $\sigma_t = \sqrt{\sigma_o^2 + \sigma_p^2}$, so α depends only on μ and σ_t. Alternatively, for a probit model with link $\Phi(.)$, the standard Gaussian cumulative distribution function, we have $\alpha = E[\Phi(\mu + \sigma_t Z)]$, where $Z \sim G(0, 1)$, with the well-known closed form $\alpha = \Phi(\mu/\sqrt{1 + \sigma_t^2})$ (McCulloch et al. 2008, pp. 237). With the probit model, if μ is negative ($\alpha < 1/2$), the misclassification probability decreases as σ_t decreases. For a logistic link, there is no simple closed form for α but we see in Fig. 2 that for $\alpha < 1/2$, again the misclassification probability decreases as σ_t decreases. We conclude that by reducing σ_t, we reduce the misclassification probability α and hence improve the performance of the measurement system.

As a simple alternative to the conditional model, we use a marginal model specified directly by α and other marginal parameters. To define the model, suppose we have $m = 2$ parts and $n = 2$ operators sampled randomly from the populations of interest and each operator measures each part r times. We arrange the observed number of passes s_{ij}, $i = 1, 2$, $j = 1, 2$ or the corresponding proportion of passes $\hat{\alpha}_{ij} = s_{ij}/r$ as in Table 3 with rows corresponding to parts and columns to operators. In Table 3 and its generalization to $L(m, n, r)$, each entry is an unbiased estimate of α and each has the same variance denoted by λ. Proportions in the same row are dependent with

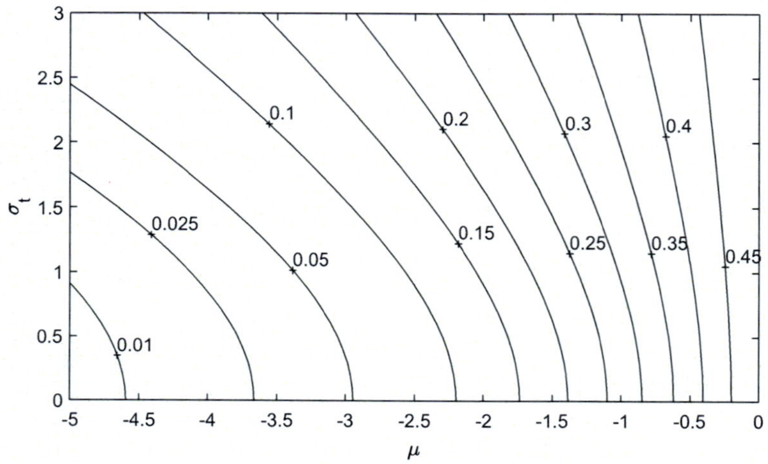

Fig. 2 Contour plot of the misclassification probability, α, from a logistic model as a function of μ and σ_t for $\alpha < 1/2$

Table 3 Sample proportions from $L(2, 2, r)$

	Operator 1	Operator 2
Part 1	$\hat{\alpha}_{11}$	$\hat{\alpha}_{12}$
Part 2	$\hat{\alpha}_{21}$	$\hat{\alpha}_{22}$

covariance δ because they share the same part effect. Similarly, proportions in the same column are dependent with covariance γ. We expect δ and γ to be positive. Entries such as $\hat{\alpha}_{11}$ and $\hat{\alpha}_{22}$ in different rows and columns are independent because they share neither an operator nor a part. If we have a $L(m, n, r)^T$ plan with operators and parts selected at random, then the same four parameters, i.e. α, λ, δ, γ describe the means, variances and covariances among the table entries. Two proportions that differ in both row and column within a table are independent as are entries among the T independent copies of $L(m, n, r)$.

We interpret α as the probability that the measurement of a randomly selected (nonconforming) part by a randomly selected operator is misclassified.

As in the fixed operator effects case, here the marginal model can stand on its own or may arise from the integration of a latent conditional model. The conditional model has three parameters and the marginal model has four so they are not in 1–1 correspondence. If we assume $g(.)$ is a logistic or probit function, then as long as the misclassification probability is reasonably small, say $\alpha < 0.2$, given α, δ and γ, we can find the corresponding values for μ, σ_O and σ_P. Then we can determine the corresponding λ. Alternately, we can add a part by operator interaction effect (OP) in the conditional model so that (7) becomes

$$\alpha = P(Y = 1) = E[g(\mu + O + P + OP)] \tag{8}$$

where $OP \sim G(0, \sigma_{OP})$. Now the two models each have four parameters and for feasible values, it appears from a numerical investigation, that the two sets of parameters are in 1-1 correspondence. We make use of this equivalency in the analysis and planning steps.

3.2 Estimation

Given the results of an assessment study $L(m, n, r)^T$, we can estimate the parameters of the conditional or marginal models as defined in the Sect. 3.1.

Estimation with the conditional model (7) or its extension to multiple tables is complicated. Due to the number of realizations of the latent random effects, a single calculation of the likelihood involves an $m + n$ dimensional integral for each of the T tables. For example, when $m = n = 2$ and $T = 1$, we have the following likelihood involving a four dimensional integral:

$$L(\mu, \lambda, \rho, \gamma) = E\left[\prod_{i,j} \binom{r}{s_{ij}} \eta_{ij}^{s_{ij}} (1 - \eta_{ij})^{1-s_{ij}}\right]$$

where $\eta_{ij} = g(\mu + o_i + p_j)$ and the expectation is over the joint distribution of the four random effects P_1, P_2, O_1, O_2. To make the evaluation of the log-likelihood and its derivatives even more difficult, the parameters of interest α, λ, δ and γ are defined implicitly by μ, σ_P and σ_O. The conditional model uses several untestable assumptions (e.g., additive independent Gaussian effects, choice of link function). Based on the study of other latent class models Albert and Dodd (2004, 2008), Akkerhuis et al. (2017), we suspect that the maximum likelihood estimates will not be robust to deviations from these assumptions.

Since the marginal model specifies only the first two moments, we use moment estimates. The estimates are simple and robust to the underlying latent structure. Suppose we have a $L(m, n, r)^T$ plan with observed proportions of passes $\hat{\alpha}_{ikt}$, $i = 1, \ldots, m, k = 1, \ldots, n, t = 1, \ldots, T$. We estimate α directly by the overall average

$$\hat{\alpha} = \frac{\sum_{i,j,t} \hat{\alpha}_{ijt}}{mnT} \tag{9}$$

For this estimator, we have

$$\begin{aligned} Var[\hat{\alpha}] &= \frac{mn\lambda + mn(n-1)\delta + nm(m-1)\gamma}{m^2 n^2 T} \\ &= \frac{\lambda + (n-1)\delta + (m-1)\gamma}{mnT} \end{aligned} \tag{10}$$

since for a single $L(m, n, r)$, each of the m rows has $n(n-1)$ correlated ordered pairs and each of n columns has $m(m-1)$ such pairs. Entries in different tables are independent since they arise from different parts and operators.

To find a standard error for $\hat{\alpha}$, we have

$$\lambda = E[\hat{\alpha}_{ijt}^2] - E\big[\hat{\alpha}_{ijt}\big]^2 = E[\hat{\alpha}_{ijt}^2] - \alpha^2 ,$$
$$\delta = E[\hat{\alpha}_{ijt}\hat{\alpha}_{ikt}] - \alpha^2 , \ \gamma = E[\hat{\alpha}_{ijt}\hat{\alpha}_{ljt}] - \alpha^2 \tag{11}$$

and $Var[\hat{\alpha}] = E[\hat{\alpha}^2] - \alpha^2$ or equivalently $\alpha^2 = E[\hat{\alpha}^2] - Var[\hat{\alpha}]$. Substituting in (10) and rearranging terms, we have

$$Var[\hat{\alpha}] = \frac{E[\hat{\alpha}_{ijt}^2]+(n-1)E[\hat{\alpha}_{ijt}\hat{\alpha}_{ikt}]+(m-1)E[\hat{\alpha}_{ijt}\hat{\alpha}_{ljt}]-(m+n-1)E[\hat{\alpha}^2]}{mnT-(m+n-1)} \tag{12}$$

We get unbiased estimates of each of the first three expectations in the numerator of (12) using the corresponding averages over all tables and of the fourth expectation using the estimate $\hat{\alpha}^2$. Substituting in the right side of (12) gives an unbiased estimate of $Var[\hat{\alpha}]$ and hence a standard error for $\hat{\alpha}$ that is close to being unbiased. For large tables, it is useful to note that the sum of all cross products in a row is the sum of the row entries squared minus the sum of the squares of the row entries, so we have the estimate

$$\hat{E}[\hat{\alpha}_{ijt}\hat{\alpha}_{ikt}] = \frac{\sum_{i,t}\left[\left(\sum_j \hat{\alpha}_{ijt}\right)^2 - \sum_j \hat{\alpha}_{ijt}^2\right]}{mTn(n-1)}$$

and a similar expression for the estimate $\hat{E}[\hat{\alpha}_{ijt}\hat{\alpha}_{ljt}]$.

As in the previous section, we carried out a study using simulated data to compare the maximum likelihood estimate to the moment estimate of α. We used the MATLAB (2012) function *fitgmle* to maximize the likelihood from the conditional model. In the simulation, we tried to match as closely as possible the plan and results obtained from the inter-laboratory assessment example that follows in Sect. 3.3. However, since it was not straightforward to adapt the Matlab function *fitgmle* to allow for two separate tables of results, we restricted attention to the case where $T = 1$. We assumed a $L(8, 20, 2)^1$ plan where there are 8 operators and 20 parts. Note that in this case, the log-likelihood for the conditional model involves 28 integrals and it is challenging and slow to maximize. We generated data from the conditional model with parameter values $(\mu, \sigma_p, \sigma_o, \sigma_{op}) = (-4.30, 3.22, 1.97, 0)$. We selected these particular parameter values for the simulation since they match the model parameter estimates obtained from the inter-laboratory example (once we translated the estimates for the marginal model to the corresponding estimates for the expanded conditional model (8)—since we were unable to fit the conditional model directly). Figure 3 shows the comparison for the estimates of α derived from both the conditional and marginal models for 100 simulation runs. We see that for the random operators model, the correspondence between the marginal and conditional models is fairly strong, but

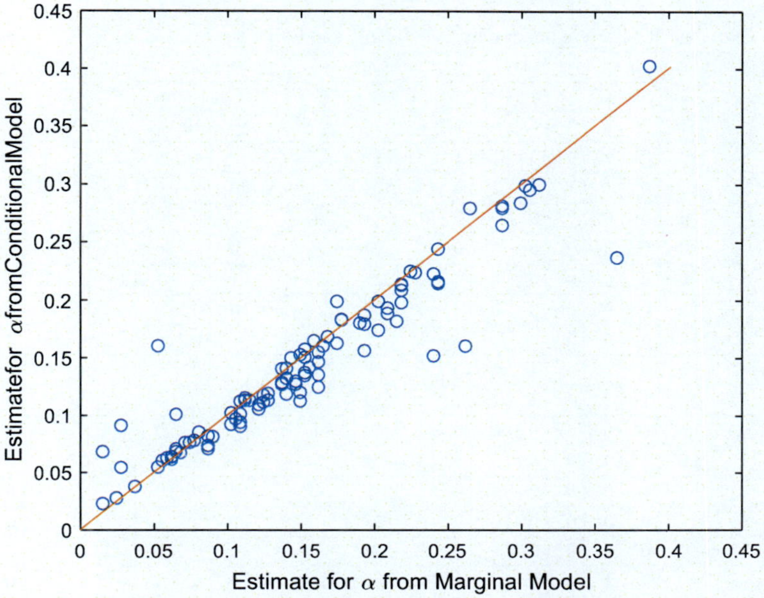

Fig. 3 Scatterplots for the conditional model MLEs and marginal model estimates for α assuming random operators

not nearly as close as with the fixed operator effect model described in the Sect. 2. Further simulation studies (not shown) suggest that the larger discrepancies between the conditional and marginal model results arise due to convergence issues with fitting the conditional model. The marginal model provides unbiased estimates of α and, for the utilized parameter values and assessment plan, is preferred.

3.3 Example

The example is artificial but the context is realistic based on Bashkansky et al. (2012). In an inter-laboratory assessment to estimate misclassification errors when measuring a binary property, 40 samples with the property and 40 without were prepared under a range of conditions that mirrored actual practice. Each sample was split into 8 subsamples giving a total of $80 \times 8 = 640$ subsamples. Sixteen laboratories were selected at random from cooperating units and divided into two groups A and B. Every lab in group A was sent 20 subsamples of each category and were requested to measure each subsample twice in random order. The technicians carrying out the measurements were blind to the true status of each subsample. We consider subsamples as parts (rows) and labs as operators (columns). In our notation, we have two $L(8, 20, 2)^2$ plans, one plan for assessing each misclassification probability. We

Table 4 Number of Misclassification by Sub-sample and Lab from $L(20, 8, 2)^2$

Subsample	Group A								Subsample	Group B							
	Lab number									Lab number							
	1	2	3	4	5	6	7	8		9	10	11	12	13	14	15	16
1	0	0	2	0	0	2	1	2	21	2	0	0	0	0	0	0	1
2	0	0	1	1	0	2	0	2	22	0	0	0	0	0	0	0	0
3	0	0	0	0	0	0	0	1	23	0	0	0	0	0	0	0	0
4	0	0	0	0	0	0	0	0	24	0	0	0	0	0	0	0	0
5	0	0	0	0	0	0	0	2	25	1	0	1	1	2	1	2	2
6	0	0	0	0	0	0	0	1	26	0	0	0	0	0	0	0	0
7	0	0	0	0	0	0	0	0	27	1	0	0	0	1	0	1	1
8	0	0	1	1	1	2	0	2	28	0	0	0	0	0	0	0	0
9	2	1	2	2	1	2	2	2	29	0	1	0	0	0	0	0	1
10	1	0	0	0	0	0	0	1	30	0	0	0	0	0	0	0	0
11	0	0	0	0	0	0	0	1	31	0	0	0	0	0	0	0	0
12	0	0	0	0	0	0	0	1	32	0	0	0	0	0	1	0	0
13	0	0	0	0	0	0	0	0	33	0	0	1	0	0	0	0	1
14	0	0	0	0	0	1	0	2	34	0	0	0	0	0	0	0	0
15	0	0	1	0	0	2	0	2	35	0	0	0	0	0	0	0	1
16	0	0	1	0	0	0	0	2	36	0	0	0	0	1	0	0	0
17	0	0	0	0	0	0	0	1	37	0	0	0	0	0	0	0	0
18	0	0	0	0	0	0	0	1	38	0	0	0	0	0	0	0	1
19	2	2	2	2	2	2	2	2	39	0	0	0	1	0	0	0	0
20	0	0	0	0	0	0	0	0	40	0	0	0	0	0	0	0	0

present the data in Table 4 only for those subsamples without the property of interest. Using the results for the marginal model, we have $\hat{\alpha} = 0.152$ and by substitution $\hat{E}(\hat{\alpha}_{ijt}^2) = 0.123$, $\hat{E}(\hat{\alpha}_{ijt}\hat{\alpha}_{ikt}) = 0.040$, $\hat{E}(\hat{\alpha}_{ijt}\hat{\alpha}_{ljt}) = 0.070$, so the standard error of $\hat{\alpha}$ is 0.048.

In a GR&R assessment of a continuous measurement system, we may decompose the estimated variation of the measurement system into components due to operators, part by operator interaction and repeatability, the variation in repeated measurements made by the same operator on the same part. If the variation due to a particular source is large relative to the overall variation, we use different strategies to improve the measurement system.

The estimates from the marginal model do not tell us how much improvement is available by reducing the variation among operators. Here we consider a "what if" analysis to assess the effect of variation among the operators on the misclassification probability. We require both the marginal model and conditional model with the logistic link function. First, we use the marginal model to estimate $\hat{\alpha}$, $\hat{\lambda}$, $\hat{\delta}$ and $\hat{\gamma}$ as above. Next, we find the corresponding estimates $\hat{\mu}$, $\hat{\sigma}_p$, $\hat{\sigma}_o$ and $\hat{\sigma}_{op}$ in the expanded conditional model (8) with an interaction effect. Then to see the effect of the operators on the misclassification probability, we set $\hat{\sigma}_o = 0$, $\hat{\sigma}_{op} = 0$ and translate back to the marginal model. Doing this gives $\hat{\alpha} = 0.121$. Here it seems we can make only a small improvement in the system by making the labs (operators) more homogeneous.

3.4 Choice of Plans

We consider $L(m, n, r)^T$ plans under a variety of constraints. Recall that the main goal of the assessment study is to estimate the misclassification probability α. As such, better plans will result in a smaller standard error for $\hat{\alpha}$. Looking at $Var(\hat{\alpha})$ given by (10), if the total number of distinctive part by operator combinations mnT is fixed, then $L(1, 1, r)^T$ generates the smallest standard deviation but gives no information about the covariances δ or γ, useful for a "what if" analysis, as we illustrated for the example. The next best plan that provides estimates of the three parameters is $L(2, 2, r)^{T/4}$. Note also that multiple copies of small plans are much easier to manage than a large plan with $T = 1$.

From the marginal model and (9), we see that $Var[\hat{\alpha}]$ does not depend explicitly on r. To study the effect of changing r, we revert to the corresponding conditional model. Consider making r measurements on a part and operator selected at random. By conditioning on the random effects and using conditional independence of the repeated measurements, the variance of the proportion of passes is

$$\lambda = E\left[\frac{g(1-g)}{r}\right] + Var[g] = \frac{\alpha(1-\alpha)}{r} + \frac{r-1}{r} Var[g] \tag{13}$$

where $Var[g]$ in (13) is calculated over the distribution of the random effects P and O and depends only on μ and σ_t. Note that as the number of repeated measurements (r) goes to infinity, λ does not approach zero. We can easily show that ρ and γ do not depend on r so $Var[\hat{\alpha}]$ depends on r only through λ.

To compare plans, we need some idea of the unknown parameters α, λ, δ and γ or equivalently α, $Var[g]$ and

$$\theta_O = \frac{P(Y_{11} = 1 \mid Y_{12} = 1)}{\alpha} = \frac{\delta + \alpha^2}{\alpha^2}, \quad \theta_P = \frac{P(Y_{11} = 1 \mid Y_{21} = 1)}{\alpha} = \frac{\gamma + \alpha^2}{\alpha^2}$$

Note that θ_O quantifies the relative increase of the probability that the first operator misclassifies a part given that the second operator has done so. The ratio θ_P can be similarly interpreted replacing operators with parts. For planning purposes, we can elicit reasonable guesses for α, θ_O and θ_P.

Note from above we have

$$\delta = \alpha^2(\theta_O - 1), \quad \gamma = \alpha^2(\theta_P - 1) \tag{14}$$

However, there is no direct way to suggest a value for $Var(g)$. Again we resort to the conditional model. If we assume no interaction effect (i.e. $\sigma_{op} = 0$), given values for α, θ_O and θ_P, we can calculate the corresponding conditional model parameters μ, σ_o and σ_p and hence $Var(g)$. In Table 5, we give values of $Var(g)$ when $\alpha = 0.05, 0.10, 0.15$ for a range of values of θ_O and θ_P.

Suppose the goal of the assessment is to estimate α to a specified precision. We set the variance of $\hat{\alpha}$ to a specified value A giving

$$A = \frac{\alpha(1-\alpha)/r + Var(g)(r-1)/r + (n-1)\rho + (m-1)\gamma}{mnT} \tag{15}$$

We choose $L(m, n, r)^T$ to minimize some objective function subject to the constraint (15) and any other constraints due to cost or logistics.

As an example of how to investigate possible plans, we consider a situation where we believe $\alpha = 0.10$. As well, we choose $\theta_O = 2$ and $\theta_P = 4$. Solving, using (14), we have $\delta = 0.01$, $\gamma = 0.03$ and from Table 5, $Var[g] = 0.056$. As well, we set the desired precision $A = (0.02)^2$. Substituting in (15) we have

$$0.0004 = \frac{0.09/r + 0.056(r-1)/r + 0.01(n-1) + 0.03(m-1)}{nmT}$$

Table 5 Value of $Var(g)$ when $\alpha = 0.01, 0.05, 0.10$

$\alpha = 0.01$	θ_O			
θ_P	1.0	1.5	2.0	3.0
1.5	0.0001	0.0001	0.0002	0.0003
2.0	0.0001	0.0002	0.0003	0.0005
3.0	0.0002	0.0003	0.0004	0.0007
4.0	0.0003	0.0005	0.0006	0.0009
5.0	0.0004	0.0006	0.0008	0.0011
$\alpha = 0.05$	θ_O			
θ_P	1.0	1.5	2.0	3.0
1.5	0.0013	0.0029	0.0046	0.0077
2.0	0.0025	0.0046	0.0065	0.0101
3.0	0.0050	0.0077	0.0101	0.0148
4.0	0.0075	0.0107	0.0137	0.0194
5.0	0.0100	0.0138	0.0173	0.0244
$\alpha = 0.10$	θ_O			
θ_P	1.0	1.5	2.0	3.0
1.5	0.0050	0.0114	0.0176	0.0298
2.0	0.0100	0.0176	0.0248	0.0395
3.0	0.0200	0.0298	0.0395	0.0627
4.0	0.0300	0.0425	0.0563	0.0907
5.0	0.0400	0.0566	0.0893	0.0920

We can use the above expression to compare various choices for $L(m, n, r)^T$. Consider the extreme case $L(1, 1, r)^T$ where each operator measures a single part r times. If $r = 1$, we need $T = 225$ part/operator pairs to attain the desired precision. By increasing r, we can reduce T so, for example, with $r = 5$, we need $T = 157$. The limiting minimum value as r increases is $T = 140$. Note with these plans, we cannot estimate θ_o or θ_p but we minimize the number of part/operator pairs. Alternately, we can consider a series of small sub-plans, say $L(2, 2, r)^T$. With $r = 1$, we require $T = 81$ copies and, increasing r to 5, yields $T = 64$. If we are constrained to using a single crossed design with no repeated measurements (i.e. $r = 1, T = 1$), then among other choices we have $m = 50$ parts and $n = 150$ operators. Increasing r in this plan has no material effect. We can also consider minimizing functions such as the total number of measurements $rmnT$ subject to the constraint (15).

These plans may seem infeasible. However, because we have a binary response and two random effects, we need a large number of operators, parts and measurements to get the required precision. Plans such as those described above will be complicated to manage and expensive to execute. Great care should be taken in choosing the assessment plan.

4 Discussion

We consider the assessment of a binary measurement system with random part effects. If the system involves only a few operators, we treat the operator effects as fixed; otherwise if the assessment study involves a sample of operators, we treat their effects as random. In either case, we develop simple estimates of the misclassification probabilities and their standard errors that can be used for inference and for planning the assessment study.

When operator effects are fixed, increasing the number of repeated measurements r can dramatically reduce m, the number of parts required to achieve the required precision. For random operator effects, increasing r was found to have less effect and we generally require many more parts and operators in the study.

For random operator effects, plans such as $L(2, 2, r)^T$ are much easier to manage than a single replicate plan $L(m, n, r)^1$. If we are using human subjects then it is likely that there is an upper bound of the number of times a subject can be assessed. Such constraints make plans such as $L(2, 2, r)^T$ attractive since each subject needs to be assessed by only two operators. However, the $L(2, 2, r)^T$ plan requires more parts and operators than the corresponding plan $L(m, n, r)^1$ with the same precision.

One weakness of the approach is that we require a large population of both conforming and non-conforming parts (or samples) as verified by a gold standard. In the examples of a high volume process such as credit card blanks and in the inter-laboratory test, it was feasible to find or create such samples. Severn et al. (2016) use partial verification in the assessment of a BMS with no operator effects to reduce the burden of using the gold standard on every part while maintaining almost all of the precision available from full verification. We plan to pursue this idea further in the case of a BMS with multiple operators.

We provide plans for assessing α (or α_j), the probability of misclassifying a non-conforming part. In practice we can simultaneously have the same operators measure a sample of conforming parts to estimate β (or β_j), the probability of misclassify a conforming part. Combining the two assessment plans would help to blind the operators as to the true status of any part.

References

AIAG. (2010). *Measurement system analysis. Reference Manual.* 4th Edition Detroit MI. Automotive Industry Action Group.

Akkerhuis, T., de Mast, J., & Erdmann, T. (2017). The statistical evaluation of binary tests without gold standard: Robustness of latent variable approaches. *Measurement, 95*, 473–479.

Albert, P. S., & Dodd, L. E. (2004). A cautionary note on the robustness of latent class models for estimating diagnostic error without a gold standard. *Biometrics, 60*, 427–435.

Albert, P. S., & Dodd, L. E. (2008). On estimating diagnostic accuracy from studies with multiple raters and partial gold standard evaluation. *Journal of the American Statistical Association, 103*, 61–73.

Bashkansky, E., Gadrich, T., & Kuselman, I. (2012). Inter-laboratory comparison of test results of an ordinal or nominal binary property: Analysis of variation. *Accreditation and Quality Assurance, 17*, 239–243.

Beavers, D. P., Stanley, J. D., & Bekele, B. N. (2011). A bayesian model to assess a binary measurement system when no gold standard system is available. *Journal of Quality Technology, 43*, 16–27.

Burdick, R. K., Borror, C. M., & Montgomery, D. C. (2005). In *Design and Analysis of Gauge R&R Studies: Making Decisions with Confidence Intervals in Random and Mixed ANOVA Models.* Philadelphia PA, ASA Alexandria VA: SIAM. ASA-SIAM Series on Statistics and Applied Probability.

Danila, O., Steiner, S. H., & MacKay, R. J. (2012). Assessing a binary measurement system with varying misclassification rates using a latent class random effects model. *Journal of Quality Technology, 44*, 179–192.

Danila, O., Steiner, S. H., & Mackay, R. J. (2013). Assessing a binary measurement system with varying misclassification rates when a gold standard is available. *Technometrics, 55*, 335–345.

De Boeck, P. (2008). Random item IRT models. *Psycometrika, 73*, 533–559.

De Mast, J., & Erdmann, T. P. (2011). Measurement system analysis for binary inspection: Continuous versu dichotomous measurands. *Journal of Quality Technology, 43*, 99–112.

Erdmann, T. P., Akkerhuis, T. S., de Mast, J., & Steiner, S. H. (2016). The statistical evaluation of a binary test based on combined samples. *Journal of Quality Technology, 48*, 54–67.

Gwet, K. L. (2014). *Handbook of inter-rater reliability.* Gaithersburg: Advanced Analytics.

MATLAB and Statistics Toolbox Release. (2012b). *The MathWorks Inc.* Massachusetts, United States: Natick.

McCulloch, C. E., Searle, S. R., & Neuhaus, J. M. (2008). *Generalized, linear, and mixed models.* 2nd edn. Hoboken: Wiley.

Pepe, M. S. (2003). *The statistical evaluation of medical tests for classification and prediction* (1st ed.). New York: Oxford University Press.

Severn, D. E., Steiner, S. H., & MacKay, R. J. (2016). Assessing binary measurement systems: A cost-effective alternative to complete verification. *Journal of Quality Technology, 48*, 128–138.

Steiner, S. H., Stevens, N. T., Browne, R., & Mackay, R. J. (2011). Planning and analysis of measurement reliability studies. *Canadian Journal of Statistics*, *39*, 344–355.

Van Wieringen, W. N., & de Mast, J. (2008). Measurement system analysis for binary data. *Technometrics*, *50*, 468–478.

Concepts, Methods, and Tools Enabling Measurement Quality

Antonio Possolo

Abstract This contribution provides an overview, illustrated with examples, of applications of statistical methods that support measurement quality and guarantee the intercomparability of measurements made worldwide, in all fields of commerce, industry, science, and technology, including medicine. These methods enable a rigorous definition of measurement uncertainty, and provide the means to evaluate it quantitatively, both for qualitative measurands (for example, the sequence of nucleobases in a DNA strand) and for quantitative measurands (for example, the mass fraction of arsenic in rice). Measurement quality is its trustworthiness and comprises several attributes: reliable calibration involving standards; traceability to the international system of units or to other generally recognized standards; measurement uncertainty that realistically captures contributions from all significant sources of uncertainty; and fitness for purpose of the measurement results, which comprise measured values and evaluations of associated uncertainties. Statistical methods play key roles in the quality system that validates the measurement services (reference materials, calibrations, reference data, and reference instruments) provided by the National Institute of Standards and Technology (NIST). And these services in turn support measurement quality in laboratories, factories, farms, hospitals, transportation, utilities, and weather and environmental monitoring stations throughout the world, contributing to ensure food safety, to manufacture reliable products, and to monitor industrial and natural processes accurately. The *NIST Uncertainty Machine* (NUM) and the *NIST Consensus Builder* (NICOB) are web-based tools freely available to metrologists everywhere, that help maintain measurement quality. The NUM serves to evaluate measurement uncertainty and the NICOB builds consensus values from measurement results obtained independently for the same measurand.

Keywords Bottom-up · Calibration · Consensus · Count · Measurand · Measurement equation · Measurement model · Measurement uncertainty · Mutual consistency · Observation equation · Qualitative · Standard · Top-down · Traceability · Trustworthiness

---

A. Possolo (✉)
NIST (National Institute of Standards and Technology), Gaithersburg, Maryland, USA
e-mail: antonio.possolo@nist.gov

S. Knoth and W. Schmid (eds.), *Frontiers in Statistical Quality Control 13*,
Frontiers in Statistical Quality Control,
https://doi.org/10.1007/978-3-030-67856-2_19

1 Preamble

This contribution provides an overview of measurement and of measurement uncertainty (Sects. 3 and 4), highlighting how statistical models and methods (Sects. 7, 8, 9) enable measurement and ensure measurement quality. The concepts of calibration (Sect. 5) and traceability (Sect. 6), which are distinctive traits of measurement, are examined in detail. In particular, Sect. 6.1 offers new proposals concerning how traceability should be established for counts and for assignments of value to qualitative properties—the latter fall within the broad understanding of measurement laid out in Sect. 3.

2 Introduction

Measurement serves to estimate values of properties of natural and man-made entities and processes, which are used to inform choices and decisions made in all sectors of the human enterprise. Measurement quality is its trustworthiness: the extent to which it tracks the true values of those properties sufficiently closely for the intended purpose, with assuredly high confidence.

Measurement tracks the truth when its results are metrologically traceable to appropriate, widely recognized standards. It does so sufficiently closely when measurement uncertainty is small enough to warrant effectively regarding the measured value as a proxy of the corresponding true value in the context where the measurement results will be used to make a decision, thus achieving fitness for purpose.

Trustworthiness requires also that there should be a strong, justified belief (that is, assuredly high confidence) that the true value lies within the reported margin of uncertainty of the measured value. This belief is strengthened considerably when multiple, fundamentally different measurement methods produce measured values that agree with one another to within their respective margins of uncertainty.

- Jean Baptiste Perrin was reassured by "the very remarkable agreement found between values derived from the consideration of such widely different phenomena" (including viscosity of gases, critical opalescence, black body radiation, and Brownian motion), as he made measurements of the Avogadro constant and established the existence of molecules (Perrin 1916; Hudson 2018);
- The measurements of the speed of light, c, made using widely different measurement methods achieved such mutual agreement toward the middle of the twentieth century, that a consensus value was deemed to be sufficiently well characterized to warrant a definitive assignment of value to c in 1983 (Quinn 2012, p. 299)—MacKay and Oldford (2000) use historical measurement results for the speed of light as a focus for a discussion of how statistical methods and best statistical practices contribute to the advancement of science;

- The marked reduction, over time, of the variability of historical determinations of the Planck constant, h, made using very different measurement methods—most recently the Kibble balance and X-ray silicon crystal density (Possolo et al. 2018) —, enabled a definitive assignment of value to h in 2018 (Mohr et al. 2018).

3 Measurement

We adopt here a broad definition of measurement, to encompass the myriad measurement services offered by, and metrological research pursued at national metrology institutes (NMIs) worldwide, and in particular at NIST, which is the NMI of the United States of America.

In science, medicine, manufacturing, agriculture, indeed in most fields of human endeavor, both quantitative and qualitative properties of material or virtual objects are of interest. The mass of a white powder in a plastic bag that was left on a seat of a bus is a quantitative property. The chemical nature of this powder (whether it is soy protein, baking soda, cocaine, etc.) is a qualitative property.

The NIST Measurement Services Council has recently affirmed the view implicit in the *NIST Quality Manual for Measurement Services* (NIST-QM-I, Version 10, 27-Dec-2016, www.nist.gov/qualitysystem/) that the concept of measurement need not be restricted to the assignment of value to quantitative properties but may also be used in relation to qualitative properties.

However, this ought not to prevent that other, either more specialized or more informal terms be employed to describe assignments of value that fall within this broad concept of measurement. For example, instead of saying that one has measured the mass of the white powder in the bag aforementioned, or that one has measured its identity, it may be more natural to say that one has weighed and identified its contents (Possolo 2018).

The NIST-QM-I defines ***measurement*** as an "experimental or computational process that, by comparison with a standard, produces an estimate of the true value of a property of a material or virtual object or collection of objects, or of a process, event, or series of events, together with an evaluation of the uncertainty associated with that estimate, and intended for use in support of decision-making."

NIST Technical Note 1900 (NIST *Simple Guide*) supplements this definition with the following clarification (Possolo 2015, Note 2.1): "The property intended to be measured (*measurand*) may be qualitative (for example, the identity of the nucleobase at a particular location of a strand of DNA), or quantitative (for example, the mass concentration of 25-hydroxyvitamin D_3 in NIST SRM 972a, Level 1, whose certified value is 28.8 ng mL^{-1}). The measurand may also be an ordinal property (for example, the Rockwell C hardness of a material), or a function whose values may be quantitative (for example, relating the response of a force transducer to an applied force) or qualitative (for example, the provenance of a glass fragment determined in a forensic investigation)."

4 Measurement Uncertainty

A measurement result comprises an estimate of the true value of the property intended to be measured (*measurand*), and an evaluation of the uncertainty associated with this estimate. For example, the value recommended most recently for the Newtonian constant of gravitation, by the Task Group on Fundamental Constants of the Committee on Data for Science and Technology (CODATA), International Council for Science (CODATA 2019), is

$$G = 6.67430 \times 10^{-11}\,\mathrm{m}^3\,\mathrm{kg}^{-1}\,\mathrm{s}^{-2}\,,$$

with associated standard uncertainty

$$u(G) = 0.00015 \times 10^{-11}\mathrm{m}^3\,\mathrm{kg}^{-1}\,\mathrm{s}^{-2}$$

Measurement uncertainty is the doubt about the true value of the measurand that remains after making a measurement. Two aspects of measurement uncertainty require characterization: the width and depth of the corresponding margin of doubt Bell (1999). In the example above, $u(G)$ is the width of the margin of doubt, and its depth (unstated but implied) is the complement of the probability (approximately $1 - 0.68 = 32\%$) that quantifies the confidence in the true value of G lying within the interval $G \pm u(G)$: the smaller the confidence, the deeper the doubt.

Measurement uncertainty may be described fully and quantitatively by a probability distribution on the set of values of the measurand. For example, one might say that the uncertainty concerning the true value of G is described by a Gaussian probability distribution with mean $6.67430\ 10^{-11}\mathrm{m}^3\,\mathrm{kg}^{-1}\,\mathrm{s}^{-2}$, and with a standard deviation $0.00015\ 10^{-11}\mathrm{m}^3\,\mathrm{kg}^{-1}\,\mathrm{s}^{-2}$. At a minimum, it may be described summarily and approximately by a quantitative indication of the dispersion (or scatter) of such distribution, for example, $u(G)$ as given above (Possolo 2015, Sect. 3).

5 Calibration

When a truck stops at a highway scale to be weighed, it applies a force to one or several load cells under the scale, which generates a potential difference between the electrical terminals that the load cells are connected to. Calibration is the procedure that establishes a relation between values of the force applied to a load cell and corresponding values of potential difference, thereby making possible to "translate" indications of voltage into values of force. These values of force, in turn, are translated into values of mass using the local value of the Earth's gravity and Newton's second law of motion.

Calibration is usually performed by presenting a set of standards representing different values of the measurand to the measuring instrument to be calibrated, pos-

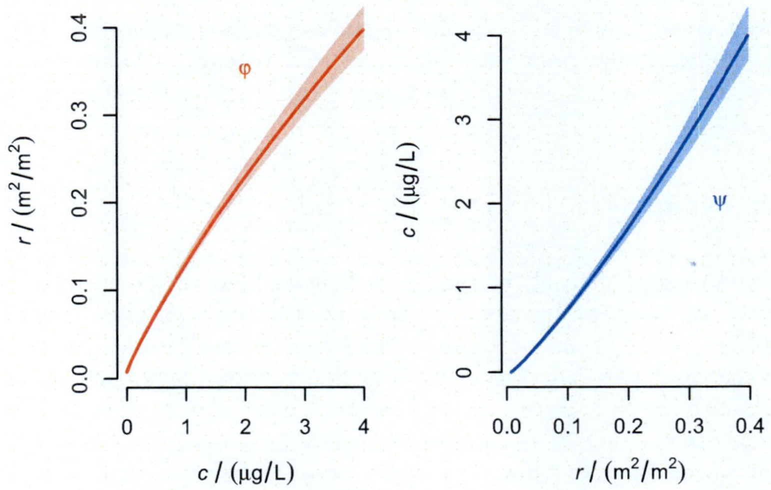

Fig. 1 Calibration (ϕ) and analysis (ψ) functions for a gas chromatography, mass spectrometry (GC/MS) system, and associated 95% coverage bands, used to measure the mass concentration *c* of chloromethane using observed values of the ratio *r* of areas of peaks corresponding to chloromethane and to an internal standard, derived from data from Lavagnini and Magno (2007), Table 2

sibly multiple times each, in an order determined by the design of the calibration experiment (Bartel 2005). This data is then reduced to define a function (*calibration function*, ϕ in Fig. 1) that, given a value of the measurand, yields the value of the corresponding indication expected to be produced by the instrument.

To use the instrument in practice one needs the mathematical inverse of the calibration function, which yields a value of the measurand when it is given an instrumental indication as input. In applications in analytical chemistry, the inverse of the calibration function is often called the *analysis function*, ψ in Fig. 1. In the field of force measurement, it is sometimes called the *measurement function* (Bartel et al. 2016).

In general, calibration is a procedure that establishes a relation between values of a property realized in measurement standards, and indications provided by measuring devices, or property values of artifacts or material specimens, taking into account the measurement uncertainties of the participating standards, devices, artifacts, or specimens (Possolo 2015, Note 3.7).

According to this definition, the entity being calibrated need not be a measuring device, but can be an artifact. Gauge blocks are an instance of such artifacts: using a mechanical comparator (Possolo 2015, Exhibit 5 on P. 34), one can calibrate a gauge block by comparing its dimensions with the dimensions of one or several master blocks that act as standards. Or the entity being calibrated may be a material specimen: for example, the Mohs hardness of a synthetic ruby may be calibrated by determining which minerals in a Mohs reference measurement standard it scratches, and which scratch it. A Mohs reference standard comprises samples of talc, gypsum, calcite, fluorite, apatite, orthoclase, quartz, topaz, corundum, and diamond, which are assigned as Mohs hardness, ordinal values 1, …, 10 (Klein and Dutrow 2007).

6 Traceability

Traceability is the property of a measurement result that asserts its comparability with a reference measurement standard hence guarantees that the measurement result also is comparable with other measurement results that are traceable to the same standard.

Measurement results are comparable when the measured values (and associated uncertainties) are expressed in the same scale, and differences or ratios between them are meaningful. Only for comparable measurement results, it is meaningful to ascertain whether the true difference (or ratio) between the values of their respective measurands lies within a particular interval with some specified probability, once allowance will have been made for their associated uncertainties.

For example, when the mass concentration of cholesterol in the blood of a particular patient is measured at different times by different laboratories and all the measurement results are traceable to the International System of Units (SI) (BIPM 2019), then the results will be comparable with one another, in particular enabling an inference about whether said concentration has been increasing over time, or not.

Suppose the volume fraction of aluminate in a Portland cement clinker was determined via X-ray powder diffraction (Stutzman et al. 2016), and also derived from the bulk chemical composition via the Bogue calculation (Bogue 1929), and that both measurements are traceable to the SI. Then it is meaningful to perform a statistical test to decide whether there is a statistically significant difference between the logit-transformed volume fractions produced by those two methods.

However, a measurement of a quantity of oranges expressed as "a dozen oranges" is not comparable to 5 kg of oranges. Similarly, while one soft gel of Bausch and Lomb's *PreserVision AREDS 2 Formula + MultiVitamin*[1] contains 300 IU (International Units) of Vitamin D and 200 IU of Vitamin E, its amount of Vitamin D is not 1.5 times larger than its amount of Vitamin E because IUs are not comparable to amounts-of-substance (WHO Expert Committee on Biological Standardization 2006).

Metrological traceability is a property of a measurement result whereby the result can be related to a reference through a documented, unbroken chain of calibrations, each contributing to the measurement uncertainty. It should be noted that traceability applies to measurement results only, and neither to measuring instruments or methods nor to institutions involved in measurement. Typically, the reference is specified that a measurement result is traceable to: for example, that (i) a measurement of mass is traceable to the SI (international system of units), or that (ii) a measurement of force,

[1]Any mention of commercial products within this article is for information only; it does not imply a recommendation or endorsement by NIST.

or (iii) of the mass concentration of cholesterol in human serum, are traceable to NIST, which is shorthand for "metrologically traceable to NIST's practical realization of the definition of a measurement unit."

In the first case, and considering the definition of the SI unit of mass in effect since May 20th, 2019 (BIPM 2019), traceability to the SI means that a sequence of properly documented comparisons, all recognizing measurement uncertainty, have been made, starting with a mass standard used to calibrate the balance employed to make the measurement of mass of interest, and ending with a primary realization of the kilogram in a Kibble balance or in a silicon sphere.

In the second case, and supposing that the force in question was measured using a load cell, traceability to NIST means that the load cell used to measure the force was calibrated in a machine that applies forces measured using a force transducer that, either directly or via a chain of calibrations involving other force transducers as transfer standards, can be related to the primary force standards maintained by NIST, which are machines capable of applying forces in discrete steps, generated by stainless steel dead-weights (Bartel 2005; Jabbour and Yaniv 2001).

In the third case, concerning the measurement of the mass concentration of cholesterol in a patient's blood sample by a clinical laboratory, traceability to NIST may be established by using NIST Standard Reference Material 1951c (Lipids in Frozen Human Serum) to prepare a series of dilutions that will be used to calibrate the instrument that the clinical laboratory uses to measure the total cholesterol in the patient's sample provided that the measurements of volume of solvent used to prepare these dilutions also are traceable to NIST.

The value of a measurand y is often determined using a measurement model as described in the *Guide to the expression of uncertainty in measurement* (GUM) (JCGM 2008), that is, as a function of values of several input quantities, $y = f(x_1, \dots, x_n)$. If the measurement results that include values of the *input quantities* are traceable to the SI, say, and the uncertainties associated with these quantities are suitably propagated to evaluate the uncertainty associated with the *output quantity* y, and any uncertainty associated with the computation of values of the function f also has been quantified and propagated to the uncertainty associated with y, then y is traceable to the SI.

A particular case of the situation just described occurs when measuring mass fractions, which take the form of ratios of mass values, or when measuring ratios of other quantities of the same kind. For example, NIST SRM 158a (Silicon Bronze) lists the certified mass fraction of zinc in the material as 2.076%, with expanded uncertainty 0.019% Gonzalez and Choquette (2018).

This means that the material comprises 0.02076 kg/kg of zinc, and that the true mass fraction of zinc in it is believed to lie in the interval 0.02076 kg/kg $\pm$ 0.00019 kg/kg with approximately 95% probability. If both the numerator and denominator of the original fraction are traceable to the kilogram, and the uncertainties associated with them have been properly propagated to their ratio, and the uncertainty in the computation of the ratio and in its conversion to a percentage is negligible, then the mass fraction is traceable to the SI.

6.1 Traceability for Counts and Qualitative Measurands

Since counting is measuring, the question naturally arises about the meaning of traceability for counts, for example, when one counts the number of neutrophils among 100 white blood cells in a patient's sample. The conventional position has been to say that counts are traceable to unit 1, which is the neutral element in the SI. Referring to counts, and somewhat cryptically, the 9th edition of the SI Brochure states that "formal traceability to the SI can be established through appropriate, validated measurement procedures" (BIPM 2019, Sect. 2.3.3).

We believe that establishing traceability for counts requires further elaboration, and that their traceability cannot be to the SI, but will have to be to other references. Counting involves two kinds of standards: one standard defines the entities that are being counted (and distinguishes them from those other entities that are not to be counted); another standard that serves to assign a value to the count.

The first standard plays the role of what, in zoology and botany, is called the holotype (of a species): in the present context, it is the paradigmatic or ideal instantiation of what is being counted—the typical neutrophil, when counting white blood cells, or the typical horse, when counting horses. The same as with biological species, some diversity often needs to be accommodated, because neutrophils are not exact copies of one another, and neither are horses.

In zoology, for example, the diversity corresponding to differences due to gender may be accommodated by designating allotypes: this makes it possible to recognize both mallard drakes and hens as members of the same species, *Anas platyrhynchos*, even though they look quite different from one another. The diversity of neutrophils may have to be accommodated by formulating a standard that comprises a sufficiently diverse collection of images of neutrophils supplemented with descriptions of identifying attributes (for example, presence of intra-cellular granules, or visible response to specific staining agents).

The second standard needed for counting finitely many entities is the unique subset of the consecutive positive integers including 1 and its successors, in the sense of Peano's axioms (Mints 2018) that can be put into one-to-one correspondence with the elements of the set whose elements are being counted. The value of the count is the largest integer in the standard. If no such subset of the positive integers exists, then we say that the count is 0.

In the case of a differential white blood cell count that yields 63 neutrophils out of 100 white blood cells, saying that the measured value 63 is traceable to the unit 1 is as trivial and as unproductive as saying that $63 \times 1 = 63$. The measurement result needs to specify what is being counted, and in addition ought to include also an evaluation of measurement uncertainty: using the Poisson model, a clinical laboratory technician doing the count manually might then report having found 63 neutrophils give or take 8 neutrophils. This margin of uncertainty accounts for sampling variability, as different blood smears from the same person typically will not yield exactly the same differential leukocyte count, and possibly also mere counting error.

Claiming that counts are traceable to the SI is too much of a simplification because it neglects the fact that counting inextricably involves the definition of what is being counted, and the standard that underlies this definition is not part of the SI. This limitation applies also to quantities like chemical amounts: when one says that a Baby Aspirin contains 0.45 mmol of aspirin, one is indeed expressing a count, of molecules in this case, qualified with a statement of what is being counted, aspirin molecules, but the definition of aspirin is not within the scope of the SI.

Establishing traceability for an assignment of value to a qualitative property, for example, that the nucleobase at a particular position of a strand of DNA is adenine, involves comparison with a standard for adenine relevant to how the nucleobase is identified: for example, a mass spectrum, as specified in the NIST Chemistry WebBook (Standard Reference Database 69, https://webbook.nist.gov/chemistry/).

7 Measurement Models

Models play a central role in measurement. Therefore, their adequacy to the measurement data, and their substantive validity, are key components of measurement quality. Be they mathematical or computational, measurement models describe the relationship between the value of the measurand (*output*) and the values of qualitative or quantitative properties (*inputs*) that either determine or provide information about its value.

7.1 Measurement Equations

The measurement models considered in the GUM express the measurand, y, as a deterministic function of several inputs: $y = f(x_1, \ldots, x_n)$. This is often referred to as a *measurement equation*: for each set of values of the inputs, it produces a single, corresponding value of the output. The function f may reflect a physical law, or merely describe an empirical relationship that is represented either analytically (that is, by means of a mathematical formula), or only algorithmically (that is, as a sequence of computational steps).

The Pitot tube affords an instance of a measurement model that reflects a physical law. A typical Pitot tube used to measure airspeed has an orifice facing directly into the airflow to measure the total pressure, and at least one orifice whose opening is parallel to the airflow, to measure static pressure (Fig. 2). Airspeed v is determined by the difference Δ between the total and static pressures, and by the mass density ρ of air, according to the measurement equation $v = \sqrt{2\Delta/\rho}$, which is a consequence of Bernoulli's equation (Anderson 2017, Sect. 3.4).

The procedure developed by Ciddor (1996), to estimate the refractive index of air as a function of the vacuum wavelength of the radiation of interest (in the range 300–1700 nm), air temperature, atmospheric pressure, air humidity, and amount frac-

Fig. 2 The steel part attached to the end of the horizontal yellow tube, and visible immediately to the left of the wheel, is a Pitot tube mounted on a helicopter (published with permission of Zátonyi Sándor, (ifj.), https://en.wikipedia.org/wiki/Pitot_tube): it has one large, forward-facing, circular orifice to measure total pressure, and several small circular orifices behind a trim ring, to measure static pressure

tion of carbon dioxide (in the range 0 to 2000 µmol/mol), is an example of an empirical measurement model that effectively is specified as an algorithm involving ten steps (Ciddor 1996, p. 1572). This procedure is implemented in an online calculator that is part of the NIST *Engineering Metrology Toolbox* (Stone and Zimmerman 2004).

7.2 Observation Equations

The curve in Fig. 3 is an example of a measurement model that describes an empirical relationship, of the form $p = \alpha \exp(-\beta/T)$, between temperature T and the vapor pressure p of gold (Paule and Mandel 1970, Table 2, Lab 9), which was derived from data (depicted as circles in Fig. 3) whose scatter around the curve reflects unpredictable contributions from uncontrolled sources of uncertainty.

Referring to exercises such as the derivation of this empirical model from the data, Bogen and Woodward (1988) state that "an important source of progress in science is the development of procedures for the systematic handling of observational and measurement error and procedures for data-analysis and data-reduction which obviate the need for a theory to account for what is literally seen."

Given any two data points, (T_i, p_i), and (T_j, p_j), one can solve a system of two simultaneous equations with α and β as unknowns, to obtain $\alpha_{ij} = \exp\{(T_i P_i - T_j P_j)/(T_i - T_j)\}$ and $\beta_{ij} = (P_i - P_j)/(1/T_j - 1/T_i)$, where $P_i = \log p_i$ and $P_j = \log p_j$. Doing this for all possible pairs of data points yields $55 \times 54/2 = 1485$ estimates of α, and the same number of estimates of β.

The problem is that these two sets of estimates span very wide ranges, indicating that the data, although obviously informative about the relationship between pressure and temperature, are mutually inconsistent. One could attempt to solve the problem

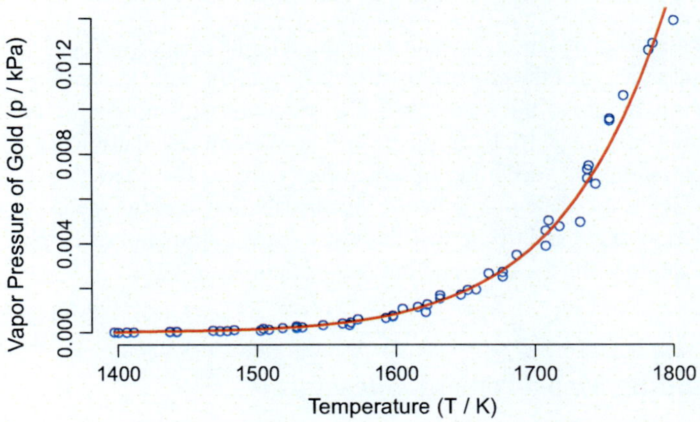

Fig. 3 Values of the vapor pressure p of gold measured at several different values of temperature T (Table 2, Lab 9, Paule and Mandel 1970), and an empirical model of the form $p = \alpha \exp(-\beta/T)$

by taking the medians of the $\{\alpha_{ij}\}$ and of the $\{\beta_{ij}\}$ as estimates of α and β. Perhaps surprisingly, these would be very good estimates of their corresponding measurands (Wilcox 2010, Chap. 11).

A more disciplined, less ad hoc approach, starts from a statistical model that defines a probability distribution for the data, where α and β figure as parameters: for example, $\log p_i = \log \alpha - \beta/T_i + \varepsilon_i$ for $i = 1, \ldots, 55$, where the $\{\varepsilon_i\}$ denote non-observable errors, assumed to be a sample from a Gaussian distribution with mean 0 and unknown standard deviation σ, and the relative measurement uncertainty associated with the values of temperature is assumed to be negligible by comparison with its counterpart for the values of pressure. The $\{\varepsilon_i\}$ may be conceived as "adjustments" that, once applied to the values of $\log p_i$, allow a single value of α and a single value of β to apply to all the "adjusted" data points.

The maximum likelihood estimates of $\log \alpha$ and β are the usual least squares estimates, whence one obtains $\widehat{\alpha} = 240.83 \times 10^6$ kPa and $\widehat{\beta} = 42\,224$ K, with associated standard uncertainties $u(\widehat{\alpha}) = 64 \times 10^6$ kPa and $u(\widehat{\beta}) = 420$ K. The estimate of σ is 0.1433.

Incidentally, the Theil-Sen estimates (the medians aforementioned), computed using R function `mblm` defined in the package of the same name Komsta (2019), are $\widetilde{\alpha} = 273.05 \times 10^6$ kPa and $\widetilde{\beta} = 42\,417$ K, with associated standard uncertainties $u(\widetilde{\alpha}) = 99 \times 10^6$ kPa and $u(\widetilde{\beta}) = 587$ K, evaluated using the non-parametric, statistical bootstrap (Efron and Tibshirani 1993).

The model just described is an *observation equation* (Possolo and Toman 2007; Forbes and Sousa 2011), where the measurand, which is the function that takes the value $p = \alpha \exp(-\beta/T)$ at T, is determined not by the data directly, but by parameters of the probability distribution of the data. In fact, according to the foregoing statistical model, each $\log p_i$ is like a realized value of a Gaussian random variable with mean $\log \alpha + \beta/T_i$ and standard deviation σ.

In an observation equation (or, statistical model), the measurand is a known function of the parameters of the probability distribution of the data. For another example, consider observations of the rupture stress of nominally identical alumina coupons under flexure (Possolo 2015, Example E14). The Weibull distribution is a reasonable model for the sampling variability of these observations. The characteristic strength of alumina, σ_C, is the scale parameter of this distribution, and the mean strength is a known function of σ_C and of the distribution's shape parameter, α: $\sigma_C \Gamma(1 + 1/\alpha)$, where "$\Gamma$" denotes the gamma function (Askey and Roy 2010).

8 Evaluating Measurement Uncertainty

The evaluation of measurement uncertainty is another key step in ensuring measurement quality. It is contingent on a model that describes how the measurand is determined by values of other properties that will have been measured previously, or how it relates to observations made in the course of a measurement experiment.

The GUM considers only measurement equations, like $y = f(x_1, \ldots, x_n)$, and provides one technique to propagate the uncertainties associated with the inputs, $\{x_j\}$, to the output, y. This assumes that the uncertainties associated with the inputs will have been characterized previously, for which the GUM contemplates the following two modalities:

- **Type A** evaluations involve the application of statistical methods to experimental data, consistently with a measurement model—observation equations typically underlie this type of evaluations;
- **Type B** evaluations involve the elicitation of expert knowledge (from a single expert or from a group of experts, also from authoritative sources including calibration certificates, certified reference materials, and technical publications), and its expression either as fully specified probability distributions, or as summary indications of the dispersion of values of such distributions (for example, standard deviations for scalar measurands).

Evaluations of measurement uncertainty may also be classified according to whether they are performed in a *bottom-up* or *top-down* fashion (Possolo and Iyer 2017):

- **Bottom-Up** evaluations involve (i) the complete enumeration of all relevant sources of uncertainty, (ii) a description of how they contribute to the uncertainty of the measurement result, and (iii) the quantification of the contributions they make to the uncertainty of the result. These elements are often summarized in an uncertainty budget, for example in Table 1.
- **Top-Down** evaluations are based on inter-comparisons of measurement results for the same measurand obtained in independent experiments, typically undertaken in different laboratories, for example to determine a consensus value for the Newtonian constant of gravitation, as depicted in Fig. 6.

Table 1 Uncertainty budget, and probability distributions used in the Monte Carlo evaluation of the uncertainty associated with a measurement of air speed using a Pitot tube

Input	Estimate	std. uncertainty	Model
Δ	1.993 kPa	0.0125 kPa	Gaussian
T	292.8 K	0.11 K	Gaussian
p	101.4 kPa	1.05 kPa	Gaussian

8.1 *NIST Uncertainty Machine*

Since the mass density ρ of air, in the measurement equation for air velocity $v = \sqrt{2\Delta/\rho}$, is usually estimated by application of the ideal gas law, the measurement equation becomes $v = \sqrt{2\Delta R_s T/p}$, where p and T denote the air pressure and temperature, respectively, and $R_s = 287.058\,\mathrm{J\,kg^{-1}\,K^{-1}}$ is the specific gas constant for dry air, whose associated relative uncertainty is negligible by comparison with the relative uncertainties associated with the other inputs.

Table 1 lists the uncertainty budget for the evaluation of the uncertainty, $u(v)$, associated with the estimate of airspeed. This may be done using the *NIST Uncertainty Machine* (NUM), available at https://uncertainty.nist.gov (Lafarge and Possolo

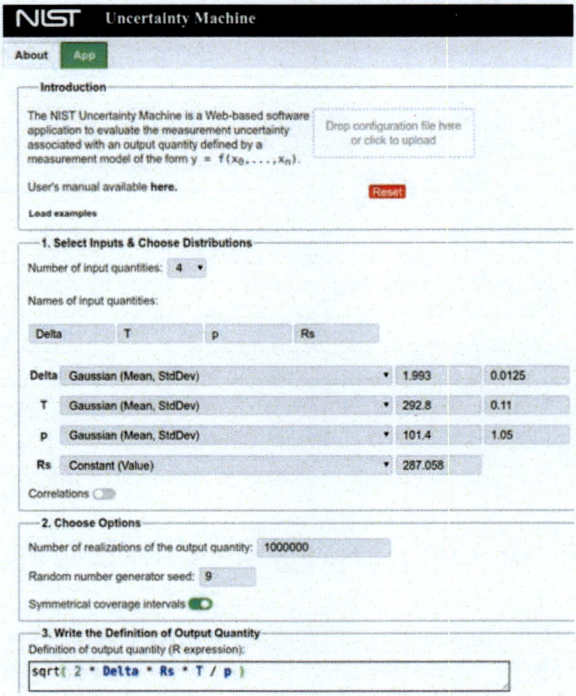

Fig. 4 Input Web page for the NIST Uncertainty Machine to evaluate the uncertainty associated with the value of airspeed measured using a Pitot tube

Fig. 5 Evaluation of the uncertainty associated with the value of airspeed measured using a Pitot tube, as produced by the NIST Uncertainty Machine

```
===== RESULTS ==============================

Monte Carlo Method

Summary statistics for sample of size 1000000

ave     = 57.483
sd      = 0.348
median  = 57.481
mad     = 0.35

Symmetrical coverage intervals

99% ( 56.5826,   58.3826)         k =      2.6
95% ( 56.8016,   58.1636)         k =        2
90% ( 56.9106,   58.0546)         k =      1.6
68% ( 57.1366,   57.8286)         k =     0.99

ANOVA (% Contributions)

            w/out Residual w/ Residual
Delta                26.89       26.89
T                     0.09        0.09
p                    73.02       73.02
Residual                NA        0.01

--------------------------------------------

Gauss's Formula (GUM's Linear Approximation)

       y   = 57.48
       u(y) = 0.348

             SensitivityCoeffs Percent.u2
Delta                   14.000     27.000
T                        0.098      0.096
p                       -0.280     73.000
Rs                       0.100      0.000
Correlations                NA      0.000
============================================
```

2015), which does it in two different ways: by application of the conventional formula in the GUM (Eq. (10), (JCGM 2008), apparently first used by Gauss (1823), often referred to as the Delta method (Casella and Berger 2002); or by application of a Monte Carlo method Morgan and Henrion (1992), Joint Committee for Guides in Metrology (2008). Figure 4 shows a screenshot of the inputs used by the NUM, and Fig. 5 shows the corresponding outputs.

9 Mutual Consistency and Consensus Building

The trustworthiness of measurement is bolstered considerably by the agreement between measurement results that essentially different measurement methods may produce for the same measurand. Such agreement can be gauged by comparing the

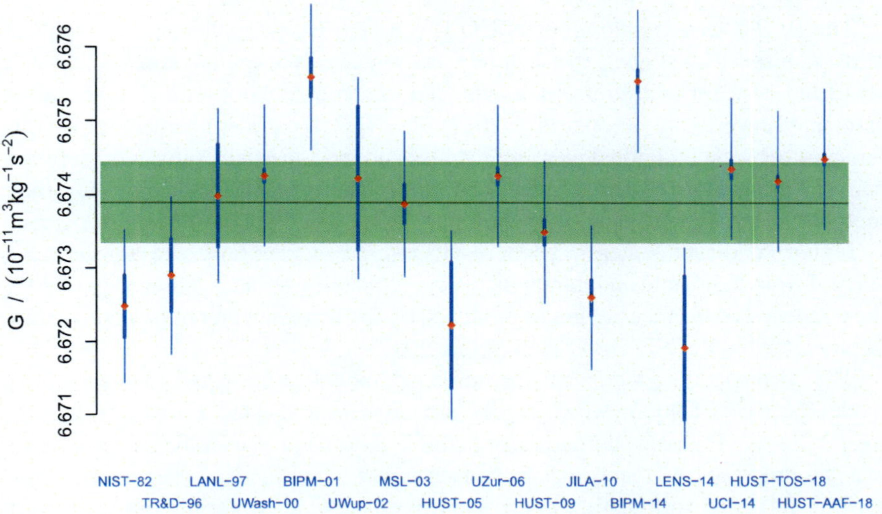

Fig. 6 The red diamonds represent the measured values $\{G_j\}$. The vertical, thick line segments represent the associated standard uncertainties (1-sigma intervals), $\{G_j \pm u(G_j)\}$. The labels along the horizontal axis describe the provenance of the measurement results and are the same used in Table 1 of Merkatas et al. (2019). The thin lines that extend the thick lines represent the contribution from dark uncertainty. The horizontal, dark green line represents the consensus value produced by the Bayesian procedure implemented in the NICOB, and the thickness of the horizontal, light green band, represents a corresponding, 95% credible interval for the true value of the measurand, G

variability between measured values produced by different experiments conducted independently, with the reported uncertainties associated with these measured values.

The collection of measurement results for the Newtonian constant of gravitation, G, that are depicted in Fig. 6 (and are listed in Table 1 of Merkatas et al. 2019) are mutually inconsistent: the standard deviation of the 16 measured values of G is about 4 times larger than the median of the reported uncertainties associated with them. That is, the measured values are much more dispersed than what their associated uncertainties suggest they should be. Cochran's Q test is often used for a formal assessment of homogeneity, even if it suffers from important shortcomings (Hoaglin 2016).

This excess variance is sometimes characterized as an expression of *dark uncertainty* (Thompson and Ellison 2011), so-called because it becomes apparent only once independent measurement results are compared. Detecting and quantifying dark uncertainty is part and parcel of the process improvement that leads to trustworthy measurement. Recognizing the presence of dark uncertainty is the first, necessary step toward eventually controlling and abating the corresponding sources.

But even when measurement results for the same measurand are mutually inconsistent, it is often useful to combine them into a consensus value, provided that the uncertainty associated with this value will reflect not only the individual reported uncertainties but also the dark uncertainty that seems to affect them all.

The *NIST Consensus Builder* (NICOB) is a Web-based application, available at https://consensus.nist.gov/, that can be used to assess the mutual consistency of a set of independent measurement results, and to compute a consensus value and its associated uncertainty (Koepke et al. 2017). The NICOB can also characterize the differences between individual measured values and the consensus value, and between pairs of individual measured values, all along taking into account not only their reported uncertainties but also any dark uncertainty that may have been uncovered.

Figure 6 depicts the measurement results under consideration, a consensus value derived from them, and an interval that, with probability 95%, is believed to include the true value of the measurand, as produced by the Bayesian hierarchical procedure implemented in the NICOB (Koepke et al. 2017).

The corresponding statistical measurement model is the usual random effects model (Searle et al. 2006), which expresses each measured value, $G_j = G + \lambda_j + \varepsilon_j$ for $j = 1, \dots, 16$, as an additive superposition of the true value of the measurand, G, of a random effect, λ_j, that is specific to each experiment, and of a measurement error, ε_j. The Bayesian formulation is particularly effective at capturing and expressing the variance component induced by dark uncertainty, which is τ^2, the variance of the $\{\lambda_j\}$, and also the fact that the estimate of τ is based on a fairly small number of degrees of freedom (15 in this case, which is one less than there are measurement results).

10 Summation and Conclusions

Measurement quality is its trustworthiness, which is achieved through reliable calibration involving well-characterized standards, traceability to the international system of units or to other generally recognized standards, rigorous evaluation of measurement uncertainty that realistically captures contributions from all significant sources of uncertainty, and fitness for purpose.

When the same measurand is measured using fundamentally different measurement methods, applied independently by different laboratories, and the measurement results are mutually consistent, then their agreement offers considerable reassurance that the measurement results indeed are trustworthy, and that all are targeting the same measurand.

The *NIST Consensus Builder* (NICOB) serves to assess whether measurement results obtained independently are mutually consistent, and will produce a consensus value even when they are not. In this case, the uncertainty associated with the consensus value includes a component of dark uncertainty that the NICOB also estimates and propagates.

Statistical models and methods are essential tools to gauge and ascertain the trustworthiness of measurements. They figure preeminently in uncertainty evaluation, for example, in the procedures implemented in the *NIST Uncertainty Machine* (Delta method and Monte Carlo method), and they provide the foundation and technical machinery for consensus building, and in particular to decide whether measurement results are mutually consistent.

The reliance on standards agreed upon by the metrological community, and the confidence derived from the mutual agreement of measurements made independently of one another, possibly also employing different methods, show that measurement in fact is a collective enterprise at multiple scales of engagement between individuals and nations worldwide, hence also is a vehicle for cooperation and an enabler of commerce and trade.

References

Askey, R. A., & Roy, R. (2010). Gamma function. In Olver, F. W. J., Lozier, D. W., Boisvert, R. F., Clark, C. W. (eds.), NIST handbook of mathematical functions. Cambridge: Cambridge University Press

Bartel, T. (2005). Uncertainty in NIST force measurements. *Journal of Research of the National Institute of Standards and Technology*, *110*(6), 589–603.

Bartel, T., Stoudt, S., & Possolo, A. (2016). Force calibration using errors-in-variables regression and Monte Carlo uncertainty evaluation. *Metrologia*, *53*(3), 965–980. https://doi.org/10.1088/0026-1394/53/3/965.

Bell, S. (1999). A Beginner's Guide to Uncertainty of Measurement, Measurement Good Practice Guide, vol. 11 (Issue 2). National Physical Laboratory, Teddington, Middlesex, United Kingdom, www.npl.co.uk/publications/guides/a-beginners-guide-to-uncertainty-of-measurement, amendments March 2001

BIPM. (2019). The international system of units (SI), 9th edn. International bureau of weights and measures (BIPM), Sèvres, France. https://www.bipm.org/en/publications/si-brochure/

Bogen, J., & Woodward, J. (1988). Saving the phenomena. *The Philosophical Review*, *97*, 303–352. https://doi.org/10.2307/2185445.

Bogue, R. H. (1929). Calculation of the compounds in portland cement. *Industrial and Engineering Chemistry Analytical Edition*, *1*(4), 192–197. https://doi.org/10.1021/ac50068a006.

Casella, G., & Berger, R. L. (2002). *Statistical inference* (2nd edn.). Pacific Grove, California: Duxbury.

Ciddor, P. E. (1996). Refractive index of air: New equations for the visible and near infrared. *Applied Optics*, *35*(9), 1566–1573. https://doi.org/10.1364/AO.35.001566.

CODATA. (2019). 2018 CODATA recommended values of the fundamental physical constants. https://physics.nist.gov/cuu/Constants/bibliography.html. Retrieved May 21st 2019. World Metrology Day

Efron, B., & Tibshirani, R. J. (1993). *An introduction to the bootstrap*. London: Chapman & Hall.

Forbes, A. B., & Sousa, J. A. (2011). The GUM, Bayesian inference and the observation and measurement equations. *Measurement*, *44*(8), 1422–1435. https://doi.org/10.1016/j.measurement.2011.05.007.

Gauss, C. (1823). Theoria combinationis observationum erroribus minimis obnoxiae. In: Werke, Band IV, Wahrscheinlichkeitsrechnung und Geometrie, Könighliche Gesellschaft der Wissenschaften, Göttingen, http://gdz.sub.uni-goettingen.de

Gonzalez, C. A., & Choquette, S. J. (2018) Standard Reference Material 158a, Silicon Bronze (chip form). Office of Reference Materials, National Institute of Standards and Technology, Department of Commerce, Gaithersburg, Maryland. www.nist.gov/srm/

Hoaglin, D. C. (2016). Misunderstandings about Q and 'Cochran's Q test' in meta-analysis. *Statistics in Medicine*, *35*, 485–495. https://doi.org/10.1002/sim.6632.

Hudson, R. (2018). The reality of jean perrin's atoms and molecules. *The British Journal for the Philosophy of Science*,. https://doi.org/10.1093/bjps/axx054.

Anderson, J. D. (2017). *Fundamentals of Aerodynamics* (6th edn.). New York: McGraw-Hill Education.

Jabbour, Z. L., & Yaniv, S. L. (2001). The kilogram and measurements of mass and force. *Journal of Research of the National Institute of Standards and Technology*, *106*(1), 25–46.

Joint Committee for Guides in Metrology. (2008). Evaluation of measurement data — Supplement 1 to the "Guide to the expression of uncertainty in measurement" — Propagation of distributions using a Monte Carlo method. International Bureau of Weights and Measures (BIPM), Sèvres, France. www.bipm.org/en/publications/guides/gum.html, BIPM, IEC, IFCC, ILAC, ISO, IUPAC, IUPAP and OIML, JCGM 101:2008

Joint Committee for Guides in Metrology (JCGM). (2008). Evaluation of measurement data — Guide to the expression of uncertainty in measurement. International Bureau of Weights and Measures (BIPM), Sèvres, France, www.bipm.org/en/publications/guides/um.html. BIPM, IEC, IFCC, ILAC, ISO, IUPAC, IUPAP and OIML, JCGM 100:2008, GUM 1995 with minor corrections

Klein, C., & Dutrow, B. (2007). *Manual of mineral science* (23rd edn.). Hoboken: Wiley.

Koepke, A., Lafarge, T., Toman, B., Possolo, A. (2017). NIST Consensus Builder — User's Manual. National Institute of Standards and Technology, Gaithersburg, MD. https://consensus.nist.gov

Komsta, L. (2019). mblm: Median-Based Linear Models. https://CRAN.R-project.org/package=mblm, r package version 0.12.1

Lafarge, T., & Possolo, A. (2015). The NIST uncertainty machine. *NCSLI Measure Journal of Measurement Science*, *10*(3), 20–27.

Lavagnini, I., & Magno, F. (2007). A statistical overview on univariate calibration, inverse regression, and detection limits: Application to gas chromatography/mass spectrometry technique. *Mass Spectrometry Reviews*, *26*(1), 1–18. https://doi.org/10.1002/mas.20100.

MacKay, R. J., & Oldford, R. W. (2000). Scientific method, statistical method and the speed of light. *Statistical Science*, *15*(3), 254–278.

Merkatas, C., Toman, B., Possolo, A., & Schlamminger, S. (2019). Shades of dark uncertainty and consensus value for the Newtonian constant of gravitation. arXiv:1905.09551

Mints, G. E. (2018). Peano axioms. In: Encyclopedia of Mathematics, Springer & European Mathematical Society. https://www.encyclopediaofmath.org/index.php/Peano_axioms. Retrieved December 1, 2018

Mohr, P. J., Newell, D. B., Taylor, B. N., & Tiesinga, E. (2018). Data and analysis for the CODATA 2017 special fundamental constants adjustment. *Metrologia*, *55*, 125–146. https://doi.org/10.1088/1681-7575/aa99bc.

Morgan, M. G., Henrion, M. (1992). *Uncertainty — a guide to dealing with uncertainty in quantitative risk and policy analysis*, first paperback edn. New York,: Cambridge University Press. 10th printing, 2007.

Paule, R. C., Mandel, J. (1970). *Analysis of interlaboratory measurements on the vapor pressure of gold (Certification of standard reference material 745)*. Washington: National Bureau of Standards. special Publication 260-19.

Perrin, J. (1916). *Atoms*. New York: D. Van Nostrand Company. Translation of 4th Revised French Edition by D. L. Hammick

Possolo, A. (2015). Simple Guide for evaluating and expressing the uncertainty of NIST measurement results. National Institute of Standards and Technology, Gaithersburg, MD. http://dx.doi.org/10.6028/NIST.TN.1900, NIST Technical Note 1900

Possolo, A. (2018). Measurement. In: Forbes, A. B., Zhang, N. F., Chunovkina, A., Eichstädt, S., Pavese, F. (eds.) Advanced mathematical and computational tools in metrology and testing: AMCTM XI, Series on advances in mathematics for applied sciences (vol. 89, pp. 273–285) Singapore: World Scientific Publishing Company. https://doi.org/10.1142/9789813274303_0027

Possolo, A., & Iyer, H. K. (2017). Concepts and tools for the evaluation of measurement uncertainty. *Review of Scientific Instruments*, *88*(1), 011301. https://doi.org/10.1063/1.4974274.

Possolo, A., & Toman, B. (2007). Assessment of measurement uncertainty via observation equations. *Metrologia*, *44*, 464–475. https://doi.org/10.1088/0026-1394/44/6/005.

Possolo, A., Schlamminger, S., Stoudt, S., Pratt, J. R., & Williams, C. J. (2018). Evaluation of the accuracy, consistency, and stability of measurements of the Planck constant used in the redefinition of the international system of units. *Metrologia*, *55*, 29–37. https://doi.org/10.1088/1681-7575/aa966c.

Quinn, T. (2012). *From Artefacts to Atoms: The BIPM and the Search for Ultimate Measurement Standards*. New York: Oxford University Press.

Searle, S. R., Casella, G., & McCulloch, C. E. (2006). *Variance components*. Hoboken, NJ: Wiley.

Stone, J. A., Zimmerman, J. H. (2004). Index of Refraction of Air: Vacuum Wavelength and Ambient Conditions based on Ciddor Equation. https://emtoolbox.nist.gov/Wavelength/Ciddor.asp. Engineering Metrology Toolbox. Retrieved on June 3rd, 2019

Stutzman, P. E., Feng, P., & Bullard, J. W. (2016). Phase analysis of Portland cements by combined quantitative X-ray powder diffraction and scanning electron microscopy. *Journal of Research of the National Institute of Standards and Technology*, *121*, 47–107. https://doi.org/10.6028/jres.121.004.

Thompson, M., & Ellison, S. L. R. (2011). Dark uncertainty. *Accreditation and Quality Assurance*, *16*, 483–487. https://doi.org/10.1007/s00769-011-0803-0.

WHO Expert Committee on Biological Standardization. (2006). Recommendations for the preparation, characterization and establishment of international and other biological reference standards. Technical Report 932, World Health Organization, Geneva, Switzerland, annex 2 of 55th Report

Wilcox, R. R. (2010). *Fundamentals of modern statistical methods: Substantially improving power and accuracy* (2nd edn.). Berlin: Springer.

Assessing Laboratory Effects in Key Comparisons with Two Transfer Standards Measured in Two Petals: A Bayesian Approach

Olha Bodnar and Clemens Elster

Abstract We propose a new statistical method for analyzing data from a key comparison when two transfer standards are measured in two petals. The approach is based on a generalization of the classical random effects model, a popular procedure in metrology. Bayesian treatment of the model parameters, as well as of the random effects is suggested. The latter can be viewed as potential laboratory effects which are assessed through the proposed analysis. While the prior for the laboratory effects naturally is assigned as a Gaussian distribution, the Berger and Bernardo reference prior is taken for the remaining model parameters. The results are presented in terms of the posterior distributions derived for the laboratory effects. From these distributions, posterior means and credible intervals are calculated. The proposed method paves the way for applying the established random effects model also for data arising from the measurement of several transfer standards in several petals. Finally, the new approach is illustrated for measurements of two 500 mg transfer standards carried out in key comparison CCM.M-K7.

Keywords Extended random effects model · Two transfer standards · Reference analysis · Non-informative prior · Credible interval · CCM.M-K7

1 Introduction

Meta-analysis is an important statistical tool which deals with drawing inferences from data that are themselves the results of analyses. Applications of meta-analysis comprise, for example, the combination of results from clinical trials (see, Sutton and

O. Bodnar (✉)
Unit of Statistics, School of Business, Örebro University, Fakultetsgatan 1, 70182 Örebro, Sweden
e-mail: olha.bodnar@oru.se

C. Elster
Physikalisch-Technische Bundesanstalt,, Abbestrasse 2-12 10587 Berlin, Germany
e-mail: clemens.elster@ptb.de

S. Knoth and W. Schmid (eds.), *Frontiers in Statistical Quality Control 13*, Frontiers in Statistical Quality Control,
https://doi.org/10.1007/978-3-030-67856-2_20

Higgins 2008; Bodnar et al. 2017), the determination of fundamental constants (see, Mohr et al. 2012; Bodnar et al. 2016a), or the analysis of interlaboratory comparisons (see, Toman 2007).

In order to establish the equivalence between national metrology institutes, interlaboratory comparison measurements known as key comparisons are carried (cf., Bureau International des Poids et Mesures 2003). The statistical analysis of data from key comparisons has been recently developed in a number of papers (see, e.g., Kacker 2004; Chunovkina et al. 2008; Elster and Toman 2010; Bodnar et al. 2013; Elster and Toman 2013; Chunovkina et al. 2016; Forbes 2016; Shirono et al. 2016; Koepke et al. 2017) for both the fixed effects model and the random effects model. Bayesian methods were established for the determination of a reference value on basis of a random effects model by Bodnar et al. (2016b), and also for the estimation of laboratory effects using a fixed effects model by Elster and Toman (2010). Recently, the estimation of laboratory effects has been explored when applying a random effects model in Rukhin and Possolo (2011); Bodnar and Elster (2018); while this has been done from the viewpoint of conventional statistics in Rukhin and Possolo (2011), a full Bayesian treatment based on a noninformative prior has been proposed in Bodnar and Elster (2018). Laboratory effects, i.e. the fixed effects in a fixed effects model or the random effects in a random effects model, are key results in such a statistical analysis (cf., Toman and Possolo 2009).

The above mentioned approaches are applicable to measurement results obtained in a single petal. However, in some key comparisons, several transfer standards are circulated among participants in separate petals, and only the pilot laboratory participates in each petal (see, e.g., Abbott et al. 2015; Lee et al. 2017). One such example is key comparison CCM.M-K7, where measurements of 5 kg, 100 g, 10 g, 5 g and 500 mg stainless steel mass standards have been carried out. In Fig. 1, the measurement data for the 500 mg stainless steel mass standard are presented which were obtained in two petals. The first petal consists of measurements provided by KRISS, NIS, VNIIM, CENAM, NIST, while the second petal shows measurements of KRISS, PTB, METAS, CEM, INRIM, NIM. A separate transfer standard was used in each petal, and the pilot laboratory in this key comparison was KRISS. Although two measurements were done by KRISS in each of the two petals, we used only the first ones in order to keep the same conditions for all participated laboratories. The data in Fig. 1 consist of the differences between the measurement result and the nominal level of the transfer standard, presented together with the corresponding standard uncertainties provided by the laboratories.

The difficulty in the analysis of the CCM.M-K7 data in Fig. 1 is to combine the measurements obtained from the two different petals in a meaningful way. The results of key comparison CCM.M-K7 reported in the BIPM KCDB (see, Lee et al. 2017) are based on the differences between the values provided by the participating laboratories and the value measured by the pilot laboratory in the same petal. However, proceeding in such a way has several drawbacks: (i) after this transformation, some important information present in the initial data may be lost; for example, no laboratory effect is obtained for the pilot laboratory; (ii) when the initial observations measured by the different laboratories can be assumed to have been obtained independently, this

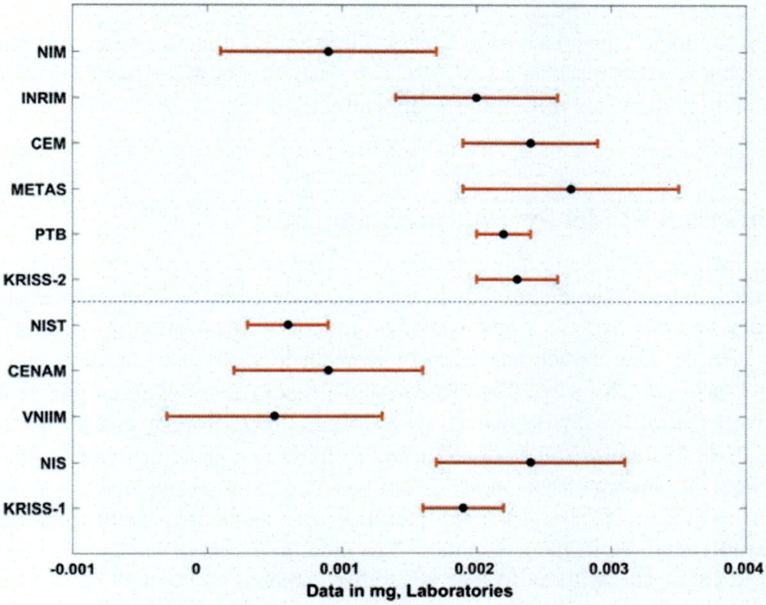

Fig. 1 Measurement data from CCM.M-K7 with two 500 mg transfer standards measured in two petals. Error bars indicate standard uncertainties. The dashed line in the figure separates the two petals. The pilot laboratory, KRISS, participated in both petals

does no longer hold for the transformed data. As a result, uncertainties are enlarged, correlations are introduced, and a statistical analysis for the transformed data may be less informative. An additional challenge appears when the measurement data in one or both petals are inconsistent. For instance, it was pointed out that the measurements are not consistent in four out of five comparisons considered in key comparison CCM.M-K7 Lee et al. (2017).

We contribute to the existing literature on the analysis of data from key comparisons by suggesting a novel statistical approach for combining results from two petals. The approach is based on an extension of the random effects model and its Bayesian treatment. Under this model, the Berger and Bernardo reference prior (see, Berger and Bernardo 1992; Bodnar and Elster 2014) for the underlying means in the two petals and the heterogeneity parameter are derived as well as the corresponding posterior for the heterogeneity parameter and the random effects is obtained. In this way, Bayesian inference procedures are provided for the underlying means, the additional between-laboratory variability, and the laboratory effects. The theoretical results are applied to reanalyze part of the key comparison CCM.M-K7 data recently published in Lee et al. (2017).

The paper is organized as follows. In Sect. 2, a new statistical model is introduced for key comparison data with two transfer standards and measurements performed in two petals. The Bayesian inference procedures proposed for the treatment of this statistical model are then derived in Sect. 3, while Sect. 4 illustrates their application to data from key comparison CCM.M-K7. Final remarks are provided in Sect. 5. Technical derivations are moved to the appendix (Sect. 6).

2 Statistical Model for Measurement Data

In order to model measurement data in the case of two transfer standards measured in two petals, we introduce a new statistical model in this section that extends the random effects model which has recently been successfully used in metrology and medicine (see, e.g., Kacker 2004; Toman and Possolo 2009; Rukhin and Possolo 2011; Turner et al. 2015; Bodnar et al. 2016b, a, 2017; Bodnar and Elster 2018; Bodnar 2019; Muhmuza and Bodnar 2020). Without such an adjustment application of the classical random effects model is not possible in the considered case.

Let $\boldsymbol{X} = (X_1, ..., X_n)^T$ and $\boldsymbol{Y} = (Y_1, ..., Y_m)^T$ denote the two vectors of measurement results obtained in the two petals. The measurements of the pilot laboratory in the two petals are, without loss of generality, denoted by X_1 and Y_1. In contrast to existing approaches which suggest to link the measurement results from the two petals by building the differences between the measurements provided by the participating laboratories and those of the pilot laboratory, we propose a new statistical model which allows to combine the measurement data in an appealing way and avoids the problems discussed in the introduction when differences are calculated. In particular, our approach arrives at an estimate of the laboratory effect also for the pilot laboratory.

More precisely, we assume that $\boldsymbol{X}$ and $\boldsymbol{Y}$ follow an extended random effects model expressed as

$$\boldsymbol{X} = \mu_X \mathbf{1}_n + \boldsymbol{\lambda}_X + \boldsymbol{\varepsilon}_X \,, \tag{1}$$

$$\boldsymbol{Y} = \mu_Y \mathbf{1}_m + \tilde{\boldsymbol{\lambda}}_Y + \boldsymbol{\varepsilon}_Y \,, \tag{2}$$

where $\mathbf{1}_k$ denotes the k-dimensional vector of ones. We further assume that the first random effects in both petals coincide, i.e., the random effects of the pilot laboratory are the same. This means that any potential laboratory effect of the pilot laboratory which is not accounted for in his uncertainty budget is assumed to remain constant.

Following the assumptions of the classical random effects model, $\boldsymbol{\lambda}_X = (\lambda_{1,X}, ..., \lambda_{n,X})^T$ and $\boldsymbol{\lambda}_Y = (\lambda_{2,Y}, ..., \lambda_{m,Y})^T$ are assumed to be independently distributed with

$$\boldsymbol{\lambda}_X | \sigma \sim N_n(\mathbf{0}_n, \sigma^2 \boldsymbol{I}_n) \,, \tag{3}$$

$$\boldsymbol{\lambda}_Y | \sigma \sim N_{m-1}(\mathbf{0}_{m-1}, \sigma^2 \boldsymbol{I}_{m-1}) \,, \tag{4}$$

where $\mathbf{0}_k$ denotes the k-dimensional vector of zeros and $\boldsymbol{I}_k$ stands for the k-dimensional identity matrix. Later on, we also use the notation $\mathbf{O}_{k,p}$ for the $k \times p$ dimensional matrix of zeros. Furthermore, the model residuals are assumed to be normally distributed, but not obviously independent, according to

$$\boldsymbol{\varepsilon} = \begin{pmatrix} \boldsymbol{\varepsilon}_X \\ \boldsymbol{\varepsilon}_Y \end{pmatrix} \sim N_{n+m}(\mathbf{0}_{n+m}, \boldsymbol{V}) \quad \text{with} \quad \boldsymbol{V} = \begin{pmatrix} \boldsymbol{V}_{11} & \boldsymbol{V}_{12} \\ \boldsymbol{V}_{21} & \boldsymbol{V}_{22} \end{pmatrix}. \tag{5}$$

The covariance matrix $\boldsymbol{V}$ is assumed to be positive definite and it is formed from the uncertainties quoted by the laboratories, together with the assessment made about their correlation. Subsequently, $\boldsymbol{V}$ is treated as known and the dependence of the results on it will be suppressed in our notation.

Summarizing (1)–(5), we obtain the following extended random effects model expressed as

$$\begin{pmatrix} \boldsymbol{X} \\ \boldsymbol{Y} \end{pmatrix} = \begin{pmatrix} \mu_X \mathbf{1}_n \\ \mu_Y \mathbf{1}_m \end{pmatrix} + \boldsymbol{L} \begin{pmatrix} \boldsymbol{\lambda}_X \\ \boldsymbol{\lambda}_Y \end{pmatrix} + \boldsymbol{\varepsilon}, \tag{6}$$

where $\boldsymbol{L} : (n+m) \times (n+m-1)$ matrix which transforms $(\boldsymbol{\lambda}_X^T, \boldsymbol{\lambda}_Y^T)^T$ into $(\boldsymbol{\lambda}_X^T, \tilde{\boldsymbol{\lambda}}_Y^T)^T$ and it is given by

$$\boldsymbol{L} = \begin{pmatrix} \boldsymbol{I}_n & \mathbf{O}_{n,m-1} \\ \mathbf{i}_n^T & \mathbf{0}_{m-1}^T \\ \mathbf{O}_{m-1,n} & \boldsymbol{I}_{m-1} \end{pmatrix} \quad \text{with} \quad \mathbf{i}_n^T = (1, \underbrace{0, \ldots, 0}_{n-1}).$$

From model (6), we arrive at the marginal model

$$\begin{pmatrix} \boldsymbol{X} \\ \boldsymbol{Y} \end{pmatrix} \sim N_{n+m} \left[\begin{pmatrix} \mu_X \mathbf{1}_n \\ \mu_Y \mathbf{1}_m \end{pmatrix}, \boldsymbol{V} + \sigma^2 \boldsymbol{L} \boldsymbol{L}^T \right]. \tag{7}$$

3 Bayesian Inference Based on the Reference Prior

In this section, we present the Bayesian inference procedures for the parameters of model (6), namely $\{\mu_X, \mu_Y, \sigma\}$, as well as for $\boldsymbol{\lambda} = (\boldsymbol{\lambda}_X^T, \boldsymbol{\lambda}_Y^T)^T$. We will mainly concentrate on the derivation of the posterior distribution of $\boldsymbol{\lambda}$, while considering $\{\mu_X, \mu_Y, \sigma\}$ as nuisance parameters. This approach will allow us directly to estimate laboratory effects, which are important outputs in the analysis of key comparison data.

In our Bayesian treatment of the random effects model (6), $\boldsymbol{\lambda} = (\boldsymbol{\lambda}_X^T, \boldsymbol{\lambda}_Y^T)^T$ are treated as parameters to be inferred, and formally our treatment is that of a fixed effects model. The fact that by assumption the random effects are drawn from the Gaussian distribution (3) and (4) lends itself naturally to taking that distribution as the prior for $\boldsymbol{\lambda}$. The prior distribution for the remaining parameters, σ, μ_X, μ_Y, is then determined by the reference prior for the marginal model (7).

3.1 Reference Prior for the Marginal Model

In the derivation of the Berger and Bernardo reference prior (see, Berger and Bernardo 1992; Bodnar and Elster 2014), the Fisher information matrix calculated for model (7) is utilized. For grouping $\{\{\mu_X, \mu_Y\}, \sigma\}$, it is given by (see Sect. 6.1 in the appendix)

$$\boldsymbol{F} = \begin{pmatrix} \boldsymbol{K}^T(\boldsymbol{V}+\sigma^2\boldsymbol{L}\boldsymbol{L}^T)^{-1}\boldsymbol{K} & \mathbf{0}_2 \\ \mathbf{0}_2^T & 2\sigma^2 tr\left((\boldsymbol{L}^T(\boldsymbol{V}+\sigma^2\boldsymbol{L}\boldsymbol{L}^T)^{-1}\boldsymbol{L})^2\right) \end{pmatrix} \tag{8}$$

with

$$\boldsymbol{K} = \begin{pmatrix} \mathbf{1}_n & \mathbf{0}_n \\ \mathbf{0}_m & \mathbf{1}_m \end{pmatrix}. \tag{9}$$

In using (8), the reference prior is obtained and it is given as the square root of the right corner of the Fisher information matrix (see, Corollary to Proposition 5.29 in Bernardo and Smith 2000) expressed as

$$\begin{aligned} \pi(\mu_X, \mu_Y, \sigma) &\propto \sigma\sqrt{tr\left((\boldsymbol{L}^T(\boldsymbol{V}+\sigma^2\boldsymbol{L}\boldsymbol{L}^T)^{-1}\boldsymbol{L})^2\right)} \\ &= \sigma\sqrt{tr\left(\left(\boldsymbol{I}_{n+m-1}+\sigma^2\boldsymbol{L}^T\boldsymbol{V}^{-1}\boldsymbol{L}\right)^{-1}\boldsymbol{L}^T\boldsymbol{V}^{-1}\boldsymbol{L}\right)^2}. \end{aligned} \tag{10}$$

3.2 Posterior for Laboratory Effects

Because the derived expression of the reference prior (10) depends on σ only, in the following we write $\pi(\sigma)$ instead of $\pi(\mu_X, \mu_Y, \sigma)$. Altogether, the prior

$$\pi(\mu_X, \mu_Y, \sigma, \boldsymbol{\lambda}_X, \boldsymbol{\lambda}_Y) = \pi(\boldsymbol{\lambda}_X, \boldsymbol{\lambda}_Y|\sigma)\pi(\sigma) \tag{11}$$

is taken, where $\pi(\boldsymbol{\lambda}_X, \boldsymbol{\lambda}_Y|\sigma) = \pi(\boldsymbol{\lambda}_X|\sigma)\pi(\boldsymbol{\lambda}_Y|\sigma)$ is given by (3) and (4), and $\pi(\sigma)$ by (10).

Let

$$\boldsymbol{R} = \boldsymbol{V}^{-1} - \boldsymbol{V}^{-1}\boldsymbol{K}(\boldsymbol{K}^T\boldsymbol{V}^{-1}\boldsymbol{K})^{-1}\boldsymbol{K}^T\boldsymbol{V}^{-1}.$$

Then, the conditional posterior for $\boldsymbol{\lambda}$ given σ is expressed as (see, Sect. 6.2 in the appendix)

$$\boldsymbol{\lambda}|\sigma, \boldsymbol{X}, \boldsymbol{Y} \sim N_{n+m-1}\left(\boldsymbol{\mu}_{\boldsymbol{\lambda}|\sigma}, \boldsymbol{V}_{\boldsymbol{\lambda}|\sigma}\right) \tag{12}$$

with

$$\boldsymbol{\mu}_{\boldsymbol{\lambda}|\sigma} = \left(\boldsymbol{L}^T\boldsymbol{R}\boldsymbol{L}+\frac{1}{\sigma^2}\boldsymbol{I}_{n+m-1}\right)^{-1}\boldsymbol{L}^T\boldsymbol{R}\begin{pmatrix}\boldsymbol{X}\\ \boldsymbol{Y}\end{pmatrix}, \quad \boldsymbol{V}_{\boldsymbol{\lambda}|\sigma} = \left(\boldsymbol{L}^T\boldsymbol{R}\boldsymbol{L}+\frac{1}{\sigma^2}\boldsymbol{I}_{n+m-1}\right)^{-1},$$

and

$$\pi(\sigma|X,Y) \propto \frac{\pi(\sigma)}{\sqrt{det\left(\sigma^2 L^T V^{-1} L + I_{n+m-1}\right)}\sqrt{det(K^T(V+\sigma^2 LL^T)^{-1}K)}} \times \exp\left(-\frac{1}{2}\left(\begin{pmatrix}X\\Y\end{pmatrix}^T R \begin{pmatrix}X\\Y\end{pmatrix} - \mu_{\lambda|\sigma}^T V_{\lambda|\sigma}^{-1}\mu_{\lambda|\sigma}\right)\right). \tag{13}$$

Finally, using the properties of the multivariate normal distribution, we get the conditional marginal posteriors for each λ_i separately as

$$\lambda_i|\sigma \sim N\left(\mathbf{e}_i^T \mu_{\lambda|\sigma}, \mathbf{e}_i^T V_{\lambda|\sigma}\mathbf{e}_i\right),$$

where $\mathbf{e}_i$ is the i-th basis vector in $I\!R^{n+m-1}$ and the marginal posterior of σ is given in (13).

In the following theorem, we derive necessary conditions under which the posterior for λ is proper and its first and second moments exist.

Theorem 1 *The posterior $\pi(\lambda, \sigma|X, Y)$ obtained from (12) and (13) for the reference prior from (10) is proper if $n + m > 3$. For the according marginal posterior $\pi(\lambda|X, Y)$ mean and variance exist if $n + m > 5$ and $n + m > 7$, respectively.*

Proof The proof of the theorem follows from (10), (12), and (13) by noting that no singularity in zero is present for (13), while the marginal posterior for σ behaves as σ^{-n-m-2} for $\sigma \to \infty$.

In order to calculate the marginal posterior for the laboratory effects, one can calculate the posterior $\pi(\lambda_i|X, Y)$ by simply calculating the one-dimensional integral

$$\pi(\lambda_i|X,Y) = \int \pi(\sigma|X,Y)\frac{1}{\sqrt{2\pi \mathbf{e}_i^T V_{\lambda|\sigma}\mathbf{e}_i}} \exp\left(-\frac{1}{2}(\mathbf{e}_i^T \mu_{\lambda|\sigma} - \lambda_i)^2/\mathbf{e}_i^T V_{\lambda|\sigma}\mathbf{e}_i\right) d\sigma \tag{14}$$

numerically.

Finally, we note that the conditional posterior mean vector $\mu_{\lambda|\sigma}$ possesses an interesting interpretation. From its structure and using that R is a projection matrix on the space determined by the matrix K, we conclude that $\mu_{\lambda|\sigma}$ is close to the zero vector if and only if the measurement results X and Y almost belong to the space determined by K. The latter statement is equivalent to the case when the measurement data are similar to each other with respect to their quoted uncertainties in both petals, i.e., when the data are consistent.

4 Analysis of CCM.M-K7 Data

In this section, we apply the developed procedure to reanalyze the measurement results for two 500 mg transfer standards obtained in two petals within key comparison CCM.M-K7 (Table 1). The data are presented in Fig. 1 where two petals are separated by the dashed line. The pilot laboratory KRISS participated in both petals. Moreover, it is pointed out by applying a χ^2-test on consistency that the measurement data from the first petal (lower part in Fig. 1) are not consistent (c.f., Table 7 from Lee et al. 2017). The data together with their standard uncertainties are provided in Table 1. The matrix $\boldsymbol{V}$ is constructed in the following way: it has the squared standard uncertainties associated with the measurement results as its diagonal entries; the only nonzero non-diagonal elements are the ones which correspond to the measurements of the pilot laboratory and which are equal to 0.3 multiplied by the product of the corresponding standard uncertainties (see, Sect. 8.2 in Lee et al. 2017). To this end, we point out that the value of the correlation coefficient may have a minor impact on the coverage properties of the credible intervals constructed for laboratory effects (see, e.g., Sect. 5 in Bodnar and Elster 2018).

Key results of the suggested approach are the joint posterior and the marginal posteriors of the laboratory effects. The marginal posteriors are presented in Fig. 2. The obtained marginal posterior distributions are roughly symmetric, indicating that their approximation by a normal distribution might provide a good fit. Moreover, the constructed 95% probabilistically symmetric credible intervals always include zero (cf. also Fig. 3), which is taken as an indication that at the chosen 95% level of significance, no significant laboratory effect is present. This result is consistent with the initial analysis as documented in Table 10 of Lee et al. (2017). Note that while none of the laboratory effects differs significantly from zero at the 95% level of

Table 1 Measurement data from CCM.M-K7 with two 500 mg transfer standards measured in two petals. The dashed line in Fig. 1 separates the two petals. The pilot laboratory, KRISS, participated in both petals with the correlation between its measurements equal to 0.3

Laboratory	Measurement in mg	Standard uncertainty in mg
KRISS-1	0.0019	0.0003
NIS	0.0024	0.0007
VNIIM	0.0005	0.0008
CENAM	0.0009	0.0007
NIST	0.0006	0.0003
KRISS-2	0.0023	0.0003
PTB	0.0022	0.0002
METAS	0.0027	0.0008
CEM	0.0024	0.0005
INRIM	0.0020	0.0006
NIM	0.0009	0.0008

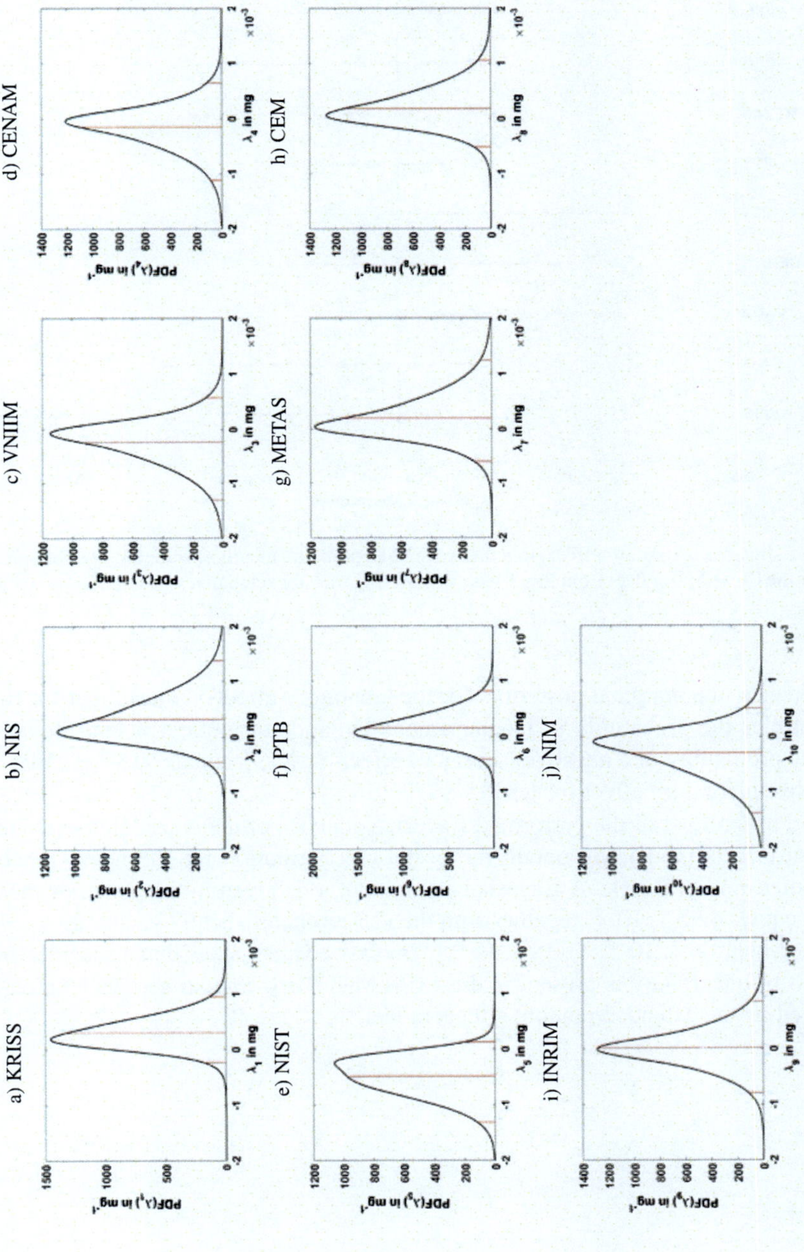

Fig. 2 Marginal posterior for laboratory effects λ_i for $i = 1, \ldots, 10$ in the CCM.M-K7 data from Fig. 1 with two 500 mg transfer standards measured in two petals. The red lines show the posterior means and the limits of 95% probabilistically symmetric credible intervals

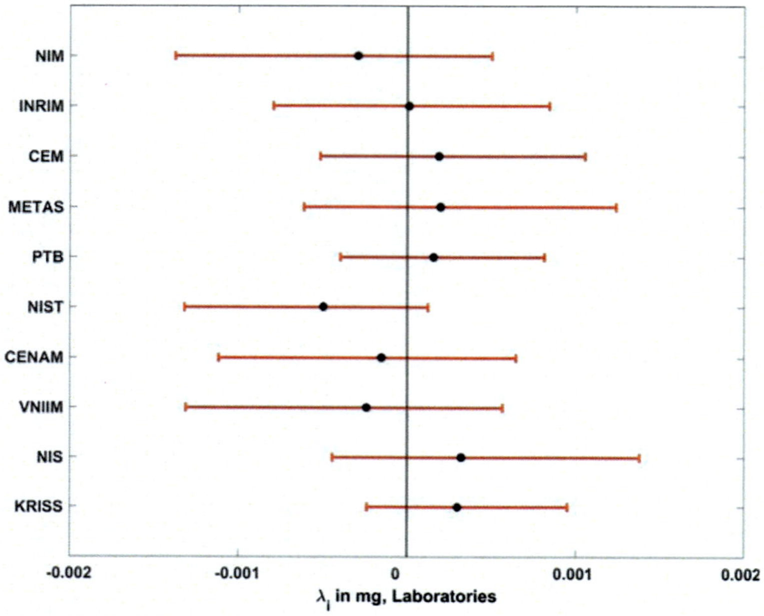

Fig. 3 Estimated laboratory effects together with 95% probabilistically symmetric credible intervals for the CCM.M-K7 data from Fig. 1 with two 500 mg transfer standards measured in two petals

significance, the marginal posteriors for the laboratory effects indicate that for two laboratories the probability of having a nonzero laboratory effect is considerable, i.e., the probability of a negative laboratory effect for NIST equals 93%, and that of a positive laboratory effect for KRISS 86%.

As a by-product of the performed Bayesian analysis of the considered extended random effects model, we obtain the posterior distribution for the heterogeneity parameter σ (see, Fig. 4). A Bayesian estimate of σ is given by its posterior mean which equals 4.95×10^{-4}, together with the 95% probabilistically symmetric credible interval $[1.06 \times 10^{-4}, 11.18 \times 10^{-4}]$. These two values show that a considerable amount of uncertainty is present in the CCM.M-K7 key comparison data which is captured by the extended random effects model.

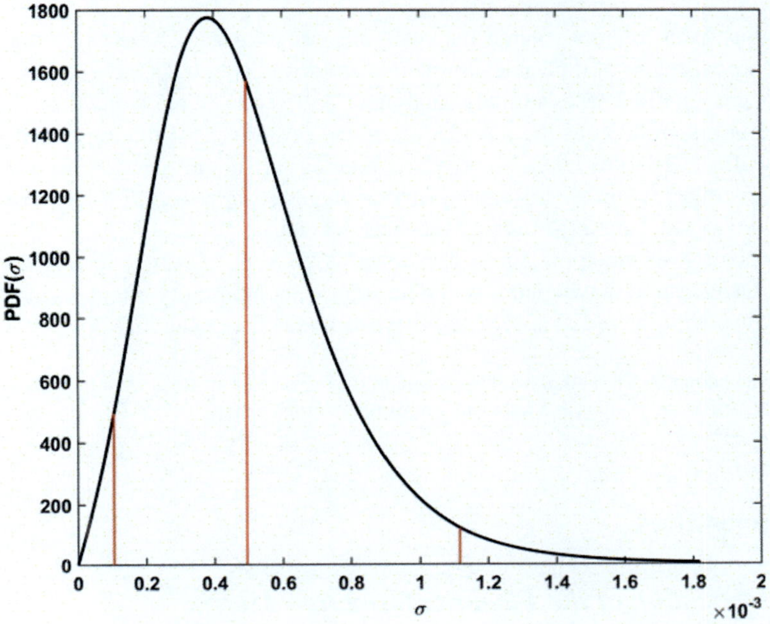

Fig. 4 Marginal posterior for heterogeneity parameter σ in the case of the CCM.M-K7 data with two 500 mg transfer standards measured in two petals. The red lines show the posterior mean and the limits of 95% probabilistically symmetric credible interval

5 Conclusions

In many key comparisons, the participating laboratories measure several transfer standards simultaneously in several petals. Linking together the measurement results obtained in the different petals is crucial in those situations. Current approaches address this task by constructing differences from the measurements of the participating laboratories and the pilot laboratory in each petal first, followed by a subsequent analysis of these differences (see, Abbott et al. 2015; Lee et al. 2017).

In this paper, an alternative approach is proposed based on an appropriate extension of the popular random effects model (see, e.g., Kacker 2004; Toman and Possolo 2009; Rukhin and Possolo 2011; Turner et al. 2015; Bodnar et al. 2016b, 2017) and its Bayesian inference. The prior distributions for the unknowns in the model were selected by utilizing the assumed Gaussian distribution of the laboratory effects, together with the Berger and Bernardo reference prior principle for the parameters of the marginal model. The Bayesian treatment results in the posterior distribution, from which the summary statistics such as the posterior means or the posterior credible intervals can be derived for the laboratory effects.

The main advantage of the novel approach is that it provides estimates for all laboratory effects, including the pilot laboratory. The method does not produce a single key comparison reference value but two consensus values. These two consensus values are estimates of the values of the two employed standards.

The new method is essentially analytical and requires only one-dimensional numerical integration for the calculation of the posterior distributions of the laboratory effects. Markov chain Monte Carlo methods are not needed for this purpose. This constitutes a further advantage of the suggested approach. Finally, the new procedure is well suited for treating inconsistent data.

Application of the approach to data from CCM.M-K7 for two 500 mg transfer standards measured in two petals found no significant laboratory effects, which is in accordance with the initial analysis of these data (see, Lee et al. 2017).

Acknowledgements The author would like to thank the Editors and the Reviewer for the thoughtful and constructive suggestions that led to an improvement of the paper.

6 Appendices

6.1 Derivation of the Fisher Information Matrix

The Fisher information matrix is given by

$$F = -E_{\mu_X,\mu_Y,\sigma}\begin{pmatrix} \frac{\partial^2 l(X,Y|\mu_X,\mu_Y,\sigma)}{\partial\mu_X^2} & \frac{\partial^2 l(X,Y|\mu_X,\mu_Y,\sigma)}{\partial\mu_X\partial\mu_Y} & \frac{\partial^2 l(X,Y|\mu_X,\mu_Y,\sigma)}{\partial\mu_X\partial\sigma} \\ \frac{\partial^2 l(X,Y|\mu_X,\mu_Y,\sigma)}{\partial\mu_X\partial\mu_Y} & \frac{\partial^2 l(X,Y|\mu_X,\mu_Y,\sigma)}{\partial\mu_Y^2} & \frac{\partial^2 l(X,Y|\mu_X,\mu_Y,\sigma)}{\partial\mu_Y\partial\sigma} \\ \frac{\partial^2 l(X,Y|\mu_X,\mu_Y,\sigma)}{\partial\mu_X\partial\sigma} & \frac{\partial^2 l(X,Y|\mu_X,\mu_Y,\sigma)}{\partial\mu_Y\partial\sigma} & \frac{\partial^2 l(X,Y|\mu_X,\mu_Y,\sigma)}{\partial\sigma^2} \end{pmatrix},$$

where

$$l(X,Y|\mu_X,\mu_Y,\sigma) \propto -\frac{1}{2}\log\left(det\left(V+\sigma^2 LL^T\right)\right)$$
$$-\frac{1}{2}tr\left(\left(V+\sigma^2 LL^T\right)^{-1}\begin{pmatrix} X-\mu_X\mathbf{1}_n \\ Y-\mu_Y\mathbf{1}_m \end{pmatrix}\begin{pmatrix} X-\mu_X\mathbf{1}_n \\ Y-\mu_Y\mathbf{1}_m \end{pmatrix}^T\right)$$
$$= -\frac{1}{2}\log\left(det\left(V+\sigma^2 LL^T\right)\right)$$
$$-\frac{1}{2}tr\left(\left(V+\sigma^2 LL^T\right)^{-1}\begin{pmatrix} (X-\mu_X\mathbf{1}_n)(X-\mu_X\mathbf{1}_n)^T & (X-\mu_X\mathbf{1}_n)(Y-\mu_Y\mathbf{1}_m)^T \\ (Y-\mu_Y\mathbf{1}_m)(X-\mu_X\mathbf{1}_n)^T & (Y-\mu_Y\mathbf{1}_m)(Y-\mu_Y\mathbf{1}_m)^T \end{pmatrix}\right).$$

Let

$$W(\sigma)^{-1} = \left(V+\sigma^2 LL^T\right)^{-1} = \begin{pmatrix} W_{11}^{(-)}(\sigma) & W_{12}^{(-)}(\sigma) \\ W_{21}^{(-)}(\sigma) & W_{22}^{(-)}(\sigma) \end{pmatrix}.$$

It then holds that

$$\frac{\partial^2 l(\boldsymbol{X},\boldsymbol{Y}|\mu_X,\mu_Y,\sigma)}{\partial \mu_X^2} = -tr\left(\left(\boldsymbol{V}+\sigma^2\boldsymbol{L}\boldsymbol{L}^T\right)^{-1}\begin{pmatrix}\mathbf{1}_n\mathbf{1}_n^T & \mathbf{0}\\ \mathbf{0} & \mathbf{0}\end{pmatrix}\right) = -\mathbf{1}_n^T\boldsymbol{W}_{11}^{(-)}(\sigma)\mathbf{1}_n\,,$$

$$\frac{\partial^2 l(\boldsymbol{X},\boldsymbol{Y}|\mu_X,\mu_Y,\sigma)}{\partial \mu_X\partial\mu_Y} = -\frac{1}{2}tr\left(\left(\boldsymbol{V}+\sigma^2\boldsymbol{L}\boldsymbol{L}^T\right)^{-1}\begin{pmatrix}\mathbf{0} & \mathbf{1}_n\mathbf{1}_m^T\\ \mathbf{1}_m\mathbf{1}_n^T & \mathbf{0}\end{pmatrix}\right)$$
$$= -\mathbf{1}_n^T\boldsymbol{W}_{12}^{(-)}(\sigma)\mathbf{1}_m\,,$$

$$\frac{\partial^2 l(\boldsymbol{X},\boldsymbol{Y}|\mu_X,\mu_Y,\sigma)}{\partial \mu_Y^2} = -tr\left(\left(\boldsymbol{V}+\sigma^2\boldsymbol{L}\boldsymbol{L}^T\right)^{-1}\begin{pmatrix}\mathbf{0} & \mathbf{0}\\ \mathbf{0} & \mathbf{1}_m\mathbf{1}_m^T\end{pmatrix}\right) = -\mathbf{1}_m^T\boldsymbol{W}_{22}^{(-)}(\sigma)\mathbf{1}_m$$

Furthermore, since $\frac{\partial l(\boldsymbol{X},\boldsymbol{Y}|\mu_X,\mu_Y,\sigma)}{\partial\mu_X}$ and $\frac{\partial l(\boldsymbol{X},\boldsymbol{Y}|\mu_X,\mu_Y,\sigma)}{\partial\mu_Y}$ are linear functions in $(\boldsymbol{X}-\mu_X\mathbf{1}_n)$ and $(\boldsymbol{Y}-\mu_Y\mathbf{1}_m)$, taking the derivative with respect to σ and then the expectation will lead to zero.

Next, we derive the (3, 3)-th block of $\boldsymbol{F}$. It holds that

$$\frac{\partial^2 \boldsymbol{W}(\sigma)}{\partial\sigma} = 2\sigma\boldsymbol{L}\boldsymbol{L}^T\,,\quad \frac{\partial^2 \boldsymbol{W}(\sigma)}{\partial\sigma^2} = 2\boldsymbol{L}\boldsymbol{L}^T\,.$$

In using the following formulas (cf., Harville 1997, p. 309)

$$\begin{aligned}\frac{\partial^2\boldsymbol{W}(\sigma)^{-1}}{\partial\sigma^2} &= -\boldsymbol{W}(\sigma)^{-1}\frac{\partial^2\boldsymbol{W}(\sigma)}{\partial\sigma^2}\boldsymbol{W}(\sigma)^{-1}\\ &\quad+2\boldsymbol{W}(\sigma)^{-1}\frac{\partial\boldsymbol{W}(\sigma)}{\partial\sigma}\boldsymbol{W}(\sigma)^{-1}\frac{\partial\boldsymbol{W}(\sigma)}{\partial\sigma}\boldsymbol{W}(\sigma)^{-1}\,,\\ \frac{\partial^2\log(det(\boldsymbol{W}(\sigma)))}{\partial\sigma^2} &= tr\left(\boldsymbol{W}(\sigma)^{-1}\frac{\partial^2\boldsymbol{W}(\sigma)}{\partial\sigma^2}\right)\\ &\quad- tr\left(\boldsymbol{W}(\sigma)^{-1}\frac{\partial\boldsymbol{W}(\sigma)}{\partial\sigma}\boldsymbol{W}(\sigma)^{-1}\frac{\partial\boldsymbol{W}(\sigma)}{\partial\sigma}\right)\end{aligned}$$

together with

$$E_{\mu_X,\mu_Y,\sigma}\left(\begin{pmatrix}\boldsymbol{X}-\mu_X\mathbf{1}_n\\ \boldsymbol{Y}-\mu_Y\mathbf{1}_m\end{pmatrix}\begin{pmatrix}\boldsymbol{X}-\mu_X\mathbf{1}_n\\ \boldsymbol{Y}-\mu_Y\mathbf{1}_m\end{pmatrix}^T\right) = \boldsymbol{V}+\sigma^2\boldsymbol{L}\boldsymbol{L}^T\,,$$

we get

$$E_{\mu_X,\mu_Y,\sigma}\left(\frac{\partial^2 l(\boldsymbol{X},\boldsymbol{Y}|\mu_X,\mu_Y,\sigma)}{\partial\sigma^2}\right) = -\frac{1}{2}\frac{\partial^2\log(det(\boldsymbol{V}+\sigma^2\boldsymbol{L}\boldsymbol{L}^T))}{\partial\sigma^2}$$

$$
\begin{aligned}
& -\frac{1}{2}tr\left(\frac{\partial^2(\boldsymbol{V}+\sigma^2\boldsymbol{L}\boldsymbol{L}^T)^{-1}}{\partial\sigma^2}E_{\mu_X,\mu_Y,\sigma}\left(\begin{pmatrix}\boldsymbol{X}-\mu_X\mathbf{1}_n\\ \boldsymbol{Y}-\mu_Y\mathbf{1}_m\end{pmatrix}\begin{pmatrix}\boldsymbol{X}-\mu_X\mathbf{1}_n\\ \boldsymbol{Y}-\mu_Y\mathbf{1}_m\end{pmatrix}^T\right)\right)\\
&= -tr((\boldsymbol{V}+\sigma^2\boldsymbol{L}\boldsymbol{L}^T)^{-1}\boldsymbol{L}\boldsymbol{L}^T)\\
&+ 2\sigma^2 tr((\boldsymbol{V}+\sigma^2\boldsymbol{L}\boldsymbol{L}^T)^{-1}\boldsymbol{L}\boldsymbol{L}^T(\boldsymbol{V}+\sigma^2\boldsymbol{L}\boldsymbol{L}^T)^{-1}\boldsymbol{L}\boldsymbol{L}^T)\\
&+ tr((\boldsymbol{V}+\sigma^2\boldsymbol{L}\boldsymbol{L}^T)^{-1}\boldsymbol{L}\boldsymbol{L}^T)\\
&- 4\sigma^2 tr((\boldsymbol{V}+\sigma^2\boldsymbol{L}\boldsymbol{L}^T)^{-1}\boldsymbol{L}\boldsymbol{L}^T(\boldsymbol{V}+\sigma^2\boldsymbol{L}\boldsymbol{L}^T)^{-1}\boldsymbol{L}\boldsymbol{L}^T)\\
&= -2\sigma^2 tr((\boldsymbol{V}+\sigma^2\boldsymbol{L}\boldsymbol{L}^T)^{-1}\boldsymbol{L}\boldsymbol{L}^T(\boldsymbol{V}+\sigma^2\boldsymbol{L}\boldsymbol{L}^T)^{-1}\boldsymbol{L}\boldsymbol{L}^T)\\
&= -2\sigma^2 tr\left((\boldsymbol{L}^T(\boldsymbol{V}+\sigma^2\boldsymbol{L}\boldsymbol{L}^T)^{-1}\boldsymbol{L})^2\right).
\end{aligned}
$$

Using the notation

$$
\boldsymbol{K}=\begin{pmatrix}\mathbf{1}_n & \mathbf{0}_n\\ \mathbf{0}_m & \mathbf{1}_m\end{pmatrix},
$$

we get the expression of the Fisher information matrix as given in (8).

6.2 Derivation of the Posterior

Let

$$
\boldsymbol{L}=\begin{pmatrix}\boldsymbol{L}_1\\ \boldsymbol{L}_2\end{pmatrix} \text{ and } \boldsymbol{\lambda}=\begin{pmatrix}\boldsymbol{\lambda}_X\\ \boldsymbol{\lambda}_Y\end{pmatrix}.
$$

The posterior for $\{\mu_X, \mu_Y, \boldsymbol{\lambda}_X, \boldsymbol{\lambda}_Y, \sigma\}$ is expressed as

$$
\begin{aligned}
&\pi(\mu_X, \mu_Y, \boldsymbol{\lambda}_X, \boldsymbol{\lambda}_Y, \sigma|\boldsymbol{X}, \boldsymbol{Y})\\
&\propto \exp\left(-\frac{1}{2}\begin{pmatrix}\boldsymbol{X}-\boldsymbol{L}_1\boldsymbol{\lambda}-\mu_X\mathbf{1}_n\\ \boldsymbol{Y}-\boldsymbol{L}_2\boldsymbol{\lambda}-\mu_Y\mathbf{1}_m\end{pmatrix}^T\boldsymbol{V}^{-1}\begin{pmatrix}\boldsymbol{X}-\boldsymbol{L}_1\boldsymbol{\lambda}-\mu_X\mathbf{1}_n\\ \boldsymbol{Y}-\boldsymbol{L}_2\boldsymbol{\lambda}-\mu_Y\mathbf{1}_m\end{pmatrix}\right)\\
&\times \exp\left(-\frac{1}{2\sigma^2}\boldsymbol{\lambda}^T\boldsymbol{\lambda}\right)\frac{\pi(\sigma)}{\sigma^{n+m-1}}\\
&= \exp\left(-\frac{1}{2}\begin{pmatrix}\boldsymbol{X}-\boldsymbol{L}_1\boldsymbol{\lambda}\\ \boldsymbol{Y}-\boldsymbol{L}_2\boldsymbol{\lambda}\end{pmatrix}^T\boldsymbol{V}^{-1}\begin{pmatrix}\boldsymbol{X}-\boldsymbol{L}_1\boldsymbol{\lambda}\\ \boldsymbol{Y}-\boldsymbol{L}_2\boldsymbol{\lambda}\end{pmatrix}\right)\exp\left(-\frac{1}{2}G\right)\\
&\times \exp\left(-\frac{1}{2\sigma^2}\boldsymbol{\lambda}^T\boldsymbol{\lambda}\right)\frac{\pi(\sigma)}{\sigma^{n+m-1}}
\end{aligned}
$$

with

$$
\begin{aligned}
G &= \begin{pmatrix} \mu_X \mathbf{1}_n \\ \mu_Y \mathbf{1}_m \end{pmatrix}^T \boldsymbol{V}^{-1} \begin{pmatrix} \mu_X \mathbf{1}_n \\ \mu_Y \mathbf{1}_m \end{pmatrix} - 2 \begin{pmatrix} \boldsymbol{X} - \boldsymbol{L}_1 \boldsymbol{\lambda} \\ \boldsymbol{Y} - \boldsymbol{L}_2 \boldsymbol{\lambda} \end{pmatrix}^T \boldsymbol{V}^{-1} \begin{pmatrix} \mu_X \mathbf{1}_n \\ \mu_Y \mathbf{1}_m \end{pmatrix} \\
&= \begin{pmatrix} \mu_X \\ \mu_Y \end{pmatrix}^T \boldsymbol{K}^T \boldsymbol{V}^{-1} \boldsymbol{K} \begin{pmatrix} \mu_X \\ \mu_Y \end{pmatrix} - 2 \begin{pmatrix} \boldsymbol{X} - \boldsymbol{L}_1 \boldsymbol{\lambda} \\ \boldsymbol{Y} - \boldsymbol{L}_2 \boldsymbol{\lambda} \end{pmatrix}^T \boldsymbol{V}^{-1} \boldsymbol{K} \begin{pmatrix} \mu_X \\ \mu_Y \end{pmatrix} \\
&= \left[\begin{pmatrix} \mu_X \\ \mu_Y \end{pmatrix} - \boldsymbol{A}^{-1} \boldsymbol{M}(\boldsymbol{\lambda}) \right]^T \boldsymbol{A} \left[\begin{pmatrix} \mu_X \\ \mu_Y \end{pmatrix} - \boldsymbol{A}^{-1} \boldsymbol{M}(\boldsymbol{\lambda}) \right] - \boldsymbol{M}(\boldsymbol{\lambda})^T \boldsymbol{A}^{-1} \boldsymbol{M}(\boldsymbol{\lambda})
\end{aligned}
$$

where

$$
\boldsymbol{A} = \boldsymbol{K}^T \boldsymbol{V}^{-1} \boldsymbol{K} \text{ and } \boldsymbol{M}(\boldsymbol{\lambda}) = \boldsymbol{K}^T \boldsymbol{V}^{-1} \begin{pmatrix} \boldsymbol{X} - \boldsymbol{L}_1 \boldsymbol{\lambda} \\ \boldsymbol{Y} - \boldsymbol{L}_2 \boldsymbol{\lambda} \end{pmatrix}.
$$

Hence,

$$
\begin{aligned}
&\pi(\mu_X, \mu_Y, \boldsymbol{\lambda}_X, \boldsymbol{\lambda}_Y, \sigma | \boldsymbol{X}, \boldsymbol{Y}) \\
&\propto \exp\left(-\frac{1}{2} \left[\begin{pmatrix} \mu_X \\ \mu_Y \end{pmatrix} - \boldsymbol{A}^{-1} \boldsymbol{M}(\boldsymbol{\lambda}) \right]^T \boldsymbol{A} \left[\begin{pmatrix} \mu_X \\ \mu_Y \end{pmatrix} - \boldsymbol{A}^{-1} \boldsymbol{M}(\boldsymbol{\lambda}) \right] \right) \\
&\times \exp\left(-\frac{1}{2} \begin{pmatrix} \boldsymbol{X} - \boldsymbol{L}_1 \boldsymbol{\lambda} \\ \boldsymbol{Y} - \boldsymbol{L}_2 \boldsymbol{\lambda} \end{pmatrix}^T \boldsymbol{V}^{-1} \begin{pmatrix} \boldsymbol{X} - \boldsymbol{L}_1 \boldsymbol{\lambda} \\ \boldsymbol{Y} - \boldsymbol{L}_2 \boldsymbol{\lambda} \end{pmatrix} \right) \\
&\times \exp\left(\frac{1}{2} \boldsymbol{M}(\boldsymbol{\lambda})^T \boldsymbol{A}^{-1} \boldsymbol{M}(\boldsymbol{\lambda}) \right) \exp\left(-\frac{1}{2\sigma^2} \boldsymbol{\lambda}^T \boldsymbol{\lambda} \right) \frac{\pi(\sigma)}{\sigma^{n+m-1}}.
\end{aligned}
$$

Because $\boldsymbol{A}$ does not depend on $\boldsymbol{\lambda}$ and σ, integrating over μ_X and μ_Y, we get

$$
\begin{aligned}
&\pi(\boldsymbol{\lambda}_X, \boldsymbol{\lambda}_Y, \sigma | \boldsymbol{X}, \boldsymbol{Y}) \\
&\propto \exp\left(-\frac{1}{2} \begin{pmatrix} \boldsymbol{X} - \boldsymbol{L}_1 \boldsymbol{\lambda} \\ \boldsymbol{Y} - \boldsymbol{L}_2 \boldsymbol{\lambda} \end{pmatrix}^T \boldsymbol{R} \begin{pmatrix} \boldsymbol{X} - \boldsymbol{L}_1 \boldsymbol{\lambda} \\ \boldsymbol{Y} - \boldsymbol{L}_2 \boldsymbol{\lambda} \end{pmatrix} \right) \exp\left(-\frac{1}{2\sigma^2} \boldsymbol{\lambda}^T \boldsymbol{\lambda} \right) \frac{\pi(\sigma)}{\sigma^{n+m-1}},
\end{aligned}
$$

where

$$
\boldsymbol{R} = \boldsymbol{V}^{-1} - \boldsymbol{V}^{-1} \boldsymbol{K} (\boldsymbol{K}^T \boldsymbol{V}^{-1} \boldsymbol{K})^{-1} \boldsymbol{K}^T \boldsymbol{V}^{-1}.
$$

We consider

$$
\begin{aligned}
&\begin{pmatrix} \boldsymbol{X} - \boldsymbol{L}_1 \boldsymbol{\lambda} \\ \boldsymbol{Y} - \boldsymbol{L}_2 \boldsymbol{\lambda} \end{pmatrix}^T \boldsymbol{R} \begin{pmatrix} \boldsymbol{X} - \boldsymbol{L}_1 \boldsymbol{\lambda} \\ \boldsymbol{Y} - \boldsymbol{L}_2 \boldsymbol{\lambda} \end{pmatrix} + \frac{1}{\sigma^2} \boldsymbol{\lambda}^T \boldsymbol{\lambda} \\
&= \begin{pmatrix} \boldsymbol{X} \\ \boldsymbol{Y} \end{pmatrix}^T \boldsymbol{R} \begin{pmatrix} \boldsymbol{X} \\ \boldsymbol{Y} \end{pmatrix} + \boldsymbol{\lambda}^T \boldsymbol{L}^T \boldsymbol{R} \boldsymbol{L} \boldsymbol{\lambda} - 2 \begin{pmatrix} \boldsymbol{X} \\ \boldsymbol{Y} \end{pmatrix}^T \boldsymbol{R} \boldsymbol{L} \boldsymbol{\lambda} + \frac{1}{\sigma^2} \boldsymbol{\lambda}^T \boldsymbol{\lambda} \\
&= \begin{pmatrix} \boldsymbol{X} \\ \boldsymbol{Y} \end{pmatrix}^T \boldsymbol{R} \begin{pmatrix} \boldsymbol{X} \\ \boldsymbol{Y} \end{pmatrix} + \boldsymbol{\lambda}^T \left(\boldsymbol{L}^T \boldsymbol{R} \boldsymbol{L} + \frac{1}{\sigma^2} \boldsymbol{I}_{n+m-1} \right) \boldsymbol{\lambda} - 2 \begin{pmatrix} \boldsymbol{X} \\ \boldsymbol{Y} \end{pmatrix}^T \boldsymbol{R} \boldsymbol{L} \boldsymbol{\lambda} \\
&= (\boldsymbol{\lambda} - \boldsymbol{\mu}_{\boldsymbol{\lambda}|\sigma})^T \boldsymbol{V}_{\boldsymbol{\lambda}|\sigma}^{-1} (\boldsymbol{\lambda} - \boldsymbol{\mu}_{\boldsymbol{\lambda}|\sigma}) + \begin{pmatrix} \boldsymbol{X} \\ \boldsymbol{Y} \end{pmatrix}^T \boldsymbol{R} \begin{pmatrix} \boldsymbol{X} \\ \boldsymbol{Y} \end{pmatrix} - \boldsymbol{\mu}_{\boldsymbol{\lambda}|\sigma}^T \boldsymbol{V}_{\boldsymbol{\lambda}|\sigma}^{-1} \boldsymbol{\mu}_{\boldsymbol{\lambda}|\sigma},
\end{aligned}
$$

where

$$\boldsymbol{\mu}_{\boldsymbol{\lambda}|\sigma} = \left(\boldsymbol{L}^T\boldsymbol{R}\boldsymbol{L} + \frac{1}{\sigma^2}\boldsymbol{I}_{n+m-1}\right)^{-1} \boldsymbol{L}^T\boldsymbol{R}\begin{pmatrix}\boldsymbol{X}\\ \boldsymbol{Y}\end{pmatrix}$$

and

$$\boldsymbol{V}_{\boldsymbol{\lambda}|\sigma} = \left(\boldsymbol{L}^T\boldsymbol{R}\boldsymbol{L} + \frac{1}{\sigma^2}\boldsymbol{I}_{n+m-1}\right)^{-1} .$$

Rewriting the formula of $\pi(\boldsymbol{\lambda}_X, \boldsymbol{\lambda}_Y, \sigma|\boldsymbol{X}, \boldsymbol{Y})$, we get

$$\begin{aligned}
&\pi(\boldsymbol{\lambda}_X, \boldsymbol{\lambda}_Y, \sigma|\boldsymbol{X}, \boldsymbol{Y})\\
&\propto \exp\left(-\frac{1}{2}\left(\boldsymbol{\lambda}-\boldsymbol{\mu}_{\boldsymbol{\lambda}|\sigma}\right)^T \boldsymbol{V}_{\boldsymbol{\lambda}|\sigma}^{-1}\left(\boldsymbol{\lambda}-\boldsymbol{\mu}_{\boldsymbol{\lambda}|\sigma}\right)\right)\\
&\quad \exp\left(-\frac{1}{2}\left(\begin{pmatrix}\boldsymbol{X}\\ \boldsymbol{Y}\end{pmatrix}^T \boldsymbol{R}\begin{pmatrix}\boldsymbol{X}\\ \boldsymbol{Y}\end{pmatrix} - \boldsymbol{\mu}_{\boldsymbol{\lambda}|\sigma}^T\boldsymbol{V}_{\boldsymbol{\lambda}|\sigma}^{-1}\boldsymbol{\mu}_{\boldsymbol{\lambda}|\sigma}\right)\right)\frac{\pi(\sigma)}{\sigma^{n+m-1}} .
\end{aligned}$$

Hence

$$\boldsymbol{\lambda}|\sigma \sim N_{n+m-1}\left(\boldsymbol{\mu}_{\boldsymbol{\lambda}|\sigma}, \boldsymbol{V}_{\boldsymbol{\lambda}|\sigma}\right) \tag{15}$$

and the marginal posterior for σ is given by

$$\begin{aligned}
\pi(\sigma|\boldsymbol{X}, \boldsymbol{Y}) &\propto det(\boldsymbol{V}_{\boldsymbol{\lambda}|\sigma})^{1/2}\frac{\pi(\sigma)}{\sigma^{n+m-1}}\\
&\quad\times \exp\left(-\frac{1}{2}\left(\begin{pmatrix}\boldsymbol{X}\\ \boldsymbol{Y}\end{pmatrix}^T \boldsymbol{R}\begin{pmatrix}\boldsymbol{X}\\ \boldsymbol{Y}\end{pmatrix} - \boldsymbol{\mu}_{\boldsymbol{\lambda}|\sigma}^T\boldsymbol{V}_{\boldsymbol{\lambda}|\sigma}^{-1}\boldsymbol{\mu}_{\boldsymbol{\lambda}|\sigma}\right)\right) .
\end{aligned}$$

Next, we simplify the posterior for σ using the fact that

$$det(\boldsymbol{V}_{\boldsymbol{\lambda}|\sigma}) = 1/det\left(\boldsymbol{L}^T\boldsymbol{R}\boldsymbol{L} + \frac{1}{\sigma^2}\boldsymbol{I}_{n+m-1}\right)$$

and

$$\begin{aligned}
&det\left(\boldsymbol{L}^T\boldsymbol{R}\boldsymbol{L} + \frac{1}{\sigma^2}\boldsymbol{I}_{n+m-1}\right)\\
&= det\left(\boldsymbol{L}^T\boldsymbol{V}^{-1}\boldsymbol{L} + \sigma^{-2}\boldsymbol{I}_{n+m-1} - \boldsymbol{L}^T\boldsymbol{V}^{-1}\boldsymbol{K}(\boldsymbol{K}^T\boldsymbol{V}^{-1}\boldsymbol{K})^{-1}\boldsymbol{K}^T\boldsymbol{V}^{-1}\boldsymbol{L}\right)\\
&= det\left(\boldsymbol{L}^T\boldsymbol{V}^{-1}\boldsymbol{L} + \sigma^{-2}\boldsymbol{I}_{n+m-1}\right)\\
&\times det\left(\boldsymbol{I}_2 - (\boldsymbol{K}^T\boldsymbol{V}^{-1}\boldsymbol{K})^{-1/2}\boldsymbol{K}^T\boldsymbol{V}^{-1}\boldsymbol{L}(\boldsymbol{L}^T\boldsymbol{V}^{-1}\boldsymbol{L} + \sigma^{-2}\boldsymbol{I}_{n+m-1})^{-1}\boldsymbol{L}^T\boldsymbol{V}^{-1}\boldsymbol{K}(\boldsymbol{K}^T\boldsymbol{V}^{-1}\boldsymbol{K})^{-1/2}\right)\\
&= det\left(\boldsymbol{L}^T\boldsymbol{V}^{-1}\boldsymbol{L} + \sigma^{-2}\boldsymbol{I}_{n+m-1}\right)\\
&\times \frac{det(\boldsymbol{K}^T(\boldsymbol{V}^{-1} - \boldsymbol{V}^{-1}\boldsymbol{L}(\boldsymbol{L}^T\boldsymbol{V}^{-1}\boldsymbol{L} + \sigma^{-2}\boldsymbol{I}_{n+m-1})^{-1}\boldsymbol{L}^T\boldsymbol{V}^{-1})\boldsymbol{K})}{det(\boldsymbol{K}^T\boldsymbol{V}^{-1}\boldsymbol{K})}\\
&= \sigma^{-2(n+m-1)}det\left(\sigma^2\boldsymbol{L}^T\boldsymbol{V}^{-1}\boldsymbol{L} + \boldsymbol{I}_{n+m-1}\right)\frac{det(\boldsymbol{K}^T(\boldsymbol{V} + \sigma^2\boldsymbol{L}\boldsymbol{L}^T)^{-1}\boldsymbol{K})}{det(\boldsymbol{K}^T\boldsymbol{V}^{-1}\boldsymbol{K})} ,
\end{aligned}$$

where the second equality follows from Sylvester's determinant theorem and the fourth one is obtained by applying Woodbury's matrix identity.

Hence

$$\pi(\sigma|\boldsymbol{X},\boldsymbol{Y}) \propto \frac{\pi(\sigma)}{\sqrt{det\left(\sigma^2\boldsymbol{L}^T\boldsymbol{V}^{-1}\boldsymbol{L}+\boldsymbol{I}_{n+m-1}\right)}\sqrt{det(\boldsymbol{K}^T(\boldsymbol{V}+\sigma^2\boldsymbol{L}\boldsymbol{L}^T)^{-1}\boldsymbol{K})}}$$
$$\times \exp\left(-\frac{1}{2}\left(\begin{pmatrix}\boldsymbol{X}\\\boldsymbol{Y}\end{pmatrix}^T\boldsymbol{R}\begin{pmatrix}\boldsymbol{X}\\\boldsymbol{Y}\end{pmatrix}-\boldsymbol{\mu}_{\lambda|\sigma}^T\boldsymbol{V}_{\lambda|\sigma}^{-1}\boldsymbol{\mu}_{\lambda|\sigma}\right)\right).$$

References

Abbott, P. J. et al. (2015). Final report of CCM key comparison of mass standards CCM. M-K6, 50 kg. *Metrologia 52*(1A), 07,004

Berger, J., & Bernardo, J. M. (1992). On the development of reference priors. In: Bernardo, J. M., Berger, J., Dawid, A. P., Smith, A. F. M. (eds) *Bayesian statistics* (vol. 4, pp. 35–60). Oxford: University Press.

Bernardo, J., & Smith, A. (2000). *Bayesian theory*. Chichester: Wiley.

Bodnar, O. (2019). Non-informative Bayesian inference for heterogeneity in a generalized marginal random effects meta-analysis. *Theory of Probability and Mathematical Statistics, 100*, 7–23.

Bodnar, O., & Elster, C. (2014). Analytical derivation of the reference prior by sequential maximization of Shannon's mutual information in the multi-group parameter case. *Journal of Statistical Planning and Inference, 147*, 106–116.

Bodnar, O., & Elster, C. (2018). Assessment of vague and noninformative priors for Bayesian estimation of the realized random effects in random-effects meta-analysis. *AStA Advances in Statistical Analysis, 102*(1), 1–20.

Bodnar, O., Link, A., Klauenberg, K., Jousten, K., Elster, C. (2013). Application of Bayesian model averaging using a fixed effects model with linear drift for the analysis of key comparison CCM. P-K12. *Measurement Techniques 56*(6), 584–590

Bodnar, O., Elster, C., Fischer, J., Possolo, A., & Toman, B. (2016a). Evaluation of uncertainty in the adjustment of fundamental constants. *Metrologia, 53*, S46–S54.

Bodnar, O., Link, A., & Elster, C. (2016b). Objective Bayesian inference for a generalized marginal random effects model. *Bayesian Analysis, 11*, 25–45.

Bodnar, O., Link, A., Arendacká, B., Possolo, A., & Elster, C. (2017). Bayesian estimation in random effects meta-analysis using a non-informative prior. *Statistics in Medicine, 36*(2), 378–399.

Bureau International des Poids et Mesures. (revision 2003). Mutual Recognition of National Measurement Standards and of Calibration and Measurement Certificates issued by National Metrology Institutes. CIPM.

Chunovkina, A., Elster, C., Lira, I., & Wöger, W. (2008). Analysis of key comparison data and laboratory biases. *Metrologia, 45*(2), 211.

Chunovkina, A., Stepanov, A., & Burmistrova, N. (2016). Evaluation of inconsistent data: Comparison of two adjustment algorithms. *Measurement, 91*, 707–712.

Elster, C., & Toman, B. (2010). Analysis of key comparisons: Estimating laboratories' biases by a fixed effects model using bayesian model averaging. *Metrologia, 47*, 113–119.

Elster, C., & Toman, B. (2013). Analysis of key comparison data: Critical assessment of elements of current practice with suggested improvements. *Metrologia, 50*(5), 549.

Forbes, A. B. (2016). A hierarchical model for the analysis of inter-laboratory comparison data. *Metrologia, 53*(6), 1295.

Harville, A. D. (1997). *Matrix algebra from a statistician's perspective*. New York: Springer.

Kacker, R. N. (2004). Combining information from interlaboratory evaluations using a random effects model. *Metrologia*, *41*, 132–136.

Koepke, A., Lafarge, T., Possolo, A., & Toman, B. (2017). Consensus building for interlaboratory studies, key comparisons, and meta-analysis. *Metrologia*.

Lee, S et al. (2017). The final report for CCM. M-K7: Key comparison of 5 kg, 100 g, 10 g, 5 g and 500 mg stainless steel mass standards. *Metrologia*, *54*(1A)07,001

Mohr, P. J., Taylor, B. N., & Newell, D. B. (2012). CODATA recommended values of the fundamental physical constants: 2010. *Journal of Physical and Chemical Reference Data*, *41*(043), 109.

Muhmuza, R., Bodnar, O. (2020). On modeling the correlation as an additional parameter in random effects model. *Theory of Probability and Mathematical Statistics* (to appear)

Rukhin, A. L., & Possolo, A. (2011). Laplace random effects models for interlaboratory studies. *Computational Statistics & Data Analysis*, *55*, 1815–1827.

Shirono, K., Shiro, M., Tanaka, H., & Ehara, K. (2016). Proficiency tests with uncertainty information: Detection of an unknown random effect. *Measurement*, *83*, 144–152.

Sutton, A. J., & Higgins, J. (2008). Recent developments in meta-analysis. *Statistics in Medicine*, *27*, 625–650.

Toman, B. (2007). Bayesian approaches to calculating a reference value in key comparison experiments. *Technometrics*, *49*, 81–87.

Toman, B., & Possolo, A. (2009). Laboratory effects models for interlaboratory comparisons. *Accreditation and Quality Assurance*, *14*, 553–563.

Turner, R. M., Jackson, D., Wei, Y., Thompson, S. G., & Higgins, J. (2015). Predictive distributions for between-study heterogeneity and simple methods for their application in Bayesian meta-analysis. *Statistics in Medicine*, *34*, 984–998.

Quality Control Activities Are a Challenge for Reducing Variability

Ken Nishina

Abstract It is well known that reducing variability is the basis of quality control activities. The production process can be regarded roughly as a value chain, which is composed of customer voice, product planning, product design, and manufacturing the product. In the outcome of the value chain, three kinds of variability, which are the variability before shipping to market, the variability after shipping to market, and the variability of satisfaction of market, can be considered. Quality control activities can be regarded as thinking about what can be done to reduce the three variabilities and taking actions, then ensuring quality for customers by implementing them. In the value chain, many proposals and improvements have been implemented to reduce the variabilities. In this paper, a structure of the three variabilities above is shown; then activities to reduce the variabilities are discussed. As a result, the activities can be classified into four approaches and they can be systematized as the four approaches to reduce the three kinds of variability.

Keywords Value chain of production process · Three kinds of variability · Four approaches to reduce the variabilities

1 Introduction

A column which at that time American Sony's vice-president wrote was placed in a Japanese newspaper on April 17, 1979. Its contents explained the reasons why the quality of Japanese products had been better than US products. Figure 1 expresses the reasons plainly. It shows the differences in the distribution of a property (color density) of comparable color televisions manufactured at a Sony plant in San Diego and a Sony plant in Japan. According to the article, Fig. 1 shows one of the reasons

This paper is based on Chaps. 1 and 2 on Nishina et al. (2018).

K. Nishina (✉)
Aich Institute of Technology, 2-49-2, Jiyugaoka, Chikusa-ku, Nagoya 464-0044, Japan
e-mail: ken-nishina@aitech.ac.jp

S. Knoth and W. Schmid (eds.), *Frontiers in Statistical Quality Control 13*,
Frontiers in Statistical Quality Control,
https://doi.org/10.1007/978-3-030-67856-2_21

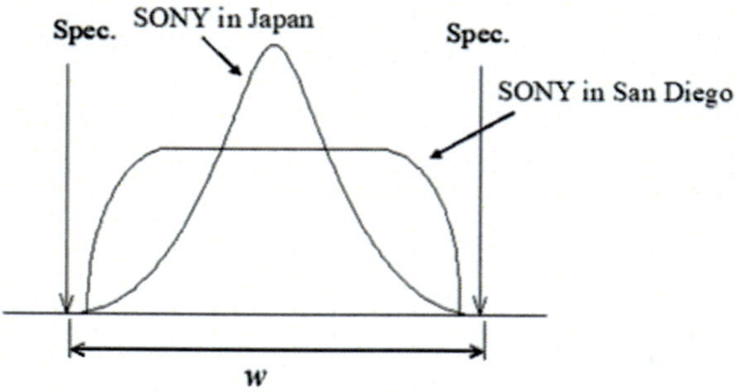

Fig. 1 Distribution of a color TV property (difference between San Diego Sony plant and Japan Sony plant)

why the quality of Japanese-made products is superior to that of American-made products.

The period following the publication of this article, the 1980s, may be considered a period of transfer of Japanese quality control technology to the USA. The trigger was a television show called "If Japan Can, Why Can't We?" broadcast by American network NBC in 1980. In 1987, the Malcolm Baldrige National Quality Award, which benchmarks Japan's Deming Prize, was established.

In Japan, operation standards are followed thoroughly and work is done with the aim of making process values agree as much as possible with target values. However, variability occurs as a result of many factors that, while their contribution to the characteristic is small, cannot be controlled. As shown in Fig. 1, in such cases, the distribution can approach a normal distribution. At the San Diego plant, on the other hand, work was carried out with the attitude that it was acceptable as long as it remains within the specification values. As a result, properties have a mixed distribution consisting of several distributions.

The differences in the two distributions in Fig. 1 can be quantified as the differences in the size of the variability. Assuming a normal distribution in Japan and a uniform distribution in San Diego, let us take σ_{jp} and σ_{san} to be the respective standard deviations of the distributions. With "W" as the allowable tolerance, we get

$$\sigma_{jp} = \frac{W}{6}, \quad \sigma_{san} = \frac{W}{2\sqrt{3}} \left(> \sigma_{jp}\right)$$

Taguchi (1993). The above demonstrates that the level of quality can be quantified according to variability and that the smaller the variability, the better the quality.

It is well known that reducing variability is the basis of quality control activities. Then, how should we understand the variabilities and reduce them in practical quality control activities of production processes?

The production process can be regarded roughly as a value chain, which is composed of customer voice, product planning, product design, and manufacturing the product. In the outcome of the value chain, three kinds of variability can be considered. One is the variability before shipping to market. Another is the variability after shipping to market and the last is the variability of satisfaction of market. Quality control activities can be regarded as thinking about what can be done to reduce the three variabilities and taking actions, then ensuring quality for customers by implementing them. In the value chain, many proposals and improvements have been implemented to reduce the variabilities.

In this paper, a structure of the three kinds of variability in the value chain is shown; then a classification and a systematization of approaches to reduce the variabilities is discussed.

2 Value Chain in Production Processes

The process of making a product can be roughly divided into the three stages of planning, design, and manufacturing, which is shown in Fig. 2. It is clear that quality is not built in the manufacturing stage alone. Quality is built with activities in each stage.

The input to the product planning stage is the voice of customers. In this stage, the voice of the customer is transferred to a concept of a product. That is the production activity answers the questions "What are we going to sell, and to sell to whom?". Then, the grade of the product including the cost planning is determined. That is the production activity answers the question "What grade of product are we going to sell?".

The next stage is the product design, where the product planning is transferred to information for manufacturing the product, that is, design is the production activity of creating specifications and drawings to produce the quality that is promised to the customer in the product planning stage. Product design also serves to provide information with consideration of ease of work in manufacturing, which is the process downstream of the product design stage.

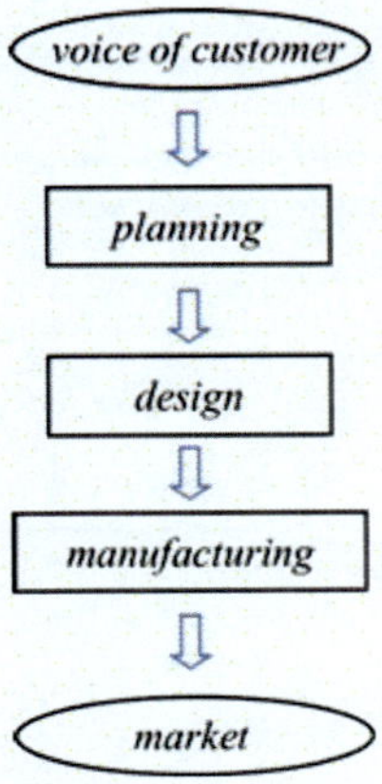

Fig. 2 Value chain of production process

Manufacturing is the production activity of transforming drawings received from design into a product. The quality properties, that are the outcome of the production activities of manufacturing, are measured, and the outcome is evaluated by the variability from the target. This is the process capability. These stages shown in Fig. 2 can be regarded as a value chain.

As mentioned in Sect. 1, quality can be quantified by variability. In this paper, on the assumption of the value chain above, three kinds of variability are considered and we discuss how to reduce the variabilities in each stage of the value chain.

3 Three Kinds of Variability of Outcome in the Value Chain

Figure 3 shows the three kinds of variability of outcomes in the value chain, the variability before shipping to market, the variability after shipping to market, and the variability of satisfaction in the market. As shown in Fig. 3, the variability before shipping to market is that of the distribution with which the mean is the target value provided by the product design. The causes of variation in the manufacturing process can be summarized with 5M1E (man, machine, method, material, measurement, and environment). The product manufacturing department has a responsibility to reduce the variability because most of the causes are internal noises in the 5M1E. But it should be noted that it is not "all" but "most". For example, "material" may come from a supplier. In addition, the product design department also has a responsibility, because some causes in 5M1E are determined by the product design.

A customer purchases a product from the outcome of the production manufacturing and uses it under various conditions and over a long period, then the function of the outcome has a variability. This is the variability after shipping to market. There

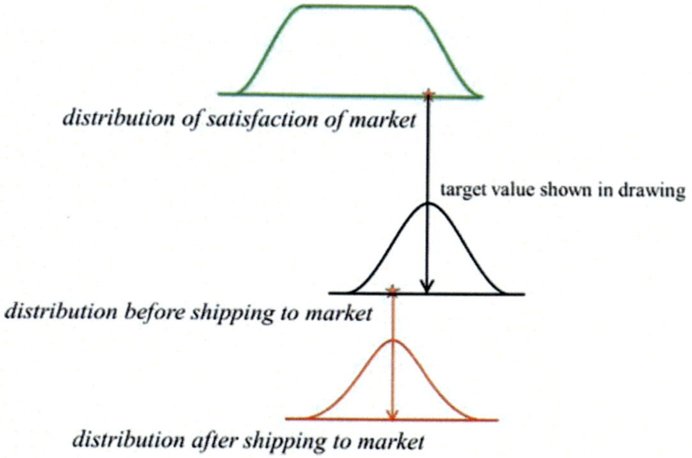

Fig. 3 Three kinds of variability of outcomes in value chain

is a responsibility to reduce it in the product design department. Unlike reducing the variability before shipping to market, in the case of the variability after shipping to market, the causes of the variability are external noises, for example, outside temperature and atmospheric pressure. Robust design for the external noises is required to reduce this variability.

The variability of satisfaction of market may be the most difficult problem to take direct actions against its causes because the causes of the variability are customer needs. The distribution shape of the satisfaction of market is trapezoidal because the distribution is composed of a mixture of distributions with different means. Because there is a divergence of values in the market needs.

From the discussions above, it should be noted that there are two cases. One is that the causes of variability are internal noises for the manufacturer. The other is that the causes of variability are external noises for the manufacturer. Roughly speaking, in the former cases, an action can be taken against the cause; on the other hand, in the latter cases, an action cannot be taken against the causes.

4 Four Approaches to Reduce the Variability

4.1 Four Approaches in Causal Model

The variability before shipping to market, the variability after shipping to market, and the variability of satisfaction of the market are caused by noises in the manufacturing processes, noises in the usage environment, and the individual feelings of customers, respectively. There are many possible situations, but their structure can be modeled as consisting of cause, effect, and a causal relationship between cause and effect. Figure 4 shows a simple causal relationship between cause and effect. The activities to reduce the variabilities can be classified from the perspective of "what should we take action against?". The subjects which we can take action against are three elements mentioned above, that is, cause, effect, and their causal relationship. The activities can be classified into four approaches as in Tatebayashi (2004). The four approaches are shown in Fig. 4. The approaches are as follows:

Approach A: Taking action against effect,
Approach B: Taking action against cause,
Approach C: Observing the situation of causes and taking action against effect according to the situation,
Approach D: Taking action against causal relationship.

A related study is the excellent book (Steiner and MacKay 2005). They proposed seven approaches to reducing variability. Our study proceeded independently of the one by Steiner and MacKay (2005). The four approaches above are very similar to the seven approaches by Steiner and MacKay (2005), although there are slightly different

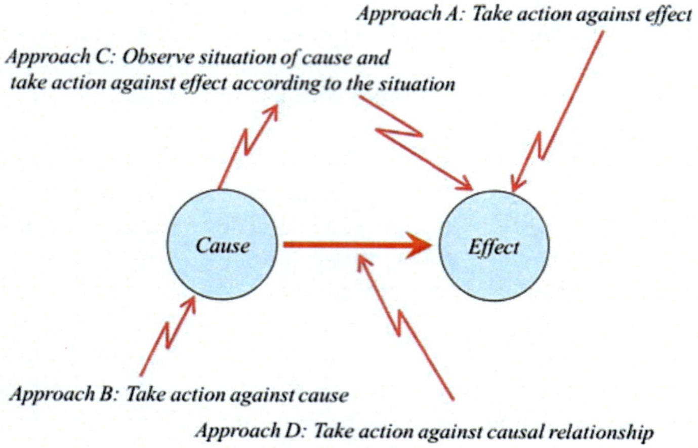

Fig. 4 Four approaches to reduce the variability

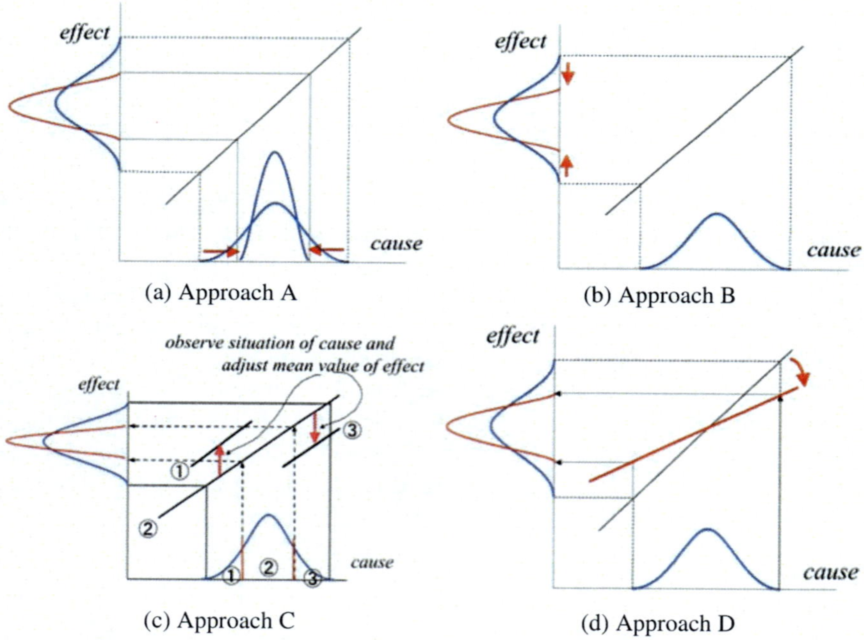

Fig. 5 Four approaches by assuming a linear model between cause and effect

explanations. Our study can be characterized by systematization of the measures for reducing the above three variabilities when looking at the whole production process.

The four approaches can be explained by assuming a simple linear model which represents the cause, the effects, and their causal relationship. In Fig. 5, the four approaches are expressed graphically.

Note that "taking action against effect" means directly or indirectly reducing the variability of the effect. "Taking action against cause" means reducing the variability of the cause and stopping the variability from reproducing. "Taking action against causal relationship" means decreasing the change ratio of the causal relationship.

Next, we will explain the four approaches in detail and include some examples.

4.2 Approach A: Taking Action Against Effect

Approach A can be classified as direct and indirect actions against the effects. The former is "selection and sorting". A specific example is 100% inspection. The latter is "adjustment" after observing the effect. As the adjustment is performed by operating a control variable, it is an indirect action against the effect. A typical example is feedback control. In these approaches, an action is implemented after observing the effect and no actions are taken against its cause as shown in Fig. 5, so the approach is not a recurrence prevention but a spillage prevention. Note that in Approach A, there is no need to identify the cause.

In the case of selection and sorting, high precision measurements are needed, and actions must be standardized. The actions are incorporated in tact time, and implementation obviously takes man-hours. Therefore, automation, for example, using image processing is helpful to reduce labor cost. In the case of adjustment, the actions must be standardized. A precondition is that the relationship between the control variable and the effect stays constant. If the precondition fails, the action may introduce larger variability and the production process may be in an out of control condition. Measuring the output and making adjustment to the process may require additional cost. Therefore, it is necessary to reduce man-hours for measurement and adjustment.

4.3 Approach B: Taking Action Against Cause

In Approach B, there are two kinds of actions depending upon when the action is taken. One is recurrence prevention. That is so-called "kaizen activity". After a problem is identified, its root cause is determined and then a corrective action is taken against the cause as shown in Fig. 6a. It is well known that the QC seven tools and SQC methods are useful for identifying a problem, searching for the root cause, and confirming the effect by the action. The process of searching for a root cause can create improvement. So this approach is fundamental and traditional in quality control. However, it should be noted that the approach may increase cost. Considering cost-effectiveness, a different approach may be better.

The other possibility with Approach B is prevention. If the cause is known in advance, the actions can be taken against the cause as shown in Fig. 6b. A typical example is the determination of tolerance in the product design stage. A severe

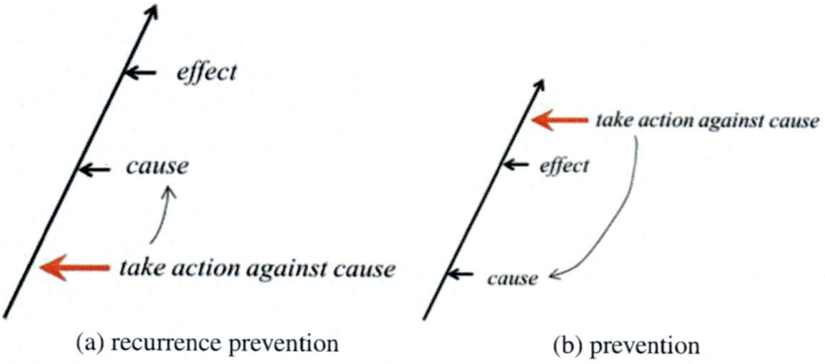

Fig. 6 Two kinds of action depending upon when the action is taken

tolerance may be required. The design of experiments may be useful for the determination of the tolerance. But in general, this approach is more costly than Approach D because manufacturing costs increase.

4.4 Approach C: Observing the Situation of Causes and Taking Action Against Effect According to the Situation

Suppose the cause is known but no action can be taken against it, unlike with Approach B. For example, the cause is the variability of the material from a supplier, but no action can be taken against the cause. In such cases, after measuring the value of the cause, a condition of the production process is adjusted. An example is shown in Fig. 5c where feedforward control is used. The action is taken after the causes have already occurred but before the effect occurs as shown in Fig. 7. Therefore, it can be called "adaptive prevention".

In Approach C, the real-time observation and the adjustment at high precision are required. Like with the adjustment in Approach A, the cost of the observation and the adjustment should be considered. However, with modern IoT technologies, these

Fig. 7 Adaptive prevention according to the situation of cause

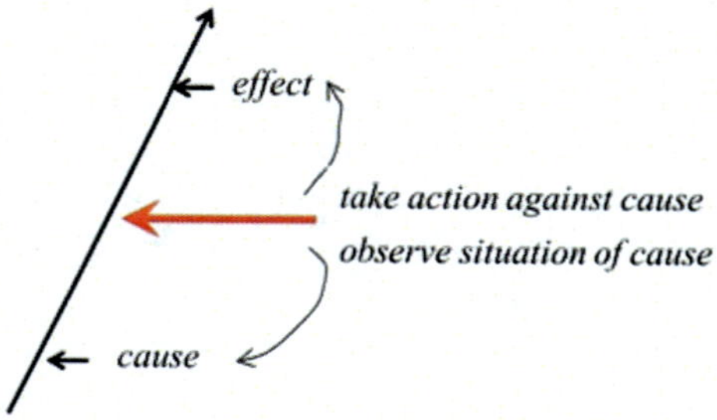

conditions are more easily satisfied. As a result, in the future, Approach C will likely become more useful for reducing variability.

4.5 Approach D: Taking Action Against Causal Relationship

Approach D means using robust design, which is the origin of the Taguchi method. The variability due to the noise, which a supplier cannot take direct action against, can be reduced by using Approach D. In Approach D, an action is taken before the cause occurs; therefore, Approach D is preventive. This approach does not require us to take action against the cause; therefore, the cost can be reduced (see Taguchi 1993). But in Approach D, it is necessary to adjust the process mean by finding another factor, which can shift the process mean, so-called adjustment factor.

There are two methods in Approach D. One is an application of the non-linear relationship. The other is an application of the interaction between a design variable and a noise factor.

The application of the non-linear relationship is a measure for the noise of inner factors. Figure 8a shows an example of a process design in a semiconductor wafer manufacturing process. In this case, it is known that the cash register strike film thickness which is an effect in the just before process is a cause of the variability of the cash register strike dimensions which is an effect in the just after process. Then the target value of the cash register strike film thickness is determined as shown in Fig. 8a in order to reduce the variability of the cash register strike dimensions.

Figure 8b shows an example concerning the optimization of plasticity of a candy by Kackar (1985). In this case, the formulate of the candy is optimized by analyzing the interaction between the external noise (temperature) and a formulate element of the candy. As shown in Fig. 8b, the variability of plasticity due to temperature after improvement is less than before the improvement.

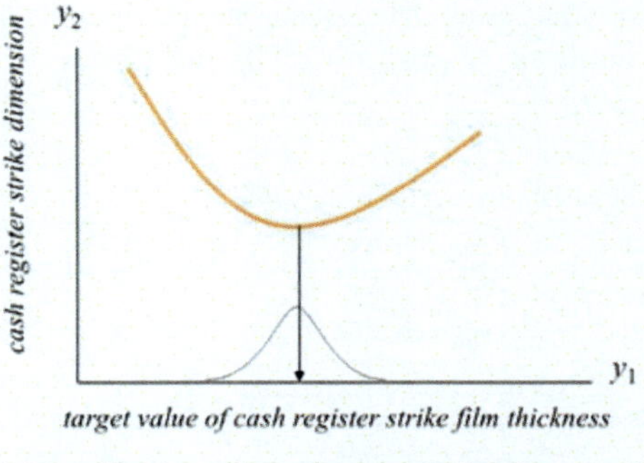

(a): application of non-linear relationship

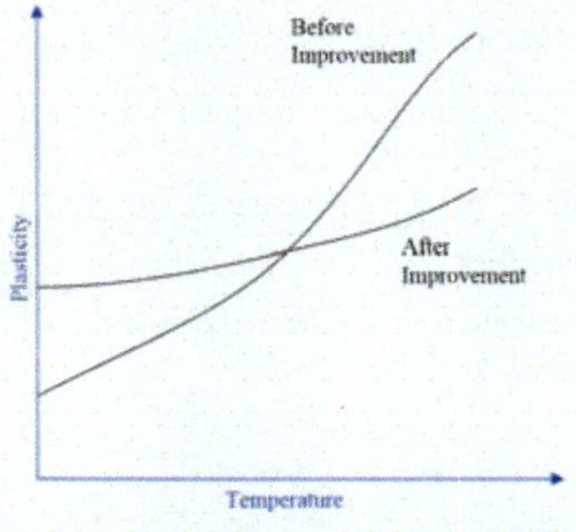

(b) application of interaction (Kackar, 1985)

Fig. 8 Two methods of Approach D

Approach D is similar to Approach C from the viewpoint of a preventive measure; however, the point in time to take an action is different. In Approach D, the action has been determined before the cause occurs. On the other hand, in Approach C, the action is taken after the cause has occurred, a condition of the production process is adjusted depending on the situation of the cause. As mentioned earlier, Approach C will likely become much more useful for reducing the variability under the IoT condition. That means the usability of Approach C will be enhanced as a supplemental approach of Approach D.

5 Four Approaches to Reduce the Three Kinds of Variability Overlooking the Value Chain

In Sects. 2 and 3, we have explained the three kinds of variability in the value chain and the four approaches to reduce the variabilities. In this section, they are reviewed from the viewpoint of the value chain.

Table 1 shows the utilization of the four approaches to the three kinds of variability. Double circle (◎) and circle (○) mean highly effective and moderately effective utilization; however, the point in time to take an action is different. In Approach D, the action has been determined before the cause occurs. On the other hand, in Approach C, the action is taken after the cause has occurred, a condition of the production process is adjusted depending on the situation of the cause. As mentioned earlier, Approach C will have been much more useful for reducing the variability under the IoT condition. That means the usability of Approach C will be enhanced as a supplemental approach of Approach D, respectively, while Black circle (●) means the approach is undesirable.

First, we discuss the reduction of the variability in market satisfaction. The product planning department is responsible for reducing this variability. The department, however, cannot take direct actions to reduce customer variability. Market segmentation is well known as an effective activity in marketing. Market segmentation means stratification of the cause. This is an example of Approach C. The most essential

Table 1 Four approaches to reduce three kinds of variability overlooking values

Variability	Approach A	Approach B	Approach C	Approach D
Of satisfaction of market	● Withdrawal from the market	○ Awareness in the market	◎ Market segmentation	
After shipping to market	● Recall	◎ Tolerance design	○ Adaptive specification	◎ Robust design for product
Before shipping to market	◎ Inspection feedback control	◎ Corrective action for process	◎ Feedforward control	◎ Robust design for process

activity of marketing is market research. This is nothing but observing the situation of the cause because different customer needs lead to the variability of satisfaction of market.

Approach B is an effective approach in this case, where the target of action is customers. Its typical example is the advertisement for a new product, where the action is directly taken against customer's consciousness that is a cause of the variability of satisfaction of market. For example, customer's awareness of the new product with some merits can provide the reduction of the variability. However, the effectiveness of Approach B is typically less than that of Approach C.

Approach A is an undesirable approach in this case because the action is too late and provides extensive cost. It may be a worst story for the supplier. The ultimate action is withdrawal from the market.

Second, we discuss the reduction of the variability after shipping to market. A responsible department for reducing this variability is the product design department. Like the variability of satisfaction of market, the responsible department cannot take actions directly against root causes of the variability because the causes are external noises for the supplier. In this case, Approach D, Taguchi's robust design, of which the target of action is the causal relationship, is most effective to reduce the variability. As mentioned in Sect. 4.3, this approach is a preventive measure.

The other preventive measure, Approach B, can be useful if the cause is known in advance. As shown in Fig. 6b, the determination of severe tolerance provides a reduction in the variability after shipping to market. However, this approach sometimes increases cost (see Taguchi 1993). In the Taguchi method, this approach is called Tolerance design.

Approach C is an adaptive specification. A typical example is that a specification is adaptively determined according to the market. For example, a vehicle has a special specification for a cold location. In this case, it is essential to know in the early stage of product design which specification can be useful for the adaptation.

Like the variability in market satisfaction, Approach A is also an undesirable approach in this case. A typical action is recall. Recalls are usually very expensive.

Lastly, we discuss the variability before shipping to market. The activity which greatly contributes to reducing this variability is Statistical Process Control (SPC). The activities lifecycle of SPC are comprised of the following four stages, mass production preparation, pilot mass production, early-stage mass production, and routine mass production. The four approaches must be utilized in the right place at the right time in these four stages to reduce this variability.

The main mission in the mass production preparation is determining the optimal conditions of each machine to assure the highest machine performance. In this stage, Approach D is applied because some noise factors in the stage of mass roduction are assumed. The engineers hope that this stage of the SPC lifecycle leads to the stage of routine mass production as soon as possible. In this sense, Approach D is required in this stage. In the stage of pilot mass production, the production line is formed and the elements in the production line are standardized. In this stage, the short time process capability is assured.

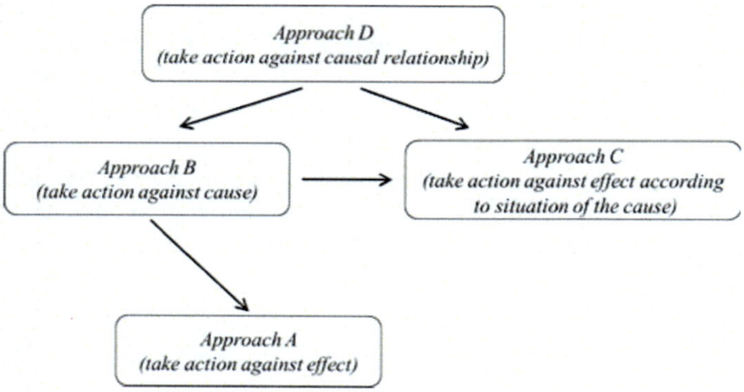

Fig. 9 Relationship among four approaches to reduce variabilities

Approach B is compulsory in the early-stage mass production because the variability is increasing due to many noise factors related to time in this stage, for example, the variability among the material lots and the deterioration of parts. That is so-called "kaizen activity". Action is taken against the root cause and the related standards are revised. However, there is a case where action cannot be taken against the root cause even if the cause is clear. In such a case, Approach C is effective in a sense of making up for Approach B.

Approach A also is included as the standards to reduce the variability. As mentioned in Sect. 4.2, in Approach A no action is taken against the cause; therefore, Approach A is effective as the action of spillage prevention.

From the discussion above, in the case of reduction of the variability before shipping to market, the relationship among the four approaches is shown in Fig. 9. Figure 9 shows activity steps in the SPC lifecycle for reducing this variability. At first, a sufficient implementation of Approach D in the stage of mass production preparation is needed to shift more quickly to the routine mass production stage. Next, Approach B or Approach C is implemented for process improvement. As a last resort, Approach A plays a role in spill prevention.

Finally, we suggest not only the product manufacturing department but also the product design department is responsible for reducing the variability before shipping to market. The product manufacturing department is a kind of customer for the product design department. As mentioned in Sect. 2, the design department should provide information with consideration of the ease of work in manufacturing. Without the collaboration of both departments, the reduction of the variability cannot be realized.

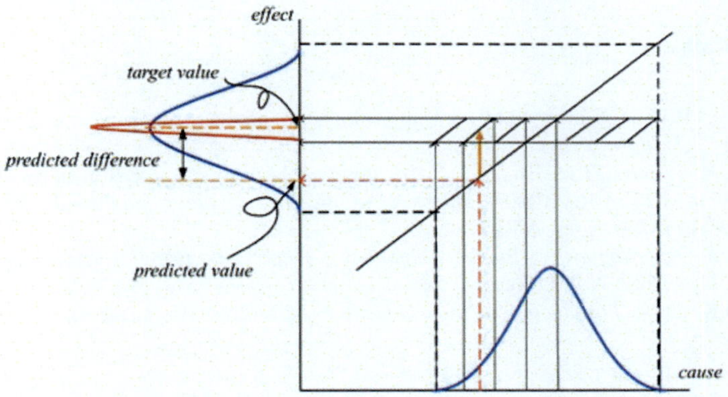

Fig. 10 Advancement of Approach C

6 Further Discussions and Conclusive Remarks

It is well known that production outcome has three values, that is, Quality, Cost, and Delivery. They are so-called "QCD". However, Q has a quite different character from C and D. The values of C and D are decided before shipping to market; on the other hand, the value of Q is decided after shipping to market. The customers evaluate Q after their purchase. But this does not mean that the variability before shipping to market is not very important. Figure 3 says that it is necessary to reduce the variability before shipping to market in order to reduce the variability after shipping to market.

The prevention approach and the recurrence prevention (Approach D and Approach B) have been conventionally emphasized in Quality Control. Of course, that should be done from now on, also. In addition to that, the adaptive prevention, that is, Approach C may be regarded as more important in the IoT era. As mentioned earlier, the real-time observation of the cause at high precision and the adjustment with high precision are required so that Approach C functions effectively. Much more advancement is possible in Approach C. If the real-time observation of the cause and the adjustment can be more precise using advanced IT technology, the variability can be drastically reduced. The mechanism of advancement of Approach C is shown in Fig. 10, which is an extension to a more precise mechanism than the fundamental one of Approach C shown in Fig. 5.

In this paper, the quality control activities have been discussed as a challenge for reducing the variability in the value chain. As a result, the scheme, which consists of the three kinds of variability and the four approaches to reduce them, is only one viewpoint of the quality control activities; however, the scheme can not only systematize the quality control activities in the value chain of production but also can lead to better management of the manufacturing process.

Acknowledgements The author would like to thank Prof. Steiner of the University of Waterloo for providing useful information.

References

Kackar, R. N. (1985). Off-line quality control, parameter design, and the Taguchi method. *Journal of Quality Technology*, *17*(4), 176–188.

Nishina, K., Kawamura, H., & Ishii, N. (2018). *Standard quality control*. Baifukan (in Japanese).

Steiner S. H., & MacKay R. J. (2005). *Statistical engineering: An algorithm for reducing variation in manufacturing processes*. ASQ Printing.

Taguchi, G. (1993). *Taguchi on robust technology development: Bringing quality engineering upstream*. New York: ASME Press.

Tatebayashi K. (2004) *Introduction to Taguchi method*. JUSE Press (in Japanese).

Is the Benford Law Useful for Data Quality Assessment?

Wolfgang Kössler, Hans-J. Lenz, and Xing D. Wang

Abstract Data quality and data fraud are of increasing concern in the digital world. Benford's Law is used worldwide for detecting non-conformance or data fraud of numerical data. It says that the first non-zero digit D_1, of a data item from a universe, is not uniformly distributed. The shape is roughly logarithmically decaying starting with $P(D_1 = 1) \cong 0.3$. It is self-evident that Benford's Law should not be applied for detecting manipulated or faked data before having examined the goodness of fit of the probability model while the business process is free of manipulations, i.e. 'under control'. In this paper, we are concerned with the goodness-of-fit phase, not with fraud detection itself. We selected five empirical numerical data sets of various sample sizes being publicly accessible as a kind of benchmark, and evaluated the performance of three statistical tests. The tests include the chi-square goodness-of-fit test, which is used in businesses as a standard test, the Kolmogorov–Smirnov test, and the MAD test as originated by Nigrini (1992). We are analyzing further whether the invariance properties of Benford's Law might improve the tests or not.

Keywords Benford's Law · Invariance properties · Goodness-of-fit tests · Data quality · Data fraud · Data manipulation

1 Introduction

Benford's Law describes an astonishing phenomenon. In many data sets, the first non-zero digit D_1 is not uniformly distributed but obeys a logarithmic law, $P(D_1 =$

W. Kössler · X. D. Wang
Institut für Informatik, Humboldt Universität zu Berlin, Rudower Chaussee 25, 12489 Berlin, Germany
e-mail: koessler@informatik.hu-berlin.de

H.-J. Lenz (✉)
Institut für Statistik und Ökonometrie, Freie Universität Berlin, Boltzmannstr. 20, 14195 Berlin, Germany
e-mail: hans-j.lenz@fu-berlin.de

S. Knoth and W. Schmid (eds.), *Frontiers in Statistical Quality Control 13*,
Frontiers in Statistical Quality Control,
https://doi.org/10.1007/978-3-030-67856-2_22

$d) = \log(1 + \frac{1}{d})$[1] for all $d \in \{1, \ldots, 9\}$, cf. Newcomb (1881), Benford (1938). It is used worldwide for assessing the data quality or detecting data fraud of business and economic data. Auditors of big companies use Benford's Law for detecting data manipulation mostly by applying the χ^2-goodness-of-fit test. No doubt, not every empirical or artificial data set follows Benford's Law. The question arises as to which conditions data sets follow that law, and how this can be tested in practice.

More than 50 years after Newcomb's detection (Pinkham 1961) proved that scale invariance leads to Benford's Law, Nigrini (1992) as well as Berger and Hill (2011) proved further the base and significand-sum invariance. The latter means that the sum of all the 1-significands is equal to each sum of the significands of the remaining digits 2, ..., 9.

In this paper, the authors use five publicly available data sets of various fixed-sample sizes as a benchmark for testing whether they are Benford distributed or not. Observational data has the advantage of a given sample size, n, opposite to experimental data where n is to be planned. The χ^2 test is used as a yardstick, the Kolmogorov–Smirnov test and the MAD test, a test based on the mean absolute deviation, are further applied. The multiple test problem does not arise in our study, because the χ^2 test and the other two tests used behave in conformance if the critical values of the KS and MAD tests were revised. While the scale and base invariance properties don't contribute too much to improve the test power, there is evidence that the patterns of the significant-sums can help to reduce the error of applying the Benford hypothesis in practice when it is false.

2 Mathematical Basics of Benford's Law

Benford's Law makes claims about the leading digits of a number regardless of its scale. Hence, we begin by introducing the formal notation of significands and significant digits.

Definition 1 (*Significant Digits and the significand,* Berger and Hill 2015) The first significant digit $D_1(x) = d$ of $x \in \mathbb{R}$ is given by the unique integer $d \in \{1, 2, \ldots, 9\}$ where $10^k d \leq |x| < 10^k(d+1)$ with an integer $k \in \mathbb{R}$.

The mth significant digit $D_m(x) = d$ with $m \geq 2$ can recursively be determined by

$$10^k \left(\sum_{i=1}^{m-1} D_i(x)10^{m-i} + d\right) \leq |x| < 10^k \left(\sum_{i=1}^{m-1} D_i(x)10^{m-i} + d + 1\right)$$

where $d \in \{0, 1, \ldots, 9\}$ and $k \in \mathbb{Z}$. The significand $S(x)$ of $x \in \mathbb{R}$ is defined as $S(x) = t$ with $t \in [1, 10)$ where $|x| = 10^k t$ if $x \neq 0$, else $S(x) := 0$.

Examples are $D_1(e) = 2, D_2(\pi) = 1, D_1(88) = 8, S(\frac{1}{2}) = 5, S(10^8) = 1,$ *and* $S(67594) = 6, 7594$. Next, we can state when the first significant digit and the significand of a random variable X are distributed according to Benford's Law.

[1] $log(x) = log_{10}(x)$.

Table 1 Benfords' Law of the first significant digit D_1

d	1	2	3	4	5	6	7	8	9
$P(d)\%$	30.1	17.6	12.5	9.7	7.9	6.7	5.8	5.1	4.6

Definition 2 (*Benford's Law*) The probability of the first significant digit $d \in \{1, 2, 3...9\}$ is $P(D_1(X) = d) = \log(1 + d^{-1})$.

In Table 1, we give the distribution of the leading digit D_1. The corresponding statement for significands reads as follows:

$$P(S(X) \leq t) = \log t \text{ for all } t \in [1, 10). \tag{1}$$

In the following, we say that a random variable X is Benford iff (1) is satisfied. Such variables own some interesting properties. We want to give a brief overview here, and in the upcoming sections we will use them for checking the Benford hypothesis. For further explanations and proofs, we refer to the literature (Berger and Hill 2011; Pinkham 1961; Nigrini 1992).

A first characterization of Benford's Law involves the uniform distribution between 0 and 1.

Theorem 1 *A random variable X is Benford if and only if* $\log|X| \mod 1$ *is uniformly distributed between* 0 *and* 1.

Applying this theorem and reminding that a random variable X is uniformly distributed mod 1 if and only if $kX + b$ is uniformly distributed mod 1 for all integers $k \neq 1$ and all real numbers b, we can derive that the random variables αY^k for a given Benford random variable Y and for all reals α, integers k and $\alpha \cdot k \neq 0$ are Benford too. This motivates both the scale invariance as well as the base invariance property of Benford random variables. The scale invariance implies that multiplying a Benford random variable with a scalar is still Benford and the base invariance implies that the Benford property is kept under exponentiation as well.

Theorem 2 (Scale Invariance Pinkham 1961) *A random variable X is Benford if and only if X has scale-invariant significant digits, i.e. for all $\alpha > 0, t \in [1, 10)$*

$$P(S(\alpha X) < t) = P(S(X) < t).$$

Theorem 3 (Base Invariance Nigrini 1992; Berger and Hill 2011) *A random variable X is Benford if and only if X has base-invariant significant digits, i.e. for all $m \in \mathbb{N}, t \in [1, 10)$*

$$P(S(X^m) < t) = P(S(X) < t).$$

Nigrini (1992) and Berger and Hill (2011) showed another property of a Benford random variable, the sum invariance. Summing all significands with the first digit 1 yields the same sum as summing all significands with the first digit 2, 3, or 4, etc.

Theorem 4 (Sum Invariance Nigrini 1992; Allart 1997) *A random variable X is Benford if and only if X has sum-invariant significant digits, i.e. the sum of all significands,* $V_d = \sum_{j=1, D_1=d}^{n} S(X_j)$*, is the same for all* $d = 1, \ldots, 9$.

In Sect. 5, we shall make use of these properties. The scale and base invariance, X is Benford iff aX^b is Benford $(a \neq 0), b \in \mathbb{Z}^+$, and the sum invariance, $E_0(V_d) = n/ln10 =: V_0$ for all $d = 1, \ldots, 9$, where $V_d = \sum_{j=1, D_1=d}^{n} S(X_j)$, are applied to detecting deviations between the empirical and Benford distributions.

3 Statistical Goodness-of-Fit Tests

The χ^2 test is the most popular goodness-of-fit test and was originated by Pearson (1900). It is assumed that the sample $(x_1, x_2, \ldots, x_n)$ of size n is of the simple random type, i.e. $X_1, X_2, \ldots, X_n$ are random variables. In most practical cases in industry and business, $n = ALL$ is true. This means that no proper random sampling is executed, and all the data of a given period is captured. However, conceiving annual bookkeeping figures coming from the super-population of all past and present annual data sets, the necessary independence assumption can be justified. We are testing
H_0 : X is Benford against H_1 : X is not Benford.
The alternative hypothesis H_1 reveals the first weakness of the χ^2-goodness-of-fit test. It is unspecified as it includes all alternative probability distributions different from Benford's Law with domain $\{1, 2, \ldots, 9\}$. The χ^2 test statistic measures the relative distance between the relative frequencies n_j/n and the probabilities $p_j = P(D_1 = d_j)$ for all $j = 1, 2, \ldots, 9$ under Benford's Law, and is defined by

$$\chi^2 = n \sum_{j=1}^{9} \frac{(n_j/n - p_j)^2}{p_j} = \sum_{j=1}^{9} \frac{(n_j - np_j)^2}{np_j}. \tag{2}$$

The χ^2 test rejects the test hypothesis H_0 if $\chi^2 > \chi^2_{1-\alpha,8}$. The significance level α is usually set equal to 0.01, 0.05 or 0.1 when no prior knowledge about the utility or probability related to the application area is available. Note that the sample $(x_1, x_2, \ldots, x_n)$ itself does not enter into formula (2), and the nine frequencies $n_1, n_2, \ldots, n_9$ are used as sufficient statistics instead. In the era of *Big Data*, the sample size may become very large, for instance, larger than $n \cong 10^5$. This implies that the χ^2 test rejects the null hypothesis already for very small, perhaps purely random deviations from the logarithmic law, cf. Definition 1. In the section on performance analysis of the three tests, we shall find this effect later. Göb (2007) showed that the power function of the test given $\alpha = 0.01$ has steep ascents near H_0 for a distorted alternative of the Benford distribution, i.e. $P_q(D_1 = d) = \log(q^{d-5}(1 + 1/d)$ for $d = 1, 2, \ldots, 9$, and $q \in [0.98, 1.05]$. This implies that the test hypothesis is almost always rejected for very large sample sizes even if the deviation from H_0 'is small'. This effect is called the 'Excess power problem', cf. Nigrini (2000). Conse-

Table 2 Critical values of the χ^2 and the KS test

	χ^2	Kolmogorov–Smirnov	
α	$c_{1-\alpha,8}$ Pearson (1900)	$d_{1-\alpha}$ Miller (1956)	$d_{1-\alpha}$ Morrow (2014)
0.01	20.09	1.628	1.420
0.05	15.51	1.358	1.148
0.10	13.36	1.224	1.012

quently, he recommends not to use the test if $n > 1000$, (Nigrini 2012). Furthermore, the χ^2 test is an approximate test, and a necessary condition is that the sample size is 'sufficiently large', i.e. $np_j > 5$ for all j. The critical values $c_{1-\alpha,8}$ of the χ^2 test are tabulated in the second column of Table 2.

An alternative goodness-of-fit test is the Kolmogorov–Smirnov (KS) test, cf. Kolmogorov (1933), Smirnov (1948), and Darling (1957). Its idea is to compare the empirical distribution $F_n(x)$ with a fully specified theoretical, continuous one, $F_0(x)$. In our case, F_0 is the Benford distribution which is, of course, a discrete one. Therefore the critical region of the test has to be adapted. While the χ^2 test statistic is the sum of all single deviations as a distance between F_n and F_0, the KS test uses the supremum norm

$$d_{max} = sup_{x \in \mathbb{R}} |F_n(x) - F_0(x)|. \tag{3}$$

The critical values of the KS test were completely tabulated by Miller (1956) for continuous distributions. Morrow (2014) computed appropriate tighter bounds by MC simulation of the distribution given in Definition 1. The KS test rejects the H_0 hypothesis, if $\sqrt{n}\, d_{max} > d_{1-\alpha}$ for all sample sizes $n > 40$; see columns at the right in Table 2.

Finally, we include in our test set the MAD test as originated by Nigrini (2012). It is based on the mean absolute deviation between the relative frequencies and the Benford probabilities.

$$MAD = \sum_{j=1}^{9} \frac{|n_j/n - p_j|}{9}. \tag{4}$$

Originally, this was not a statistical test in its proper sense with acceptance and rejection domains. Nigrini (2012) only gave the linguistic terms *good*, *acceptable*, *weak acceptable*, and *no conformance*. Therefore we simulated (simulation size N = 10000) the critical values of the MAD statistic for all five data sets and $\alpha \in \{0.01, 0.05, 0.10\}$; see Table 3.

Table 3 Critical values of the MAD test for sample size n and significance level α

n	$\alpha = 0.1$	$\alpha = 0.05$	$\alpha = 0.01$
197	0.0227	0.0248	0.0288
601	0.0132	0.0144	0.0167
1456	0.0084	0.0091	0.0105
3998	0.0051	0.0055	0.0064
7022	0.0038	0.0042	0.0048

4 Our Benchmark Data

In this section, we describe the benchmark data set used for our performance evaluation based on the three tests mentioned in Sect. 3. The relative frequencies n_i/n for $i = 1, 2 \ldots, 9$ represent the real data, and the meta data give some background information about the domain of concern. The values of the first significant digit, D_1, the Benford probabilities, and the relative frequencies of the five data sets together with their sample size n are displayed in Table 4.

Data set #1:News ($n = 601$)

The collection includes five numeric files from publicly accessible sources. The data set was extracted from the Internet on Dec 27, 2017, accessing German online websites like Die Zeit, WirtschaftsWoche, and Sportschau. The relative frequencies of this data set are displayed in line 3 of Table 4. The numbers represent a mixture of topics from politics, economics, sport, and content. Berger and Hill (2011) argue that such mixtures are sufficient for Benford's Law becoming true. We shall check this conjecture in the subsequent Sect. 5.

Data set #2: Financial Report of Deutsche Bank ($n = 1456$)
This medium-sized or even 'large' data set in Nigrini's sense corresponds to the quarterly financial reports of Deutsche Bank, Germany, and was retrieved from the German Bundesanzeiger in September 2017.[2] The relative frequencies are shown in line 4 of Table 4. The bank is the largest one in Germany, and is internationally active

Table 4 Benford probabilities and frequencies of all five data sets

Data set	1	2	3	4	5	6	7	8	9	n
Benford	0.301	0.176	0.125	0.097	0.079	0.067	0.058	0.051	0.046	
News	0.261	0.171	0.143	0.113	0.113	0.058	0.050	0.048	0.042	601
Bank	0.273	0.166	0.134	0.129	0.080	0.078	0.041	0.049	0.049	1456
MBB SE	0.305	0.213	0.102	0.076	0.086	0.086	0.046	0.041	0.046	197
Reddit	0.493	0.191	0.109	0.069	0.043	0.029	0.025	0.022	0.019	7022
Population	0.526	0.194	0.088	0.062	0.041	0.034	0.019	0.019	0.017	3998

[2] http://www.bundesanzeiger.de/ebanzwww/wexsservlet.

in fields like banking and other financial services for private, commercial, corporate clients, and start-ups. Other fields of business activities are financing, brokerage, digital banking, and the management of investments, deposits, loans, and mortgages. The company is a prime standard of Deutsche Börse, Frankfurt.

Data set #3: Financial Report of MBB SE ($n = 197$)
MBB stands for Messerschmitt-Bölkow-Blohm. The relative frequencies are shown in line 5 of Table 4. The data set is part of the quarterly financial report of MBB SE Berlin. The firm is a medium-sized family business, and has worldwide known technological and engineering expertise. Sales in the year 2018 were about 500 Mio Euro. The data set is very small, and was retrieved online in Sep 2017 from the Deutsche Bundesanzeiger[2].

Data set #4: No. of subscribers of Reddit ($n = 7022$)
The relative frequencies are displayed in line 6 of Table 4. The website Reddit is a US online platform of news for all topics like sport, politics (world news), science, movies, and includes, of course, ads. The platform is divided into subreddits devoted to special topics like sport, movies, etc. Altogether Reddit has about 10.000 subreddits with each up to 20 Mio users, the data was sampled in Dec 2017,[3] and includes the 7022 most popular subreddits, i.e. the subreddits with the largest amount of subscribers, ranging 21 million–10,000 at that time. It is the largest data set we analyzed.

Data set #5: Population in large cities ($n = 3998$)
The final data set of the benchmark consists of the number of inhabitants in a mixture of cities worldwide larger than 100.000 people. The relative frequencies are displayed in the last line of Table 4. The figures correspond to the year 2016. The data set was downloaded and integrated, accessing the Internet servers UNStat[4] and Worldpopulation Report.[5] As the data is coming from two sources representing many countries, one may expect Benford' Law to hold, according to the conjecture of Berger and Hill (2015).

5 Performance Analysis

In the following, we apply the three tests of Sect. 3 to the Benchmark data discussed above. The test significance level is uniformly fixed to $\alpha = 0.01$. In each study, we start by displaying the empirical and the Benford distribution, present the results of the three goodness-of-fit tests,[6] and close by applying the three invariance properties.

Let us consider the data set #1: News ($n = 601$); see Table 4. The empirical and the Bendford distribution suggest a reasonable good fit; see Fig. 1. This evidence is

[3] http://redditmetrics.com/top.

[4] https://unstats.un.org/unsd/demographic-social/products/dyb/documents/dyb2016//table08.pdf.

[5] http://worldpopulationreview.com/countries/china-population/cities/.

[6] Note that all $c_{1-\alpha}$ of the KS test are chosen according to Morrow (2014).

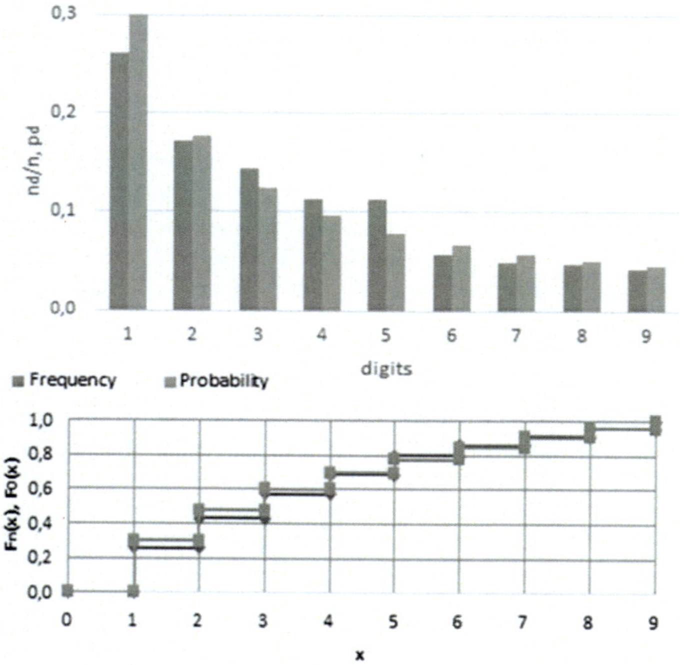

Fig. 1 Results of #1:News (n = 601)

confirmed by the χ^2, Kolmogorov–Smirnov, and MAD tests tabulated in Table 5. All the three tests are signaling 'accept'.

In Sect. 2, three important characteristics of Benford's Law were presented, i.e. the scale, base, and the sum invariance. The results of applying them to the *News* data are given in Table 6.

The frequency distributions of the first digit D_1 related to $Y = aX^b$ for $a, b \in \{2, 7\}$ deviate from the Benford distribution in a similar way as the frequencies of D_1 corresponding to X do. This evidence supports the Benford hypothesis. Furthermore, the sums of the significands $\sum_{j=1, D_1=d}^{n} S(x_j)$ have an arithmetic mean $\bar{v} = 250$ which is similar to the expected value $E_0(V_d) = V_0 = 261$. We close by accepting Benford's Law being true for *News*. This finding is in accordance with the conjecture of Berger and Hill (2011) that mixed data lead to Benford's Law.

Table 5 Test results for #1:News, $\alpha = 0.01$

$\sqrt{n}d_{max}$	$1.103 < 1.42$	Accept
χ^2	$16.9 < 20.09$	Accept
MAD	$0.015 < 0.017$	Accept

Table 6 Transformations $Y = aX^b$ and invariant sums of data set #1:News

	d	1	2	3	4	5	6	7	8	9	Sum
Relative frequencies	X	0.261	0.171	0.143	0.113	0.113	0.058	0.050	0.048	0.042	1
	$2X$	0.311	0.173	0.088	0.113	0.058	0.103	0.111	0.017	0.025	1
	$7X$	0.263	0.228	0.140	0.088	0.070	0.047	0.113	0.022	0.030	1
	X^2	0.316	0.195	0.082	0.135	0.043	0.075	0.027	0.038	0.090	1
	X^7	0.416	0.203	0.058	0.088	0.025	0.075	0.078	0.045	0.012	1
Benford probability	X	0.301	0.176	0.125	0.097	0.079	0.067	0.058	0.051	0.046	1
Invariant sums		205	234	280	284	352	217	218	236	232	$n/ln10 \approx 261$

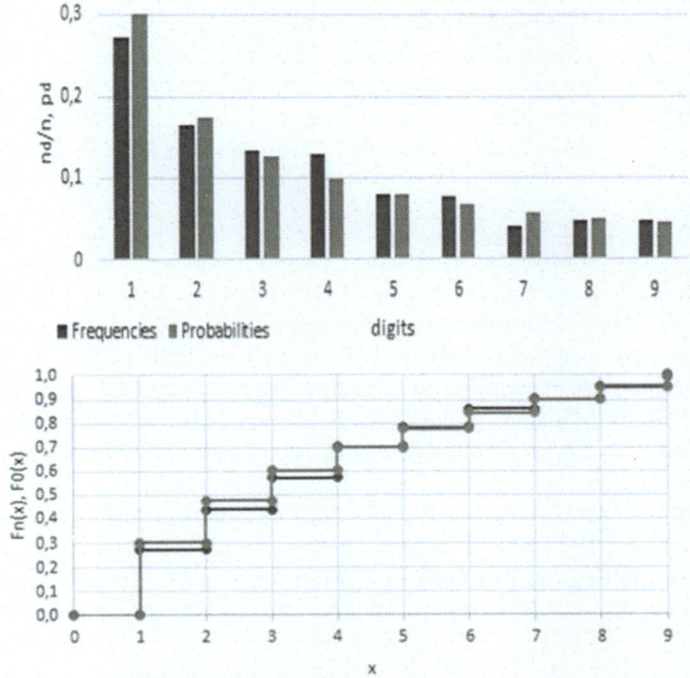

Fig. 2 Results of #2:Bank DB (n = 1456)

Next, we turn to the bank data set #2 : Bank ($n = 1456$), the quarterly financial report of Deutsche Bank; see Table 4. On the first glimpse, the frequencies n_i/n and probabilities p_i seem to fit well, especially, because the deviations for $d = 1$ and $d = 3$ are smaller than those of the data set *News* (Fig. 2).

However, the χ^2, KS, and MAD tests lead to a rejection of H_0, cf. Table 7. Notice the increased value of the test statistic, $\chi^2 = 31.72$, compared with its value of data

Table 7 Test results for #2:Bank DB, $\alpha = 0.01$

$\sqrt{n}d_{max}$	1.46>1.42	Reject
χ^2	31.72>20.09	Reject
MAD	0.013>0.011	Reject

Table 8 Transformations $Y = aX^b$ and invariant sums of data set #2:Bank DB

	d	1	2	3	4	5	6	7	8	9	Sum
Relative frequencies	X	0.273	0.166	0.134	0.129	0.080	0.078	0.041	0.049	0.049	1
	$2X$	0.298	0.163	0.109	0.096	0.070	0.071	0.070	0.071	0.051	1
	$7X$	0.271	0.190	0.154	0.103	0.066	0.066	0.064	0.040	0.046	1
	X^2	0.322	0.166	0.121	0.128	0.045	0.058	0.052	0.052	0.057	1
	X^7	0.337	0.191	0.096	0.082	0.073	0.084	0.060	0.041	0.036	1
Benford probability	X	0.301	0.176	0.125	0.097	0.079	0.067	0.058	0.051	0.046	1
Invariant sums		562	582	676	832	638	736	449	600	681	$n/ln10 \approx$ 632

set #1:*News*, and observe sample size $n = 1456 > 1000$, too. Further information about the Bank data is gained by applying the invariance characteristics, especially by comparing the significand sums with their expected value. Table 8 shows the details.

The relative frequency distributions of X and aX^b resemble each other, while the significand sums have evidently large deviations from the expected value, $E_0(V_d) = V_0 = 632$, and around their mean, $\bar{v} = 640$. Therefore, the sum invariance assumption seems doubtful, and keeping the test results in mind we classify the data *Bank* as 'non-Benford'.

The third data set, #3: Financial report MBB SE, is depicted in Table 4. It is a small data set (n = 197<1000). Comparing the plots of the relative frequencies and Benford probabilities, our first vote is to accept the Benford hypothesis despite the frequencies $n_d/n, d = 2, 3$; see Fig. 3. All three tests confirm our intuitive perception, cf. Table 9.

Table 10 gives evidence that the original data and the transformed data, especially, for $2X$ and X^7, conform to Benford's Law. Furthermore, the invariant sums have small dispersion around the mean $\bar{v} = 86$, which is equal to the expected value $V_0 = 86$. Therefore, we classify data set *MBB SE* as obeying Benford's Law.

Data set #4: Reddit is the largest one, $n = 7022$, and is tabulated in Table 4. The frequencies clearly deviate from the related Benford probabilities; see Fig. 4.

This evidence is supported by the three goodness-of-fit tests in Table 11 which uniformly flag 'reject' H_0. Observe again the effect of a large sample size on the excessive value of the χ^2 statistic. Moreover, the test statistic of the KS test, $\sqrt{n}d_{max} = 17.347$, is influenced, too.

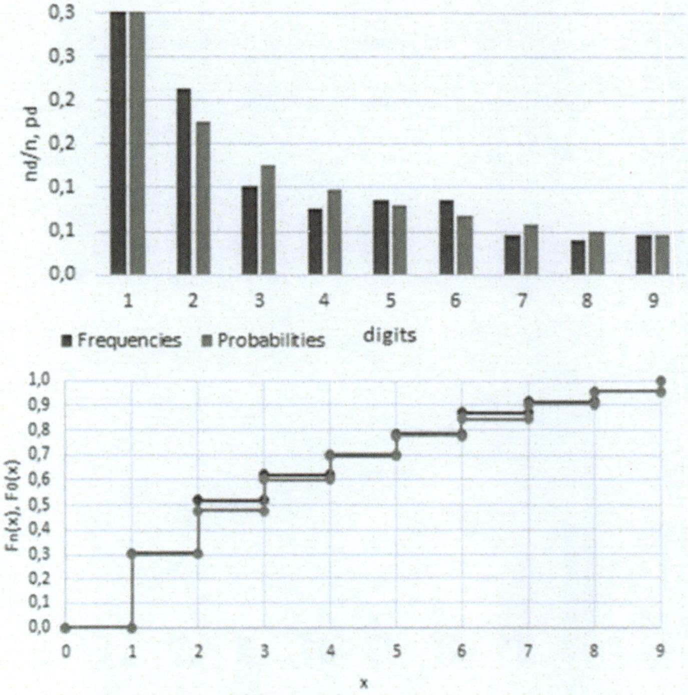

Fig. 3 Results of #3:MBB, (n = 197)

Table 9 Test results for #3:MBB, $\alpha = 0.01$

$\sqrt{n}d_{max}$	0.575<1.42	Accept
χ^2	5.46<20.09	Accept
MAD	0.015<0.029	Accept

Next, we examine whether or not this data set conforms to the invariance properties and to the sum invariance characteristic. The results are presented in Table 12. We observe large deviations of the relative frequencies from Benford's probabilities for $Y = 2X$, and of the invariant sums from their expected value, $V_0 \approx 3050$, and the mean, $\bar{v} = 2205$, too. Especially, the sums of digits, v_d, $d = 1, 9$, are quite far from their expected value. We conclude that the data set *Reddit* is not Benford.

Finally, we analyze the data set #5: Population, tabulated in Table 4. It consists of $n = 3998$ records. As it is a mixed data set, we may assume that the Benford hypothesis becomes true according to Berger and Hill (2011).

However, the large deviation between the relative frequency, n_1/n, and probability, p_1, evident from Fig. 5, causes the first doubt. This is confirmed when applying the three tests, cf. Table 13. They deliver the conforming result 'reject'. Evidently, the 'large' sample size, $n = 3998$, increases again the values of $\sqrt{n}d_{max}$ and χ^2.

Table 10 Transformations $Y = aX^b$ and invariant sums of data set #3:MBB SE

	d	1	2	3	4	5	6	7	8	9	Sum
Relative frequencies	X	0.305	0.213	0.102	0.076	0.086	0.086	0.046	0.041	0.046	1
	$2X$	0.305	0.188	0.117	0.112	0.102	0.020	0.081	0.003	0.046	1
	$7X$	0.345	0.132	0.091	0.147	0.066	0.061	0.046	0.061	0.051	1
	X^2	0.254	0.183	0.147	0.147	0.056	0.076	0.056	0.051	0.030	1
	X^7	0.325	0.178	0.117	0.051	0.076	0.076	0.076	0.003	0.071	1
Benford probability	X	0.301	0.176	0.125	0.097	0.079	0.067	0.058	0.051	0.046	1
Invariant sums		86	101	72	68	97	111	67	68	86	$n/ln10 \approx$ 86

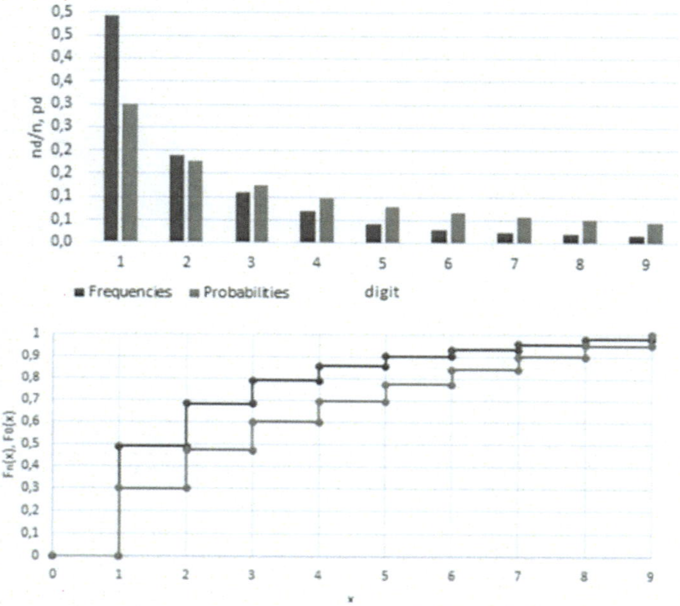

Fig. 4 Results of #4:Reddit (n = 7022)

Table 11 Test results for #4:Reddit, $\alpha = 0.01$

$\sqrt{n}d_{max}$	17.347>1.42	Reject
χ^2	1567.88>20.09	Reject
MAD	0.046>0.0053	Reject

Table 12 Transformations $Y = aX^b$ and invariant sums of data set #4:Reddit

	d	1	2	3	4	5	6	7	8	9	Sum
Relative frequencies	X	0.493	0.191	0.109	0.069	0.043	0.029	0.025	0.022	0.019	1
	$2X$	0.138	0.305	0.188	0.114	0.078	0.063	0.047	0.036	0.033	1
	$7X$	0.392	0.152	0.078	0.045	0.036	0.025	0.104	0.094	0.074	1
	X^2	0.387	0.189	0.127	0.081	0.062	0.050	0.038	0.034	0.032	1
	X^7	0.321	0.185	0.120	0.093	0.074	0.067	0.056	0.046	0.038	1
Benford probability	X	0.301	0.176	0.125	0.097	0.079	0.067	0.058	0.051	0.046	1
Invariant sums		4921	3270	2646	2167	1669	1320	1332	1281	1236	$n/ln10 \approx$ 3050

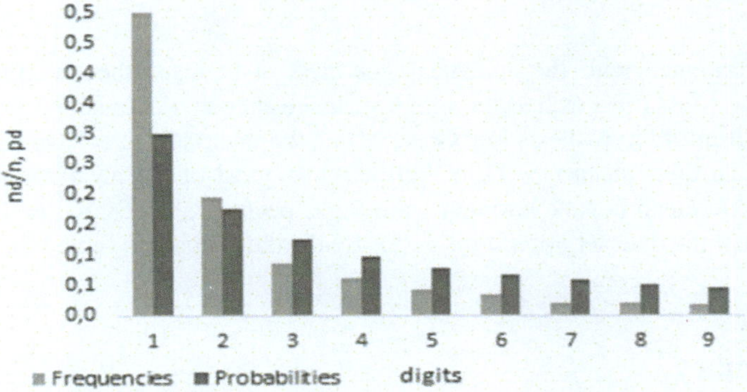

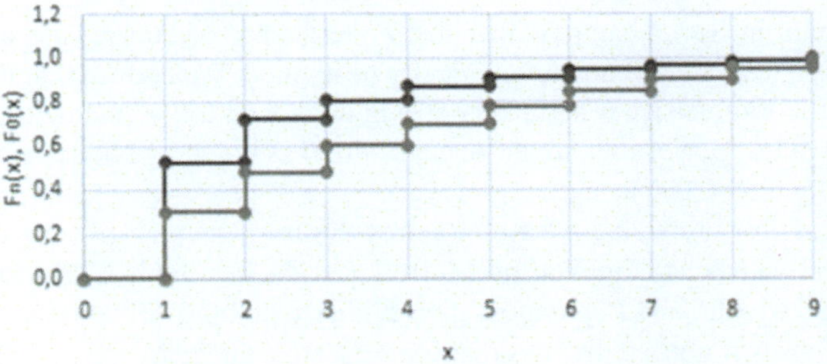

Fig. 5 Results of #5:Population (n = 3998)

Table 13 Test results for #5: Population, $\alpha = 0.01$

$\sqrt{n}d_{max}$	15.365>1.42	Reject
χ^2	1168.33>20.09	Reject
MAD	0.054>0.0064	Reject

Table 14 Transformations $Y = aX^b$ and invariant sums of data set #5:Population

	d	1	2	3	4	5	6	7	8	9	Sum
Relative frequencies	X	0.526	0.194	0.088	0.062	0.041	0.034	0.019	0.019	0.017	1
	$2X$	0.130	0.352	0.174	0.117	0.077	0.049	0.039	0.034	0.028	1
	$7X$	0.383	0.130	0.073	0.046	0.027	0.026	0.132	0.107	0.076	1
	X^2	0.412	0.191	0.110	0.085	0.058	0.045	0.040	0.033	0.027	1
	X^7	0.334	0.183	0.126	0.095	0.074	0.056	0.050	0.039	0.044	1
Benford probability	X	0.301	0.176	0.125	0.097	0.079	0.067	0.058	0.051	0.046	1
Invariant sums		2907	1887	1218	1104	901	866	574	654	649	$n/ln10 \approx$ 1736

In conformance with the statistical goodness-of-fit tests, the transformations $Y = 2X$ and $Y = X^2$ as well as the sum invariance characteristics support the rejection of the Benford hypothesis too. Observe that the invariant sums have large deviations around their mean $\bar{v} = 1196$. Furthermore, most of the invariant sums and the arithmetic mean deviate strongly from the expected value $E_0(V_d) = 1736$. We conclude that the data set *Population* is not Benford distributed (Table 14).

6 Summary

The test results in Sect. 5 suggest that the χ^2 test, and, of course, the other two goodness-of-fit tests should not automatically be applied. We deduce that its dependency upon the sample size n has a very strong impact on the χ^2 statistic and leads to an increased test power for sample size $n > 1000$ as noted by Nigrini (2012) and exemplified by Göb (2007).

We advocate for preferring the Kolmogorov–Smirnov test, cf. (3), but using the tighter critical values as compiled by Morrow (2014); see Table 2. The large data sets, i.e. Bank, Reddit, and Population, give evidence that the KS test is sensitive to large sample sizes if non-continuous distributions are analyzed.

The MAD statistic, cf. (4), as proposed by Nigrini (2000), should not be used in its original form. It is not a statistical test in its proper sense. The boundaries correspond to linguistic terms and cause vagueness of interpretation. If critical values for fixed α-values are determined by the Monte Carlo simulation as done above, the MAD test can be conceived as a test competitor.

The scale and base invariance properties gave only weak evidence of accepting or rejecting Benford's Law. The sum invariance, however, became helpful. It seems worthwhile considering the invariant sums together with their expected value, $E_0(V_d)$, and the corresponding arithmetic mean, $\bar{v}$.

There remains one point of interest. The data sets *News*, *Bank*, and *MBB* may be conceived visually belonging to the same cluster having a 'small' distance between the frequency and probability distributions of D_1. However, *Bank* data is a kind of outlier opposite to the other two data sets because our three tests uniformly lead to a rejection of the Benford Law. This means purity of this cluster of about 66% only. The remaining data sets, *Reddits* and *Population*, form a cluster with identical (negative) test results and similar visual deviance of frequencies and probabilities. This implies purity of 100%.

QQ-plots based on F_0 and F_n may be an option for further exploratory visualization of data assumed to be obeying Benford's Law.

Acknowledgements The authors thank an anonymous referee for helping to improve the paper substantially.

References

Allart, P. C. (1997). An invariant-sum characterization of Benford's law. *Journal of Applied Probability34*(1), 288–291.

Benford, F. (1938). The law of anomalous numbers. *Proceedings of the American Philosophical Society*, *78*(4), 551–572.

Berger, A., & Hill, T. P. (2015). *An introduction to Benford's Law*. Princeton: Princeton University Press.

Berger, A., & Hill, Th P. (2011). A basic theory of Benford's Law. *Probability Surveys*, *8*, 1–126.

Darling, A. D. (1957). The Kolmogorov-Smirnov, Cramér-von-Mises Tests. *Annals of Mathematical Statistics*, *28*(4), 823–838.

Deutsche Bank Aktiengesellschaft, Quartalsfinanzbericht zum 30. September 2017. http://www.bundesanzeiger.de/ebanzwww/wexsservlet.

Göb, R. (2007). Data conformance testing by digital analysis - A critical review and an approach to move appropriate testing. *Quality Engineering*, *19*(4), 281–297.

Kolmogorov, A. N. (1933). Sulla determinazione empirica di una legge di distribuzione. *Giorn. dell'Inst. Ital. degli Att.*, *4*, 83–91.

Miller, L. H. (1956). Table of percentage points of Kolmogorov statistics. *Journal of the American Statistical Association*, *51*(273), 111–121.

Morrow, J. (2014). Benford's Law, families of distributions and a test basis. Discussion Paper No 1291, Centre for Economic Performance, LSE, London.

Newcomb, S. (1881). Note on the frequency of use of the different digits in natural numbers. *American Journal of Mathematics*, *4*(1), 39–40.

Nigrini, M. (1992). The detection of income evasion through an analysis of digital distributions. PhD dissertation, University of Cincinnati.

Nigrini, M. J. (2000). *Digital analysis using Benford's Law: Tests and statistics for auditors*. Vancouver: Global Audit Publication.

Nigrini, M. (2012). *Benford's Law: Applications for forensic accounting, auditing, and fraud detection*. Hoboken: Wiley.

Pearson, K. (1900). On the criterion that a given system of deviations from the probable in the case of a correlated system of variables is such that it can be reasonably supposed to have arisen from random sampling. *Philosophical Magazine and Journal of Science*, *5*(50), 157–175.

Pinkham, R. S. (1961). On the distribution of first significant digits. *Annals of Mathematical Statistics*, *32*(4), 1223–1230.

Smirnov, N. V. (1948). Table of estimating goodness of fit of empirical distributions. *Annals of Mathematical Statistics*, *19*(2), 279–281.

UNStats Report. (2016). https://unstats.un.org/unsd/demographic-social/products/dyb/documents/dyb2016//table08.pdf.
Worldpopulation Report. (2016). http://worldpopulationreview.com/countries/china-population/cities/.